Energy Saving and Carbon Reduction

Tony A. Chen

Energy Saving and Carbon Reduction

Approaches for Energy and Chemical Industries

Springer

Tony A. Chen
Chevron Canada Ltd. (retired)
Calgary, AB, Canada

ISBN 978-981-19-5297-5 ISBN 978-981-19-5295-1 (eBook)
https://doi.org/10.1007/978-981-19-5295-1

This Springer imprint is published by the registered company Springer Nature Singapore Pte Ltd.
The registered company address is: 152 Beach Road, #21-01/04 Gateway East, Singapore 189721, Singapore

About This Book

The book provides an integrated energy/exergy analysis method to identify the energy utilization issues and systematically proposes the cost-effective energy-saving and CO_2 mitigation/capture solutions, meeting the increasing market needs for energy-saving and GHG (greenhouse gas) gas reduction. CO_2 mitigation/capture will achieve the economic benefit of fuel, power, and carbon tax saving as well as the environmental benefit of GHG reduction.

The book is a professional reference book for energy saving and GHG gas mitigation technology in oil and gas, oil refining, and chemical process. It is an integrated technical book that combines energy utilization theory and practical methods including thermodynamic analysis for unit operation and process units; energy and exergy calculation for various process streams and utilities; three-link energy/exergy analysis model; energy/exergy balance of equipment, process units, and entire plant; approach and technology of energy saving; optimization of pipeline and equipment; pinch energy-saving technology and its application; CO_2 capture and utilization with ten case studies incorporated for all different scenarios; key energy-saving technologies such gas turbine, FCCU regeneration CO combustion and energy recovery, flue gas turbine system optimization, and low-grade heat recovery and utilization.

The book is intended for engineers and professional personnel who are working in process engineering, production, energy management, chemical and petrochemical plants, refineries, oil and gas production facilities, and power generation plants; this book can also be a professional reference or textbook for undergraduate or graduate-level university students and teaching personnel of chemical, energy, and process engineering faculties of universities.

About the Author

Tony A. Chen is a registered professional engineer in Canada with over 40 years of experience in the petrochemical and oil and gas industries. He is an expert in energy integration, GHG mitigation, process engineering, process simulation, and optimization. He received his Master's Degree in petroleum engineering from Beijing Petroleum University. Most recently, he served as a senior process advisor for Chevron Corporation worked on Chevron LNG plant, gas processing, TCO oil and gas facilities and received Chevron TCO Management Recognition Award in 2020 for initiating and implementing an ethane gas flaring reduction project that achieved significant cost savings. While at Chevron, Tony also led the energy integration and GHG reduction program for an LNG project with significant cost saving and made a presentation for Chevron FE 2014 World Conference in Houston on Energy Integration and GHG Mitigation for the LNG Project.

He is also the author of Methods and Technology of Energy Conservation on Petrochemical Process, written two textbooks used in the continuing engineering training for SINOPEC. He delivered an integrated simulation model on H&M balance and BFD development using HYSYS for heavy oil upgrading/oil refining and presented it on 2009 Aspen Worldwide User Conference in Houston. He received four Science and Technology Awards from SINOPEC and was granted 1 invention patent by Patent Bureau of China.

Prior to Chevon, as a principal process engineer, he also worked for Jacobs and other EPC companies in Canada, experienced whole range of process engineering, including FEED and detail engineering, and completed 50+ projects with cost-effective solutions. He was the manager of environmental engineering department of LPEC/SINOPEC before year 2001. In this role, he completed multiple oil refinery energy integration and GHG mitigation projects and process design and energy integration for multiple process units.

Preface

With the deepening of energy conservation work and greenhouse gas (GHG) reduction in the oil and gas industry, it is more and more urgent to establish a set of scientific energy consumption analysis, evaluation, and improvement methods to make a practical evaluation of the energy consumption of enterprises, find out the energy-saving potential, propose improvement measures, reduce energy consumption, and improve economic efficiency. In the energy-saving books published in recent years, there are not many practical books on energy analysis combined with GHG mitigation, in identifying the issues and providing solutions. Therefore, it is extremely important and urgent to establish a systematic energy utilization analysis method integrated with CO_2 capture, from shallow to deep, from outside to inside, to diagnose process energy utilization and CO_2 emission problems, and to propose improvement solutions.

How to carry out the energy balance work of process plant, how to carry out the energy and exergy balance analysis of the equipment and process units deeply, and analyze and diagnose the energy consumption of process units or complex plants according to the results of energy balance and exergy balance, and which energy-saving technologies to adopt to reduce the energy consumption of the equipment or process units, all these are the practical issues faced by enterprise energy conservation workers. Based on this, in this book, the author has engaged in petrochemical equipment and the process plant energy consumption analysis, energy-saving projection, process engineering, heat integration, and CO_2 capture experience for more than forty years with systematic summarization and development.

This book has the following features:

- The book provides an approach and technologies to smartly use fossil oil with efficiency improved significantly and CO captures and utilization, which meets the strong market needs on GHG gas reduction and achieves the economic benefit of fuel, power, and carbon tax saving.
- The book provides an integrated energy/exergy analysis method to identify the energy utilization issues and systematically proposes the cost-effective energy-saving and CO_2 mitigation/capture solutions.

- The book incorporated nine case studies with process simulation models that support heat integration and CO_2 capture/utilization.
- The book provides energy and exergy calculation method for all process streams and utilities used in production, systematically energy and exergy balance method using three-link energy analysis model, analyzes and identifies the energy utilization issues from three-level of equipment, process unit, and the entire plant.
- Energy analysis on ambient condition, plant size to identify its actual impact on energy consumption, provided energy optimization for single equipment, pipeline as well as system heat integration.

This book is divided into four parts. Part I (Chaps. 1–4) is the basis of thermodynamic analysis, mainly solving the calculation method of the energy and exergy for petrochemical process, unit operation, and thermodynamic analysis method of complex petrochemical process.

Part II (Chaps. 5–8) is the energy balance and exergy balance method based on the three-link energy analysis model, which proposes the calculation and evaluation methods for process equipment, process units, auxiliary systems, and the energy balance of the entire plant.

Part III (Chaps. 9–11) provides energy consumption analysis, including the impact analysis on ambient temperature and plant size, approach of energy saving, CO_2 capture, and utilization with nine case studies, technical and economic evaluation of energy-saving measures.

Part IV (Chaps. 12–14) discusses the optimization method for energy-saving measures, Pinch energy-saving technology, and its application with a case study of using Aspen Energy Analyzer, some key energy-saving technologies such as gas turbine combined cycle, FCCU regeneration gas CO combustion, and recovery. Flue gas turbine system optimization, low-temperature heat recovery with three approaches of same level utilization, refrigeration, power generation, etc.

This book avoids complexity, simplifies, focuses on practicality, and is easy for engineering and technical personnel to master and apply. It is a tool book and reference book for the engineering technical personnel who are working on process engineering, oil gas and production, energy management in oil refining, oil and gas production and processing, petrochemical, chemical, and pharmaceutical enterprises, and the power industry. This book can also be a professional reference or textbook for undergraduate or graduate-level university students and teaching personnel of chemical, energy, and process engineering faculties of universities.

The author of this book has provided both English and Chinese manuscripts; the English version and Chinese version of this book are published at the same time; the Chinese version of this book is《能源化工行业节能和碳减排》, 作者, 陈安民, 由中国石化出版社出版, 版权所有.

Calgary, Canada
June 2022

Tony A. Chen

Acknowledgements

I would like to express my deep and sincere gratitude to Dr. Haoming Li in Li's Process Consulting Service Inc. for his detailed review and valuable input to the CO_2 Capture and Utilization section, and for his help in checking and reviewing the simulation cases of this book.

I would like to express my heartfelt thanks to Dr. Lu Chunxi, Professor at Beijing University of Petroleum, for reviewing the manuscript proposal of this book and providing valuable review comments, and for providing strong support for the preparation and publication of this book.

I am very grateful to Dr. Mengchu Huang, Sridevi Purushothaman, Satish S. A. and his team of Springer Nature Press for their hard work, professionalism, help, support, and coordination for the timely publication of the English edition of this book.

I would also like to sincerely thank my employers Chevron, Jacobs, and LPEC/SINOPEC for providing me with a large, great working platform to enhance my knowledge, broadened my experience and allowed me to grow into an expert in process engineering, process simulation, energy-saving/carbon reduction.

Contents

List of Figures

List of Tables

Part I
Thermodynamic Basis of Energy Utilization Analysis

Chapter 1
Thermodynamics Fundamentals

Abstract To introduce the thermodynamic to process energy utilization analysis and improvement approach, this chapter provides the energy basic terms, energy forms to be used, and their reference states, which are frequently used in process energy utilization analysis and improvement; the first thermodynamic law and second thermodynamic law have been briefly introduced, which is the basis of entire energy utilization analysis and improvement.

Keywords Energy form · Energy utilization · Reference state · Greenhouse gas · Thermodynamic law

The petrochemical process is the process of processing petroleum raw materials into products through physical and chemical changes. Energy is the source and driven force to promote the process. Completing a series of processes from raw materials to products consumes energy. Energy conversion, energy process use, and energy recovery after using constitute the characteristics and regular pattern of energy use in the chemical and petrochemical process. Studying these characteristics and regular patterns, analyzing the process of energy use, identifying the potential, and proposing improvement measures is a realistic topic for our energy conservation workers.

Use thermodynamic methods to calculate and analyze the energy conversion and transfer, utilization and loss, recovery and rejection of the process energy utilization, reveal the energy demand magnitude, causes behind the energy consumption, and point out the direction for improving the process and increasing energy utilization efficiency, and use technical and economic optimization methods to analyze and screen, to identify efficiency improvement opportunities. This is the thermodynamic analysis method of the petrochemical process. At present, in the fields of petroleum, chemical industry, metallurgy, thermal engineering, etc., a systematic effective energy analysis method has been gradually formed, that is, the exergy analysis method, which has played an import role in the analysis and diagnosis of energy saving.

The so-called exergy analysis method relies on the first and second laws of thermodynamics to combine energy quantity and energy quality to calculate, analyze, diagnose and improve the energy utilization of processes and systems. The energy conversion and conservation law pointed out by the first law of thermodynamics has

T. A. Chen, *Energy Saving and Carbon Reduction*,
https://doi.org/10.1007/978-981-19-5295-1_1

been widely used in industrial processes. It has shown great advantages in the energy quantity balance of equipment, process units, and entire plant in industrial enterprises. It is the basis of energy analysis. However, only an energy quantity balance is not enough, the concept of energy quality still needs to be introduced to make the energy analysis more relevant to the essence, otherwise, incorrect conclusions could be drawn. For example, according to the thermodynamics first law, the energy entering and exiting the process equipment is conserved, and the quantity remains unchanged, so there is no "consumption", energy conversion and energy recovery that serve it both have the consumption of energy in quantity (flue gas, heat dissipation, cooling, rejection, etc.) conceal the essence of energy use in the process.

This chapter briefly introduces the thermodynamics first and second law based on introducing the basic concepts of thermodynamics.

1.1 Basic Terms

1.1.1 System and Environment

In the process of energy utilization, to clarify the scope of the research object, the research object is usually separated from the surrounding objects by the interface. This kind of artificially divided separately as the research and analysis object is called the system, and the system has a certain space and clear boundaries. The selection of the system is determined following the convenience and needs of our analysis and research, as small as a piece of equipment or even a part of a piece of equipment, as large as a process unit, a plant, an enterprise, or a region, all can be selected as a system for our energy analysis.

For an energy utilization analysis system, if the substance constituting the system is composed of a large number of particles with the macroscopic thermal phenomenon shown by the irregular movement of these particles, which determines the nature and condition of the system, it is usually called a thermodynamic system.

Most of the processing media involved in the chemical and petrochemical process are fluids, and their thermodynamic properties (T, p, V) are closely related to the processing process itself and energy utilization status, so they can all be regarded as thermodynamic systems.

Once the system is determined, all spaces and objects outside the system are called the "outside". The earth's surface (lower atmosphere, ground surface, and shallow water surface) on which humans live and production activities—the environment—is an important part of the outside.

The environment has an infinite capacity, its composition and the "strength properties" such as temperature and pressure are uniforms, and each part is relatively static, and will not change due to the input of the system or the energy absorbed from it.

Since the chemical and petrochemical process system is surrounded by the environment, the focus of our energy analysis research is the relationship between the system and the environment.

Under normal circumstances, there may be substances and energy exchanges between the system and the environment. A system without substance exchange with the environment is called a closed system. The substance exchange between the system and the environment is called an open system, also known as a flowing system. The system without heat exchange between the system and the environment is also called an adiabatic system. If a system has neither energy exchange nor substance exchange with the environment, it is called an isolated system.

Intermittent operations in petrochemical processes, such as closed reactors and simple distillation, are all closed systems. However, most continuous and stable processing processes belong to a flowing system.

The closed system has a constant quality, and the change of the system is a function of time.

The flowing system has a relatively constant spatial scope or boundary. The change of the processed materials in the system is a function of spatial position, not a function of time, and most chemical petrochemical processes belong to a flowing system.

1.1.2 State and State Parameters [4]

All the properties of the system determine the state of the system. There are many properties of the system, such as mass, pressure, temperature, volume, refractive index, density, heat capacity, composition, etc.

Under certain environmental conditions, if these properties of the system do not change with time, the system is in a state of thermodynamic equilibrium. The thermodynamic equilibrium state refers to the dynamic equilibrium of heat, force, and chemical composition. The temperature of each part of the system is equal and reaches the thermal equilibrium; when the system is composed of more than one substance, there is an equilibrium state of chemical reaction, and the composition of the system does not change with time, the chemical equilibrium is reached; the relative displacement between the system and the environment is reached force balance or mechanical balance.

Once the state of the system is determined, all its properties are also determined. Since the properties of the system are related to each other, when determining the state of the system, as long as a few of them are specified, the rest of the properties will be determined accordingly. When the property of the system changes, the state of the system also changes. It can be seen that there is a dependency relationship between the properties of the system and its state. For this reason, the nature of the system is often referred to as the state parameter of the system.

State parameters have characteristics that are not related to the process, but related to the beginning and end states of the system. It is a tool used by thermodynamics to study the properties, changes and interactions of the system. Various objective change regular pattern of the thermodynamic system is expressed by the correlation equations between state parameters.

State parameters include three types: strength quantities, such as temperature, pressure, chemical potential, etc.; volumetric quantities (extensive quantities), such as volume, electric current, mass, etc.; energy potential quantities, such as enthalpy, free energy, and free enthalpy, etc.

Energy potential (energy potential function) refers to the energy relationship between the system and the environment in the process, and it is additive with all the basic strength and volume parameters. But it is different from volumetric quantities in the following points:

(1) Both have energy dimensions; (2) they have an inclusive relationship between each other (such as $H = U + pV, U = F + TS\ldots$); (3) they can be expressed as a function of intensity and extension, and have the property of total differentiation, and (4) the amount of extension is different from the strength, they have only relative differences instead of absolute values.

1.1.3 Dead State and Reference State

The dead state refers to a specific state that the system finally reaches in the process of change, that is, the system reaches a thermodynamic equilibrium state with the surrounding environment, and then the strength parameters of the system and the environment are the same.

Because the surrounding environment of the system is different in different regions, even very different, such as north region and south region, winter and summer. However, to facilitate the analysis of energy use, the energy utilization analysis of chemical and petrochemical plants is established on the same basis, artificially stipulating a unified dead state—a reference state as a benchmark for energy balance and exergy balance calculations. The principle of determining the reference state is: as close as possible to the actual dead state and practical and convenient in engineering. For example, the CO_2 concentration of the environment is 0.03%. If the concentration is specified in the dead state, it is obvious that the energy and exergy between the concentration of the CO_2 in the system and the dead state cannot be used, but it brings difficulties to practical calculations.

1.1.4 Energy

Energy refers to the ability of the system to give or receive work and heat from the environment.

Work and heat are process quantities, which are always connected with specific energy transfer and conversion processes. The energy of a system is related to the starting and ending state of the system in a certain process. The final state of the energy use process is the dead state. In engineering, it is calculated to the specified reference state.

When the state of the system and the environment are different, energy transfer occurs between the system and the environment, and work and heat are the basic forms of energy transfer.

When the change between the system and the environment is caused by the temperature difference, energy transfer between the system and the environment will occur. The energy transferred from the system to the environment or from the environment to the system only due to the temperature difference is called heat. Heat is a process quantity, and it is stipulated that heat absorption is positive and heat release is negative.

When the changes between the system and the environment are driven by potentials other than the temperature difference, the energy changes caused are collectively called work. However, due to the difference of impulse potential, there are various forms of work. For example, the work using mechanical potential as the driving force is called mechanical work; the work using chemical potential as the driving force is called chemical work. In principle, work can be completely and continuously transformed into another form of work. Any form of work can be completely converted into heat. Heat cannot be completely converted into work. The degree of conversion is limited by the second law of thermodynamics and stipulates that the system delivering work to external is positive and the work obtained is negative.

1.1.5 Reversible and Irreversible Processes

When the system changes from one state to another, we say that the system has gone through a certain process. The process has three elements: initial state, final state, and path. There can be different paths between the states, which constitute different processes. Such as: constant temperature process, constant pressure process, constant volume process, adiabatic process etc.

If the system starts from a certain state, goes through a series of changes, and then returns to this state, the process in which the initial state and the final state are the same is called a cyclic process.

The reversible process is ideal in thermodynamics. It requires no frictional resistance during mechanical movement and no temperature difference during heat transfer. At every moment of system change, the medium or working fluid is in thermal equilibrium at any time, and the system changes by a series of equilibrium process of the infinitely small difference in intensity, and the process can be restored to the original state without leaving any changes in the outside of the system. In the actual process, the difference in strength is the driving force for change, the temperature difference promotes heat transfer, the pressure difference causes expansion, compression and flow, and the concentration difference causes mass transfer. For a closed system, an infinitely small difference in intensity means that the process is infinitely slow; for a flowing system, an infinitely small difference in intensity means that infinite space is required. For example, the process of reversible fluid transportation requires an infinite pipe diameter, and reversible heat transfer requires an infinite heat transfer area. Therefore, it is also an unrealistic process.

The key to the reversibility of the process lies in two conditions, one is frictionless (energy dissipation), and the other is a "quasi-static process". System changes consist of a series of equilibrium processes with the infinitely small driving force.

There is no real reversible process, and no process in nature is a reversible process. But under certain conditions, certain processes are close to reversible processes, such as the evaporation of pure liquid at its boiling point. The significance of studying the reversible process lies in: firstly, comparing the actual process with the reversible process as a benchmark to measure the efficiency of the actual process; secondly, for a changing process of the given system state, the change of certain state functions is only when the state change assumes a reversible process is easy to calculate. Since the state function only depends on the initial and final states, even though the actual process is irreversible, it does not affect the calculation result of the system state change based on the hypothetical reversible process.

In the actual petrochemical process, in order to complete the established production task within a unit of time, it is necessary to use a larger process driving force (intensity difference), which is far from the infinitely small quasi-equilibrium process of the driving force, therefore, all are irreversible processes; but the study of reversible process still has very important significance, first of all, as the direction of process improvement, the minimum energy consumption limit is compared with the actual process, and the actual process and the reversible process are compared with the efficiency. Secondly, for thermodynamic functions that depend on the initial and final states, the change value of the function can be determined by designing a reversible process.

1.1.6 Energy Use Process, Energy Use Process Thermodynamics

As a basic subject, thermodynamics has different meanings in different fields.

In the field of thermal engineering, engineering thermodynamics focuses on the expansion and compression of the working fluid in the thermal engine process, with the purpose of obtaining mechanical work from the internal energy of the working fluid. There is no definite initial and final state.

Chemical thermodynamics focuses on the chemical and physical changes (phase equilibrium, chemical composition, etc.). The purpose is to clarify the direction and limits of the process and to find suitable operating conditions. There is no certainty of initial and final states.

The petrochemical energy use process focuses on the reasonable efficiency of energy use. The process is very different, and there is no uniform initial and final states. To find the utilization efficiency of energy resources, energy consumption analysis requires people to take the same reference state and calculate the energy released or absorbed by the process.

Energy source is a variety of substances or systems in nature that can produce spontaneous changes in the natural environment where humans live, and release energy in the form of work and heat in the process of change. This spontaneous change can happen because of the difference in intensity between these systems or substances and the environment. These differences in intensity form the driving force for process change and have the nature of "energy". The process by which release and use the energy of the energy source or system is called the energy-using process. The final state of the energy-using process is the dead state.

Energy is used in any production process. The process of using energy is the process by which energy substances release energy and be used. Using thermodynamics theory to study the energy use process and the rationality of energy use is called energy use process thermodynamics. The purpose of energy use process thermodynamics is to obtain the highest efficiency as possible and the greatest economic benefits for the process at the lowest possible cost. It is not only to obtain work required, but to put in less energy source to maximize the various forms of energy (heat, work, etc.) required by the process.

1.2 Energy Form and Reference State

1.2.1 The Form of Energy

Energy is a measurement of the change and interaction of the movement of a substance. Because the movement of substance has many forms, such as mechanical movement, electromagnetic movement, thermal movement, chemical change, and nuclear fission or fusion etc. Therefore, there are energy forms corresponding to the forms of motion, such as mechanical energy, electrical energy, electromagnetic energy, thermal energy, chemical energy, and nuclear energy etc. Just as movement is the basic attribute of substance, energy is also possessed by all substance, and it is the driving force that drives substance movement and interaction.

The energy of substance can be divided into two categories, one is the energy of an object relative to another, including potential energy and kinetic energy, and the other is the energy contained within the object, called internal energy, which includes all forms of energy other than potential energy and kinetic energy. There are many levels of internal energy. Internal kinetic energy, internal potential energy, chemical energy, nuclear energy, etc. are all internal energy. Generally speaking, internal energy refers to the energy contained in the substance without chemical and nuclear changes. Internal energy is an important thermodynamic function, and its absolute value is difficult to determine. Generally, selected reference state is used to determine it.

Although there are many different forms of energy, the energy can be divided into work and heat according to the characteristics of substance particles movement-orderly and disorderly movement. Correspondingly, exergy can also be divided into work exergy and thermal exergy. Chemical energy belongs to the category of work due to the orderly movement of electrons within the molecular structure. But once the compound is burned, it changes from orderly movement to disorderly movement, generating heat.

The petrochemical and chemical process is a process that uses energy, but the most direct forms of energy used in the process are work and heat. Under the current technical and economic conditions, energy source is first converted into heat, and then electricity is generated as needed, or heat is directly supplied to the process system. For small and medium petrochemical companies that do not have their power plants, electricity is often purchased from outside the company. Therefore, most of the petrochemical energy used is used for heating (including steam supply), and of course there are also gas turbines and/or steam turbines for generating power. Work and heat are the most basic forms of energy use. Determining the calculation methods of process stream work and heat in various processes in petrochemical processes under different conditions is the basis for energy use analysis. Generally speaking, the energy used in petrochemical processes mainly has the following forms: process stream heat energy (sensible heat, phase change heat, chemical reaction heat, etc.), flowing energy (work), and fuel chemical energy, and there are also corresponding exergy forms.

1.2.2 Determination of the Reference State

The energy and exergy of process streams are relative to the dead state of the system. According to the dead state of different systems, we can determine a reference state for calculating energy and exergy. Only a certain reference state can have a certain value of energy and exergy. Theoretically, the reference state is a state in which the properties of the system and the thermodynamic properties in the surrounding environment are in equilibrium, and it is the end point and limit of system changes. However, how to determine the parameters of the intensity of the environment is a concern for practical use, because the system is located in different regions and the environmental condition is also different. The atmospheric temperature varies to a certain extent with seasons, day and night, latitude and topography. What temperature can be taken as the reference temperature, and what is the most stable compound of various elements on the surface of the earth? What is its concentration? It is a matter of determining the reference status. This makes the determination of the reference state a complicated issue.

The key to the reference state of physical energy (exergy) is to determine the reference temperature T_0 and pressure p_0; the reference state of chemical energy (exergy) (including diffusion exergy) is to determine the most stable chemical composition and concentration of various elements or compounds on the earth surface.

However, whether it is a physical reference state or a chemical reference state, a unified opinion that is generally accepted by everyone has not yet formed. Generally, the determination of the reference state should follow two principles.

(1) The object of our research is the system, and the final state of system change is the surrounding environment where the system is located. Therefore, we determine the dead state (reference state) to determine the reference state based on the environmental state around the system.
(2) The processing energy analysis we conducted in the engineering energy analysis. Therefore, it is not necessary to demand a high degree of unification of theoretical concepts, but to simplify the environmental state from the perspective of practicality and convenience. In order to make the energy consumption analysis of different systems comparable, it is even possible to take an average and unified reference environment state acceptable to everyone in engineering applications.

1.2.2.1 Physical Reference State

1. Reference Temperature

The reference temperature is determined based on the principle that it is as close as possible to the ambient temperature of the system. In order to facilitate mutual comparison in the energy project, combined with the annual average temperature of different regions and seasons, the oil refining industry generally uses 15 °C (288 K), and 25 °C 298 K in the thermal and chemical industries. The value should be determined or selected according to the specific conditions of the region.

However, most of the physical property data in the chemical manuals are physical properties at 25 °C. It is more convenient to use 25 °C as the reference temperature and easy to be accepted. For the energy balance, although the energy value has slightly error if different T_0 is selected, the mass quality of the stream in and out of the system is the same under stable conditions, so the error will also be eliminated in the energy balance. But on some occasions when calculating the efficiency of the first law, choosing a different reference temperature will affect its value.

But the calculation of exergy is different. The value of thermal exergy is proportional to the ambient temperature, $EX = \left(1 - \frac{T_0}{T}\right)E$, the higher system temperature T, the smaller the influence of T_0; but when the stream temperature is lower, close to T_0 (below 100 °C), the influence is greater. Therefore, in the analysis of the general process, the temperature is relatively high, and most of the incoming and outgoing stream mass difference are ignorable. T_0 has little impact and a unified reference T_0 can be used. For the low temperature process, especially the refrigeration process, the system temperature is close to T_0, and the T_0 has a great influence on the calculation results. It should be based on the actual atmospheric temperature of the environment where the system is located, and cannot be blindly unified.

2. Reference Pressure

Reference pressure is generally taken as 1 atm (0.1 MPa); although there are slight differences in atmospheric pressure for differentcountriesy or regions, they can usually be ignored. p_0 has little effect on analysis and calculation, because the pressure in our studying system is far away from the reference pressure. However, for specific research systems, it should also be dealt with specifically.

3. Reference Phase State

The reference phase state is also an important content of the physical reference state. The determination of the reference phase state is the phase state of the system stream at the reference temperature and pressure. For example, the reference phase of water is liquid (boiling point 100 °C), and the reference phase of CH_4 is gaseous (boiling point at atmospheric pressure is -161.5 °C). The determination principle is that the normal pressure boiling point of the stream is higher than the reference temperature as the gas phase, and the atmospheric boiling point lower than the reference temperature is the liquid phase.

1.2.2.2 Chemical Reference State

The chemical energy (exergy) is caused by the difference between the constituent substance and components of the system and the environment. In order to calculate the chemical energy (exergy) of a compound, the chemical reference state must first be determined.

Determining the chemical reference state is actually to determine the reference substance of the chemical reaction, which includes the composition and phase of the reference compound, the environmental area (atmosphere, sea water, and the surface of the earth) and concentration. The selection of reference substance is very complicated, and there are different arguments.

The so-called reference substance is the most stable substance that is in equilibrium in the environmental state, that is, the substance that can no longer release energy through chemical reactions and concentration changes under ambient temperature and pressure.

1. Features of the environmental reference substance

   (1) Each element has its corresponding reference substance. The reference substance system should include reference substance of all elements in the research system.
   (2) Various reference substance should be the substances that can exist in the environment (atmosphere, ocean, and earth surface), and they can be continuously supplied by the environment without consuming useful work.
   (3) It is impossible for any spontaneous chemical changes to occur between reference substances.
   (4) Various reference substances should be the most stable substances of corresponding elements, and their energy (exergy) value is 0.

Table 1.1 Chemical reference state of the main elements (note 1)

Element	Reference phase state	Reference substance	Environmental field	Reference concentration (2)	Chemical exergy of element (kJ/kmoI)
Al	Solid	Al_2SiO_5	Earth crust	2×10^{-3}	887.89
C	Gas	CO_2	Atmosphere	0.00003	410.53
C_a	Ion	Ca^{+2}	Seawater	4×10^{-4}	717.4
Cl	Ion	Cl^-	Seawater	19×10^{-3}	117.52
Fe	Solid	Fe_2O_3	Earth crust	2.7×10^{-4}	377.74
H	Liquid	H_2O	Seawater	~1	235.049
N	Gas	N_2	Atmosphere	0.07583	720
Na	Ion	Na	Seawater	10.56×10^{-3}	343.83
O	Gas	O_2	Atmosphere	0.0204	3.97
P	Ion	$HPO_4{}^{-2}$	Seawater	5×10^{-3}	859.6
S	Ion	$SO_4{}^{-2}$	Seawater	8.84×10^{-4}	598.85
Si	Solid	SiO_2	Earth crust	4.72×10^{-1}	803.51

Note 1 All data in this table except H_2 adopts Szargut's data
Note 2 The reference concentration is mole fraction in the earth surface and partial pressure (MPa) in the atmosphere

2. Reference System

The reference systems proposed by different scholars are different, and there are certain human factors or practical considerations. The calculation of chemical energy and exergy is not required very precisely in energy analysis of petrochemical process, and it has been simplified to a certain extent. In many cases, the calculation of chemical exergy can be avoided. Table 1.1 lists the process related chemical reference state of the main elements [1] for reference.

1.3 The First Law of Thermodynamics

The thermodynamics first law is about the application of energy conversion and conservation laws in thermodynamics. All thermodynamic processes within the studying scope of thermodynamics should obey the first law of thermodynamics.

1.3.1 General Expression of the First Law of Thermodynamics

Any object has a certain amount of energy; energy has different forms, under certain conditions one form of energy can be transformed into another form of energy; in an isolated system, the sum of various forms of energy remains unchanged.

As mentioned earlier, energy can be divided into two categories: heat and work. The energy transfer driven by the temperature difference is heat. Except for the temperature difference, the energy transfer driven by other potential differences is collectively called work.

For a closed system without mass exchange outside, the first law expression is:

$$\Delta U = Q - W \tag{1.1}$$

It shows that in any change process of the closed system, the reduction of the internal energy of the system is equal to the algebraic sum of the outward work and heat release of the system.

In the above formula, ΔU is the state parameter. When the state of the system at beginning and ends is determined, ΔU is determined accordingly. However, there are countless paths and processes. The work and heat release of each process are different, but the sum of the two is constant and equal to the internal energy change value ΔU. The conversion and conservation of energy constitute the basic content of the first law of thermodynamics. Any heat engine that violates the first law of thermodynamics is called the first-type perpetual motion machine and can't be realized.

1.3.2 Energy Balance Equation of Stable Flow System [2]

Applying the first law expression of the closed system to the stable flowing system, the corresponding expression of the flowing system is obtained.

Take a stable flow system as shown in Fig. 1.1, which contains different size pipes for conveying fluids, and the fluid flows continuously and stably from the left end to the right end in the pipe. A heat exchanger is installed in the pipeline to supply heat to the fluid. When the fluid passes through the turbine in the pipeline, it pushes the turbine to rotate and perform work outward. We take the pipe inlet and outlet cross-sections 1.1 and 2.2 as the boundary to study the energy changes of the system.

The above-mentioned stable flow process means: (1) The quality of the materials entering and exiting the system is equal; (2) The state of the material at any point in the pipeline and the equipment does not change with time.

Take *mkg* of fluid in the system as the calculation basis. When it flowing into 1-1 interface Suppose that the internal energy per unit mass and specific volume are respectively U_1, V_1, the pressure is p_1, the position height is Z_1, and the flow velocity is u_1. When exiting the 2-2 interface, it becomes U_2, V_2, p_2, Z_2, and u_2 accordingly.

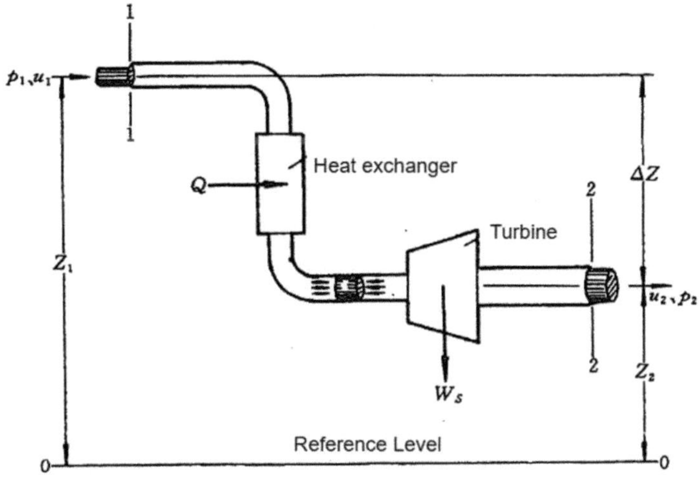

Fig. 1.1 Stable flow process

Obviously, during the flow process, the energy change of the fluid, the heat supplied to the fluid by the heat exchanger, and the shaft power output by the turbine should obey the basic relationship of the first law of thermodynamics, namely:

$$E_1 + Q = E_2 + W_s \tag{1.2}$$

where:

E_1, E_2—m kg fluid total energy at the 1-2 and 2-2 sections;

Q-*mkg*—fluid flows through the heat exchanger obtained total heat;

W_s-*mkg*—fluid flows through the turbine output total shaft power.

When fluid flows into and out of the system, its total energy includes the following

(1) **The internal energy** It reflects the energy of the movement of the microscopic particles inside the molecule is called internal energy, which is different from the macroscopic potential energy and the macroscopic kinetic energy due to the change of the position or movement of the whole substance. Including the kinetic energy of molecular motion, the potential energy of molecular interaction in various force fields, the energy of atom and electron motion, etc. Both heat and work obtained by the system can increase the internal energy of the substance.

Although it is not yet possible to determine the absolute value of the internal energy of a substance, when the transfer of heat and work occurs between the system and the environment, the internal energy changes accordingly, and only the change value needs to be calculated for thermodynamic analysis.

Since U represents the internal energy of a unit mass material, the internal energy of the import and export unit mass material is U_1 and U_2, and the total internal energy is mU_1 and mU_2.

(2) **Potential energy** The material is affected by the gravitational field and has different gravitational potential energy at different height positions, which is equal to the work required when the material is lifted from a selected reference surface to this height. In Fig. 1.1, the 0-0 plane is used as the reference plane for calculating the height (horizontal plane). The height of any point in the pipeline or equipment is represented by Z as the vertical distance from the point to the reference plane. The value of Z above the reference plane is positive, the below is negative. The potential energy of a material with a mass of mkg at a height of Z m is mgZ_1. g is the acceleration of gravity, and its unit is $m \cdot s^{-2}$, and the unit of potential energy is J. Potential energy of material import and export are $E_{p1} = mgZ_1$ and $E_{p2} = mgZ_2$

(3) **Kinetic energy** the flow velocity of a certain section of the material flowing through the pipeline or equipment is expressed in u, and its unit is m/s, so the input and output kinetic energy are (1/2) mu_1^2 and (1/2) mu_2^2 separately.

(4) **Flow work** in the continuous flow process, the work exchanged by the internal pushing of the fluid is the flow work. The reason why a certain amount of fluid can squeeze into the interface 1-1 is because it is pushed by the fluid behind, and thus receives the flow work. Similarly, when a certain amount of fluid flows out of the interface 2-2, the front fluid must be pushed to perform flow work on the front fluid. It is known that the specific volume of the fluid at the inlet interface 1-1 is V_1, and the unit is m^3/kg; the cross-sectional area of the pipe at this point isA_1, and the unit is m^2; then the distance of the m kg fluid squeezed into the interface 1-1 is $l = (mV_1/A_1)$; the driving force is$p_1 A_1$. Therefore, the input flow work is: $(p_1 A_1)(mV_1/A_1) = mp_1 V_1$. Similarly, when mkg fluid leaves the system, the output flow work is $mp_2 V_2$.

(5) **Shaft work** The work exchanged between the system and the environment through the rotating shaft of the mechanical equipment in the process of fluid flow is called shaft work, expressed in W_s.

(6) **Heat** is the heat Q supplied to the system through heat exchange equipment.

From the above, the total energy of the system at the inlet and outlet interfaces are respectively:

$$E_1 = U_1 + mgZ_1 + \left(\frac{1}{2}\right)mu_1^2 + mp_1 V_1$$

$$E_2 = U_2 + mgZ_2 + \left(\frac{1}{2}\right)mu_2^2 + mp_2 V_2$$

Substituting into the basic relationship (1.2) of the first law, we get

$$mU_1 + mgZ_1 + (1/2)mu_1^2 + mp_1 V_1 + Q$$

$$= mU_2 + mgZ_2 + (1/2)mu_2^2 + mp_2V_2 + W_s \quad (1.3)$$

Equation (1.3) is the total energy equation of the stable flow system. For unit mass fluid $m = 1$, the total energy balance is

$$U_1 + gZ_1 + (1/2)u_1^2 + p_1V_1 + q = U_2 + gZ_2 + (1/2)u_2^2 + p_2V_2 + w_s \quad (1.4)$$

where: $q = Q/m$, $w_s = W_s/m$.

Equation (1.4) can also be written as follows:

$$\Delta U + g\Delta Z + \left(\frac{1}{2}\right)\Delta u^2 + \Delta(pV) = q - w_s \quad (1.5)$$

According to the definition of enthalpy: $H = U + pV$

Therefore $\Delta H = \Delta U + \Delta(pV)$, then (1.5) can be written as

$$\Delta H + g\Delta Z + \left(\frac{1}{2}\right)\Delta u^2 = q - w_s \quad (1.6)$$

For a small thermodynamic change process, then

$$dH + gdZ + udu = \delta q - \delta w_s \quad (1.7)$$

Formula (1.4) to formula (1.7) are the mathematical expressions of the first law of thermodynamics in a stable flow system. Each term in the formula represents the energy per unit mass of fluid. According to the SI unit system, the unit of each term is kg·m^2 s^{-2} kg^{-1} = m^2 s^{-2} , that is J/kg.

1.3.3 The Form of the Total Energy Balance Equation Under Different Conditions

In the total energy balance equation of the stable flow system obtained above, there are more energy items included, but not all items need to be considered under any circumstances. In most cases, one kind of energy can be omitted to make the equation greatly simplified. Below we discuss this special form of energy balance in combination with various specific situations encountered in petrochemical production.

(1) Static system: When formula (1.5) is used in a static system, because the material does not flow and the position height does not change, $\left(\frac{1}{2}\right)\Delta u^2 = 0$; $g\Delta Z = 0$. Then the formula becomes:

$$\Delta U + \Delta(pV) = q - w_s$$

Because we are talking about a static system, for this reversible non-flowing process, its work refers to the volume work (expansion work or compression work), namely $\int_{v1}^{v2} pdV = w_s + \Delta(pV)$, therefore:

$$\Delta U = q - \int_{v1}^{v2} pdV \tag{1.8}$$

This is the formula of the first law of thermodynamics in a static system. This formula is suitable for reversible non-flow processes. For example, when it is used in the actual expansion or compression process, the pressure in the formula refers to the external pressure that the system resists.

(2) When fluid flows through pipelines, valves, heat exchangers, absorption towers, mixers, reactors and other equipment, it does not perform shaft work, and the macroscopic potential and kinetic energy of fluids often change little and can be ignored when flowing through this equipment, that is: $w_s = 0, g\Delta Z = 0, (1/2)\Delta u^2 = 0$; then the Eq. (1.6) can be simplified to:

$$\Delta H = q \tag{1.9}$$

This formula shows that the enthalpy change of the system is equal to the heat exchanged between the system and the environment, which is the basic relational formula for the heat balance of the steady flowing system.

(3) Under normal circumstances, the density of gas is very small, and the height of the pipeline does not change much, so the macroscopic potential energy change of the gas is very small and can be ignored. For example, in the flowing process, the gas does not perform shaft work, and there is no heat exchange with the outside world, then the Eq. (1.6) can be written as

$$\left(\frac{1}{2}\right)\Delta u^2 = -\Delta H \text{ or } udu = -dH \tag{1.10}$$

It can be seen that the increase in kinetic energy of the working fluid is equal to the decrease in the enthalpy of the fluid during the adiabatic flow without shaft work. Equation (1.10) is called the adiabatic stable flow process equation.

(4) When the system has neither shaft power nor heat exchange with the outside world, for example, under adiabatic conditions, passing through a mixer, reactor (assuming good insulation), or high-pressure fluid passing through a throttle valve, then: $g\Delta Z = 0$, $\left(\frac{1}{2}\right)\Delta u^2 = 0$, $w_s = 0, q = 0$; thus

$$\Delta H = 0 \text{ or } H_1 = H_2 \tag{1.11}$$

That is, when entering the equipment, the enthalpy value of the fluid is equal to the enthalpy value of the fluid when leaving the equipment. According to this relationship, the temperature change of the system in the above process can be easily obtained.

(5) When the system exchanges shaft work with the outside under adiabatic conditions, for example, when the fluid passes through compressors, blowers, pumps and other equipment with slow heat dissipation, $q \approx 0$, which can be obtained:

$$\Delta H = -w_s \tag{1.12}$$

That is, the enthalpy change of the system is equal to the shaft work exchanged between the system and the outside. Using this relationship, as long as the enthalpy value of the inlet and outlet when the working fluid passes through the equipment, the shaft work of the equipment can be obtained.

(6) Mechanical energy balance formula

The mechanical energy balance formula can be further derived from the total energy equation. The differential form of the total energy balance equation is

$$dU + pdV + Vdp + gdZ + udu = \delta q - \delta w_s \tag{1.13}$$

Because: $\delta q = dU + \delta w_s$

For the reversible process, in the above formula: $\delta w_s = pdV$, formula (1.13) becomes:

$$Vdp + gdZ + udu = -\delta w_s$$

When the friction loss cannot be ignored, the friction loss should be added to the left end of the above formula. Take δF representing the friction loss, the mechanical energy balance of the steady flow system must be used.

$$Vdp + gdZ + udu + \delta F + \delta w_s = 0 \tag{1.14}$$

If there is no shaft work the input and output in the steady flow system, the friction loss is negligible, and the fluid is incompressible, that is,

$$\int_{p_1}^{p_2} vdp = V(p_2 - p_1) = (1/\rho)\Delta p$$

where ρ—fluid density.

Then the integral form of mechanical energy balance is

$$\frac{\Delta p}{\rho} + g\Delta Z + \left(\frac{1}{2}\right)\Delta u^2 = 0 \tag{1.15}$$

This is the famous Bernoulli equation.

(7) If we don't consider the changes in macro potential energy and macro kinetic energy, $mg\Delta Z = 0$; $\left(\frac{1}{2}\right)m\Delta u^2 = 0$, then from the frmular (1.6), the first law of thermodynamics in a steady flow system becomes

$$\Delta H = q - w_s \tag{1.16}$$

This is a very important formula in the calculation of enthalpy balance.

1.3.4 Application of the First Law in the Petrochemical Process

The first law can be applied to the chemical and petrochemical process, but a certain transformation is still needed. This is because:

(1) For a process system operating under continuous and stable flow conditions, there is often more than one stream that enters and exits the system; the number of exchanges of work and heat between the system and the outside may be more than one.
(2) The entry and exit of system materials reflects the change from raw materials to products. The energy in and out are often reflected in the conversion and utilization of energy for the purpose of achieving material changes. Therefore, the work, heat and process Stream Energy (enthalpy difference) are often not expressed as the total result of the relationship between the system and the outside, but expressed as the input side and the output side respectively.
(3) Due to engineering and economic factors, the use of energy in the process generally cannot be "to the end". In other words, most of the streams or heat leaving the system carries out energy at pressure and temperature higher than the ambient conditions. Therefore, the energy balance mode for the general process system is:

For the energy balance relationship shown in Fig. 1.2, there are obviously:

$$\sum(H_i - H_{i0}) + Q_i + W_i = \sum(H_e - H_{e0}) + Q_e + W_e$$

Fig. 1.2 Schematic diagram of the energy balance of the process

$(H - H_0)$ is the energy of the Process streams, represented by E

$$\sum E_i + Q_i + W_i = \sum E_e + Q_e + W_e \tag{1.17}$$

In the formula, the subscript i represents the input streams, the subscript e represents the output streams, and the subscript 0 represents the reference state. Equation (1.17) shows that the sum of the energy carried by all the streams entering the process unit and the work and heat provided by the outside world, which is equal to the sum of the energy carried by all the streams leaving the process unit plus the work discharged and heat output from the process. Which reflects the energy conservation. The values of work and heat in the formula can be positive or negative according to regulations.

Formula (1.17) is widely used in single equipment and local systems, and is the basis of energy balance. It is inconvenient for complex chemical or petrochemical processes composed of many process units and operating equipment. In Chap. 5, we will introduce the three-link energy utilization analysis model and balance relationship in chemical and petrochemical process based on the thermodynamics first law, the process energy use characteristics and the energy evolution law.

1.4 The Second Law of Thermodynamics

The first law of thermodynamics explains the conservation of energy in the process of conversion and transfer, and cannot point out the direction, conditions and limits of energy transfer. It is not that all processes that do not violate the first law of thermodynamics can be realized, but the second law of thermodynamics clarifies this problem scientifically.

In 1956, Yugoslav scholar Rant proposed that "the part of energy that can be converted into technical work under given natural environmental conditions is called Exergy". This formulation has been recognized by the international academic community [3]. Since then, people's understanding of the principles and laws of energy utilization has entered a new stage, and has promoted the development of various new energy-using technologies and the rapid improvement of energy utilization efficiency. At the same time, thermodynamics has made new developments in new applications. The second law of thermodynamics is applied to the process energy use and is a milestone in the development of thermodynamics.

1.4.1 The Expression of the Second Law of Thermodynamics

There are many ways to express the second law of thermodynamics. For example, heat cannot be transferred spontaneously from a colder object to a hotter object; it is impossible to take out heat from a single heat source and turn it into work completely without producing other changes. In fact, heat can only be transferred from a high-temperature object to a low-temperature object, indicating the direction of the heat transfer process; taking out heat from a single heat source can only partially perform work, indicating the conditions and limits of heat conversion into work. Therefore, the second law of thermodynamics is a law that describes the direction, conditions, and limits of spontaneous processes. However, the most basic expression of the second law of thermodynamics is: the entropy of an isolated system increases constantly.

$$\Delta S_{iso} = \Delta S_{system} + \Delta_{outside} \geq 0 \tag{1.18}$$

The equal sign in the formula indicates the reversible situation, and the greater than sign indicates the actual process.

Entropy, as a measure of the chaotic degree in the system, has its statistical laws. The behavior of tiny particles (gas molecules) that always move randomly is in accordance with statistical laws. Statistical thermodynamics points out that the temperature T is a measure of the average movement energy of a large number of particles in the system, and the entropy S is a function of the probability Ω of the microscopic state of the particles in the system, expressed as Boltzmann's equation.

$$S = k.ln\Omega \tag{1.19}$$

In the formula, k is Boltzmann's constant, $k = \frac{R}{N_0}$ is the ratio of the gas constant to the Alvogadro constant $N_0(6.022045 \times 10^{23}\text{mol}^{-1})$.

Ω represents the chaotic degree of the disordered particle movement of the system. Boltzmann's equation links the macroscopic thermodynamic functions with the physical quantities of the microscopic state, thus building a bridge from the macroscopic to the microscopic.

Combining the entropy function with the first law of thermodynamics, a new state function-free energy G can be defined:

$$G = H - TS$$

Under isothermal and isostatic conditions, the maximum useful work performed by the reversible process is equal to the reduction in the free energy of the system.

$$W_{useful} = -\Delta G_{TP} \tag{1.20}$$

It shows the limit of the system's work under isothermal and isostatic conditions.

In the case of not doing useful work, such as the general chemical reaction process is carried out under constant temperature, constant pressure, and not doing useful work, the direction and limit of the spontaneous progress of the process can be judged:

$(dG)_{pT} < 0$ Indicates that the process can be carried out spontaneously;

$(dG)_{pT} = 0$ Indicates that the process has reached an equilibrium and has reached the maximum.

For the phase change process, because it is also under constant temperature and pressure, this criterion can also be used.

1.4.2 The Maximum Limit of Thermal Variable Work, Carnot Factor

As shown in Fig. 1.3 a, if a heat engine realizes a certain reversible cycle between A and two heat sources ($T > T_0$), in order to make the reversible cycle proceed, each step of the cycle must meet the requirements of the above reversible process requirements, therefore, select the following process to form a cycle.

Heat absorption and heat release process: According to the given conditions, since there is only one heat source T and one cold source T_0, in order to make the heat absorption and release process between the two heat (cold) sources meet the reversible conditions, the temperature difference between the heat source and the cold

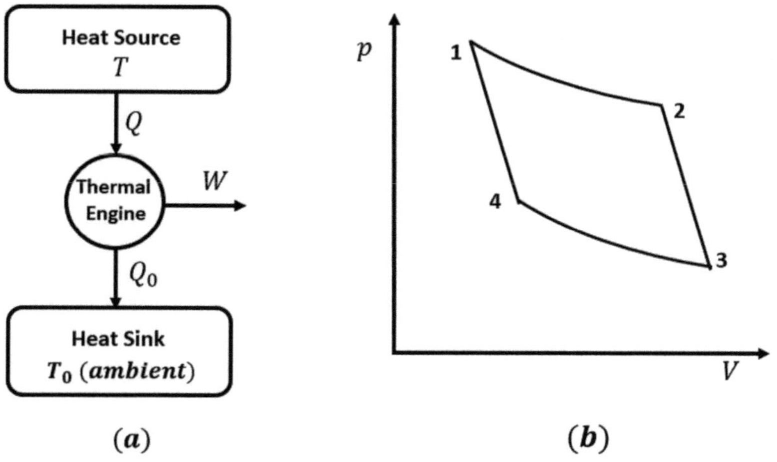

Fig. 1.3 Heat engine cycle process

source requires infinitely small, which means that constant temperature exotherm and constant temperature endothermic must be achieved, but the constant temperature lines of endothermic heat and exothermic heat cannot intersect on the pV diagram. Only two constant temperature processes cannot form a cycle, therefore, adiabatic expansion and adiabatic compression are inserted in the process of heat absorption and heat release, which constitutes a cyclic process. As shown in Fig. 1.3b, 1.2 is the constant temperature endothermic process at temperature T; 2-3 is the adiabatic expansion process; 3-4 is the constant temperature exothermic process at temperature T_0; 4-1 It is an adiabatic compression process.

The cycle formed by the above process is called the Carnot reversible cycle. When the heat absorption is Q and the heat release is Q_0, the completed work is $W = Q - Q_0$; its efficiency $\eta = \frac{W}{Q} = 1 - Q_0/Q$, due to $\left(\frac{Q_0}{Q}\right) = (\frac{T_0}{T})$

$$\eta = 1 - T_0/T \tag{1.21}$$

It can be seen from formula (1.21) that the thermal efficiency of the Carnot reversible cycle is only related to the temperature of the heat source and the cold source, and has nothing to do with the nature of the working fluid and the type of heat engine; it is impossible to absorb all the heat from the heat source to convert it all into work in any cycle, the conversion efficiency between heat and work is always less than 1; $T_0 = T$, $\eta = 0$, which indicates: if there is no temperature difference, it is impossible using a single heat source to cycle work, even an ideal cycle such as the Carnot reversible cycle. To realize the conversion between heat and work, there must be more than two heat sources with different temperatures. This is an indispensable thermodynamic condition for all heat engines.

The Carnot cycle is a reversible cycle without power consumption. The work obtained cannot be reached by any actual cycle. It is the maximum work, that is, the maximum limit of heat conversion into work.

If the Carnot cycle is reversed, it is the cycle of a reversible refrigerator cycle or a reversible heat pump cycle. At this time, the coefficient of refrigeration is:

$$\varepsilon = \frac{1}{\frac{T}{T_0} - 1} = T_0/(T - T_0) \tag{1.22}$$

The coefficient of performance of the heat pump should be:

$$cop = \frac{1}{(1 - \frac{T_0}{T})} = T/(T - T_0) \tag{1.23}$$

1.4.3 The Expression and Significance of the Energy Use Process of the Second Law

The entropy increase principle of the isolated system can be used in the energy use process, as shown in Fig. 1.4a. For future convenience, a flow system with a controlled volume is given. The characteristics of the energy use process are embodied here as follows:

(1) The final state of the process is a dead state, and the system intensity p_0, T_0 etc. are the same as the environmental intensity.
(2) The heat released by the system in the process must eventually be transferred to the environment. This is because the environment is the only final heat sink for the energy use process of natural energy. If there is a certain energy-using process that heat Q' is transferred to another system instead of directly to the environment, then the heat Q' will eventually be released to the environment during the change process of that system (Fig. 1.4b). At this time, a complete energy use process must include these two systems at the same time. It can also be seen from Fig. 1.4b that if the entire energy use process is reversible, the total work done will be the largest, that is, the heat released to the environment Q will be the smallest.

Therefore, when formula (1.18) is used in the energy use process, there are

$$S_{system} = S_0 - S$$

$$S_{outside} = S_{envor} = Q/T_0$$

Because the outside only obtains the shaft work w, its entropy remains unchanged, and the environment always exchanges heat at the temperature of T_0, which can always be considered reversible. Substituting into Eq. (1.18), we get

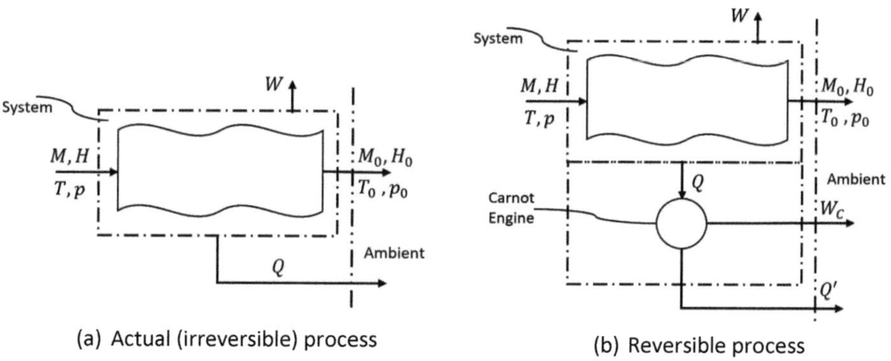

Fig. 1.4 Fluid flow system versus environment

$$S_{iso} = S_0 - S + \left(\frac{Q}{T_0}\right) \geq 0$$

After finishing, we have to use the expression of the second law of the energy use process:

$$Q \geq T_0(S - S_0) \tag{1.24}$$

In general, this formula can be expressed as follows: all the energy released by the system $S \neq S_0$ in the energy use process will always be transferred to the environment in the form of heat$Q$; for the reversible process, $Q = T_0(S - S_0)$; For the irreversible process$Q > T_0(S - S_0)$. Under certain conditions, such as a low-temperature system, both ends of formula (1.24) are negative. The system absorbs heat from the environment and performs work on the outside, but the above formula still be right.

Equation (1.24) actually points out the relationship between the relative quantity of work and heat in the energy conversion process. It shows that this relative quantity is mainly determined by the change of the entropy S of the system itself in addition to the environmental conditions (T_0).

1.4.4 Concept and Calculation of Exergy

Exergy is a thermodynamic state function proposed by people in recent years to evaluate the capability of work performing of the system in the reversible process. It is defined as: The energy that can theoretically be converted into work to the maximum extent based on the equilibrium environmental state (dead state) is called Exergy (EX), and the energy that cannot be converted into work in theory is called waste energyA_N. In theory, the only condition for maximum conversion into work is that the process is reversible. According to this definition, ordered energy such as mechanical energy and electrical energy can be completely converted into work in theory, so all of them is Exergy. For thermal energy, it cannot be completely converted into work, Limited by Carnot factor.

As shown in Fig. 1.5, the material system is in an initial state (state parameters are$P$, T, H, S, etc.) through several reversible processes to reach a state of equilibrium with the environment (state parameters are represented by subscript0). These processes are collectively referred to as process A. Suppose total work w_1 and heat release Q_1. These heat Q_1 may not necessarily be at ambient temperature, so they can also form a Carnot cycle, denoted as process B, perform workw_2, and heat Q_0 is discharged to environment. In this way, it is considered that the system has made the maximum work.

Combine processes A and B to establish an energy balance:

$$H = H_0 + w_1 + w_2 + Q_0$$

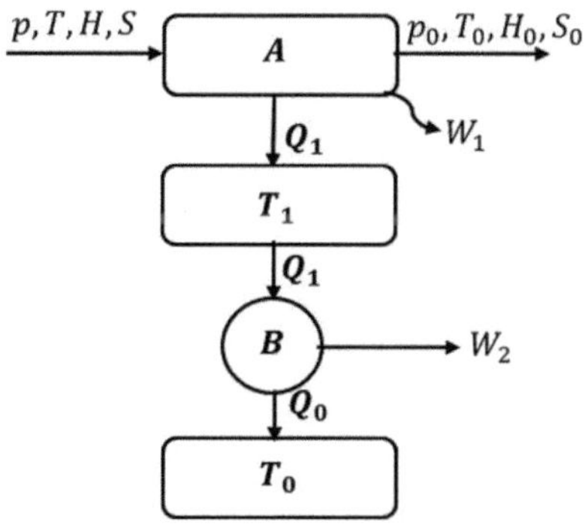

Fig. 1.5 Stream exergy calculation

In the output item, H_0 is the enthalpy value in the ambient state, and Q_0 is the heat in the ambient temperature, neither of which has any Exergy. So, by definition:

$$EX = w_1 + w_2 = H - H_0 - Q_0$$

All the above processes are reversible, and according to the second law of thermodynamics, the total entropy change (the sum of the entropy change of the system and the entropy change of the environment) of the reversible process is zero. Here, the entropy of the system changes from S to S_0; the working medium of the Carnot machine forms a closed cycle, and the entropy becomes zero; the environment receives heat Q_0, and its entropy increases to (Q/T_0) according to the definition of entropy. Then, the total entropy becomes:

$$S_0 - S + \frac{Q_0}{T_0} = 0$$

Substitute Q_0 into the above:

$$EX = (H - H_0) - T_0(S - S_0) \tag{1.25}$$

This formula is the basic formula for calculating the exergy of fluid flowing system in a certain state. In the formula, H and S are the enthalpy and entropy of the system, and H_0, S_0 and T_0 are the enthalpy, entropy and temperature of the system when it changes to the environmental state. It can be seen that, like enthalpy, entropy and other thermodynamic properties, they are all state functions.

Most of the chemical and petrochemical processes we are studying are steady fluid flowing systems, therefore, Exergy is the maximum work performed by the system:

$$EX = W_{max} = H - H_0 - T_0(S - S_0) \tag{1.26}$$

$$A_N = T_0(S - S_0) \tag{1.27}$$

The enthalpy difference in the preceding term at the right end of the formula (1.26) is the energy of the system ($E = H - H_0$); therefore, it can be seen that the energy minus the waste energy that cannot be performed for work is the Exergy.

$$E = EX + A_N \tag{1.28}$$

Therefore, the formula (1.24) and formula (1.25) can be expressed as:

$$Q \geq A_N W \leq EX$$

Write the differential formula of exergy, waste energy and energy state parameters as:

$$\delta EX = dH - T_0 dS \tag{1.29}$$

$$\delta A_N = T_0 dS \tag{1.30}$$

$$\delta E = \delta EX + \delta A_N \tag{1.31}$$

Its physical meaning is: the maximum work (under reversible conditions) that a fluid system may make during the energy use process is equal to the exergy of the system, and the minimum heat released to the environment is equal to the waste energy A_N of the system, and the sum of work and heat release, no matter whether it is reversible or not, it is equal to the energy E of the system.

The upper limit of the work that a fluid system can make in the energy utilization process is Exergy, and the low limit of heat release to the environment is the waste energy (all under reversible conditions). They are all determined by the state of the system (H, S) and the dead state condition (T_0, H_0, S_0).

In different states of the system, in its total energy ($E = H - H_0$), the ratio of Exergy to waste energy is different, and the ratio depends on the entropy S of the system. Exergy is the capable performing work part of energy and the "potential" that contributes to the ability of the system to change. Therefore, evaluate the value of energy, except the energy quantity, and the energy quality- Exergy should also be included, usually defines the ratio of exergy in energy as the energy grade ε:

$$\varepsilon = EX/E \tag{1.32}$$

Obviously, for the heat energy of a constant temperature heat source, its energy grade is the Carnot efficiency($\varepsilon = 1 - \frac{T_0}{T}$), and for the energy grade of a variable temperature heat source($\varepsilon = 1 - T_0/T_m$), T_m is the average temperature of the variable temperature heat source.

The work of orderly movement, such as electricity, mechanical energy, potential energy, etc., has an energy grade of 1, and chemical energy is also the result of the orderly movement of electrons outside the nucleus of the molecular structure, and its energy grade is also approximately 1.

Energy grade is also an important parameter to characterize energy quality in energy analysis.

Generally, the energy and exergy based on the intensity difference (T, p, phase state) of the physical state are called physical energy and physical exergy, and the energy and exergy based on the intensity difference (chemical composition, concentration) in the chemical are called chemical energy and chemical exergy.

According to the first law of thermodynamics, energy is always conserved in any energy-using process (regardless of whether it is reversible or not). However, exergy is conserved only in the reversible process, and is not conserved in the actual process. The exergy leaving the system is less than the exergy entering the system. The difference between them is the irreversible exergy loss of the actual process. The commonly referred to as exergy balance is to attribute the irreversible exergy loss that has not left the system to the outgoing exergy flow, and artificially "balance" the exergy.

Energy is the basic property of substance system and a universal concept. Exergy is a specific concept that represents the part of energy that can be converted into work under given natural environmental conditions. It has many different characteristics compared with exergy. The properties of energy are: (1) obeys the law of conservation; (2) it is the state function of the matter under consideration; (3) it can be calculated from any given reference state; (4) its value increases with increasing temperature; (5) in an ideal gas state, it does not depend on pressure. The characteristics of exergy are: (1) It is conserved in the reversible process and transformed into waste energy in the spontaneous process; (2) It is the state function of the substance under consideration and the environmental substance; (3) Its reference state is the state of the surrounding natural environment; (4) It has a minimum value when the ambient temperature is reached in the isobaric process, and its value increases with the decrease of temperature at low temperature; (5) It depends on the pressure.

1.4.5 The Exergy Balance Equation of the Energy Use Process [1]

As mentioned earlier, the energy utilization in the actual process is not "to the end", that is, the final state of the process cannot be equal with the environment. The formula (1.25) is suitable for the evaluation of the utilization limit of the energy system. For any specific process, you can still use the model in Fig. 1.2 as the basis, and use the method of deriving formula (1.17) to derive practical formulas including the first and second laws.

As shown in Fig. 1.6, apply the second law Eq. (1.18) to an isolated system that includes a controlled volume flow system and the entire outside world.

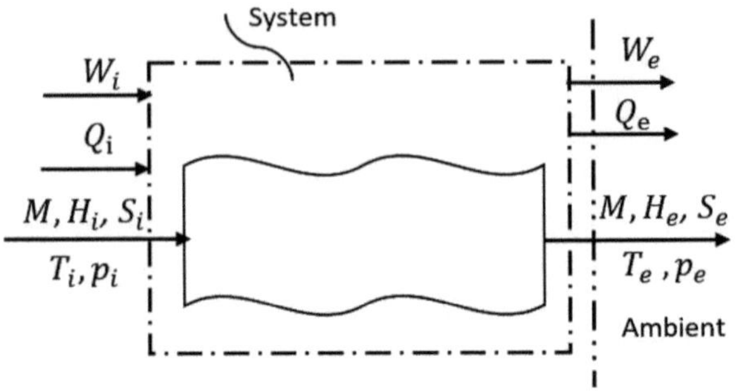

Fig. 1.6 Controlled volume flow system Schematic diagram

Considering more than one streams, the entropy changes of the system $S_{System} = S_0 - S_i$ can be written as

$$S_{system} = \sum (S_e - S_0) - \sum (S_i - S_0) \tag{1.33}$$

(S_e and S_i are two different states of the same streams, so they have the same dead state entropy S_0).

The external entropy change can be written as

$$S_{outside} = \left(\frac{Q_e}{T_e}\right) - (Q_i/T_i) \tag{1.34}$$

Substituting the formula (1.33), formula (1.34) into formula (1.18), you can get

$$\sum (S_e - S_0) - \sum ((S_i - S_0) + \frac{Q_e}{T_e} - Q_i/T_i \geq 0$$

Just move items and organize:

$$\sum (S_i - S_0) + Q_i/T_i \leq \sum ((S_e - S_0) + Q_e/T_e \tag{1.35}$$

Equation (1.35) is actually the application of the thermodynamics second law to the relationship between the energy using process of the process, and can be approximately expressed as: for any flow process system, the total entropy of the entering stream (the entropy of the stream is estimated as the difference to the entropy of the dead state) is always equal to (reversible process) or less than (actual process) the total entropy of the leaving stream.

Combining Eq. (1.35) with Eq. (1.17), we can get an equation that is easier to understand and apply. Multiply both ends of formula (1.35) byT_0, then subtract Eq. (1.35) from formula (1.17), we have:

$$\sum (H_i - H_0) - T_0 \sum (S_i - S_0) + Q_i - T_0\left(\frac{Q_i}{T_i}\right)$$

$$+ w_i \geq \sum (H_e - H_0) - T_0 \sum (S_e - S_0) + Q_e - T_0\left(\frac{Q_e}{T_e}\right) + w_e$$

where: $\sum(H_i - H_0) - T_0 \sum(S_i - S_0) = \sum (EX_m)_i$

$\sum (EX_M)_i$—The sum of exergy of all streams entering the system, the subscript m of EX represents Streams

$$Q_i\left(1 - \frac{T_0}{T_i}\right) = (EX_H)_i$$

$(EX_H)_i$—The exergy contained in the total heat of the supply system, the subscript H represents the heat.

$$W_i = (EX_W)_i$$

$(EX_W)_i$—The exergy corresponding to the work of the supply system is the same amount as work, and the subscript W represents work. You can also summarize the items listed here:

$$\begin{aligned} &\sum (EX_M)_i + (EX_H)_i + (EX_W)_i \\ &\geq \sum (EX_M)_e + (EX_H)_e + (EX_W)_e \end{aligned} \tag{1.36}$$

In fact, the heat and work of the in and out sides are more than one, and they should all be prefixed with the sum symbol "$\sum$", which is simplified here to avoid cumbersomeness and make it easy to use.

The equal sign indicates the reversible process, and the ">" sign indicates the actual process. Equation (1.36) shows that the actual process exergy is always non-conservative, and always tends to decrease, that is, it is constantly lost in the process.

If D_K represents the exergy loss in the actual process, the above formula can be written as

$$\sum (EX_M)_i + (EX_H)_i + (EX_W)_i = \sum (EX_M)_i + (EX_H)_i + (EX_W)_i + D_K \tag{1.37}$$

It is a practical formula of the basic equation of energy use process, and it can also be called a practical exergy balance equation of process. The exergy loss D_K in the formula is the sum of the exergy loss caused by the internal entropy increase in the actual process:

$$D_K = T_0 \sum \Delta S_{pj} \tag{1.38}$$

It can be seen that exergy loss and entropy increase essentially reflect the same natural law from different perspectives. However, the calculation methods and approaches of the two are not the same in practice. The calculation of entropy increase often requires a lot of physical data, which is difficult to complete and the calculation is more cumbersome. The calculation of Exergy loss can often avoid the way of entropy increase, and it is easier to understand the application. Equation (1.37) is a practical equation used to actually calculate the exergy loss of equipment, energy utilization links and even the entire process.

1.5 Energy Saving and GHG Gas Mitigation

Climate change is one of the most important environmental issues of our time. Climate change is caused by the increase in concentrations of greenhouse gases (GHGs) in the atmosphere. These increases are primarily due to human activities such as the use of fossil fuels.

GHGs include emissions for 7 greenhouse gases under the Kyoto Protocol:

- carbon dioxide (CO_2)
- methane (CH_4)
- nitrous oxide (N_2O)
- hydrofluorocarbons (HFCs)
- perfluorocarbons (PFCs)
- Sulphur hexafluoride (SF_6)
- nitrogen trifluoride (NF_3).

Among the GHGs, the CO_2 account for dominate proportion due to human activities using of fossil fuels, reducing CO_2 will contribute to GHGs reduction greatly.

CO_2 production most comes from the Carbon element combustion of fossil fuels, CO_2 producing rate is highly depend the ration of C:H in the fossil fuels. For example, burning 1 ton Fuel oil will produce about 2.5–3 tons CO_2 depending on C:H ratio fuel oil.

Therefore, it is concluded that saving energy is equivalent to CO_2 Emission reduction.

Carbon Peak value—Carbon peaking means that at existing human activities such as using fossil fuels, with human continually effort to mitigate CO_2 production, the CO_2 emission will reach the peak point at a certain point in time, carbon dioxide emissions no longer increase to a peak, and then gradually fall back.

Carbon neutrality refers to achieving net-zero carbon dioxide emissions in overall. It may be achieved in a region, a country, a continent or globally. It is used in the context of carbon dioxide-producing and utilization processes associated with transportation, energy production, agriculture, and industry. This can be done mainly by the followings

Mitigation carbon dioxide emissions by improvement of fuel efficiency; substitute fossil fuels with renewable energy, nuclear energy etc. CO_2 capture, enhance oil and gas production or storage; CO_2 chemical utilization.

References

1. H. Ben, *Analysis and Synthesis of Energy Consumption in Processing Process* (Hydrocarbon Processing Press, 1989), p. 1
2. Z. Lianke, *Chemical Thermodynamics (for inorganic chemical industry)* (Chemical Industry Press, 1980)
3. W. Wenhu, Q. Yanlong, Refining Des. **3**, 79 (1983)
4. Z. Hongxin, *Physical Chemistry* (Chemical Industry Press, 1980), pp. 39–40

Chapter 2
Calculation of Thermophysical Energy and Exergy

Abstract Based on the thermodynamic first law and second law, developed the energy and exergy calculation method for all the energy analysis related process streams, utilities and heat dissipation; discussed the calculation of energy and exergy of process thermal effect including process stream sensible heat, phase change latent heat, reaction heat effect, mixed heat effect, etc. According to petroleum enthalpy and its fitting correction, developed a calculation method for the liquid phase and vapor phase of petroleum and its fractions at different reference phases, make the oil refining, oil & gas industry exergy balance, and analysis a simple and effective; developed energy and exergy calculation method of light hydrocarbon and its mixtures, including calculation of enthalpy, entropy, and exergy of ideal gas hydrocarbon and its mixture; this chapter also provides energy and exergy calculation method for utility streams, and heat dissipation measuring and calculation with a simple and more accurate method of convection and radiation combined heat transfer coefficient. The petrochemical process involves many process streams and utilities, and the calculation of its energy and exergy is the basis of energy analysis.

Keywords Energy calculation · Exergy calculation · Thermal effect · Petroleum exergy · Stream exergy · Heat dissipation

2.1 Calculation of Energy and Exergy of Process Thermal Effect

Heat is a macroscopic manifestation of the irregular movement of a large number of molecules in substance. As a phenomenon, heat generally exists in the production process, such as heat release or absorption caused by temperature change of objects, the latent heat of phase change, the thermal effect of the material mixing process, the heat of chemical reaction, etc., which are collectively referred to as the thermal effect of the process.

T. A. Chen, *Energy Saving and Carbon Reduction*,
https://doi.org/10.1007/978-981-19-5295-1_2

2.1.1 Calculation of Process Stream Sensible Heat Energy and Exergy

2.1.1.1 Process Streams Sensible Heat Energy

The process stream does not undergo chemical change and phase change, but the heat absorbed or released by the change in the temperature of the process stream during the heating or cooling process is called sensible heat. The sensible heat of process stream can be calculated by the specific heat capacity of the process streams, or can be calculated also by looking up the thermodynamic property table.

The heat absorbed (or released) for each degree of temperature increase (or decrease) of an object is called heat capacity. Usually, the heat capacity expressed in unit mass or unit volume is called specific heat capacity. The heat capacity is related to the characteristics of the process. The heat capacity of a constant volume process is called the isovolumic heat capacity at constant volume $C_V = \left(\frac{\partial U}{\partial T}\right)_V$, and the heat capacity of a constant pressure process is called the isobaric heat capacity $C_P = \left(\frac{\partial H}{\partial T}\right)_P$.

According to the thermodynamics first law, the heat of the constant volume process is equal to the internal energy difference between the initial state and the final state of the system; the heat of the constant pressure process equals the enthalpy difference between the initial state and the final state of the system, namely: $Q_V = \Delta U$, $Q_P = \Delta H$.

Therefore, the thermal effect of the process can also be obtained separately from the change of internal energy and the change of enthalpy.

Generally speaking, the heat capacity of a constant volume process is only a function of temperature, so when calculating the heat absorption of a continuous constant volume heating process, the heat capacity relationship $C_V = C_V(T)$ should be substituted into the Eq. (2.1) to obtain the integral constant volume heating rate, when the lower limit of the integral is the reference temperature, the heat is the required process Stream Energy:

$$Q_V = \Delta U.m = m \int_{T_0}^{T} C_V(T)dT \tag{2.1}$$

where: m—Mass flow rate of stream, kg/s;

Q_v—Isovolumic process heat, kW;

C_V—Isovolumic heat capacity, kJ/(kg.k).

Isobaric heat capacity is a function of temperature and pressure. But when the pressure is not too high, the pressure has little effect on the specific heat capacity of the incompressible liquid phase, which can be considered as a function of temperature. The isobaric heat capacity of an ideal gas in an ideal state is only a function of temperature, and the thermal energy of the isobaric process is:

$$Q_P = \Delta H.m = m\int_{T_0}^{T} C_P(T)dT \tag{2.2}$$

where: Q_P—sobaric process heat,kW;

C_P—Heat capacity of isobaric process, kJ/(kg K).

Isobaric heat capacity is one of the commonly used physical properties in process calculations. It is mostly expressed as a function of temperature. The heat capacity correlation formula varies with different process stream calculation methods. Simply expressed as a linear relationship (such as air):

$$C_P = a + bT \tag{2.3}$$

The complicated ones are expressed as multiple nonlinear equations: such as hydrocarbon ideal gas, the correlation equation of equal pressure specific heat capacity [1]:

$$C_P = B + CT + DT^2 + ET^3 + FT^4 \tag{2.4}$$

Another correlation [2] is:

$$C_P = A\left(\frac{1.8T}{100}\right) + B\left(\frac{1.8T}{100}\right)^2 + C\left(\frac{1.8T}{100}\right)^3 + \frac{100D}{1.8T} \tag{2.5}$$

In formulas (2.3) to (2.5), a, b, A–F are coefficients of the specific heat capacity equation, which can be obtained from the petroleum refining or petrochemical Databook.

The correlation equation of specific heat capacity of process stream is mostly determined by experimental regression. Different experimental methods and regression methods lead to different specific heat correlations. The coefficients in the correlation equations are different, therefore, be careful when using them.

By substituting the specific heat correlation into Eq. (2.2) for different process streams, the energy value can be accurately calculated. For the complex form of $C_P(T)$, the numerical integration method should be used to calculate it.

The heat capacity mentioned above is all true heat capacity, and the result obtained by this integration is also relatively accurate. In order to avoid cumbersome integration, when the calculation accuracy is not required very high, the average specific heat between two temperatures is often used to calculate the average specific heat in the process calculation. Many manuals give the average specific heat. The so-called average specific heat capacity refers to the heat capacity within a certain temperature range. The average heat capacity value should be distinguished and used correctly. For average specific heat:

$$Q_V = m(C_{V2}T_2 - C_{V1}T_1) \tag{2.6}$$

$$Q_p = m(C_{p2}T_2 - C_{p1}T_1) \tag{2.7}$$

If the system that absorbs or releases heat is a mixture, assuming the mixed components do not affect each other when the heat changes, the average constant pressure molar heat capacity C_{pm} of the mixture is equal to the sum of the mole fraction Y_i of each component multiply by the average constant pressure molar heat capacity C_{pi} of each component at same temperature range:

$$C_{pm} = \sum Y_i(C_{pi}) \tag{2.8}$$

Then calculate the sensible heat of the mixture in the same way as above.

2.1.1.2 Sensible Heat Exergy of Process Stream

When there is no chemical change, the thermodynamic differential equations of enthalpy H and entropy S

$$dH = C_p dT + Vdp - T\left(\frac{\partial V}{\partial T}\right)_p dp \tag{2.9}$$

$$dS = \frac{C_p}{T} dT - \left(\frac{\partial V}{\partial T}\right)_p dp \tag{2.10}$$

Substitute into the basic definition equation of exergy:

$$\delta EX = C_p\left(1 - \frac{T_0}{T}\right)dT + Vdp - (T - T_0)\left(\frac{\partial V}{\partial T}\right)_p dp \tag{2.11}$$

The physical exergy in any state (T, p) relative to the reference state (T_0, p_0)can be expressed as

$$EX = \int_{T_0}^{T}\left(1 - \frac{T_0}{T}\right)C_p dT + \int_{p_0}^{p} Vdp - \int_{p_0}^{p}(T - T_0)\left(\frac{\partial V}{\partial T}\right)_p dp \tag{2.12}$$

The first item in formula (2.12) is physical exergy due to temperature changes, called thermal exergy

$$EX_h = C_p\left(1 - \frac{T_0}{T}\right)dT \tag{2.13}$$

The second and third item are the physical exergy due to pressure changes. It is called pressure exergy

$$EX_p = \int_{p_0}^{p} Vdp - \int_{p_0}^{p} (T - T_0)\left(\frac{\partial V}{\partial T}\right)_p dp \tag{2.14}$$

The sensible heat exergy of process streams is the physical exergy that changes with temperature.

From the Eq. (2.13), we can obtain the process stream thermal exergy according to the temperature of the stream and the specific heat capacity integration of the streams. For a constant specific heat capacity (the specific heat capacity of most streams can be approximately regarded as a constant), the thermal exergy calculation formula obtained by integration is:

$$EX_h = \int_{T_0}^{T} C_p\left(1 - \frac{T_0}{T}\right)dT = C_p(T - T_0) - C_pT_0\ln(\frac{T}{T_0})$$

Due to $C_p(T - T_0) = \Delta H$

$$\begin{aligned} EX_H &= C_p(T - T_0) - C_pT_0\ln\left(\frac{T}{T_0}\right)\left(\frac{T - T_0}{T - T_0}\right) \\ &= \Delta H - \Delta H\frac{T_0}{\frac{T-T_0}{\ln\left(\frac{T}{T_0}\right)}} = \Delta H(1 - \frac{T_0}{T_m}) \end{aligned} \tag{2.15}$$

$$T_m = \frac{T - T_0}{\ln\left(\frac{T}{T_0}\right)}.$$

Where T_m—The logarithmic average temperature of the streams at the temperature T and the reference temperature T_0.

In fact, the specific heat capacity of the stream mostly changes with the temperature change of the stream. For those with a small degree of change, the above formula can be used, and the arithmetic average temperature can be taken, and the result is more accurate, that is, $T_m = (T + T_0)/2$.

What has been discussed above is the case of variable temperature heat sources. Many processes can be regarded as constant temperature change processes, such as pure component phase change processes, and heat dissipation on the surface of high-temperature process equipment, the Eq. (2.15) becomes:

$$EX = \Delta H(1 - \frac{T_0}{T}) \tag{2.16}$$

where T—Heat source temperature, K;

Cold exergy is also a kind of thermal exergy, the difference is that the temperature of the system is below the ambient temperature. Cold energy is usually obtained by refrigeration equipment through the circulation process of the refrigerating medium (refrigerant), the heat is moved from the low-temperature system to the "high-temperature" environment. To achieve such a process must consume energy, mainly high-quality electrical energy.

Different from thermal exergy, the ambient temperature of the refrigeration process is higher than the temperature of the system, and cold source exergy at constant temperature and specific heat capacity is [3]:

$$EX = \Delta H(\frac{T_0}{T} - 1) \tag{2.17}$$

where ΔH—Cooling capacity.

In Eq. (2.17), generally $T_0 > T$, exergy is directly proportional to the temperature difference $(T_0 - T)$ and inversely proportional to the system temperature. Therefore, when the temperature of the refrigeration system is lower, the exergy value is higher, and the cold energy quality is higher. For the same system temperature difference $|T - T_0|$, the thermal exergy T is $T_0 + |T - T_0|$, and the cold exergy T is $T_0 - |T - T_0|$, so the cold exergy value is higher.

Cold energy is the result of heat transfer in a system below ambient temperature. The direction of cold flow is opposite to that of hot exergy.

For a variable temperature and constant specific heat capacity cold source, it is the same as heat, and the specific heat capacity relationship of the working fluid needs to be integrated to obtain the cold exergy.

$$\begin{aligned}
EX &= \int_T^{T_0} C_p(\frac{T_0}{T} - 1)dT \\
&= T_0 C_p \int_T^{T_0} \frac{dT}{T} - C_p \int_T^{T_0} dT \\
&= C_p T_0 \ln\left(\frac{T_0}{T}\right) - C_p(T_0 - T) \\
&= C_p T_0 \frac{\ln(\frac{T_0}{T})(T_0 - T)}{T_0 - T} \\
&= C_p(T_0 - T)\frac{T_0}{\frac{(T_0 - T)}{\ln\left(\frac{T_0}{T}\right)}} - 1 \\
&= \Delta H(T_0/(T_m^{'}) - 1)
\end{aligned} \tag{2.18}$$

where T_m'—Average temperature of cold stream energy.

When T_0 and T are not much different, the arithmetic average of the two can also meet the engineering error requirements, that is, $T'_m = (T + T_0)/2$. In the actual process, the specific heat capacity of a stream often changes with the change of the system temperature, and the specific heat capacity of some streams changes considerably, for example, the specific heat of the petroleum fraction changes between 1.6 and 3.6 kJ/(kg.C). At this time, the streams specific heat capacity correlation equation must be substituted into the Eq. (2.13) integral calculus to obtain exergy.

2.1.2 Energy and Exergy Calculation of Phase Change Latent Heat

When a substance changes from one phase to another under constant temperature and pressure, the heat released or absorbed is the phase change latent heat of the substance.

There are four types of phase change processes in petrochemical production [4]:

$$\text{Liquid Phase}\,\frac{\text{Solidification}}{\text{Melt}}\,\text{Solid Phase}$$

$$\text{Liquid Phase}\,\frac{\text{Vaporazation}}{\text{Condensation}}\,\text{vapor Phase}$$

$$\text{Solid Phase}\,\frac{\text{Sublimation}}{\text{Desublimation}}\,\text{vapor Phase}$$

$$\text{Solid Phase I}\,\frac{\text{Crstal Transition}}{}\,\text{Solid Phase II}$$

The corresponding latent heat of phase change is heat of melting (or solidification), heat of vaporization (or condensation), heat of sublimation (or desublimation), and heat of crystal transformation. The latent heat of phase change can also be calculated using empirical formulas or using Databook.

The process of vaporization and condensation is the most common phase change process in petroleum and chemical plants, and it is very important to understand the calculation of the latent heat of phase change.

There are three commonly used empirical formulas for calculating the latent heat of vaporization in engineering:

(1) Kistiakovsky formula

$$\Delta H_V = 4.1868 T_b (8.75 + 4.517 lg T_b) \tag{2.19}$$

where ΔH—Molar latent heat of vaporization, J/mol;

T_b—Normal boiling point of the substance, K;

This formula can be used to calculate the heat of vaporization of non-polar or weakly polar compounds at the normal boiling point, with a maximum error of less than 3%.

(2) Nernst formula

$$\Delta H_V = 4.1868 T_b (4.5 lg T_b - 0.007 T_b) \tag{2.20}$$

(3) Wasson's empirical formula:

$$(\Delta H_V)_2/(\Delta H_V)_1 = \left(\frac{1 - T_{r2}}{1 - T_{r1}}\right)^{0.38} \tag{2.21}$$

In the formula, the subscripts 1 and 2 refer to the heat of vaporization and the reduced temperature T_r at two temperatures. This formula is relatively simple and quite accurate. For below the critical temperature 10 K, the average error is only 1.8%, so it is suitable for calculating the change in heat of vaporization with temperature in engineering.

The heat of fusion of many elements and compounds can be estimated as follows:

$$\frac{\Delta H_F}{T_F} = constant \tag{2.22}$$

For elements, the constant at the right end of the above formula is 2–3, for inorganic compounds, it is 5–7; for organic compounds, it is 9–11.

Where ΔH_F—Molar heat of fusion of solid substance, cal/mol;

T_F—Melting point, K;

The most typical example of a phase change heat data table is the steam table, which lists the value of the heat of vaporization. there are similar tables or Mollie diagrams for refrigerants such as ammonia and freon, as well as oxygen and nitrogen.

In general, the phase change heat data of the phase change process of the streams can be found in the relevant manuals, and the manuals are mostly the atmospheric boiling point. The phase change heat value is different at different temperatures. Therefore, we should try to find the correlation formula of the phase change heat with temperature. For example, the correlation formula of the vaporization heat of hydrocarbons can use the Wasson's empirical formula:

$$\Delta H_T = \Delta H_b \left(\frac{T_c - T}{T_c - T_b}\right)^{0.38} \tag{2.23}$$

where: ΔH_T—Heat of vaporization at any temperature;

ΔH_b–Heat of vaporization at atmospheric boiling point;

T_c–Material critical temperature, K;

T_b–Atmospheric boiling point of the substance, K;

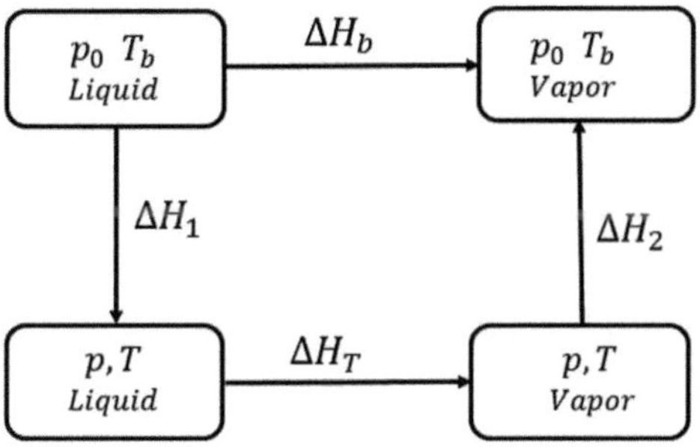

Fig. 2.1 The substance heat of vaporization under different temperature

In the absence of experimental data and correlations, the heat of vaporization at the required temperature may be obtained by heat of vaporization at normal boiling temperature. The design reversible path is shown in Fig. 2.1.

$$\Delta H_T = \Delta H_b + \Delta H_1 + \Delta H_2 \tag{2.24}$$

$$\Delta H_1 = \int_{T_b}^{T} C_{p(l)} dT$$

$$\Delta H_2 = \int_{T}^{T_b} C_{p(v)} dT$$

The above method assumes: (1) Ignore the influence of pressure on the enthalpy of the liquid; (2) The gas can be considered to be in an ideal gas state. It is a rough treatment method, so

$$\Delta H_T = \Delta H_b + \int_{T_b}^{T} C_{p(l)} dT - \int_{T}^{T_b} C_{p(v)} dT \tag{2.25}$$

It is the formula for calculating the heat of vaporization at any temperature, which uses the features of the state function to design different paths to simplify the calculation. This method is called the thermodynamic state function method.

For pure compound, the phase change process is a constant temperature process, and the phase change thermal exergy is:

$$EX = \left(1 - \frac{T_0}{T}\right) \Delta H_b \tag{2.26}$$

The phase change process of the mixture is the temperature change process between the bubble point and the dew point of the mixture. In particular, the composition of petroleum fractions is complex, and the bubble and dew point are far apart. The temperature-enthalpy relationship of the petroleum fraction phase change process is approximately a straight line [5]. Generally speaking, for the mixture phase change process, using the arithmetic mean temperature T_m of its bubble point and dew point temperature instead of T in the above formula can meet the engineering error requirements.

2.1.3 *Reaction Heat Effect and Reaction Exergy*

2.1.3.1 Heat of Reaction

During the chemical reaction process, the chemical bonds between the reactant molecules and atoms are recombined to complete the transformation from reactant to product process. At this time, some chemical bonds are broken and new chemical bonds are formed. In the process of breaking and generating, a change in chemical energy takes place, and a thermal effect is produced. The heat of chemical reaction is the heat released or absorbed by the chemical reaction between substances under the condition of constant temperature and pressure. The reaction heat usually exchanges energy with the outside system in the form of heat.

1. Reaction heat effect of known composition

Calculating the heat of chemical reaction is always based on the enthalpy of the product minus the enthalpy of the reactant in thermodynamics, that is

$$\Delta H_R = \sum H_y - \sum H_{Rec} \tag{2.27}$$

where ΔH_R—Chemical reaction heat;

$\sum H_{Rec}$—Enthalpy of reactant;

$\sum H_y$—Enthalpy of reaction product.

(1) **Guess's law**

Regardless of whether a chemical reaction or physical change is completed in one step or several steps, the total thermal effect remains the same. In fact, Guess's law is an extension of the first law of thermodynamics. Because enthalpy or internal energy are state functions, as long as the initial state and the final state of a chemical reaction are fixed, ΔH or ΔU is a fixed value; and it doesn't really matter how the reaction is completed.

Applying Guess's law, the indirect method can be used to calculate the chemical reaction heat that is difficult to control and difficult to measure the thermal effect. For example, it is difficult to directly measure the reaction of carbon monoxide by incomplete combustion of carbon, because we burn carbon in the experiment, there is always CO production, and it is impossible to produce 100% O_2. And the two reaction $C(s) + O_2(g) = CO_2(g)$ and $CO(g) + \left(\frac{1}{2}\right)O_2(g) = CO_2(g)$ reaction heats at 298.16 K are experimentally determined to be -97.81 kcal and 68.11 kcal. According to Guess's law, the reaction of incomplete combustion of carbon to carbon monoxide can be combined with the above two reactions as shown in the figure below:

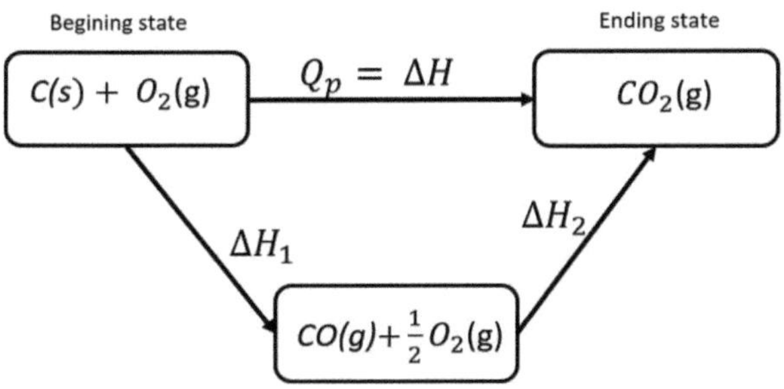

So, get: $\Delta H_1 = \Delta H - \Delta H_2 = -97.81 + 68.1 = -29.70\text{kcal} = -124.3\text{kJ}$

According to Guess's law, it is also possible to calculate the reaction heat of some other reactions through the measured reaction heat data, instead of measuring the thermal effect of each reaction through experiments. In addition, you can also use the experimental results and calculated data perform comparison and analysis to check the accuracy of reaction heat data.

In the actual production process, a variety of thermal effects often occur at the same time, and the conditions are changeable, the heat balance is more complicated, and sometimes the necessary thermochemical data is not available; according to Guess's law, a way can be designed using the existing data and the chart for calculating the required data to complete the heat balance of the whole process.

The heat of chemical reaction can be calculated based on the formation heat and the combustion heat of substance.

The thermal effect of directly synthesizing 1 mol of a compound from a stable element is called the formation heat of the compound. Specifying that the formation reaction is carried out under the conditions of 298.16 K, 1 atm, the reaction heat is called the standard formation heat, expressed as ΔH_f^0.

Using the standard formation heat data of the substances, the difference between the sum of the standard formation heat of the reaction products and the sum of the standard formation heat of the reactants can be used to find the standard reaction heat of the chemical reaction.

$$\Delta H_R^0 = \sum \left(\Delta H_f^0\right)_y - \sum \left(\Delta H_f^0\right)_{Reactants} \tag{2.28}$$

If a substance can completely react with oxygen to generate the corresponding oxide, this kind of reaction is called a combustion reaction of a substance. Therefore, the thermal effect when 1 mol of a substance is completely burned to produce CO_2(g) and H_2 O (l) is called the combustion heat of the substance. The value measured at 25 °C, 1 atm is called the standard heat of combustion and is represented by ΔH_C^0. For example, the standard heat of combustion of methane is the reaction heat of the following reaction:

$$CH_4(g) + 2O_2(g) \xrightarrow[\rightarrow]{25°C,\ 1\ atm} CO_2(g) + H_2O$$

$$\left(\Delta H_C^0\right)_{CH_4(g)} = \Delta H_R^0 = -212.798\ \text{kcal}$$

From the above discussion, it can be seen that the combustion heat is different from the formation heat. The former considers the reactants and the latter considers the products. Only when the reactant of the combustion reaction is a single substance (such as carbon), the value of the combustion heat is equal to the heat of generation.

When the reactants and products before and after the reaction can "burn" react with oxygen, and the standard combustion heat data of the substance is used in the calculation, the sum of the standard combustion heat of the reactants minus the sum of the standard combustion heat of the products, which is the standard reaction heat of this reaction:

$$H_R^0 = \sum \left(\Delta H_C^0\right)_{Reactants} - \sum \left(\Delta H_C^0\right)_y \tag{2.29}$$

(2) **Kirchhoff's Law**

Most reactions in chemical production are not carried out at 298.16 K. For example, the water gas shift reaction is carried out at 450–500 °C, the reaction in the gas generator is carried out above 900 °C, and the ammonia synthesis reaction is carried out at 400–500 °C. How to use the existing reaction heat data at 298.16 K to find the thermal effect of the reaction at any temperature has become a issue to be solved.

As shown in Fig. 2.2, there are two reaction materials A and B entering the reactor under $T(K)$, and the chemical reaction of $aA + bB = eE + gG$ is carried out in it, and the products E and G leave the reactor at same temperature $T(K)$.

This is a chemical reaction process with stable flow. Since the reaction system does not perform axial work $w_s = 0$, if the potential energy and kinetic energy changes of the system are not considered, that is, $mg\Delta Z = 0$, $\frac{1}{2}m\Delta u^2 = 0$, then According to the thermodynamics first law of the steady flow system; $Q = \Delta H$. As shown in Fig. 2.3, that is

$$Q = \Delta H_{RT} = \Delta H_R^0 + \Delta H_P - \Delta H_R = \Delta H_R^0$$

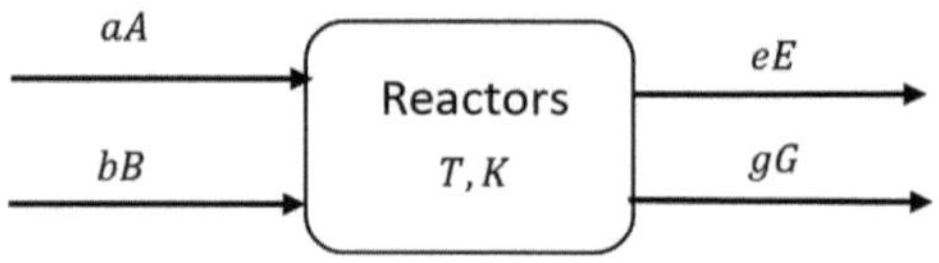

Fig. 2.2 Chemical reaction process at high temperature

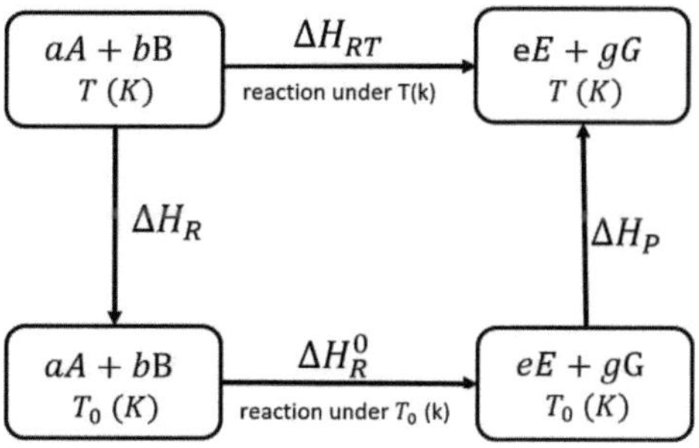

Fig. 2.3 Thermal effects of chemical reactions at high temperatures

$$+\int_{T_0}^{T}\left((eC_{pE}+gC_{pG}) - (aC_{pA}+bC_{pB})\right)dT$$

Make $\int_{T_0}^{T}((eC_{pE}+gC_{pG}) - (aC_{pA}+bC_{pB})) = \int_{T_0}^{T}\Delta C_p$ (The sum of the heat capacity of the product − the sum of the heat capacity of the reactant), Then

$$\Delta H_{RT} = \Delta H_R^0 + \int_{T0}^{T}\Delta C_p dT \tag{2.30}$$

The above formula is Kirchhoff's law. In the formula, ΔH_{RT} refers to the thermal effect of the reaction when the temperature is T. C_p indicates the constant pressure heat capacity of a substance. The rest of the symbols are the same as above.

Because the empirical formula for the heat capacity of general materials is $C_p = a + bT + CT^2$;

$aC_{pA} = a(a_A + b_A T + C_A T^2)$

$bC_{pB} = b(a_B + b_B T + C_B T^2)$

$eC_{pE} = e(a_E + b_E T + C_E T^2)$

$gC_{pG} = g(a_G + b_G T + C_G T^2)$

$\Delta a = ((ea_E + ga_G) - (aa_A + ba_B))$

$\Delta bT = T((eb_E + gb_G) - (ab_A + bb_B))$

$\Delta CT^2 = T^2((eC_E + gC_G) - (aC_A + bC_B))$

Then $\Delta C_p = \Delta a + \Delta bT + \Delta CT^2$, Substitute above to Eq. (2.30) to get:

$$\begin{aligned}\Delta H_{RT} &= \Delta H_R^0 + \int_{T_0}^{T}(\Delta a + \Delta bT + \Delta CT^2)dT \\ &= \Delta H_R^0 + \Delta a(T - T_0) + \frac{\Delta b}{2(T^2 - T_0^2)} + C/3(T^3 - T_0^3)\end{aligned} \tag{2.31}$$

Applying the above formula, as long as you know the heat of reaction ΔH_R^0 at T_0, the thermal effect of this reaction ΔH_{RT} at $T(K)$ can be calculated.

2.1.3.2 Thermal Effects of Complex Reactions

There are quite a lot of reactions in the petrochemical process. The exact composition of the reactants and products is unknown, and the type and number of reactions are not known. Therefore, it is difficult to obtain the reaction heat at the reference temperature from the Guess's law. For example, in the oil refining industry, the reactants and products are mostly complex petroleum fractions composed of thousands of molecules. The calculation of the reaction heat can only be determined by the empirical formula summed up in practice or through the calculation of the heat balance of the equipment. The empirical formula has different methods and formulas for different devices and reaction processes. Such as catalytic cracking reaction heat, the reaction heat generated per kilogram of catalytic carbon is 2180 kcal/kg coke, and Catalytic Carbon = Total Carbon − Additioanl Carbon – Strippable Carbon. There are similar empirical formulas for other reaction processes, which will not be listed one by one. Here, we will only briefly discuss the method of calculating the reaction heat through the heat balance of the reaction equipment.

In the energy analysis method of the petrochemical process, the chemical energy calculation of reactants and products is not considered in the system energy balance calculation, thus avoiding the complicated calculation of the chemical energy of hydrocarbons and the errors caused by it. The energy carried by the stream entering the reaction equipment is mostly physical energy that satisfies the operating conditions of the reaction, and it is also physical energy when leaving the reaction equipment. The difference between the chemical energy of the exiting streams and the entering streams is just the reaction heat of the process. In addition, there is also the heat dissipation of the equipment. Sometimes, in order to meet the high equilibrium conversion rate and stop the reaction, some heat removal and chill facilities are set up, and no shaft work is performed externally. The heat balance of general reaction equipment can be expressed as shown in Fig. 2.4.

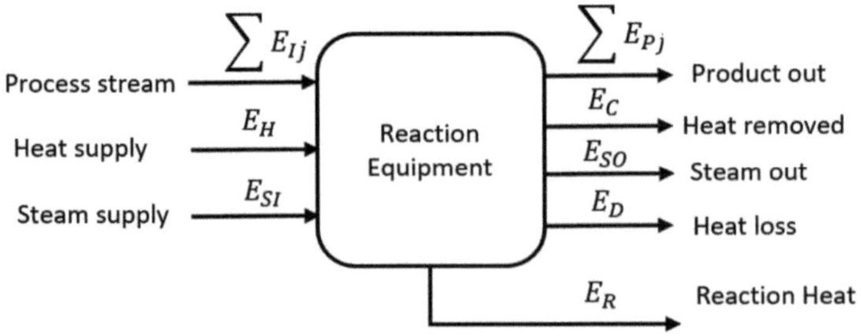

Fig. 2.4 Heat balance diagram of reaction equipment

For m raw material streams and n product streams, the reaction equipment is thermally balanced, and the reaction heat is:

$$E_R = \sum\nolimits_{j=0}^{m} E_{Ij} + E_H + E_{SI} - \sum\nolimits_{j=0}^{n} E_{pj} - E_C - E_D - E_{SO} \quad (2.32)$$

When the energy of the reactant stream entering the equipment is measured at the entrance of the reaction equipment, the energy brought by the stream has included the process heat supply. Sometimes, when the process stream parameters are included in the energy before the process heat supply, the process heat supply item should be listed separately, and attention should be paid to deduct the heat dissipation of the pipeline.

When the calculated reaction heat is positive, indicating that the reaction is endothermic. When the heat of reaction is negative, it indicates that the process is an exothermic reaction. The heat balance of the reaction equipment will be specifically introduced in Chap. 6.

2.1.3.3 Calculation of Reaction Exergy

Exergy's Balance view is that the chemical reaction process also changes exergy. The so-called chemical reaction exergy refers to the difference between the exergy value contained in the product and the reactant. That is, during the reaction process, due to the change of molecular structure, the chemical exergy difference between the product and the reactant is produced. The chemical reaction exergy for endothermic reactions comes from the heat energy provided by the outside system. Under the action of the catalyst, the thermal energy of disordered movement enters into the product molecules to become orderly energy. The reaction exergy of exothermic reaction is to release chemical energy under the conditions of the system, so that the chemical exergy of the reactant is reduced, so that the released part of the ordered chemical energy is transformed into the heat energy of disordered movement.

1. **Calculation of chemical reaction exergy with known composition**

It can also be calculated by the basic calculation formula (1.26) of the steady flowing system:

$$EX = \Delta H - T_0 \Delta S$$

With the reaction: $aA + bB = cC + dD$, the ΔH and ΔS can be calculated from the Databook manual, and then the reaction exergy can be calculated:

$$EX = (ch_C + dh_D - ah_A - bh_B) - T_0(cS_C + dS_D - aS_A - bS_B) \quad (2.33)$$

In the formula, h_i, S_i are the partial molar enthalpy and partial molar entropy of the reaction system, and when the temperatures before and after the reaction are equal and both are equal to the ambient temperature, the reaction exergy is equal to the free enthalpy change ΔG of the reaction.

$$\Delta G = -RT_0\ln\left(\frac{K_p}{J_p}\right) = \Delta G^0 + RT_0 ln J_p \tag{2.34}$$

When the reaction substance is in the standard state (25 °C, 1 atm, pure substance), the reaction exergy is equal to the standard free enthalpy change ΔG^0.

Where: ΔG, ΔG^0-free enthalpy and standard free enthalpy of the reaction process;

J_p—is the ratio of product of reaction products partial pressure to product of reactants partial pressure under reference temperature and reaction conditions.

$$J_p = (p_C^c.P_D^d)/(p_A^a \cdot p_B^b)$$

K_p—equilibrium constant at T_0 temperature.

For an isothermal isobaric reaction at temperature T, the free enthalpy changes

$$\Delta G = \Delta H - T\Delta S \tag{2.35}$$

Substitute above to the formula (1.26) to get the relationship between the chemical reaction exergy and the free enthalpy.

$$EX = \Delta G + (T - T_0)\Delta S \tag{2.36}$$

That is, the chemical reaction exergy equals to the free enthalpy change ΔG under the reaction conditions plus a correction term $(T - T_0)\Delta S$, substituting Eq. (2.34) into Eq. (2.36), we get:

$$EX = -RT_0\ln\left(\frac{K_p}{J_p}\right) + (T - T_0)\Delta S \tag{2.37}$$

Because of the entropy change of the reaction process, ΔS is approximately the thermal entropy (ratio of the reaction heat to temperature). The formula (2.37) can be written as

$$EX = -RT_0 ln\left(\frac{K_p}{J_p}\right) + (1 - \frac{T_0}{T})\Delta H \tag{2.38}$$

where: T—Reaction temperature;

ΔH—Heat of reaction under operating conditions.

In the above formula, ΔS, ΔH are the values per unit of raw materials and products, including the conversion factor.

2. **Reaction exergy estimation for complex processes**

Due to the complexity of the reactants and products in the petrochemical process, it is difficult to accurately measure the composition of the reactants and products, so the above method cannot be used for calculation, but according to the general definition of $EX = \Delta H - T_0 \Delta S$, ΔH is the reaction heat which can be estimated from the heat balance of the reaction equipment or empirical data. The key is that it is difficult to obtain the ΔS change. However, most $T_0 \Delta S$ comparing to reaction heat ΔH is much smaller, and there is a certain error in the data of reaction heat under T_0. Therefore ignoring $T_0 \Delta S$ will not have a big impact; please note that it is only relative to T_0, P_0, the reaction exergy can be handled like this. When at the operating temperature, the analysis is as follows:

(1) Both exothermic and endothermic reactions change the chemical structure of the reactants. In the exothermic reaction, the original chemical bonds are broken, and chemical energy (such as a combustion reaction) is released from the reactants, and the heat of reaction is equal to the change in chemical energy. The value of chemical energy is close to that of chemical exergy. Therefore, the reaction exergy can be considered to be approximately equal to the reaction heat. The exergy loss caused by the reaction exothermic heat entering the system should belong to the category of process exergy loss.
(2) For the endothermic reaction, the heat of reaction comes from a heat source provided by the outside world, which transforms the disordered thermal energy into ordered chemical energy with the help of a catalyst. The transition from a low energy level to a high energy level is completed, so the heat of reaction is completely converted into the chemical energy of the product, which can be approximated as reaction exergy

2.1.4 Mixed Heat and Exergy

The mixing of two or more substances is often accompanied by a thermal effect. This thermal effect is called heat of mixing. It can release heat when mixing, such as the mixing of sulfuric acid and water; it can also absorb heat, such as ammonium nitrate soluble in water. The heat of mixing can be found in the corresponding manual.

The heat of mixing of an ideal gas is zero, and the heating effect when the real gas is mixed is generally small and can be ignored. When liquid and liquid, soluble solid materials (such as NaOH, NH_4HCO_3) and liquid are mixed, the heat of mixing is relatively large and must be considered; when gas (such as NH_3, CO_2, SO_2 etc.) is dissolved in liquid, and the catalytic cracking compressed rich gas absorbed in the absorption tower, the latent heat of condensation is also carried into the solution because the gaseous state turns into the liquid state, and the mixing heat is very significant.

The heat effect produced when gases, liquids and soluble solid substances are dissolved in a solvent to form various solutions is called the heat of solution. The

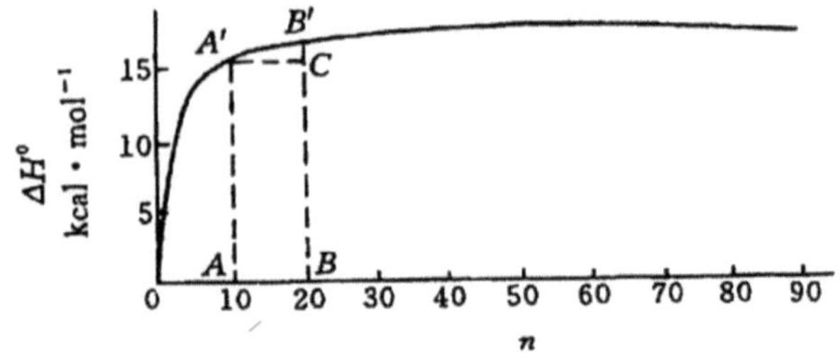

Fig. 2.5 Integral dissolving heat curve of sulfuric acid aqueous solution

heat of solution is similar to the heat of chemical reaction in many respects. When a chemical reaction occurs, due to the rearrangement of atoms, the energy of the product is different from that of the reactant at the same temperature and pressure; when the solution is formed, the difference in attraction between molecules of the same type and different types of molecules also produces energy changes, but this intermolecular gravitational force is smaller than the chemical bond force, and the resulting thermal effect is also smaller. The heat of solution of a substance is related to the temperature, pressure and concentration of the solution during dissolution. At a certain temperature and pressure, for a certain solute and solvent, the size of the heat of solution is determined by the concentration of the solution. For example, at 298.16K and 1 atm, when 1 mol of sulfuric acid is dissolved in water to become an aqueous solution of sulfuric acid, if the generated heat of dissolution and the concentration of the solution are plotted on the coordinate axis, the curve shown in Fig. 2.5 can be obtained.

The figure uses the water moles n mixing with H_2SO_4 in $H_2SO_4(nH_2O)$ to represent the concentration of the solution. It can be seen from the figure that the dilute the concentration of the solution, that is, the greater the n, the greater the heat effect when the solution is formed, but the cumulative heat of solution as the concentration becomes thinner, the increase becomes slower and slower, that is, the heat released by adding the same amount of water becomes smaller and smaller. Finally, the cumulative heat of solution approaches a certain value of 22.99 kcal/mol heat. At this time, the value of n approaches infinity compared with 1 mol. For sulfuric acid, the solution has become infinitely diluted.

When 1 mol of the solute is dissolved in a certain amount of solvent and becomes a solution of a certain concentration, the thermal effect produced is called the integral heat of solution. The cumulative heat effect of an infinitely diluted solution is the infinitely diluted integral heat of solution, expressed by ΔH_{∞}^{0}. Like the chemical reaction process, the dissolution process can also be expressed by the following equation:

$$H_2SO_4 + 5H_2O \overset{25^{\circ}C,1atm}{\Longleftrightarrow} H_2SO_4(5H_2O)$$

$$\Delta H = -13.87 \text{ kcal/mol}$$

In the formula, $H_2SO_4(5H_2O)$ means that the product of the dissolution process is a solution of 1 mol sulfuric acid in 5 mol water, and the thermal effect accompanying the dissolution process at 1 atm, 298·16 K is -13·87 kcal, which is also the enthalpy of a solution is 13.87 kcal less than that of 1 mol pure H_2SO_4 and 5 mol pure H_2O.

When 1 mol of solute is dissolved in a large amount of solution (or a small amount of solute is added to a quantitative solution), the thermal effect for each mole of solute is called differential heat of dissolution. At this time, the concentration of the solution can be regarded as constant, and differential dissolution heat is often used in processes where a large amount of solution is continuously circulated to absorb a small amount of solute.

Adding a solvent to the solution to dilute the solution produces a thermal effect called the heat of dilution. Dilution heat is related to temperature, pressure, concentration before and after dilution, and the amount of solution. Generally, the dilution heat refers to the amount of the original solution, that is, the solution containing 1 mol of solute. If the concentration after dilution is not indicated, it means that it is diluted to infinite dilution. If different amounts of solvents are used to dissolve 1 mol of solute, the difference between the two integral dissolution heats is called integral dilution heat, which is also equal to the thermal effect that occurs when a solution of one concentration is diluted into a solution of another concentration. As shown in Fig. 2.5, AA' is the integral heat of dissolution when 1 mol of H_2SO_4 is added to 10 mol of water, BB' is the integral heat of dissolution when 1 mol of H_2SO_4 is added to 20 mol of water, and BB'—AA' = B'C is equivalent to the thermal effect when A' is diluted to B'.

The differential dilution heat is similar to the differential dissolution heat. If 1 mol of pure solvent is dissolved in a very large amount of solution, the concentration of the solution can be considered unchanged, and the resulting heat of dilution is called the differential dilution heat. If ΔH is plotted against the number of moles of solvent n, the tangent of any point on the curve is the differential dilution heat of the solution.

The exergy of mixing heat can be calculated from the temperature of the mixing process according to the calculation method of physical exergy:

$$EX = \Delta H(1 - \frac{T_0}{T}) \tag{2.39}$$

where: ΔH—Mixed heat;

T—Mixing process temperature, when there is a change, the average value can be taken.

2.2 Calculation of Energy and Exergy of Petroleum and Its Fractions

Petrochemical process processing streams used mainly include petroleum distillate hydrocarbons and their mixtures, air, water, steam, etc. This section mainly discusses the calculation of petroleum distillate energy and exergy [9].

The crux of the calculation of petroleum fraction thermal energy and heat exergy is to establish the functional relationship between $C_p = f(t)$ and $F(p, V, T) = 0$. For the former, the Nelson enthalpy temperature correlation equation or graph (with relative density and Watson characteristic factor as parameters) [6] is generally used as the main characteristic formula of petroleum fractions under low pressure. This section also uses this correlation formula as a basis to derive the practical equation for heat exergy calculation; For the latter, the ideal gas equation of state can be used in the low-pressure gas phase, and the liquid phase can be simplified separately according to actual conditions.

2.2.1 *Expansion of Nelson Enthalpy Diagram Fitting Correlation*

The fitting correlation formula of Nelson petroleum fraction enthalpy diagram [6] is:

$$Z = \sum_{i=0}^{2} \sum_{k=0}^{2} A_{ik} X^i Y^k \tag{2.40}$$

where X, Y—Independent variable;

Z—Enthalpy values obtained under different X and Y, kJ/kg;

A_{ik}—coefficient;

Z, X, Y& A_{ik} coefficient values are shown in Table 2.1.

API degree is calculated by the following formula.

Table 2.1 Enthalpy diagram fitting formula coefficient table

Z	X	Y	A_{00}	A_{01}	$A_{01} \times 10^2$	A_{10}	$A_{11} \times 10^2$	$A_{20} \times 10^3$
H_L^*	t	API	15.990	1.0396	−1.1329	1.5566	0.8256	1.99
H_1	t	API	327.08	1.6442	−0.6925	1.2807	0.4094	1.9385
H_2	t	K	101.407	−86.0877	663.9	3.6103	−31.61	0.2814

$$API = \frac{141.5}{0.9942d_4^{20} + 0.009181} - 131.5 \quad (2.41)$$

where API—Petroleum fraction API degree;

d_4^{20}—Relative density of petroleum fraction at 20 °C.

Liquid phase enthalpy

$$H_L = H_L^*(0.0533K + 0.3604) \quad (2.42)$$

Gas phase enthalpy

$$H_V = H_1 - H_2 + \Delta H \quad (2.43)$$

where H_1—Uncorrected enthalpy of low-pressure gas phase petroleum fraction, kJ/kg;

H_2—Low pressure gas phase fraction enthalpy characteristic factor correction term, kJ/kg;

ΔH—Correction term for the influence of pressure on the enthalpy of gas phase fraction under high pressure, kJ/kg;

H_L—Liquid phase petroleum fraction enthalpy, kJ/kg;

H_V—Enthalpy of gas phase petroleum fraction, kJ/kg;

H_L^*—Liquid enthalpy without characteristic factor correction, kJ/kg;

K—Petroleum Fraction Characteristic Factor.

The computer program of petroleum distillate gas and liquid phase enthalpy compiled based on this has been widely used in oil refining process calculation. In order to change the above-mentioned correlation formula into a practical form of $H = H(T)$ or $C_p = f(T)$, expand the summation formula of Eq. (2.40), and the explicit function formula of gas and liquid enthalpy temperature respectively. Liquid phase enthalpy:

$$H_L = \left\{C_2 + C_3(T - 273) + 1.99 \times 10^{-3}(T - 273)^2\right\}C_1 \quad (2.44)$$

where C_1—Characteristic factor correction factor,

$$C_1 = 0.0533K + 0.3604$$

C_2, C_3—Constant,

$$C_2 = 15.99 + 1.0396API - 1.1329 \times 10^{-2}API^2$$

$$C_3 = 1.5566 + 8.256 \times 10^3 API$$

Due to $C_p = \left(\frac{\partial H}{\partial T}\right)_p$, Therefore, the partial derivative of Eq. (2.44) has:

$$C_{pL} = \left\{1.5566 + 8.256 \times 10^{-3} API + 3.98 \times 10^{-3}(T - 273)\right\} C_1 \tag{2.45}$$

Gas phase enthalpy

$$H_V = C_4 + C_5(T - 273) + 1.657 \times 10^{-3}(T - 273)^2 \tag{2.46}$$

where: T—Petroleum fraction temperature, K;

C_4, C_5—constant,

$$C4 = 225.67 + 1.644API + 86.088K - 6.925 \times 10^{-3} API^2 - 6.639K^2$$

$$C_5 = 4.17 \times 10^{-3} API + 0.3161K - 2.3295$$

Same approach, from $C_p = \left(\frac{\partial H}{\partial T}\right)_p$ get gas phase heat capacity:

$$C_{pV} = 4.17 \times 10^{-3} API + 0.3161K - 2.3295 + 3.314 \times 10^{-3}(T - 273) \tag{2.47}$$

The reference temperature of enthalpy in formula (2.44), (2.45) is -17.8 °C, and formula (2.46) is only applicable to calculation of the enthalpy of petroleum fractions gas phase at low pressure (p ≤0.5 MPa) $p \ll p_c$ or $T > T_c$.

In the energy analysis and calculation, the petroleum fraction physical exergy usually takes the reference state temperature and pressure (T_0, p_0), which are 288 K and 0.1 MPa, and the reference phase state is the phase state presented under the above (T_0, p_0). Therefore, the calculation of the physical exergy of petroleum and petroleum fractions can be roughly divided into three situations: (1) Calculation of liquid phase petroleum fraction exergy whose reference phase is liquid phase; (2) Calculation of gas phase petroleum fraction exergy whose reference phase is gas phase; (3) Calculation of gas phase petroleum fraction exergy whose reference phase is liquid phase.

2.2.2 Calculation on Liquid Petroleum Fraction Exergy with Liquid Reference Phase

In the physical exergy of petroleum fractions that are still liquid at T and pressure p, the proportion of pressure exergy is extremely small. For example, a certain petroleum fraction with a temperature of 200 °C and pressure of 0.6 MPa has a heat exergy of 102.9 kJ/kg and a pressure exergy of only 0.706 kJ/kg, accounting for 0.7% of the physical exergy. Therefore, in the general physical exergy calculation, the process stream pressure exergy is negligible, that is, the heat exergy is regarded as the physical exergy, and it is calculated according to formula (2.13).

Substituting formula (2.45) into formula (2.13) to integrate, obtain the practical calculation formula for the heat exergy of liquid petroleum fraction:

$$EX_{Lh} = \left\{A_0(T - 288) + 1.99 \times 10^{-3}\left(T^2 - 82944\right) - A_1 \ln\left(\frac{T}{288}\right)\right\}C_1 \quad (2.48)$$

Among them, $A_0 = 8.256 \times 10^{-3} API - 0.6767$

$$A_1 = 135.3 + 2.3778API$$

The energy of liquid petroleum fraction at temperature T (the enthalpy difference to 15 °C) is

$$E_{Lh} = \left\{C_3(T - 288) + 1.99 \times 10^{-3}\left((T - 273)^2 - 225\right)\right\}C_1 \quad (2.49)$$

The exergy difference and energy difference of the liquid petroleum fraction between T_1 and T_2 are respectively:

$$\Delta EX_{Lh} = \left\{A_0(T_2 - T_1) + 1.99 \times 10^{-3}\left(T_2^2 - T_1^2\right) - A_1 ln(\frac{T_2}{T_1})\right\}C_1 \quad (2.50)$$

$$\Delta E_L = \{C_3(T_2 - T_1) + 1.99 \times 10^{-3}((T_2 - 273)^2 - (T_1 - 273)^2)\} \quad (2.51)$$

where EX_{Lh}—Liquid petroleum fraction heat Exergy, kJ/kg;

ΔEX_{Lh}—heat Exergy difference of liquid petroleum fraction, kJ/kg.

When the pressure exergy or the flowing exergy of the liquid petroleum fraction is the research object, for example, the pressure exergy is recovered by a hydraulic turbine, the pressure exergy cannot be ignored, and can be simplified based on the formula (2.14), that is, (1) When the pressure change range is not too large, take $\left(\frac{\partial V}{\partial p}\right)_T = 0$, then V = constant (incompressible); (2) when the temperature change range is not too large, take $\left(\frac{\partial V}{\partial T}\right)_p = 0$ (ignoring the volume expansion coefficient). Then the formula (2.14) can be used to calculate the liquid phase petroleum fraction pressure exergy

$$\Delta EX_{Lp} = V(p - p_0) \times 10^{-3} \quad (2.52)$$

where EX_{Lp}—Liquid petroleum fraction pressure exergy, kJ /kg;

p—Petroleum fraction pressure, MPa;

p_0—Reference state pressure, MPa.

2.2.3 Calculation of Gas Petroleum Fraction Physics Exergy at Gas Reference Phase

Under low pressure and non-critical conditions, gas phase petroleum fractions can be regarded as ideal gases. Equation (2.47) and the ideal gas state equation $pV = m^{-1}RT$ relationship (for each kilogram of petroleum fraction) can be substituted into Eq. (2.13) and Eq. (2.14) respectively to obtain thermal exergy and pressure exergy. Thermal Exergy is

$$EX_{oh} = A_3(T - 288) - A_4 \ln\left(\frac{T}{288}\right) + 1.657 \times 10^{-3}(T^2 - 82944) \tag{2.53}$$

where $A_3 = 4.17 \times 10^{-3} API + 0.3161K - 4.1888$

$$A_4 = 1.201API + 91.04K - 931.48$$

The thermal energy of gas phase petroleum fraction is

$$E_{Oh} = C_5(T - 288) + 1.657 \times 10^{-3}\{(T - 273)^2 - 225\} \tag{2.54}$$

The exergy difference and thermal energy difference between T_1 and T_2 are

$$\Delta EX_{Oh} = A_3(T_2 - T_1) + A_4 \ln(\frac{T_2}{T_1}) + 1.657 \times 10^{-3}(T_2^2 - T_1^2) \tag{2.55}$$

$$\Delta E_{Oh} = C_5(T_2 - T_1) + 1.657 \times 10^{-3}\{(T_2 - 273)^2 - (T_1 - 273)^2\} \tag{2.56}$$

where EX_{Oh} Gas phase petroleum fraction thermal exergy (no phase change), kJ/kg;

ΔEX_{Oh}—Thermal exergy difference of gas phase petroleum fraction, kJ/kg;

E_{Oh}—Thermal energy of gas phase petroleum fraction, kJ/kg;

ΔE_{Oh}—Thermal energy difference of gas phase petroleum fraction, kJ/kg.

And the pressure exergy per kilogram of gas phase petroleum fraction is

$$EX_{Op} = \left(\frac{R}{m}\right) T_0 \ln\left(\frac{p}{p_0}\right) \tag{2.57}$$

where: EX_{Oh}—Gas phase petroleum fraction pressure exergy (no phase change), kJ/kg;

R—Gas constant, R = 8.314 J / (kmol K);

m—Petroleum fraction molecular weight;

T—Petroleum fraction temperature, K.

The physical exergy of gas phase petroleum fraction is the sum of heat exergy and pressure exergy.

2.2.4 Calculation of Gas Phase Petroleum Fraction Exergy at Liquid Reference Phase

The integration interval for low-pressure gas phase petroleum fractions exergy calculated whose reference phase is liquid phase includes three sections: liquid phase temperature change, gas–liquid phase change, and gas phase superheating. In addition, there is also gas phase pressure change exergy, such as catalytic cracking reaction oil and gas. For some tower overhead oil and gas that do not contain non-condensable gas, since it leaves the tower under the dew point state, it generally only contains two parts of liquid phase temperature change and gas liquid phase change.

Exergy is a state parameter, and its value is only determined by the final state or reference state of the system. In a certain reference state, when the final state parameter of the system is determined, its value is also uniquely determined and has nothing to do with the process. Therefore, for the gas phase petroleum fraction at temperature T and pressure p, the process shown in Fig. 2.6 can be designed to calculate the exergy value in four steps.

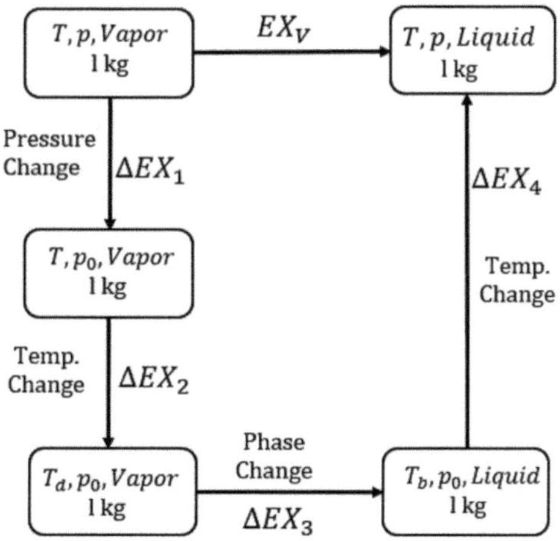

Fig. 2.6 Exergy calculation model of gas phase petroleum fraction

(1) Gas phase pressure change exergy calculation

The gas phase petroleum fraction under low pressure and non-critical state can be regarded as ideal gas, as shown in Fig. 2.6, the pressure exergy of 1 kg gas phase petroleum fraction from pressure p to reference pressure p_0 can be calculated by (2.58), namely

$$\Delta EX_1 = \frac{RT}{m}.\ln(\frac{p}{p_0}) \tag{2.58}$$

(2) Calculation of atmospheric gas phase variable temperature exergy from superheated temperature T to dew point T_d.

$$\Delta EX_2 = A_3(T - T_d) - A_4.\ln\left(\frac{T}{T_d}\right) + 1.657 \times 10^{-3}(T^2 - T_d^2) \tag{2.59}$$

(3) Calculation of gas–liquid phase change exergy from bubble point T_b to dew point T_d

The phase change process of petroleum fractions is a variable temperature phase change process within the range from bubble point to dew point temperature. The temperature-enthalpy relationship of the phase change of petroleum fractions at atmospheric pressure (under the condition of real boiling point distillation), is approximately a linear relationship [5], that is, if there is a virtual specific heat capacity in the phase change process, the specific heat capacity can be regarded as a constant; so the phase change exergy of petroleum fractions can be calculated using this constant specific heat capacity with variable temperature heat source integral formula under atmosphere pressure.

$$\Delta EX_3 = \left(\frac{1 - T_0}{T_m}\right)\Delta H_V \tag{2.60}$$

where, T_m—Arithmetic mean temperature of bubble point and dew point, K;

ΔH_V—Heating from bubble point to dew point and phase change heat, kJ/kg.

$$\Delta H_V = H_V(T_d) - H_L(T_b) \tag{2.61}$$

From formula (2.46)

$$H_V(T_d) = C_4 + C_5(T_d - 273) + 1.657 \times 10^{-3}(T_d - 273)^2 \tag{2.62}$$

From formula (2.44):

$$H_L(T_b) = \{C_2 + C_3(T_b - 273) + 1.99 \times 10^{-3}(T_b - 273)^2\}C_1 \tag{2.63}$$

It can be seen from Fig. 2.6 that T_d and T_b in the formulas (2.62) and (2.63) are the dew point and bubble point temperature of the petroleum fraction under the reference pressure p_0(0.1 MPa). In the absence of dew point and bubble point data, replacing with the initial boiling point and final boiling point of Engel's distillation can also meet the calculation requirements.

(1) Liquid phase temperature exergy calculation

When the reference temperature T_0 is 288 K, the liquid petroleum fraction temperature at the bubble point temperature T_b, its exergy can be calculated by the following formula (2.48):

$$\Delta EX_4 = \left\{A_0(T_b - 288) + 1.99 \times 10^{-3}\left(T_b^2 - 82944\right) - A_1 ln\left(\frac{T_b}{288}\right)\right\}C_1 \tag{2.64}$$

The physical exergy EX_V of the gas-phase petroleum fraction with liquid reference phase is the sum of the above four parts

$$EX_V = \Delta EX_1 + \Delta EX_2 + \Delta EX_3 + \Delta EX_4 \tag{2.65}$$

Obviously, the energy relative to the reference state T_0, p_0 (liquid) is

$$E_V = C_4 + C_5(T - 288) + 1.657 \times 10^{-3}(T - 288)^2 - (C_2 + 15C_3 + 0.488)C_1 \tag{2.66}$$

where: EX_V—Gas phase petroleum fraction physical exergy, kJ/kg;

E_V—Energy of gas phase petroleum fraction under high pressure, kJ/kg.

Example 2.1 Calculate the exergy of petroleum fraction under low pressure: (1) The relative density d_4^{20} is 0.9227, the temperature is 341 °C, the liquid phase atmospheric heavy oil exergy and the energy with the characteristic factor of 12.4; (2) The relative density d_4^{20} is 0.734, and the characteristic factor is 11.8, the exergy difference and energy difference of the gas phase components of gasoline heated from 200 to 500 °C; (3) The relative density d_4^{20} is 0.8058, the temperature is 400 °C, and the characteristic factor is 12.0, Gas phase diesel fraction exergy difference and energy. The fraction is a liquid phase at ambient temperature, the initial boiling point of Engel's distillation is 200 °C; the final boiling point is 350 °C.

The calculation results are listed in Table 2.2.

Table 2.2 Example 2.1 calculation result table unit: kJ/kg

Item			Atm heavy oil (l)	Gasoline Oil (g)	Diesel (g)
Basic data	Relative density		0.9227	0.7340	0.8058
	Characteristic factor		12.4	11.8	12.0
	Reference temperature, °C		15	15	15
	Stream temperature, °C		341	200-500	400
	Initial boiling point, °C				200
	final boiling point °C				350
Calculation results	Liquid phase	Energy Exergy difference	812· 5 287.2		433·0 102·9
	Phase change	Energy Exergy difference			636·3 301.9
	Overheating vapor	Energy Exergy difference		843· 2 451.3	144·3 80·2
	Total	Energy Exergy	812.5 287·2	843.2 451·3	1213·6 485·0

2.3 Calculation of Energy and Exergy of Light Hydrocarbon and Its Mixture

Light hydrocarbon mixture is an important processing medium in petrochemical plants, and its composition can be accurately determined by analytical instruments such as gas chromatography. The composition of light hydrocarbon mixtures is complex, containing multiple or even dozens of monomeric hydrocarbons, which brings difficulties to the calculation of its physical properties. The physical properties of monomeric hydrocarbons (enthalpy, entropy) have mature methods, such as calculating hydrocarbon ideal gas enthalpy and entropy using The Seven Coefficient equation, and the correction of the real gas deviation function, but the calculation reference state is not completely consistent with the reference state used in energy analysis, it is difficult to directly calculate the energy and exergy of hydrocarbons and their mixtures from the enthalpy and entropy data.

In calculations for the purpose of analyzing process energy, the energy and exergy of hydrocarbons and their mixtures are relative to the dead state of the process, that is, the reference state. The petrochemical process usually determines the reference state as 15 °C (288 K) 1 atm (0.1 MPa), and the reference phase state is determined as the phase state of the hydrocarbon substance under the reference state. Therefore, the calculation of enthalpy, entropy and physical exergy of light hydrocarbons and their mixtures is divided into four scenarios [10]:

- Calculation of enthalpy, entropy and exergy of gas-phase hydrocarbons whose reference phase state is gas phase;
- Calculation of enthalpy, entropy and exergy of gas-phase hydrocarbons whose reference phase state is liquid phase;
- Calculation of the enthalpy, entropy and exergy of liquid hydrocarbons whose reference phase is liquid phase;
- Calculation of enthalpy, entropy and exergy of hydrocarbon mixture.

2.3.1 Ideal Gas Hydrocarbon Enthalpy, Entropy and Exergy Calculation

For thermodynamic properties of hydrocarbon ideal gas, the Seven-coefficient equation [1] can be used, its relational formula of enthalpy and entropy is:

$$H^0 = 2.326A + 4.1868BT + 7.5362CT^2 + 13.565DT^3 + 24.417ET^4 + 43.95FT^5 \tag{2.67}$$

$$S^0 = 2.46B + 4.1868Bln(T) + 15.07CT + 20.35DT^2 + 32.56ET^3 + 54.94FT^4 + 4.1868G \tag{2.68}$$

where: H^0—Enthalpy of ideal gas, kJ/kg;

S^0—Entropy of ideal gas, kJ/ (kg K);

T—Ideal gas temperature;

$A \sim G$—The coefficients of the one-to-one Eq. (2.67) and (2.68) vary with different substances. For some light hydrocarbon coefficients, see Table A-4 in Appendix A.

The correlation formula of the specific heat capacity of the ideal hydrocarbon gas determined by the formula (2.67) is:

$$C_p = 4.1868B + 15.07CT + 40.69DT^2 + 97.667ET^3 + 219.75FT^4 \tag{2.69}$$

The enthalpy determined in the formula (2.67) and formula (2.68) is the enthalpy of the pure saturated liquid at −129 °C as zero, and the entropy is 1 kcal/kg K when the temperature is −273 °C and the pressure is 0.070307 kg/cm^2 calculated. The reference pressure of entropy deviates from the reference pressure used in energy analysis, and the calculation of the absolute value of entropy causes a deviation. The difference of the reference pressure is mainly manifested in the different value of the coefficient G of the Eq. (2.68), and the other coefficients B–F remain unchanged. Someone also given another set of coefficients G value of −273 °C and 1 atm ideal gas entropy S = 0. The method of calculating the entropy difference here is used to make the coefficient G eliminated in the calculation. Therefore, the change of the coefficient G does not affect the calculated value of the entropy difference.

2.3.1.1 Enthalpy, Entropy and Exergy of Gas-Phase Hydrocarbons at Gas Reference Phase

From Eqs. (2.67) and (2.68), we can find the enthalpy and entropy from the ideal gas reference temperature T_0 to the operating temperature T:

$$\Delta H = 4.1868B(T - T_0) + 7.5362C(T^2 - T_0^2) + 13.565D(T^3 - T_o^3) + 24.417E(T^4 - T_0^4) + 43.95F(T^5 - T_0^5) \tag{2.70}$$

$$\Delta S = 4.1868Bln\left(\frac{T}{T_0}\right) + 15.07C(T - T_0) + 20.35D(T^2 - T_0^2) + 32.56E(T^3 - T_0^3) + 54.94F(T^4 - T_0^4) \tag{2.71}$$

The influence of ideal gas pressure on entropy is:

$$\Delta S_p = \frac{R}{m}.ln\left(\frac{p_0}{p}\right) \tag{2.72}$$

Calculate physical exergy from the basic definition of exergy:

$$\Delta EX = \Delta H - T_0(\Delta S - \frac{R}{m}.\ln(\frac{p_0}{p}) \tag{2.73}$$

where: p—Operating pressure;

p_0—reference pressure。

2.3.1.2 Enthalpy, Entropy, and Exergy of Liquid Hydrocarbon at Liquid Reference Phase

The correlation formula of liquid phase hydrocarbon specific heat capacity can be divided into ten types, and the correlation formula of liquid phase specific heat capacity is shown in Table 2.3.

Substituting the specific heat capacity correlation formula of hydrocarbon substances for the integral definition formulas of enthalpy, entropy and exergy; enthalpy, entropy, and exergy can be obtained by integrating or using numerical integration methods.

Table 2.3 Correlation equation of liquid specific heat capacity of various hydrocarbon substances

Substance type	Specific heat capacity C_p correlation
Light alkanes ($C_1 - C_4$)	$(14.7486 - 9.8198T_{rb} + 12.9505T_{rb}^2 - 4.8349T_{rb}^3)^{(4-n)/3} \times (13.183 + 54.3924T_{rb} - 72.841T_{rb}^2 + 37.1562T_{rb}^3)^{(n-1)/3}$
C5 plus alkanes	$0.84167 - 1.4704T_{rb} + 1.67165T_{rb}^2 - 0.59198T_{rb}^3 + n(-0.003826 - 0.000747T_{rb} + 0.041126T_{rb}^2 - 0.01395T_{rb}^3)$
Ethylene	$0.71418 - 0.29525T_{rb} + 0.15407T_{rb}^2$
Acrylic	$0.64766 - 0.43638T_{rb} + 0.31228T_{rb}^2$
Butene-1	$0.63192 - 0.46311T_{rb} + 0.33869T_{rb}^2$
Isobutylene	$0.32575 + 0.22823T_{rb} - 0.14714T_{rb}^2 + 0.12786T_{rb}^2$
Olefin	$0.14510 + 1.26233T_{rb} - 2.03775T_{rb}^2 + 1.06953T_{rb}^3 + n(0.067176 - 0.304272T_{rb} + 0.478026T_{rb}^2 - 0.221610T_{rb}^3)$
n- cyclohexane	$0.26796 + 0.04311T_{rb} - 0.06164T_{rb}^2 + n(0.004331 - 0.003439T_{rb} + 0.039338T_{rb}^2)$
n-cylcyclopentane	$(0.3201 - 0.018T_{rb} + 0.03292T_{rb}^2 + 0.12834T_{rb}^3)^{\frac{ln\left(\frac{n}{8}\right)}{ln\left(\frac{5}{8}\right)}} \times (0.44974 - 0.53041T_{rb} + 0.96861T_{rb}^2 - 0.31487T_{rb}^3)^{\frac{ln\left(\frac{5}{n}\right)}{ln\left(\frac{5}{8}\right)}}$
Isoalkanes	$(-0.03083 + 1.77589T_{rb} + 2.50661T_{rb}^2 + 1.20202T_{rb}^3) + n(0.07967 - 0.335664T_{rb} + 0.499756T_{rb}^2 - 0.22174T_{rb}^3)$

Note n—Number of carbon atoms; T_{rb}—Reduced temperature, $T_{rb} = \frac{T}{T_b}$; T_b—boiling point at normal pressure; T—operating temperature

$$\Delta H = \int_{T_0}^{T} C_p dT \tag{2.74}$$

$$\Delta S = \int_{T_0}^{T} C_p/T dT \tag{2.75}$$

$$\Delta EX = \int_{T_0}^{T} C_p(1 - \frac{T_0}{T}) dT \tag{2.76}$$

Pressure has little effect on the enthalpy, entropy and exergy of liquid hydrocarbons and can be ignored.

2.3.1.3 Enthalpy, Entropy, and Exergy of Gas Hydrocarbons at Liquid Reference Phase

Enthalpy, entropy, and exergy are all thermodynamic state functions, and their values are only determined by the initial and final states of the system. Under a certain reference state, the enthalpy, entropy, and exergy calculation of the system in the operating state can be designed in four steps as follows: (see Fig. 2.7).

(1) Liquid phase temperature change process

The calculation methods of enthalpy, entropy and exergy of the liquid phase temperature change process are the same as those of liquid hydrocarbons whose reference phase is liquid phase. At this time, the operating temperature T should be substituted by the atm pressure boiling point T_b of the liquid hydrocarbon.

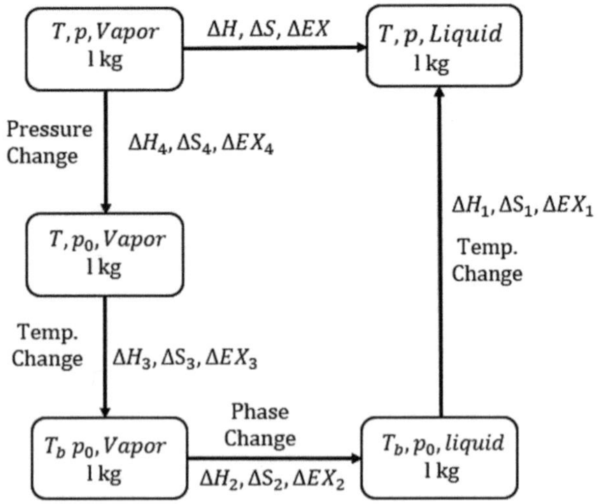

Fig. 2.7 Gas phase hydrocarbon enthalpy, entropy, exergy calculation

$$\Delta H_1 = \int_{T_0}^{T_b} C_p dT$$

$$\Delta S_1 = \int_{T_0}^{T_b} C_p/T dT$$

$$\Delta EX_1 = \int_{T_0}^{T_b} C_p(1 - \frac{T_0}{T})dT \tag{2.77}$$

(2) Vapor–liquid phase change process

The phase change process at the atm pressure boiling point is a reversible phase change process, and the phase change enthalpy is the latent heat of vaporization ΔH_V at the atm pressure boiling point, namely:

$$\Delta H_2 = \Delta H_V \tag{2.78}$$

Phase change entropy and phase change exergy:

$$\Delta S_2 = \Delta H_2/T \tag{2.79}$$

$$\Delta EX_2 = \Delta H_V(1 - \frac{T_0}{T_b}) \tag{2.80}$$

(3) Gas phase temperature change process

The calculation method is the same as the calculation method of the enthalpy, entropy and exergy of the gas phase hydrocarbon with the gas reference phase state, but the reference temperature T_0 should be replaced by the atm pressure boiling point T_b.

$$\begin{aligned}\Delta H_3 &= 4.1868B(T - T_b) + 7.5362C(T^2 - T_b^2 + 13.565D(T^3 - T_b^3) \\ &+ 24.417E(T^4 - T_b^4) + 43.95F(T^5 - T_b^5)\end{aligned} \tag{2.81}$$

$$\begin{aligned}\Delta S_3 &= 4.1868Bln(T/T_b) + 15.07C(T - T_b) + 20.35D(T^2 - T_b^2) \\ &+ 32.56E(T^3 - T_b^3) + 54.94F(T^4 - T_b^4)\end{aligned} \tag{2.82}$$

$$\Delta EX_3 = \Delta H_3 - 288\Delta S_3 \tag{2.83}$$

(4) Gas phase pressure change process

The enthalpy of an ideal gas has nothing to do with pressure, and the influence of pressure on entropy calculated according to the equation of state of ideal gas (p, V, T) belongs to the ideal gas content.

$$\Delta S_4 = -\left(\frac{R}{m}\right)\ln(\frac{p}{p_0}) \tag{2.84}$$

$$\Delta EX_4 = \left(\frac{R}{m}\right)T_0\ln(\frac{p}{p_0}) \tag{2.85}$$

The total enthalpy, entropy, and exergy of the process are the sum of the above four parts.

$$\Delta H = \Delta H_1 + \Delta H_2 + \Delta H_3 + \Delta H_4 \tag{2.86}$$

$$\Delta S = \Delta S_1 + \Delta S_2 + \Delta S_3 + \Delta S_4 \tag{2.87}$$

$$\Delta EX = \Delta EX_1 + \Delta SEX_2 + \Delta EX_3 + \Delta EX_4 \tag{2.88}$$

2.3.1.4 Enthalpy, Entropy and Exergy of Hydrocarbon Mixture

In practical applications, we have encountered a large number of hydrocarbon mixtures. The calculation of enthalpy, entropy and exergy is carried out in two steps. Firstly, the enthalpy, entropy and exergy of the monomer hydrocarbon from the operating state to the reference state are calculated according to the above three scenarios.

The reference phase state can be determined by comparing the atm pressure boiling point with the reference temperature. $T_b \geq T_o$ is the liquid phase, and $T_b < T_0$ is the gas phase; then the enthalpy, entropy, and exergy of the mixture are calculated according to this addition rule. Because under the reversible condition, the separation entropy and mixing entropy of the same value exist between the mixture and the monomer hydrocarbon, which cancel each other out, so there is no need to consider the mixing entropy.

$$\Delta H = \sum X_{Wi}\Delta H_i \tag{2.89}$$

$$\Delta S = \sum X_{Wi}\Delta H_i - (R/m ln(\frac{p}{p_0}) \tag{2.90}$$

$$\Delta EX = \sum X_{Wi}\Delta EX_i \tag{2.91}$$

Usually, the operating data is mostly volume fraction, which can be corrected to weight fraction by the following formula:

$$X_{Wi} = X_i M_i / M \tag{2.92}$$

$$M = \sum X_i M_i \tag{2.93}$$

where: X_{Wi}—Weight fraction of component i;

M—Average molecular weight;

X_i—Volume fraction of component i.

In engineering applications, there are often situations where there are two reference phases (gas and liquid coexist at the reference temperature and pressure) of gas-phase hydrocarbon mixtures. At this time, the number of components in the gas and liquid reference phases can be determined, and calculate the enthalpy, entropy and exergy using gas reference phase and the liquid reference phase separately, then the enthalpy, entropy and exergy of the gas-phase hydrocarbon mixture can be calculated by the sum rule.

For other scenarios, such as the scenario where the reference phase state is liquid phase and the operating state is gas–liquid two-phase, etc., specific analysis and treatment can also be performed according to the above method.

2.4 Calculation of Steam, Water and Air Energy and Exergy

2.4.1 *Steam*

Steam is the most widely used working fluid, and its energy and exergy can be found directly from the steam graph and table. Energy and exergy are:

$$E = \Delta H = H_T - H_0 \tag{2.94}$$

$$EX = \Delta H - T_0 \Delta S \tag{2.95}$$

It can also be calculated by the following empirical formula

$$\Delta H_T = 2417.7 - 87.9 p_s + (2.01 + 0.188 p_s) t \tag{2.96}$$

$$t_s = 179.5 p_s^{0.254} \tag{2.97}$$

where: t—Steam temperature, °C;

t_s—Steam saturated temperature, °C;

p_s—Steam pressure, MPa.

Application range: 100°C $< t <$ 420°C, $p \leq$ 0.9MPa.
Heat of vaporization of steam [7]

$$H_R = Ap_s + \frac{B}{p_s} + Cp_s^{0.5} + Dlnp_s + Ep_s^2 + Fp_s^3 + G \quad (2.98)$$

where: p_s—Steam pressure, bar (kg/cm^2).

A~G—Equation coefficients:

$$A = -1.282172, B = -0.20919, C = -72.77536, D = -34.11784,$$

$$E = 0.0055032, F = 6.271 \times 10^{-5}, G = 2337.061.$$

The entropy difference of steam changing to the liquid phase in the reference state is approximately:

$$\begin{aligned}\Delta S = 4.1868\ln\{(t_s + 273)/T_0\} + H_R/(t_s + 273)\\ +(2.004 + 0.176p_s)ln\{(t + 273)/(t_s + 273)\}\end{aligned} \quad (2.99)$$

Its exergy is calculated by formula (2.95).

2.4.2 *Energy Item Such as Water and Air*

Within a certain temperature range, the specific heat of water, air and other streams can be regarded as a constant, and the energy can be calculated according to the average specific heat method of Eq. (2.7). Heat exergy can also be calculated according to the integral formula of constant specific heat capacity (2.15).

$$EX = \Delta H(1 - \frac{T_0}{T_m}) \quad (2.100)$$

where: ΔHstream enthalpy difference (energy);

T_0, T_m—Arithmetic mean temperature of stream, K.

2.5 Calculation of Heat Dissipation Energy and Exergy

In the energy utilization analysis of petrochemical plants, the measurement and calculation of heat dissipation on the surface of the equipment and piping has an important role. Attributing the energy difference between the incoming and outgoing to the heat dissipation method does not reflect the actual situation of heat dissipation loss positively, and often conceals the errors in the statistics and calculations of the incoming and outgoing energy. The current practical heat dissipation calculation methods are very inconsistent and have large errors, so it is necessary to unify and improve.

Surface heat dissipation includes radiation, natural convection caused by temperature differences and forced convection caused by air flow (wind). The total heat dissipation is the sum of these three parts of heat. Generally, the amount of heat dissipation is calculated by the surface heat dissipation equation:

$$Q = \alpha_T F(t_W - t_f) \tag{2.101}$$

where: α_T—Convection and radiation combined heat transfer coefficient, W/m^2·°C;

F—Heat dissipation external surface area, m^2;

t_w, t_f—Respectively the equipment surface and ambient temperature, °C.

The combined heat transfer coefficient α_T summarizes the above three heat transfer methods. Strict correlation is more tedious. Some textbooks or manuals often associate α_T as an approximate simplified function of temperature difference $(t_W - t_f)$ and wind speed $W(\mathrm{m/s})$. For example, the following formula is commonly used in thermal engineering calculations:

$$\alpha_T = (10 + 6\sqrt{W}) \times 1.163 \tag{2.102}$$

This formula only considers forced convection, and regards the radiation and natural convection caused by temperature differences as a constant, which is relatively rough. Another commonly used formula is (the symbol in the formula is the same as before):

$$\alpha_T = 1.59\left(t_w - t_f\right)^{0.33} + 4.40\left(t_w - t_f\right)^{0.11} + 6.98\sqrt{W} \tag{2.103}$$

This formula includes the effects of radiation and convection caused by temperature differences, but has been simplified to a certain extent: (1) the fourth power law of radiation is treated as an exponential relationship and based on $\left(t_w - t_f\right)$ being about 100 °C; (2) the influence of forced convection is treated according to the outer diameter of the heat dissipation surface of 300–500 mm. Therefore, when the outer diameter of the surface and the temperature difference deviate far from the above conditions, the error will be larger.

Example Find the coefficient of heat dissipation on the surface of an equipment. The outer diameter of the insulation of the equipment is 500 mm, the difference between the surface temperature of the equipment and the ambient temperature is 150 °C, and the wind speed is 2 m/s.

Solution Starting from the left vertical scale with a wind speed of 2 m/s, cross the curve with a heat dissipation outer diameter of 500 mm horizontally to the right, then cross vertically downwards on a straight line with a heat dissipation temperature difference of 150 °C, and cross horizontally to the right on the right ordinate, the heat dissipation coefficient is 21.7 kcal/(m^2 h °C). Converted to SI unit is 25.237 W/(m^2 K).

It is better to use accurate correlations instead of sticking to hand calculations and sacrificing accuracy to obtain simplifications, and use computers or graphs to simplify calculations and save time. For this reason, the correlation equation directly derived from the heat transfer coefficient correlation equations of the above three heat transfer modes: radiation, natural convection, and forced convection is as follows [8]:

$$\alpha_T = 4.33\frac{\left(\frac{T_W}{100}\right)^4 - \left(\frac{T_f}{100}\right)^4}{T_W - T_f} + 1.47\left(T_w - T_f\right)^{1/3} + 4.36W^{0.6}/D^{0.4} \tag{2.104}$$

where: T_W, T_f—Respectively the surface temperature and the ambient temperature, K;

D—equipment outer diameter of heat dissipation, mm;

Draw a quick calculation diagram (Fig. 2.8) by formula (2.104), convection and radiation combined heat transfer coefficient can be easily found by wind speed, surface and ambient temperature, and the outer diameter of heat dissipation. If possible, it can also be easily calculated with a computer program.

Figure 2.8 is drawn under the condition of $t_f = 15$ °C, when $t_f \neq 15$ °C, it can be corrected by Fig. 2.9. When the heat dissipation temperature difference $(t_W - t_f)$ is less than or equal to 60 °C (generally a well-insulated surface), there is no need to check Fig. 2.9, and adding the correction value of $\gamma = 0.04\left(t_f - 15\right)$ will be good.

Tables 2.4 shows the comparison of different methods for calculating surface heat dissipation for a catalytic cracking regenerator of an oil refinery. The results show that the calculation accuracy according to formula (2.104) is higher than the other two.

The heat dissipation amount Q calculated from the measurement can be used to calculate the heat dissipation exergy loss EX_D as follows:

$$EX_D = Q(1 - T_0/T_B) \tag{2.105}$$

where T_B(K) is the temperature of the fluid inside the equipment. When the fluid temperature changes, its characteristic temperature can be used. For the heat exchangers, take the characteristic temperature of the two fluids of the shell and the tube box respectively; for the heating furnace, the T_p temperature at the top of the radiant section can be approximated; for the fractionation tower, the characteristic temperature can be calculated according to the temperature at the bottom and the top of the tower; for the pipeline, the characteristic temperature can be calculated in

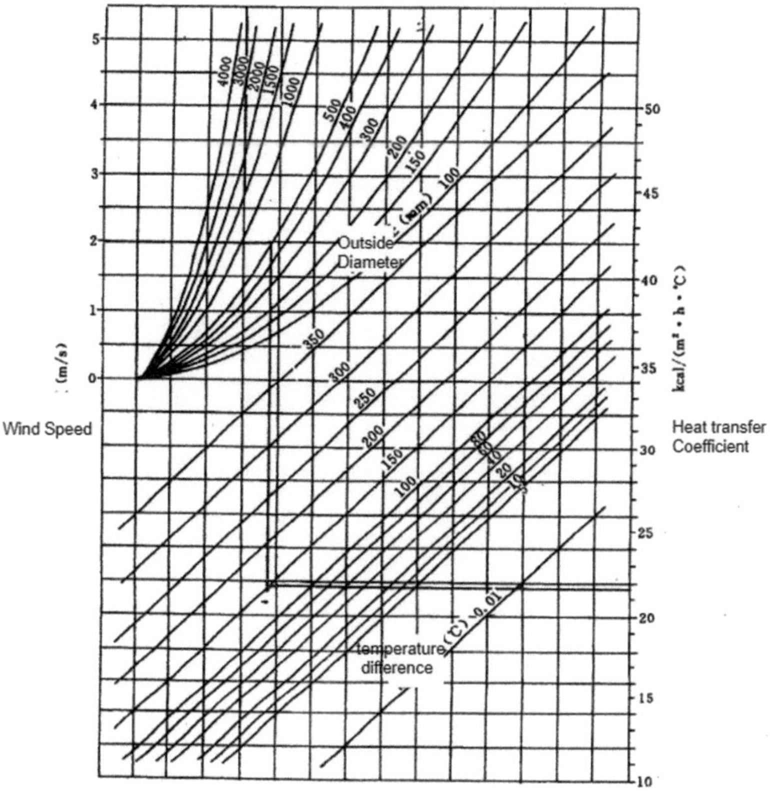

Fig. 2.8 Surface heat dissipation coefficient calculation chart

sections when the temperature drop is larger. Note that the heat dissipation exergy loss should not be calculated based on the surface temperature, because the loss of the temperature from T_B to T_W is also caused by heat dissipation.

In addition, it must be noted in the calculation of heat dissipation that the surface temperature on the windward and leeward sometimes differ greatly (Fig. 2.10). Since it is impossible to measure the wind speed at point of the temperature measurement point by point in practice, only one uniform value can be used. Therefore, the surface temperature should be the average value of multiple measurement points, and attention must be paid to the representativeness of the measurement points.

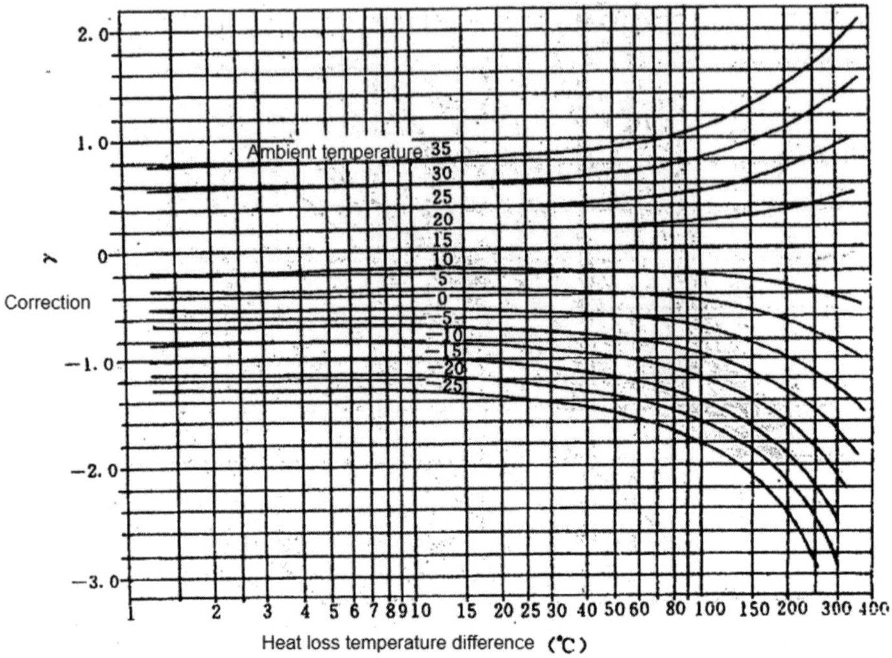

Fig. 2.9 Ambient temperature correction

Fig. 2.10 Measurement data of surface temperature of a vertical equipment

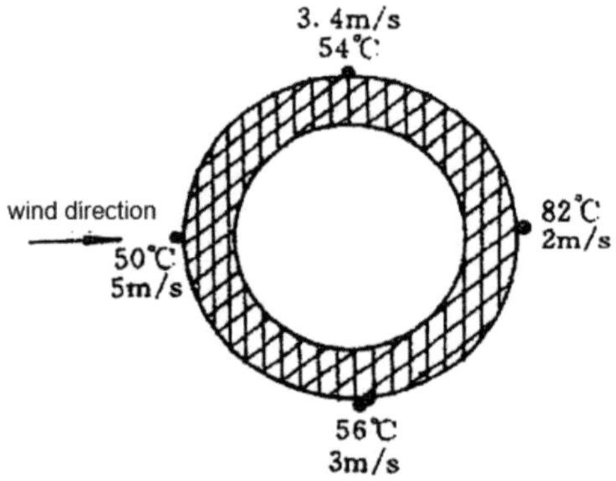

Table 2.4 Comparison of heat dissipation calculation and thermal conductivity measurement data on the surface of regenerator

Item		Double lining of regenerator	Single layer lining of regenerator
Measuring case		A	B
Ambient temperature, °C		23	18
Wind speed, (m/s)		3.2	0.2
Fluid temperature, °C		4 8 6	660
Surface temperature, °C		75	126
Insulation thickness (mm)		100	100
Measured thermal conductivity of lining, (kcal/m · h · °C)		0.18	0.26
Measured thermal conductivity (kcal/h· m^2)		740	1388
Equation (2.102)	Heat dissipation coefficient (kcal/m^2 h °C)	20.73	10.0
	Heat dissipation rate (kcal/h m^2)	1121	1127
	Error to thermal conductivity measurement (%)	51.5	18.8
Equation (2.103)	Heat dissipation coefficient (kcal/m^2 h °C)	21.6 2	12.75
	Heat dissipation rate (kcal/h m^2)	116 9	137
	Error to thermal conductivity measurement (%)	58.0	30.52
Equation (2.104)	Heat dissipation coefficient (kcal /m^2 h °C)	15. 0	12.26
	Heat dissipation rate (kcal/h m^2)	811	1382
	Error to thermal conductivity measurement (%)	9.6	0.4

(1) The "determination of thermal conductivity" in the table is calculated based on the thermal conductivity measured by Luoyang Refractory Research Institute. (2) 1 Cal = 4.184 J

References

1. API Technical Data Book—Petroleum Refining, 4th ed. (1983)
2. Oil refining teaching and research group of Shanghai Institute of Chemical Engineering, Petroleum refining design data chart and table set (part 2), p. 313 (1987)

3. Z. Xipeng, Calculation and analysis of refining process exergy. Maoming Oil Process. **4**, 25 (1988)
4. Z. Lianke Editor-in-Chief, *Chemical Thermodynamics, (inorganic chemical use)* (Chemical Industry Press, 1980), p. 12
5. C. Zuheng, *Discussion of Oil Refining Process Exergy Analysis and Energy Efficiency Economic Evaluation Methods, Refinery Design (Refinery Energy Saving Collection),* pp. 38–40 (1981)
6. Andrea Carli, Chemical Processing, **4**, 87 (1974)
7. A Correlation for Enthalpy of Petroleum Fractions, HaZim s. H Oil. & Gas Journal, Sept 4, 1989
8. H. Ben, T.A. Chen, The practical calculation of refinery heat loss and exergy economic analysis, oil refining **2** (1984)
9. C. Anmin, H. Ben, Oil and its fractional physics exergy practical calculation. Refining Eng. **3**, 35–39 (1987)
10. C. Anmin T. Hui, Energy and exergy practical calculation method of Light hydrocarbons and their mixtures. Petroleum Engineering, **6**, 52 (1990)

Chapter 3
Calculation of Mechanical Energy and Chemical Exergy

Abstract In addition to the energy and exergy calculation of process streams and utility in Chaps. 2, this chapter discusses the calculation of real gas energy and exergy based on the thermodynamic differential equations of enthalpy and entropy, deriving the enthalpy, entropy, and exergy calculation method for real gas, and the residual properties of the influence of pressure on enthalpy and entropy can be calculated in three levels according to the different calculation accuracy required. This chapter also Introduces the energy and exergy calculation of the fluid flowing process, chemical exergy, and fuel exergy calculation.

Keywords Real gas exergy · Irreversible process · Fluid flowing · Chemical exergy · Fuel exergy · Diffusion exergy

3.1 Calculation of Real Gas Energy and Exergy

In many operations, the gas stream is operated at a higher pressure. For a gaseous stream pressure greater than 0.5 MPa that has deviated from an ideal gas, the energy and exergy value of the stream cannot be calculated according to the calculation method of ideal gas physical properties. So how to calculate the real gas flow energy and exergy value is also a concern we need to solve.

3.1.1 Calculation Method of Real Gas Energy and Exergy

In the chemical engineering thermodynamics, the method of "residual properties" or "deviation function" is usually used to solve the calculation of thermodynamic properties of real gases. The so-called residual property is the thermodynamic properties difference between the real gas and the ideal gas at the same temperature and pressure [10]. The residual property is a hypothetical concept, because in the real state, under the same temperature and pressure, the streams cannot be in an ideal state. But with this method, the thermodynamic properties difference between the

T. A. Chen, *Energy Saving and Carbon Reduction*,
https://doi.org/10.1007/978-981-19-5295-1_3

real gas and ideal gas can be calculated, and then find the thermodynamic properties of real gas.

Usually, the enthalpy value of the process chart we used is mostly the enthalpy value under atm pressure. The calculation of the ideal gas entropy can also be easily conducted. Therefore, the streams exergy value can be calculated from the enthalpy, entropy and environmental parameters. The key to seeking true gas flow energy and exergy is to determine the influence of pressure on physical properties, that is, residual properties.

According to the thermodynamic differential equations of enthalpy and entropy:

$$dH = TdS + Vdp \tag{3.1}$$

$$dS = C_p dT/T - \left(\frac{\partial V}{\partial T}\right)_p dp \tag{3.2}$$

We can get:

$$dH = C_p dT + Vdp - \left(\frac{\partial V}{\partial T}\right)_p dp \tag{3.3}$$

$$\delta EX = C_p(1 - \frac{T_0}{T})dT + Vdp - (T - T_0)\left(\frac{\partial V}{\partial T}\right)_p dp \tag{3.4}$$

In formula (3.2), formula (3.3) and formula (3.4), the function that changes with pressure is called residual property, (pressure exergy includes the pressure exergy part of ideal gas)

$$\Delta H_p = \int_{p0}^{p} (V - T(\frac{\partial V}{\partial T})_p)dp \tag{3.5}$$

$$\Delta S_p = \int_{p0}^{p} (\frac{\partial V}{\partial T})_p)dp \tag{3.6}$$

$$\Delta EX_p = \int_{p0}^{p} (V - (T - T_0)(\frac{\partial V}{\partial T})_p)dp \tag{3.7}$$

Therefore, the real gas energy (enthalpy change), entropy and Exergy:

$$\Delta H = \Delta H^0 + \Delta H_p \tag{3.8}$$

$$\Delta S = \Delta S^0 + \Delta S_p \tag{3.9}$$

$$\Delta EX = \Delta EX^0 + \Delta EX_p \tag{3.10}$$

Superscript "0" represents ideal gas.

3.1.2 Calculation of Residual Properties

The key to determining the residual properties in Eqs. (3.5)–(3.7) is to find a real gas state equation that reflects the p-V-T relationship of the actual gas, so as to solve the effect of pressure on physical properties. The Virial equation of state [1] expressed in the form of a power series combined with the empirical equation of the second Virial coefficient proposed by Pitzer et al. [2] can solve the impact of petroleum fraction pressure on physical energy and exergy. The L-K three-parameter generalized correlation equation can be used to solve the influence of pressure on the energy and exergy of light hydrocarbons and their mixtures [11].

According to the different calculation accuracy required, three levels can be used to calculate the influence of pressure on enthalpy entropy.

3.1.2.1 Berthelot Equation of State Method [3]

In simple engineering calculations such as energy balance and exergy balance, high accuracy is not required, and the streams quantity (mass) entering and leaving the equipment for streams is unchanged, and the error caused by the calculation method is not very large. Therefore, it is convenient to adopt the analytical Berthelot equation of state to calculate the energy and exergy of real gas streams can also meet the engineering error requirements.

$$pV = RT(1 + \frac{9pT_c}{128p_cT}(1 - 6\frac{T_c^2}{T^2})) \tag{3.11}$$

The equation is relatively simple and takes the form of an explicit function, which is easy to solve. The terms in brackets on the right side of formula (3.11) are actually the compression factor Z in $pV = ZRT$.

From formula (3.11), we have

$$V = RT/p(1 + \left(\frac{9pT_c}{128p_cT}\right) - 54pT_c^3/(128p_cT^3)) \tag{3.12}$$

$$\left(\frac{\partial V}{\partial T}\right)_p = \frac{R}{p} + 54RT_c^3/(64p_cT^3) \tag{3.13}$$

Substituting the above Eqs. (3.12) and (3.13) into Eq. (3.5), and integral calculus from the state of p_0 to p to get

$$\Delta H = RT_c/p_c(\frac{9}{128} - 81T_c^2/(64T^2))(p - p_0) \tag{3.14}$$

Equation (3.14) is the integral calculation formula for the influence of pressure on enthalpy.

Similarly, we substitute the formula (3.12) and the formula (3.13) into Eq. (3.7) integral to get:

$$\Delta EX = RT_0\ln\left(\frac{p}{p_0}\right) + RT_c/p_c(\frac{9}{128} - \frac{81}{64T_r^2} + \frac{27T_0}{32TT_r^2})(p - p_0) \tag{3.15}$$

From Eq. (3.15), it can be seen that the former term $RT_0\ln\left(\frac{p}{p_0}\right)$ on the right side of the equation is the effect of pressure on exergy when treated as ideal gas, second term is the correction term, that is, ideal gas is used with correction term of the ideal gas pressure for the flow pressure of the real gas, let:

$$D = \frac{RT_c}{p_c} = RT_c/p_c(0.07031 - \frac{0.844}{T_r^2}\left(1.5 - \frac{T_0}{T}\right) \tag{3.16}$$

Then the formula (3.15) becomes:

$$\Delta EX = RT_0\ln\left(\frac{p}{p_0}\right) + D(p - p_0) \tag{3.17}$$

In the formula, R is for 1 kg of molecular streams. For each kilogram of the stream, the gas constant R can be replaced by R/M (M is the molecular weight of the stream).

The Berthelot equation is a two-parameter equation of state. For a gas phase process without phase change, its calculation accuracy can meet the engineering error requirements. However, it cannot be used when it is close to the critical state or the two-phase state of gas and liquid.

3.1.2.2 Virial Equation Method

The Virial equation expressed in the form of a power series is a relatively common gas state equation, and its form is:

$$Z = \frac{pV}{RT} = 1 + B'p + C'p^2 + \ldots \tag{3.18}$$

In the range where the Reduced volume of the gas is greater than 2, the second term of Eq. (3.18) can still accurately describe the relationship of the gas, so Pitzer et al. [2] believe that the Virial coefficient and its correlation equation should be used in this range which is very convenient to calculate the thermodynamic properties of gases.

$$Z = 1 + Bp/(RT) \tag{3.19}$$

Find the partial derivative of temperature under constant pressure:

$$\left(\frac{\partial Z}{\partial T}\right)_p = p/R(\frac{dB}{TdT} - \frac{B}{T^2}) \tag{3.20}$$

On the other hand, any real gas can be expressed as:

$$V = ZRT/p \tag{3.21}$$

$$\left(\frac{\partial V}{\partial T}\right)_p = R/p(Z + T\left(\frac{\partial Z}{\partial T}\right)_p) \tag{3.22}$$

Substituting formula (3.21) and formula (3.22) into (3.5) and formula (3.7), we get

$$\Delta H = RT^2 \int_{p_0}^{p} \left(\frac{\partial Z}{\partial T}\right)_p dp/p \tag{3.23}$$

$$\Delta EX = \int_{p_0}^{p} (ZRT_0 + RT\left(\frac{\partial Z}{\partial T}\right)_p (T - T_0))dp/p \tag{3.24}$$

Substituting the expressions of Z and $\left(\frac{\partial Z}{\partial T}\right)_p$, there are:

$$\Delta H = (\frac{TdB}{dT} - B)(p - p_0) \tag{3.25a}$$

$$\Delta EX = RT_0 \ln\left(\frac{p}{p_0}\right) - (\frac{TdB}{dT}(T - T_0) - B)(p - p_0) \tag{3.25b}$$

The empirical equation of the second virial coefficient [2] proposed by Pitzer et al. is:

$$B = \frac{RT_c}{p_c} - (0.33 - 0.46\omega)T_r^{-1} - (0.1385 + 0.5\omega)T_r^{-2} - (0.0121 + 0.097\omega)T_r^{-3} - 0.0073T_r^{-8}) \tag{3.26}$$

$$\frac{dB}{dT_r} = \frac{RT_c}{p_c} + (0.277 + \omega)T_r^{-3} + (0.0363 + 0.29\omega)T_r^{-4} - 0.0584T_r^{-9}) \tag{3.27}$$

Substituting B and dB/dT_r for the calculation formula can obtain the pressure enthalpy and pressure exergy value of the streams, ω is the eccentric factor of the stream (Ref to Appendix 1 Table A1).

3.1.2.3 Lee-Kesler Three-Parameter Universal Correlation Method [12]

For real gas at higher pressure, the influence of pressure on streams enthalpy and exergy can also be calculated by L-K three-parameter generalized correlation method and B-W-R Reduced equation of state. This method has high accuracy and wide application range, but it requires multiple iterative calculations. It is more convenient to apply the three-parameter generalization method to the energy consumption analysis of petrochemical plants when computers are popular today.

The eccentricity factor ω, a characteristic parameter that fascinates the size and shape of material molecules, is used to calculate the enthalpy and entropy of real gases using the L-K three-parameter generalized correlation.

$$G = G^0 + \omega/\omega^r(G^r - G^0) \tag{3.28}$$

where: G is the physical property (either compression factor, enthalpy or entropy, etc.);

ω—Eccentricity factor;

The superscript 0 represents the thermodynamic properties of simple fluids (methane, etc.);

The superscript r indicates the thermodynamic properties of the reference fluid (n-octane).

According to the above-mentioned three-parameter generalized correlation formula, the deviation functions of compression factor, enthalpy, and entropy are:

$$Z = Z^0 + \omega/\omega^r(Z^r - Z^0) \tag{3.29}$$

$$\left(\frac{H - H^0}{RT_c}\right) = \left(\frac{H - H^0}{RT_c}\right)^0 + \left(\frac{\omega}{\omega^r}\right)\left(\left(\frac{H - H^0}{RT_c}\right)^r - \left(\frac{H - H^0}{RT_c}\right)^0\right) \tag{3.30}$$

$$\frac{S - S^0}{R} = \left(\frac{S - S^0}{R}\right)^0 + \left(\frac{\omega}{\omega^r}\right)\left(\left(\frac{S - S^0}{R}\right)^r - \left(\frac{S - S^0}{R}\right)^0\right) \tag{3.31}$$

Apply the modified B-W-R Reduced equation of state to calculate the compressibility factor of simple fluid and reference fluid.

$$Z = \frac{p_r V_r}{T_r} = 1 + \frac{B}{V_r} + \frac{C}{V_r^2} + \frac{D}{V_r^5} + \frac{C_4}{V_r^2 T_r^3}(\beta + \gamma/V_r^2)\exp(-\frac{\gamma}{V_r^2})) \tag{3.32}$$

Table 3.1 Equation (3.32) coefficient table

Coefficient	Simple fluid	Reference fluid
b_1	0.1181193	0.202657
b_2	0.265728	0.331511
b_3	0.154790	0.027655
b_4	0.030323	0.203488
c_1	0.0236744	0.0313385
C_2	0.0	0.016901
c_3	0.042724	0.041577
$d_1 * 10^4$	0.155488	0.48736
$d_2 * 10^4$	0.623689	0.070336
β	0.65392	1.226
γ	0.060167	0.03754

where:

$$B = b_1 - \frac{b_2}{T_r} - \frac{b_3}{T_r^2} - \frac{b_4}{T_r^3}$$

$$C = c_1 - \frac{c_2}{T_r} + \frac{c_3}{T_r^3}$$

$$D = d_1 - \frac{d_2}{T_r}$$

The coefficients in the formula are shown in Table 3.1.

Solve the root V of the above equation by Newton's iteration method, and get the Z_0 of the simple fluid. Following this, substituting the constants of the reference fluid and the Reduced parameters p_r, T_r into the above equation, the compression factor Z^r of the reference fluid can be obtained. According to the eccentricity factor of the substance, the Z value of the substance in the operating state is obtained from the formula (3.29).

Once Z^0, Z^r are known, the dimensionless groups of residual enthalpy and residual entropy of the simple fluid and the reference fluid can be obtained by the two Eqs. (3.33) and (3.34). And then, find the residual properties of enthalpy and entropy using Eqs. (3.30) and (3.31).

$$\frac{H - H^0}{RT_c} = T\left(Z - 1 - \frac{b_2 + \frac{2b_3}{T_r} + \frac{3b_4}{T_r^2}}{T_r V_r} - \frac{c_2 - \frac{3c_3}{T_r^2}}{2T_r V_r^2} + \frac{d_2}{5T_r V_r^5} + 3E\right) \quad (3.33)$$

$$\frac{S - S^0}{R} = lnZ - \frac{\left(\frac{b_1 + b_3}{T_r^2} + \frac{2b_4}{T_r^3}\right)}{V_r} - \frac{\left(C_1 - \frac{2c_3}{T_r^3}\right)}{2V_r^2} - \frac{d_1}{5V_r^5} + 2E \quad (3.34)$$

$$E = \frac{c_4}{2T_r^3\gamma}\left[\beta + 1 - \left(\beta + 1 + \frac{\gamma}{V_r^2}\right)\exp(-\frac{\gamma}{V_r^2})\right] \tag{3.35}$$

where: T_r—reduced temperature of real gas, $T_r = T/T_c$

T_c—Hypothetic critical temperature,

$$T_c = \sum X_i/T_{ci} \tag{3.36}$$

p_c—Hypothetic critical pressure,

$$p_c = \sum X_i/p_{ci} \tag{3.37}$$

ω—Eccentricity factor,

$$\omega = \sum X_i/\omega_i \tag{3.38}$$

V_r—Reduced volume.

So far, the enthalpy of real gas can be obtained from the enthalpy and residual properties of ideal gas

$$\Delta H = \Delta H^0 + \frac{H - H_0}{RT_c}\frac{R}{M}T_c \tag{3.39}$$

The entropy of real gas can be obtained from the entropy and residual properties of ideal gas:

$$\Delta S = \Delta S^0 + [(S - S^0)/R]R/M \tag{3.40}$$

Exergy of real gas

$$\Delta EX = \Delta H + T_0\Delta S \tag{3.41}$$

Example 3.1 The compressed rich gas temperature of a catalytic cracking unit in an oil refinery is 120 °C and the pressure is 1 MPa (10 atm). The compressed rich gas composition (volume fraction) is shown in Tables 3.2 and 3.3. The results of calculating the enthalpy, entropy, and exergy of each component and mixture using the IBM-PC computer program compiled by this method are listed in Table 3.4 (take the reference temperature as 15 °C).

Table 3.2 Gas phase enthalpy, entropy and exergy at liquid reference phase

Original data		Calculation results		
Name	Volume fraction	Enthalpy (kJ/kg)	Entropy (kJ/(kg °C)	Exergy (kJ/kg)
Isopentane	0.050	542.142	1.7271	44.732
Pentane	0.008	567.204	1.7781	55.104
Cispentene-2	0.016	557.901	1.7482	54.432
Pentene-1	0.034	540.032	1.7168	45.591

Table 3.3 Gas enthalpy, entropy and exergy at gas reference phase

Original data		Calculation results		
Name	Volume Fraction	Enthalpy (kJ/kg)	Entropy (kJ/kg °C)	Exergy (kJ/kg)
Hydrogen	0.074	1512.414	4.4745	223.767
Oxygen	0.086	97.152	0.2873	14.414
Nitrogen	0.086	109.256	0.3282	16.163
Methane	0.055	247.114	0.7289	87.19
Carbon dioxide	0.012	92.985	0.2744	13.962
Ethylene	0.029	177.486	0.5223	27.053
Ethane	0.058	201.619	0.5933	30.762
Acrylic	0.217	177.601	0.5223	27.192
Propane	0.032	195.038	0.5735	29.867
Isobutane	0.089	196.862	0.5787	30.201
Isobutylene	0.036	184.628	0.543	28.234
Butene-1	0.036	179.036	0.5263	27.474
Butane	0.018	199.05	0.5854	30.46
AntiButene-2	0.036	180.525	0.5309	27.62
Maleic-2	0.028	165.329	0.4856	25.469

Table 3.4 Gas enthalpy, entropy and exergy at mixed reference phase

Item	Enthalpy (kJ/kg)	Entropy (kJ/kg °C)	Exergy (kJ/kg)
Gas reference phase	145.025	0.4268	22.102
Liquid reference phase	101.531	0.322	8.781
Ideal gas	246.555	0.7489	30.883
Pressure influence item	−10.94	−0.4833	128.243
Real gas	235.616	0.2656	159.126

3.1.2.4 Physical Properties for the Pressure Influence on Enthalpy and Exergy

When we calculate the pressure influence on the enthalpy and exergy of process streams, it involves the critical properties of streams T_c, p_c, boiling point and eccentricity factor.

For common substances, including light hydrocarbons and their mixtures and other commonly used substances. These physical properties can be found in the petrochemical physical property's manual. But for petroleum fractions, there is no ready-made data to check, which can be obtained through corresponding calculations.

The calculation method of the virtual critical properties T_c, p_c and molecular weight M of petroleum fractions, and the eccentricity factor ω reflecting the degree of deviation of the substance from the spherical shape [4].

$$T_c = 17.2164\mathrm{T}_B^{0.5848}\mathrm{d}^{0.3956} \tag{3.42}$$

$$p_c = 5.4565 \times 10^7 T_B^{-2.3125} d^{2.3201} \tag{3.43}$$

$$M = 219.06exp(0.00392T_B) \cdot exp(-3.07d)T_B^{0.118}d^{1.88} \tag{3.44}$$

$$\omega = 3/7((log_{10}p_c)/\left(\frac{T_c}{T_B} - 1\right)) \tag{3.45}$$

In the above formula, d is the relative density at 15.6 °C. If the relative density d_4^{20} of petroleum fractions at 20 °C is known, then d can be estimated by the following formula:

$$d = 0.9942d_4^{20} + 0.009181 \tag{3.46}$$

T_B is the average boiling point of petroleum fractions, which can be calculated from the data of 10–90% distillation temperature of Engel's distillation data.

$$t_V = (t_{10} + t_{30} + t_{50} + t_{70} + t_{90})/5 \tag{3.47}$$

$$SL = (t_{90} - t_{10})/80 \tag{3.48}$$

$$T_B = D_0 + D_1 t_V \tag{3.49}$$

where: t_V—Volume average boiling point, °C;

SL—The slope of Engel's distillation curve;

$$D_0 = 272.03 - 5.9915SL - 0.50067SL^3;$$

$$D_1 = 1 + 0.012809SL - 4.5286 \times 10^{-3}SL^2 + 1.17013 \times 10^{-3}SL^3$$

$t_{10} \sim t_{90}$—The 10–90% point temperature of Engel's distillation (°C). When the distillation temperature of Engel's distillation is higher than 246 °C, the cracking effect should be considered and correction should be made [4].

3.2 Calculation of Energy in the Fluid Flowing Process

Flowing energy is the main form of work. In theory, work can be transformed from one form to another. Its energy grade is 1. Therefore, the flow energy is the same as the flow exergy value in many cases.

3.2.1 Volume Work, Shaft Work and Flow Work

In the petrochemical process, the power equipment used has various power-consuming equipment (pumps, fans, compressors, etc.) and the power production equipment (thermal machinery such as gas turbines, expansion turbines and hydraulic machinery) are used to achieve exchanging of work among fluid with the environment through the rotation of the mechanical shaft.

The work exchanged between the unit mass of gas and the environment can be divided into three parts:

1. When the fluid flows into the pipeline, the work received from the gas behind is p_1V_1(the work performed to system by the environment is negative value)
2. When the fluid flows out of the pipeline, the work performed to the downstream fluid is p_2V_2;
3. The work transmitted by the fluid to the rotating shaft of the machine—shaft work.

When the unit mass of gas flows through the turbine, the volume of the gas changes. Under ideal conditions (with no friction, the driving force is infinitely small), the volumetric work performed by the gas is:

$$W = \int_{V_1}^{V_2} pdV = W_s + p_2V_2 - p_1V_1 \tag{3.50}$$

And then:

$$W_s = \int_{V_1}^{V_2} pdV - (p_2V_2 - p_1V_1) = \int_{V_1}^{V_2} pdV - \Delta(pV) \tag{3.51}$$

$$d(pV) = pdV + VdP \tag{3.51}$$

$$\int_{p_1V_1}^{p_2V_2} d(pV) = (p_2V_2 - p_1V_1) = \int_{V_1}^{V_2} pdV + \int_{p_1}^{p_2} Vdp$$

$$\int_{V_1}^{V_2} pdV = p_2V_2 - p_1V_1 - \int_{p_1}^{p_2} Vdp$$

Substitute above equation into Eq. (3.50), we get shaft work:

$$W_s = -\int_{p_1}^{p_2} Vdp \tag{3.52}$$

W_s is the ideal shaft work transmitted by the fluid to the mechanical shaft.

We usually call $\Delta(pV)$ the flow work W_f of the system, that is, the work that the fluid exchanges with the outside due to changes in the flow state during the flow of the fluid. W_f can be positive or negative. When it is a positive value, it indicates that the flow system is doing external work ($W_f > 0$, $p_2V_2 > p_1V_1$); when it is a negative value, it indicates that work must be supplied from the outside to maintain the flow condition. ($W_f < 0$, $p_2V_2 < p_1V_1$); the sum of shaft work and flow work is the total work (volume work).

$$W = W_s + W_f \tag{3.53}$$

3.2.2 *Calculation of Shaft Work and Effective Work [5, 6]*

In process energy analysis, the energy transferred from power-consuming equipment to the fluid is usually called effective work, which is mainly reflected in the change of fluid flow work. Under reversible conditions, the ideal shaft work of the pump equipment is the effective work of the fluid.

According to the integral formula of shaft work, the shaft work can be obtained from the different fluid's properties.

3.2.2.1 Calculation of Shaft Power of Liquid Conveyor Pump

$$-W_s = \int_{p_1}^{p_2} Vdp$$

The liquid is an incompressible fluid, so V = constant, there are:

$$-W_s = V(p_2 - p_1) = V\Delta p \tag{3.54}$$

where: V—Fluid specific volume, m^3/kg;

Δp—Pressure difference of fluid in and out of pump or compressor, bar (kg/cm^2).

When flow rate is G (kg/h), Δp is bar (kg/cm^2), we get:

$$-W_s = G\Delta p/367 \tag{3.55a}$$

When Q (m^3/s) is used to represent the flow rate, the meter liquid column is used to represent the pump head H, and the fluid density is γ (kg/m^3), the ideal shaft work:

$$-W_s = QH\gamma/102 \tag{3.55b}$$

where: $-W_s$—Shaft Work, kW (Negative value means system power consumption).

3.2.2.2 Calculation of Shaft Power of Gas Conveying Equipment (Compressor)

There are generally three processes for gas delivery equipment. Isothermal compression, the least power consumption (isothermal expansion does the maximum work), and the actual process is mostly not isothermal compression, but isothermal compression as a comparison benchmark has practical significance; adiabatic compression, the heat generated by the adiabatic compression gas process is used to raise the gas temperature (in fact, heat loss always exists), which is also a limit case; the variable compression process is between isothermal compression and adiabatic compression processes, and most of the actual compression processes belong to this type of compression process.

In the energy analysis that only performs energy balance, almost all of the shaft work consumed by the compressor enters the gas (in the form of flow work and thermal energy) and becomes the effective energy of the fluid. For exergy balance, the effective exergy is the sum of mechanical (compression) exergy and fluid heat exergy.

(1) Isothermal process

For 1 kg ideal gas, the compression work of isothermal process:

$$-W_s = \int_{p_1}^{p_2} Vdp = \int_{p_1}^{p_2} \frac{R'T_1}{pdp} = \left(\frac{R}{M}\right)T_1\ln(\frac{p_2}{p_1}) \tag{3.56}$$

(2) Adiabatic process

Because the adiabatic process equation is: $p_1 V_1^k = p_2 V_2^k$, So it can be exported:

$$-W_s = \left(\frac{k}{k-1}\right) R' T_1 \left(\frac{p_2}{p_1}\right)^{\frac{k-1}{k}} - 1 \tag{3.57}$$

The difference in adiabatic index is related to the properties of the gas, and strictly speaking, it is also related to the temperature. In the rough calculation of chemical thermodynamics, the ideal gas adiabatic index value can be taken: $k = C_p / C_v$

Monoatomic gas $k = 1.667$

Diatomic gas $k = 1.4$

Triatomic gas $k = 1.333$

The adiabatic index of the mixed gas can be calculated by the following formula

$$\frac{1}{k-1} = \sum Y_i / (k_i - 1)$$

where: k—Mixed gas adiabatic index;

k_i—Adiabatic index of a component i;

Y_i—Molar fraction of a component i;

R'—Gas constant, $R' = R/M$.

The theoretical shaft work of gas adiabatic compression depends on the initial temperature T_1, adiabatic index k and compression ratio (P_2/P_1).

(3) Variable compression

In the actual compression process, the gas is usually a variable compression process between isothermal and adiabatic. The −14 point temperature of Engel's relationship and theoretical power consumption of the gas are in the same form as the adiabatic process, except that the adiabatic index is replaced by the multivariable index m, $p_1 V_1^m = p_2 V_2^m$.

$$-W_s = \left(\frac{m}{m-1}\right) R' T_1 \left(\left(\frac{p_2}{p_1}\right)^{\frac{m-1}{m}} - 1\right) \tag{3.58}$$

For compressor with water jacket $1 < m < k$, such as air compressor $k = 1.4$, $m = 1.25$. In a centrifuge compressor without a jacket, all the work that overcomes the flow resistance is converted into heat, making the temperature higher than the adiabatic process, at this time $m > k$.

Isothermal compression and adiabatic compression are the special cases of the process of variable compression. When $m = 1$, it is isothermal compression, and when $m = k$, it is adiabatic compression.

For ideal gas reversible adiabatic compression or variable compression, the compression power consumption can also be calculated as follows:

$$-W_s = R' T_m \ln(\frac{p_2}{p_1}) \tag{3.59}$$

where, T_m—The logarithmic mean temperature of the import and export of the working fluid

$$T_m = (T_2 - T_1)/\ln(\frac{T_2}{T_1})$$

The verification is as follows:

$$-W_s = \frac{m}{m-1} R' T_1 \left[\left(\frac{p_2}{p_1} \right)^{\frac{m-1}{m}} - 1 \right]$$

$$= \frac{m}{m-1} R' T_1 \left(\frac{T_2}{T_1} - 1 \right) \left(\text{Because}(T_2/T_1) = \left(\frac{p_2}{p_1} \right)^{\frac{m-1}{m}} \right)$$

$$= \frac{m}{m-1} R' (T_2 - T_1)$$

$$= \frac{m}{m-1} R' T_m \ln(T_2/T_1)$$

$$= R' T_m ln((T_2/T_1)^{\frac{m}{m-1}})$$

$$= R' T_m ln(p_2/p_1)$$

So, there is

$$-W_s = R' T_m \ln(p_2/p_1)$$

It can be seen that by substituting the thermodynamic average temperature T_m with the temperature T_1, the isothermal compression formula can be used to calculate the power consumption of the variable compression process.

(4) Multi-stage and variable compression

$$-W_s = \left(\frac{Sm}{m-1} \right) R' T_1 (\left(\frac{p_2}{p_1} \right)^{\frac{m-1}{Sm}} - 1) \tag{3.60}$$

where: S—Compression stages.

3.3 Chemical Exergy and Calculation of Fuel Exergy

Both the chemical energy of a substance and the calorific value of the fuel can be checked and calculated from the manual, which establishes the calculation foundation of chemical exergy and fuel exergy. Although the calculation of chemical energy and chemical exergy is bypassed in our energy analysis system, it is still very important to consider the concept of chemical exergy and the simple calculation of fuel exergy substances exchange (physical diffusion or chemical reaction) between the stream and the environment under ambient temperature and pressure, finally, the equilibrium with the environment is reached, and the maximum work performed is chemical exergy.

The calculation of chemical exergy requires that the environmental state of each element is determined, and the chemical composition and concentration of the environment must be specified. The environmental reference state has been discussed in Chap. 1.

Chemical exergy is not very commonly used in the energy analysis of petrochemical processes. This is because in the three-link model of energy analysis (ref to Chap. 5), the chemical energy and chemical exergy of raw materials and products are not considered, and only the chemical energy difference of raw materials and products before and after processing is considered. This difference is included in the thermodynamic energy consumption of the process in the form of reaction heat effect, but for Fertilizer chemical industry that are used to calculate the chemical energy of raw materials, the chemical energy is often calculated. In the evaluation of a single process, the calculation of chemical energy (exergy) is often encountered. Therefore, it is very necessary to understand the concept of chemical exergy and master simple calculation methods.

3.3.1 Basic Concepts of Chemical Exergy [9]

The calculation of chemical exergy can be considered to consist of three sub-processes, namely, the reversible separation process of extracting certain pure substances (such as 0_2, etc.) participating in the reaction from the reference substance of the environmental state. Pure substance corresponding to environment reference substance are called "environmental reference substance". The second sub-process is a reversible reaction process between the reactant (to be studied) and the environmental reactant under the environment T_0, p_0. The product of the reaction process should be some other reference substance in the environment, called environmental products. Such a reaction process is called a reversible reference reaction, and its exergy change is the reaction exergy. In order to ensure the reversibility of the reference reaction, whether it is the substance to be studied, the environmental reactant and the environmental product must participate in the reference reaction in the form

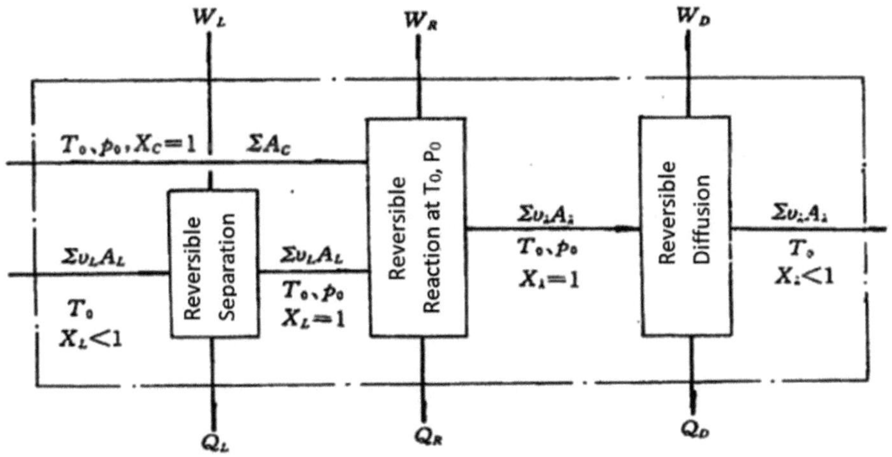

Fig. 3.1 Physical model of chemical exergy calculation

of a pure substance under T_0, p_0, and should not be mixed before entering the reaction chamber. The third sub-process is a reversible diffusion process that changes these environmental products from pure substances to the concentration of environmental reference substances. Complete these three sub-processes, that is, to achieve a complete equilibrium with the environment, as shown in Fig. 3.1.

That is, there are R environmental reactants, which are separated and purified into a pure substance with a concentration of 1 and C reactants ($Xc = 1$) under the conditions of T_0, p_0 to perform a reversible reference reaction, and P reaction products with a concentration of 1 are directed to the environment diffusion reaches the equilibrium concentration in the environment.

As can be seen from the Fig. 3.1, the calculation of chemical exergy includes two types of processes: reversible chemical reaction process and reversible concentration change process. The maximum effective work provided by these processes is the chemical exergy of the substance. Usually, we refer to the separation process of environmental reactant purification and the process of diffusion of reaction products into the environment, which are collectively referred to as the diffusion process in a broad sense. Therefore, the chemical exergy of substances is the sum of reaction exergy and diffusion exergy.

3.3.2 Calculation Method of Chemical Exergy

Combined with the basic concepts of thermodynamics, it is not difficult to derive the energy balance of the system surrounded by a dashed line in the physical model of chemical exergy calculation:

$$ex_c = -\Delta G(T_0, p_0) + (\sum_R v_L ex_{cL} + \sum_p v_\lambda ex_{c\lambda}) \quad (3.61)$$

where: ex_c—The molar chemical exergy of the element or compound (subscript L is the purification of environmental reactants, and the subscript λ is the diffusion of reactants to the environmental concentration);

ΔG—Free enthalpy of element or compound;

v—The mole fraction of environmental reactants or reaction products.

In the formula, the first term represents the reaction exergy under the reference condition, and the second term represents the diffusion exergy of the concentration change. It consists of two terms, one is the purification of environmental reactants, and the other is the diffusion of reaction products to the environmental concentration.

Because $ex_{cL} = -RT_0 lnX_L$ (X represents mole fraction)

$$ex_{c\lambda} = RT_0 lnX_\lambda$$

Therefore, the formula (3.61) can be written as:

$$ex_c = -\Delta G(T_0, p_0) - \sum_R v_L RT_0 lnX_L + \sum_p v_\lambda RT_0 lnX_\lambda \quad (3.62)$$

3.3.2.1 Chemical Exergy of Elements and Simple Substances

All un-bound elements that do not exist in a single substance state in the dead state in the natural environment on the earth's surface (reference state) have chemical energy and chemical exergy. Because they all release energy and exergy when they change from the un-bound element state to the most stable quiescent state composition.

In the manual, the standard enthalpy of formation is for pure substances (concentration is 1, the gas pressure is 1 atmosphere), but the concentration of the reference substance defining chemical energy in the environmental field is mostly less than 1. Therefore, strictly speaking, energy and enthalpy difference ΔH are conceptually different. In addition, partial pressure and concentration have little effect on enthalpy, so it can be considered as $E_{ch} \approx -\Delta H^0$.

(1) Chemical exergy of O, N, C, H elements in the reference state

The Kameyama-Yoshida system stipulates that saturated humid air at 298.15 K and 1 atm is used as the reference substance for the gas contained in the air. From this, the chemical exergy of the four elements of O, N, C, H can be determined first.

For pure gases such as O_2, N_2, CO_2, and H_20, they are only different in concentration from the reference substances and do not involve chemical reactions. Therefore, their chemical exergy is diffusion exergy.

O_2	$ex_c(O_2) = -T_0 R\ln X^0_{O_2}$
O	$ex_c(O) = -1/2T_0 R\ln X^0_{O_2}$
N_2	$ex_c(N_2) = -T_0 R\ln X^0_{N_2}$
N	$ex_c(N) = 1/2T_0 R\ln X^0_{N_2}$
CO_2	$ex_c(CO_2) = -T_0 R\ln X^0_{CO_2}$
H_2O (g)	$ex_c(H_2O) = -T_0 R\ln X^0_{H_2O}$

For C and H elements, it can be calculated according to the relevant formula of the reference reaction.

Carbon's reference reaction. $C + O_2 = CO_2$

Element C exergy $ex_c(C) = -g_f(CO_2) + ex_c(CO_2) - ex_c(O_2)$

H_2 reference reaction. $H_2 + 1/2O_2 = H_2O(l)$

Hydrogen exergy $ex_c(H_2) = -g_f(H_2O)(l) + ex_c(H_2O)(l) - 1/2ex_c(O_2)$.

2. Chemical exergy under the reference state of other elements

In order to determine the chemical exergy of element X, it is necessary to select the most stable compound $X_x A_a B_b C_c$ from all compounds containing element X as the reference substance, and consider $ex_c(X_x A_a B_b C_c) = 0$. It is believed that the reference substance is formed by the reversible reaction between element X and some other elements A, B, and C, namely

$$xX + aA + bB + \mathrm{c}C \rightarrow X_x A_a B_b C_c \quad (3.63)$$

The exergy balance of the reaction formula is listed as:

$$x.ex_c(X) + a.ex_c(A) + b.ex_c(B) + c.ex_c(C) = -\Delta G(T_0, p_o) + ex_c(X_x A_a B_b C_c) \quad (3.64)$$

Here, use $ex_c(X)$ to represent the chemical exergy of element X

$$-\Delta G(T_0, p_o) = -(g_f(X_x A_a B_b C_c) - x.g_f(X) - a.g_f(A) - b.g_f(B) - c.g_f(C) \quad (3.65)$$

Since the free enthalpy of formation of each element in the formation reaction is equal to zero, and the chemical exergy of the reference substance $X_x A_a B_b C_c$ is zero, Eq. (3.64) is sorted as

$$ex_c(X) = (1/x)(-g_f(X_x A_a B_b C_c) - a.ex_c(A) - b.ex_c(B) - c.ex_c(C)) \quad (3.66)$$

When applying the above formula to calculate chemical exergy, it should be noted: (1) the exergy of the substances A, B, and C involved in the reaction has been determined in advance; (2) the reference substance $X_x A_a B_b C_c$ has been selected.

3.3.2.2 Chemical Exergy of Compounds in the Reference State

After determining the reference substance system and the chemical exergy of the elements, as long as the free enthalpy of formation of the substance is known, the following reversible formation reactions can be used to determine the chemical exergy of any pure substance.

Suppose that the pure substance $A_a B_b C_c$ is reversibly generated by single substances or elements A, B, and C.

$$aA + bB + cC \leftrightarrow A_a B_b C_c \tag{3.67}$$

If the free enthalpy of formation $g_f(A_a B_b C_c)$ of this substance and the chemical exergy of elements A, B and C are known, the chemical exergy of this substance is:

$$ex_c(A_a B_b C_c) = g_f(A_a B_b C_c) + a.ex_c(A) + b.ex_c(B) + c.ex_c(C) \tag{3.68}$$

Corresponding chemical energy

$$E_c(A_a B_b C_c) = \Delta H_f^0(A_a B_b C_c) + a.e_A + b.e_B + c.e_C \tag{3.69}$$

where: $E_c(A_a B_b C_c)$—Chemical energy of compound.

The chemical exergy of commonly used compounds has been listed in Tables A-2, A-3 in Appendix A.

3.3.2.3 Chemical Exergy of Elements and Compounds at Other Temperatures

When performing exergy analysis, the selected ambient temperature may not be equal to the reference temperature of 298.15 K, and the aforementioned chemical exergy value under the reference state cannot be used directly. On the other hand, sometimes out of necessity, the temperature correction should be taken when calculate the chemical exergy of the substance at the temperature of T:

Suppose that the chemical exergy in the reference state isex_c, and the chemical exergy in the environmental state is $ex_c(T_0)$, if set:

$$\S = (ex_c(T_0) - ex_c)/(T_0 - 298.15)$$

Therefore

$$ex_c(T_0) = ex_c + \S(T_0 - 298.15) \tag{3.70}$$

where: § is called the temperature correction coefficient for chemical exergy.

For the calculation of the chemical exergy of a substance with a temperature of T, the above method can also be used, and T is used instead of T_0.

In the actual exergy analysis and calculation, those that are higher than the reference temperature and cannot be ignored in engineering are regarded as thermal physical exergy caused by temperature, and are calculated separately.

3.3.2.4 Diffusion Exergy [13]

Diffusion exergy is an part of chemical exergy. In engineering practice, there are often many processes without chemical changes, only the change of component concentration, that is, from the concentration in the actual stream to the concentration of the reference state, this exergy due to the concentration difference is called diffusion chemical exergy. The reverse process of diffusion is called the separation process. The diffusion exergy in the actual process is a chemical change caused by a one-way concentration difference. In the calculation of chemical exergy, diffusion exergy refers to the algebraic sum of chemical exergy in the purification of environmental reactants and the diffusion of environmental products.

The diffusion exergy discussed here mainly refers to the chemical exergy that the stream diffuses to the environment or to the low-concentration area in the actual process. The chemical exergy micro-element changes of the diffusion process shown in Fig. 3.2 (such as the flue gas CO_2 diffusion of fired furnace and boiler into the atmosphere) is:

$$EX_{cd} = \sum\nolimits_{i=1}^{m} \mu_i dn_i \tag{3.71}$$

where: EX_{cd}—Diffusion chemical exergy;

μ_i—i component chemical potential;

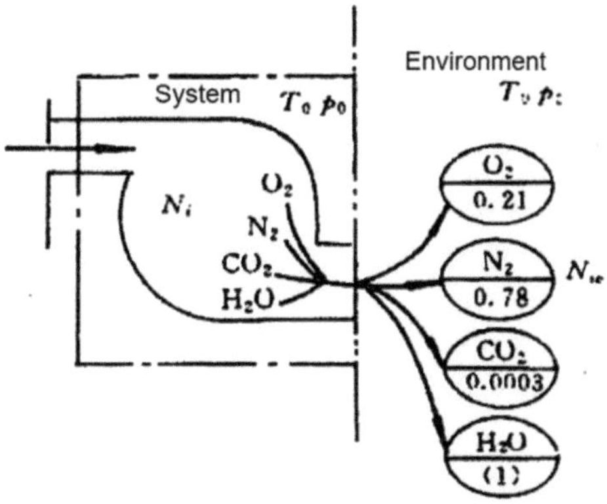

Fig. 3.2 Diffusion chemical exergy calculation model

n_i—i component kilomole.

However, in this diffusion process, all components are reference substance, and their quantity kmol has not changed, namely: n_i = constant. At the same time, the changes of the chemical potential of all components are based on the changes in the concentration of the components in the system and the environment. i component chemical potential in the system is:

$$\mu_i = \mu_i^0 + RT_0 lnN_i \tag{3.72}$$

In the formula, N_i is the concentration of component i in the system; μ_i^0 is the chemical potential of the pure component i ($N_i = 1$) under T_0, p_0.

And for the chemical potential of the i component in the environment μ_{i0}, we can write:

$$\mu_{i0} = \mu_i^0 + RT_0 lnN_{io} = 0 \tag{3.73}$$

where, N_{i0} is the concentration of the component i in the environment, which is the reference concentration.

Therefore, the integral form of the differential Eq. (3.71) should be:

$$EX_{cd} = \sum_{i=1}^{m} (\mu_i - \mu_{i0}) n_i \tag{3.74}$$

Substituting formula (3.72) and formula (3.73) into formula (3.74), we can get:

$$\begin{aligned} EX_{cd} &= \sum_{i=1}^{m} (RT_0 \ln N_i - RT_0 lnN_{i0}) n_i \\ EX_{cd} &= RT_0 \sum_{i=1}^{m} n_i \ln(N_i / N_{i0}) \end{aligned} \tag{3.75}$$

The formula (3.75) is the general formula for the calculation of diffusion chemical exergy. Among them, the concentration in the system and environment N_i and N_{i0} are mole fractions; they can also be partial pressure (for ideal gas), fugacity (for real gas) and activity (for real solution) in different situations.

Diffusion chemical exergy, mainly used for reference materials in the atmosphere and sea water. For 1 kmol pure reference material, $n_i = 1$, $N_i = 1$, diffusion chemical exergy:

$$ex_{cd} = \mu_i = RT_0 \ln N_{i0} \tag{3.76}$$

where: ex_{cd}—mole diffusion exergy.

The reference substance concentration N_i of the component i in the environmental field can be found in Table 1.1. For the, chemical, petrochemical and oil refining process, the most frequently involved diffusion chemical exergy is CO_2, O_2 and N_2.

Example 3.2 Find the diffusion chemical exergy of CO_2 and N_2 in the flue gas of the fired furnace (the content of O_2 is small and can be ignored). It is known that the partial pressures of CO_2 and N_2 in the flue gas are 0.01 MPa and 0.075 MPa, respectively.

Solution

(1) 1 kmol CO_2 diffusion chemical exergy:

$$ex_{cd} = 8.319 \times 298 ln(0.01/0.00003) = 14.41\text{MJ/kmol}$$

(2) The diffusion chemical exergy of N_2 is:

$$ex_{cd} = 8.319 \times 298 ln(0.075/0.07583) = -0.027\text{MJ/kmol}$$

According to the above calculation, the diffusion chemical exergy of the flue gas can be further calculated (assuming the average molecular weight of the flue gas is 27):

$$EX_{cd} = \frac{1}{27}[0.1 \times 14.41 + 0.75 \times (-0.027)] = 52.8\text{kJ/kg}$$

When the exhaust flue gas temperature is 200 °C and the specific heat of the flue gas is 1.05 kJ/(kg °C), the physical exergy of the flue gas is 47.4 kJ/kg.

It can be seen that the proportion of diffusion chemical exergy cannot be ignored, but with the current technical level, diffusion chemical exergy is still difficult to use, so in energy analysis, the diffusion of flue gas is not considered.

3.3.3 *Chemical Exergy of Complex Substances and Fuels*

3.3.3.1 Group Contribution Method for Complex Substances

Many substances are difficult to accurately calculate their chemical exergy due to lack of free enthalpy of formation data or complex structures.

Someone proposed to use the group contribution method to calculate the chemical energy and chemical exergy of substances, and listed the partial molar enthalpy and partial molar exergy of 127 kinds of gas and liquid organic compound groups [7, 8].

$$EX_c^0 = \sum n_j b_{cj}^0 \tag{3.77}$$

where: n_i—Number of j groups in the molecule;

b_{cj}^0—Molar chemical exergy of the j group.

3.3.3.2 Fuel Chemical Exergy

The combustion of fuel is a benchmark reaction process, so the chemical exergy of a fuel with a known composition in the reference state can be obtained from the exergy balance relationship of the benchmark reaction [9]. However, the composition of most fuels is unknown, such as heavy fuel oil for heating furnaces, but the elements that make up them are numerous, mainly C, H, S, O, N etc. It can be accurately measured through elemental analysis. On the other hand, the calorific value of fuel can be measured or calculated. Therefore, the exergy of liquid and gas fuels commonly used in petrochemical processes can be determined by their low calorific value and corresponding element composition using the following method [7].

Liquid fuel, Szargut-StyryIska equation:

$$\frac{ex_c}{LHV} = 1.0374 + 0.0159[H/C] + 0.0567[O/C] + 0.5985[S/C](1 - 0.1737\left[\frac{H}{C}\right]) \tag{3.78}$$

Nobuzawa Torao modified equation

$$\frac{ex_c}{LHV} = 1.038 + 0.1365[h/c] + 0.0308[o/c] + 0.0104[s/c] \tag{3.79}$$

Gas fuel: Szargut-Styrylska formula:

$$\frac{ex_c}{LHV} = 1.0334 + 0.0183[H/C] + 0.0694/[O/C] \tag{3.80}$$

Rant formula:

$$ex_c = 0.95HHV \tag{3.81}$$

In the above formulas, LHV is low calorific value and HHV is high calorific value. $[H/C]$, $[0/C]$, $[S/C]$ are atomic ratios, $[h/c]$, $[o/c]$, $[s/c]$ are weight ratios.

It can be seen that the chemical exergy of fuel is between low calorific value and high calorific value. The fuel exergy includes the combustion gas diffusion exergy that is difficult to recover in engineering. Therefore, in many cases, the low calorific value of the fuel directly used for exergy can also meet the requirements of energy analysis, without being restricted to strict calculations.

References

1. J.M. Smith, translated by S. Yuguang, et al., *Introduction to Chemical Engineering Thermodynamics* (Chemical Industry Press, 1982)
2. K.S Pitzer, R.F. Curl, The volumetric and thermodynamic properties of fluids. Empirical equation for the second virial coefficient. J. Amer. Chem. Soc. (1957)
3. Edited by Jiangsu Normal University, *Physical Chemistry* (People's Education Press, 1980), p.·12
4. API Technical Date Book—Petroleum Refining, 4th Ed (1983)
5. Process design information of oil refining equipment, selection of pumps and motors (Petroleum Industry Press, 1976)
6. Process design data of oil refining equipment, compressor process calculation (Petroleum Industry Press 1978), p. 4
7. H. Yaowen, *New Chemical Technology, Thermodynamic Analysis of Chemical Process*, vol. 2 (Edited by the Education Committee of China Chemical Industry Society, 1982)
8. F. Liangzheng, *Analysis and Synthesis of Chemical Process Systems from the Perspective of Thermodynamics*, vol. 1 (published by East China Institute of Chemical Technology, 1981)
9. Z. Mingshan, Exergy analysis of energy system. Tsinghua University Press **4**, 109–121 (1988)
10. L. Zanchun, X. Xun, *Principles of Chemical Energy Saving Thermodynamics* (Hydrocarbon Processing Press, 1990), p. 1
11. C. Anmin, T. Hui, Practical calculation methods for energy and exergy of light hydrocarbons and their mixtures. Refinery Des. (6), 57, 58 (1990)
12. B.I. Lee, M.G. Kesler, A generalized thermodynamic correlation based on three-parameter corresponding states. AICHE J. **21**(3), 510, 527 ·
13. H. Ben, *Analysis and Synthesis of Energy Consumption in Process* (Hydrocarbon Processing Press, 1989) pp. 1, 484–486

Chapter 4
Thermodynamic Analysis of Process Energy Utilization

Abstract The exergy loss reduction can be divided into two categories: exergy loss caused by having too much driving force (such as excessive heat transfer temperature difference), which can be eliminated; if you want to reduce the driving force further and achieve the process production goal, it is necessary to increase equipment investment. This chapter conducts the thermodynamic analysis based on the optimal energy utilization driving force, these thermodynamic analyses include (1) Thermodynamic analysis of heat transfer process, (2) Thermodynamic analysis of fluid flow process, (3) Thermodynamic Analysis of Mass Transfer Process, (4) Thermodynamic Analysis of Chemical Reaction Process and (5) Thermodynamic analysis of the combustion process.

Keywords Thermodynamic analysis · Heat transfer · Mass transfer · Chemical reaction · Combustion process · Exergy loss

Although the petrochemical process has a wide range of different types, they all process raw materials into products through a series of physical and chemical changes. The production process is carried out because of the driving effect of energy, so the production process always runs through the use and loss of energy. Energy conservation in the petrochemical process requires in-depth observation and analysis of the change law of process energy within the process, and use the technical and economic optimization methods to point out the reasonable scale of energy use and identify the efficiency improvement opportunities.

No matter how complicated the petrochemical production process is, it is composed of a series of unit operations, but the length of the process and the number of unit operations are different. In a more complex system, it is difficult to separate the integrated operations of various unit operations. The unit operation process in chemical engineering usually refers to the "three transfers and one reaction" process, that is, four types of processes: heat transfer, mass transfer, fluid flow, and chemical reaction (combustion). Research and master the energy use and energy evolution law of these four types of unit operations, is the basis for process energy analysis.

The petrochemical process is a process of using energy. In the process of converting raw materials into products, energy degenerates from high-grade to low-grade, and is finally discharged and dissipated in the environment. For a certain unit

T. A. Chen, *Energy Saving and Carbon Reduction*,
https://doi.org/10.1007/978-981-19-5295-1_4

operation process, when the heat dissipation factor is ignored, the amount of energy may not change, but the quality is decreasing.

According to the second law of thermodynamics, the motivity to drive the process is due to the existence of the difference in intensity parameters, such as temperature difference, pressure difference, concentration difference and chemical potential difference. The industrial process needs to complete the production task at a certain rate, and the intensity difference is large, which is an irreversible process. Because the process is irreversible and deviates too far from the reversible process, there is a large amount of exergy loss in the unit process. To reduce exergy losses, it is necessary to reduce the driving force.

In the current industrial process, the reduction of exergy loss can be divided into two situations: First, the process has too much driving force (such as excessive heat transfer temperature difference) with large exergy loss, which is the potential for reduction. Second, if you want to reduce the driving force and achieve the process production goal, it is necessary to increase equipment investment. At this time, technical and economic tread off should be made, and there will be an optimization result. The driving force after the result of technological and economic optimization is necessary for the process and the economy and cannot be reduced.

The irreversibility of all unit operations brings exergy loss. This exergy loss is not reflected in the amount of energy, so it is called internal process exergy loss to distinguish it from external discharge loss.

4.1 Thermodynamic Analysis of Heat Transfer Process

The heat transfer process is the main unit process, which runs through the beginning and end of the petrochemical process. It is divided into two categories. One is the direct mixing and heat transfer of the cold and hot streams, achieving equilibrium at a mixed temperature. The result of the heat transfer is that the temperature of the cold stream increases and the temperature of the heat stream decreases. The temperature reached is limited by the heat released of hot stream or heat absorbed by cold stream. Another type of indirect heat transfer is via a hear transfer areas, due to the process rate limitation, it is impossible to reach equilibrium of reversible process.

4.1.1 Heat Transfer with Ignoring the Heat Dissipation

1. Calculation of exergy of cold and hot streams in heat transfer equipment

   Heat transfer is the energy transferred from the high-temperature stream to the low-temperature stream. It is a physical process. The calculation of energy and exergy value can be found in Chap. 2. The condition for providing the heat transfer of the stream is that the stream must pass through the heat transfer equipment, which consumes fluid flowing energy. However, compared with thermal energy usage, the flow energy is very small, so it is mostly ignored. But it should be included in the exergy balance.

2. Ideal reversible heat transfer process

   The ideal reversible heat transfer process, energy and exergy are both conserved, there is no exergy loss, the heat transfer temperature difference is infinitely small at this time, and the heat transfer equipment is infinitely large. Therefore, it is actually a process that does not exist at all.

3. Actual heat transfer process

   The actual heat transfer process is irreversible, and the cold stream gets heat at a lower temperature than the hot stream. Therefore, even if the heat dissipation loss is zero, the exergy obtained by the cold stream is always less than the exergy released by the hot stream. The difference between the two is the exergy loss caused by the heat transfer process. Ignoring the heat loss and assuming that the heat transfer stream can be regarded as a constant specific heat. The heat transfer is Q, then:

   Hot stream exergy

$$EX_h = Q(1 - \frac{T_0}{T_h}) \tag{4.1}$$

Cold stream exergy

$$EX_c = Q(1 - \frac{T_0}{T_c}) \tag{4.2}$$

where: T_h, T_c—The logarithmic or arithmetic mean temperature of the hot and cold streams respectively, K.

Exergy loss for heat transfer is:

$$\begin{aligned} D_{KH} &= EX_h - EX_c = Q\left(1 - \frac{T_0}{T_h}\right) - Q\left(1 - \frac{T_0}{T_c}\right) \\ &= Q\left(1 - \frac{T_0}{T_h} - 1 + \frac{T_0}{T_c}\right) = Q\left(\frac{T_0}{T_c} - \frac{T_0}{T_h}\right) \\ &= QT_0(\frac{(T_h - T_c)}{T_h.T_c}) \end{aligned} \tag{4.3}$$

It can be seen from the above formula:

(1) The greater the temperature difference ($T_h - T_c$) between the cold and hot fluids, the greater the heat transfer exergy loss. Reducing the heat transfer temperature difference (($T_h - T_c$) is the main means to reduce heat transfer exergy loss.
(2) For the same ($T_h - T_c$), the lower the streams temperature level, the smaller the denominator ($T_h.T_c$), the greater the exergy loss, and vice versa, it can be

seen that low temperature heat transfer is greater exergy loss than high temperature heat transfer. Reducing the heat transfer temperature difference under low temperature conditions is more meaningful than reducing the heat transfer temperature difference under high temperature conditions.

For incompressible fluids, the exergy loss in the heat transfer process including the fluid flow exergy loss is:

$$D_K = T_0(Q\frac{T_h - T_c}{T_h.T_c} + \frac{V_h \Delta p_h}{T_h} + V_c\frac{\Delta p_c}{T_c}) \quad (4.4)$$

where, V_h, V_c—Respectively is the specific volume of hot and cold stream;

Δp_h, Δp_c—Pressure drop of hot and cold fluid respectively.

Considering the exergy loss of the pressure drop of the cold and hot streams is to provide conditions for heat transfer and is also the price paid for the cold stream to obtain energy. Therefore, the flow exergy loss should be considered in the accurate calculation.

(3) In the heat transfer process of equipment, the exergy loss of the process is caused by two irreversible factors, one is the temperature difference between the hot stream and the cold stream, and the other is the pressure drop of the fluid flow. To reduce the heat transfer process loss, it is necessary to reduce the influence of the two as much as possible. However, the pressure drop and heat transfer coefficient are a contradiction between each other. The pressure drop is large, the flow velocity in the equipment is large, and the heat transfer coefficient increases, which is beneficial to heat transfer, but the power consumption is large, and there is an economic optimization value.

In a heat exchanger with a certain dimension size, the fluid pressure drop can usually be regarded as a constant, and the process exergy loss is mainly a temperature difference item, which is the focus of improvement. Reducing the temperature difference increases the equipment area (investment), and is also restricted by technical and economic optimization. The heat transfer temperature difference cannot be reduced indefinitely. But under the same heat transfer temperature difference, if countercurrent heat exchange is used, the outlet temperature of the cold fluid can be close to the outlet temperature of the hot fluid, and the temperature difference along the flow direction of the heat exchanger fluid is more evenly distributed, reducing irreversibility. For multi-tube pass and multi-shell pass heat exchangers, the temperature difference correction coefficient can also be increased through adjustment of operation to reduce the exergy loss during the heat transfer process.

4.1.2 *Heat Transfer Process with Heat Loss*

In the actual process of heat transfer equipment, due to the temperature difference between the surface temperature of the equipment and the environment, there is a convection and radiation combined heat transfer from the surface of the equipment to the environment, so that the energy obtained by the cold flow is less than the energy released by the heat flow.

There are three outputs of exergy released by heat flow in the heat transfer process: exergy entering the cold flow, exergy loss for heat dissipation, and exergy loss during heat transfer. In the process of heat dissipation, the exergy loss of the heat transfer process transferred from the temperature of the internal fluid of the equipment to the temperature of the insulation surface of the equipment and the process exergy loss from the convection radiation heat loss of the temperature of the insulation surface to the environmental should be attributed to the heat dissipation exergy loss.

4.1.2.1 Exergy Loss Due to Heat Dissipation

As far as a certain point in the equipment is concerned, the fluid temperature T_b and the ambient temperature T_0 do not change with time (see Fig. 4.1), heat dissipation and exergy loss:

$$D_{KHR} = Q_R(1 - \frac{T_0}{T_b}) \tag{4.5}$$

The exergy loss for heat dissipation should not be calculated based on the surface temperature T_W of the vessel wall or insulation layer, because the exergy loss caused by the temperature drop from T_b to T_W is also caused by heat dissipation.

The heat dissipation exergy loss should also include: (1) exergy loss caused in the heat transfer process from the inside fluid temperature to the insulation surface

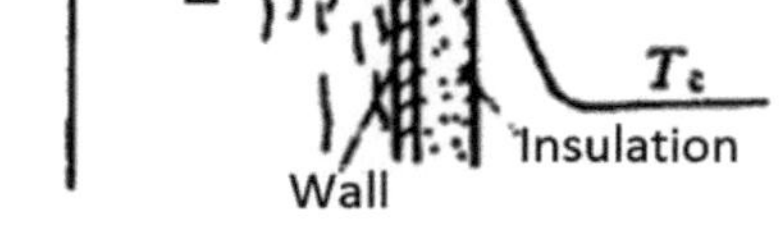

Fig. 4.1 Schematic diagram of heat dissipation on the equipment surface

temperature, and (2) the exergy loss in the process of convective radiation from the temperature of the insulation surface to the environment.

4.1.2.2 Exergy Balance Equation [1]

For a heat exchanger, whether it is evenly insulated or without insulation, the heat dissipation is mainly through the shell side fluid (of course, the tube box also has heat dissipation, but it is relatively small). When the shell-side fluid is a hot flow, the heat loss and exergy loss D_{KHR} can be represented by the area 12′60 in Fig. 4.2 (that is, T_b is between the inlet and outlet temperatures of the hot flow). At this time, the heat transfer exergy loss D_{KH} is an area of 1′234, and its correlation formula is

$$D_{KH} = (Q - Q_R)T_0(\frac{T_h - T_c}{T_h.T_c}) \tag{4.6}$$

EX_c for cold stream exergy has an area of 3465, and EX_h for hot flow exergy has an area of 1250. The relationship between these four areas can be seen from Fig. 4.2; that is, The exergy released from hot flow is the sum of the exergy received by the cold stream, exergy loss of heat transfer and heat dissipation. Reflected in the overall balance equation:

$$EX_h = EX_c + D_{KH} + D_{KHR} \tag{4.7}$$

When the cold stream travels through the shell, the relationship is still as (4.7), but D_{KH} needs to be calculated by the formula (4.3). When calculating D_{KHR}, T_b should use the cold stream average temperature.

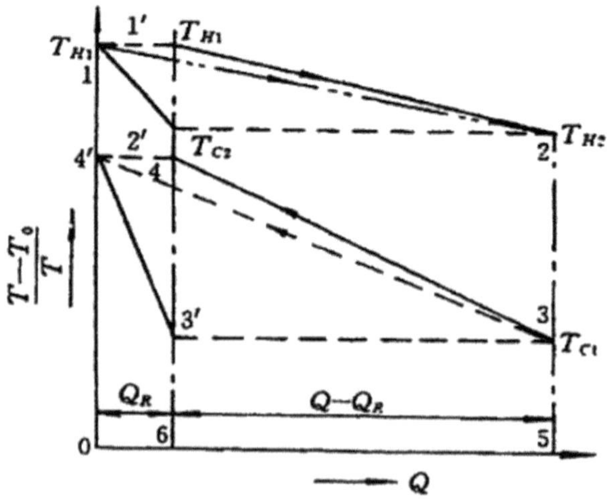

Fig. 4.2 Heat transfer exergy loss with heat dissipation

4.1.2.3 Mixing of Different Temperature Streams

Although the mixing of streams of different temperatures does not have any loss in terms of energy quantity, it produces a loss of energy quality, that is, exergy loss, which increases as the temperature difference increases between the two streams.

Example 4.1 The hot water of 45 °C in a shower room in a workshop is obtained by using 15 °C domestic water through 1 MPa (10 kg/cm^2) 260 °C steam. Find the exergy loss of this mixing process.

Solution Based on 1 ton of cold water ($T_c = T_0$, $E_c = 0$, $EX_c = 0$), Heating to 45 °C requires steam x tones.

From steam table, the 1 MPa 260 °C steam enthalpy 2964.8 kJ/kg, entropy 6.9674 kJ/kg K, and 15°C condensate entropy at 1 MPa is 0.224 kJ/kg K; 15 °C water enthalpy 62.9 kJ/kg; and 45 °C water enthalpy 188.35 kJ/kg.

$$E = (2964.8 - 62.9) \times 1000 = 2901.9\text{MJ/t}$$

$$EX = 2901.9 - 288 \times (6.9674 - 0.224) = 959.8\text{MJ/t}$$

Assuming that there is no heat loss, the heat balance (the first law) is

$$(2964.8 - 188.35)x = (188.35 - 62.9) \times 1, x = 0.04517t$$

Inlet: Steam energy and exergy:

$$E_s = 2901.9 \times 0.04517 = 131.1\text{MJ}$$

$$EX_s = 959.8 \times 0.04517 = 43.35\text{MJ}$$

Outlet: 1.04517 t 45 °C warm water exergy

$$E_m = 1.04517 \times (188.35 - 62.9) = 131.1\text{MJ};$$

$$EX_m = 131.1 \times \left(1 - \frac{288}{273 + 45}\right) = 12.37\text{MJ}$$

Therefore, the exergy loss in the mixing process of the two streams:

$$D_{kHm} = 43.35 - 12.37 = 30.98MJ$$

That is, 71% of the steam exergy is lost during the mixing process.

There are many examples of exergy loss during the process of mixing fluids at different temperatures, such as the intermediate hydrogen mixing in the hydrogenation reactor, the quenching of the cracking and viscosity breaking reaction outlets, the mixing of the 2nd vacuum withdraw line and 3rd vacuum withdraw lines as catalytic feed, and mixing of water injection with crude oil in electrical desalination, etc. For this kind only counting temperature changes, it's exergy loss can be calculated using the following simplified approximate formula.

$$D_{kHm} = QT_0(T_h^{'} - T_c^{'})/(T_h^{'}.T_c^{'}) \tag{4.8}$$

where: Q—The heat transferred from the hot flow to the cold flow during the mixing process;

$T_h^{'} = (T_{h0} + T_m)/2$—Is the arithmetic average of the hot stream from the starting temperature T_{h0} to the temperature after mixing T_m;

$T_c^{'} = (T_{c0} + T_m)/2$—Is the arithmetic average of the cold stream from the starting temperature T_{c0} to the temperature after mixing T_m.

The idea of this formula is to use the mixing process as the direct irreversible heat transfer between two unequal temperature streams to calculate the exergy loss.

Example 4.2 Calculate the mixed heat exergy loss of electric desalination and water injection. It is known that the crude oil temperature is 120 °C, the specific heat capacity is 2.09 kJ/(kg °C), the water injection temperature is 35 °C, and the water volume accounts for 5% of the crude oil.

Solution

(1) Based on 1 t of crude oil, mixing temperature obtained from heat balance:

$$(120 - t) \times 1000 \times 2.09 = (t - 35) \times 50 \times 4.1868$$

$$t = 112.3°\text{C}$$

(2) Required heat from heat balance:

$$Q = (112.3 - 35) \times 50 \times 4.1868 = 16.18\text{MJ}$$

$$T_h = \left(\frac{1}{2}\right)(120 + 112.3) + 273 = 389.15\text{K}$$

$$T_c = \left(\frac{1}{2}\right)(35 + 112.3) + 273 = 346.65\text{K}$$

Therefore

$$D_K = 16.18 \times 288\left[\frac{389.15 - 346.65}{389.15 \times 346.65}\right] = 1.47\frac{\text{MJ}}{\text{t}}crudeoil$$

It accounts for 4.6% of crude oil's exergy.

4.2 Thermodynamic Analysis of Fluid Flow Process [2]

Fluid flow is the most common type of unit operation process in petrochemical processes. For a continuous and stable production process, raw materials are processed into products in a series of flow processes. When the material flows through equipment and pipelines, it needs to overcome friction and local resistance. This kind of resistance is caused by irreversible flow and constitutes the loss of work. The power to overcome resistance is provided by fluid flow equipment such as pumps and compressors. Most of the flow work is converted into heat entering the process stream or lost in the environment. A few products leave the device under pressure, taking away the flow work.

The adiabatic throttling process is a special case of the fluid flow process. Generally, for the throttling process, there is no power and heat exchanging with the outside, so the throttling process is isenthalpic throttling, but after throttling, the entropy value increases. Therefore, the exergy (power) loss in the throttling process is $T_0\Delta S$.

From the thermodynamic differential formula:

$$dEX = dH - T_0 dS = -T_0 dS$$

Take the partial derivative of the pressure p:

$$\frac{\partial EX}{\partial p} = -T_0\frac{\partial S}{\partial p}$$

Substituting the relationship of $\frac{\partial S}{\partial p} = -\frac{V}{T}$ into the above formula, we have

$$\frac{\partial EX}{\partial p} = -T_0(-\frac{V}{T}) \tag{4.9}$$

$$\frac{\partial EX}{\partial p} = (T_0/T)Vdp \tag{4.10}$$

Equation (4.10) is the general function of exergy loss in the energy-saving process. According to formula (4.10), as long as it is substituted into the relational expressions of the streams p, V, T, it is easy to find the exergy loss:

For every kilogram of ideal gas, there is $V = RT/(mp)$

$$\Delta EX = (T_0/T)(RT/m)ln(p_2/p_1) = (R/m)T_0 ln(p_2/p_1) \quad (4.11)$$

For incompressible liquid (liquid), V can be regarded as a constant

$$\Delta EX = -(T_0/T)V(p_2 - p_1) = (T_0/T)V(p_1 - p_2) \quad (4.12)$$

There is flow resistance in the normal fluid flow process. The flow process can be approximated as a throttling process, which is regarded as an isenthalpic flow process, and because there are few phase changes during the flow process at outlet of power-consuming power equipment, for liquids and gases with little change in volume, using the average volume V_m to replace V in the formula (4.12), the exergy loss in the flow process can also be calculated.

From the above discussion, we can draw the following conclusions:

(1) It can be seen from formula (4.10) to formula (4.12) that throttling exergy loss with pressure changes is proportional to the specific volume of the fluid, which is the main reason why the throttling exergy loss of gas streams (steam, air and gas-phase process streams, etc.) is greater than the liquid throttling exergy loss.
(2) Throttling exergy loss is inversely proportional to the stream temperature. The lower the temperature, the greater the exergy loss. This is because fluid friction transforms into thermal energy at fluid temperature. The amount of thermal energy is equal to the amount of power consumption, but the energy grade of frictional thermal energy is restricted by the stream temperature.
(3) The liquid throttling exergy loss is proportional to the pressure difference, while the gas stream throttling exergy loss is proportional to the logarithmic value of the pressure ratio. For ideal gases, the average specific volume V_m can also be used to replace V in formula (4.12), and the calculation results are the same. Therefore, the ideal gas stream throttling exergy loss is also a function of the pressure difference. The reason why the gas stream throttling exergy loss is greater than the liquid stream throttling exergy loss is because the specific volume of the gas stream is far greater than that of the liquid stream, it is not caused by different functional relationships.

In the petrochemical process, the throttling exergy loss of streams can be seen everywhere. Such as the use of higher-pressure steam letdown to reduce pressure, the throttling of synthetic ammonia purge gas, and the depressurization of hydrogenation high-pressure separator to low-pressure separator, and the throttling of catalytic cracking regeneration flue gas. Therefore, reducing the exergy loss of the process throttling and recovering the expansion work of the stream is an important aspect of energy saving in the fluid flow process.

The throttling exergy loss is a function of the ambient temperature T_0, and the expansion work of the gas stream is a function of the stream temperature T. Generally, the expansion recovery work of throttling is much greater than that of throttling exergy loss. This is because when the stream is at a high temperature, the expansion process is an isentropic process, which converts the enthalpy of the stream into work. The ratio of the work performed W_e in the ideal isothermal expansion process to the exergy loss ΔEX of throttling is:

$$\frac{W_e}{\Delta EX} = \left(\frac{R}{M}\right) Tln\left(\frac{p_1}{p_2}\right) / \left[\left(\frac{R}{M}\right) T_0 ln\left(\frac{p_1}{p_2}\right)\right] = \frac{T}{T_0} \quad (4.13)$$

It is the ratio of the stream temperature to the ambient temperature. It can be seen that the throttling exergy loss of the recovery of high-temperature gaseous streams is particularly important. For example, a flue gas turbine of FCCU regeneration flue gas recovers expansion work at high temperatures. Under ideal operating conditions, the recovered work is greater than the consumed power of the main air main compressor [3].

Example 4.3 The pressure of the regeneration flue gas of a catalytic cracking unit in an oil refinery is 0.2 MPa. Through the double-acting slide valve throttling to atmospheric pressure and exhausting the chimney, the atmospheric pressure is 0.1 MPa, the flue gas flow rate is 60,000 Nm^3/h, the ambient temperature T_0 is 15 °C, and the flue gas temperature is 600 °C. Find the gas stream exergy loss via throttling valve. If a flue gas expansion turbine is installed, how much expansion work can be recovered?

Solution

(1) Throttling exergy loss

$$\Delta EX = nRTln\left(\frac{p_1}{p_2}\right) = \left(\frac{60,000}{22.4}\right) \times 8.319 \times 288\ln\left(\frac{0.2}{0.1}\right)$$
$$= 444,8281\frac{kJ}{h} = 1235.63kW$$

(2) After installing a flue gas expansion turbine, considering the enthalpy drop of the flue gas turbine, the outlet temperature is 500 °C; therefore, if the average temperature of the stream is 550 °C, the maximum expansion work is:

$$W_e = \left(\frac{60000}{22.4}\right) \times 8.319 \times (550 + 273).\ln\left(\frac{0.2}{0.1}\right) = 12711583\frac{kJ}{h} = 3531\text{kW}$$

Considering that the efficiency of the flue gas turbine is 70%, the net expansion work of 2471 kW can be recovered.

In the fluid flow process, in addition to the throttling exergy loss in the flow process, there is also a type of exergy loss, that is, the exergy loss in the flowing work conversion process. This is due to the work loss caused by friction and eddy current impact when the work conversion equipment converts electrical energy (or steam) into the flowing work required for fluid flow. This exergy loss can be calculated by calculating the effective work obtained by the fluid. It is determined by exergy balance, please refer to the pump and compressor section in Chap. 6.

4.3 Thermodynamic Analysis of Mass Transfer Process

There are a wide variety of mass transfer processes in the refining and chemical process, especially the separation process, which plays a pivotal role in the petrochemical process and forms the core process of processing and production together with chemical reactions. Performing thermodynamic analysis on the mass transfer process, pointing out the reasons and locations of energy degradation in the process, which is of great benefit to energy saving improvement. Reducing the exergy loss in the mass transfer process can improve the energy grade and quality of the process streams leaving the equipment, and it creates conditions for the energy recovery opportunities.

4.3.1 *The Minimum Exergy Consumption of the Separation Process*

4.3.1.1 Calculation of Exergy Separation of Known Components and Mixtures

Mixing and separation is a pair inverse process, the purpose of which is to change the concentration of the components, and the driving force of the mixing process is the concentration difference.

Thermodynamics stipulates that when the pure components are mixed into a solution at the same temperature and pressure, the solution whose volume and temperature do not change is called an ideal solution (in a broad sense, it includes ideal gas). For an ideal solution, volume and enthalpy exist additive. The relationship between the enthalpy and entropy of an ideal solution and the enthalpy and entropy of each pure component is:

$$H_m = X_1 H_1 + X_2 H_2 + \ldots \tag{4.14}$$

$$S_m = X_1 S_1 + X_2 S_2 + \cdots - R(X_1 \ln X_1 + X_2 \ln X_2 + \ldots) \tag{4.15}$$

In the formula, H and S are molar enthalpy and molar entropy, X_i is the molar fraction of component i, and the subscript m represents the mixture.

It can be seen from the above formula that the entropy changes before and after mixing. Since X_i is less than 1, so $\sum X_i.lnX_i < 0$; Therefore, the mixing process is always accompanied by an increase in entropy, indicating the irreversibility of this process.

The ideal work for separating an ideal solution into pure components at the same temperature and pressure. Using the previous two formulas, the separation work for a 1kmol solution is:

$$I_w = (H_m - \sum X_i H_i) - T_0(S_m - \sum X_i S_i) \tag{4.16}$$

Because the ideal solution mixing enthalpy is 0, then

$$I_w = 0 + RT_0 \sum X_i lnX_i = RT_0 \sum X_i lnX_i \tag{4.17}$$

Because $X_i < 1$, the separation work is less than 0, indicating that the separation process requires external work to be exerted on it.

The reverse process of separation is the mixing process, and the mixing process can do work outwards.

$$I_w = -RT_0 \sum X_i lnX_i \tag{4.18}$$

Under current technical conditions, it is difficult to use the actual process mixing work, and the mixing process work is irreversibly lost. Therefore, direct mixing of streams should be avoided as much as possible in the actual process.

Generally, gas mixtures can be regarded as ideal solutions, and their separation work and mixing work can be calculated. Liquids can also be treated as ideal solutions approximately. The above formula also shows that the separation work of an ideal solution is only related to the composition concentration, and has nothing to do with the physical properties of each component.

For the separation work of non-ideal solutions, it is more convenient to use concentration data, where there is a mixing heat effect. The separation work can be derived as:

$$I_w = \Delta H_m \left(1 - \frac{T_0}{T}\right) + RT_0 \sum X_i lna_i \tag{4.19}$$

where: a_i is the activity coefficient. When it is an ideal solution, $\Delta H_m = 0$, $a_i = X_i$, it is transformed into an ideal solution formula. The ideal solution can be regarded as a special case of the non-ideal solution.

4.3.1.2 Petroleum Fraction Separation Process [4]

In the petrochemical process, petroleum distillate occupies a large proportion. Petroleum fractions are fractions of multi-component mixtures. Gasoline, kerosene, diesel, etc., are the lighter fraction mixtures separated from the mixture (crude oil) in the separation process. For petroleum fractions, it is not only difficult to determine the number of compounds contained, and even more difficult to know its composition, but it can be treated as an ideal solution.

A simplified calculation method has been developed by Professor Hua Ben [4] to calculate the separation exergy of petroleum fractions. First, the total separation exergy of a system with a large number of components m is processed as a function of the number of components and the concentration distribution factor Ω:

$$EX_{TS}^{0} = -RT_0\Omega.ln.m^{-1} \tag{4.20}$$

When the value of m is large enough, the value of Ω is around 0.85.

When there is no overlap between the fractions, the separation exergy of the raw materials separating into each fraction can be calculated as the pure components of each fraction:

$$EX_{TS}^{0} = -RT_0\sum X_i ln X_i \tag{4.21}$$

where: EX_{TS}^{0} is separation exergy of raw materials into n non-overlapping fractions, kJ/kmol;

X_i—Molecular fraction of distillate fraction$i$, $\sum X_i = 1$.

When there are overlapping components (fractions), the sum of the number of pure components $p_j^{'}$ in each fraction is greater than the total number of components$m$, ($\sum p_j^{'} > m$, introducing the degree of overlapβ, let $\beta = p_j^{'}/p_j$), so the total separation exergy (EX_{TS}^{0}) of each fraction is greater than when there is no overlap. Therefore, the separation of exergy is:

$$EX_{TS}^{'} = -RT_0\sum N_j ln N_j - \Delta \tag{4.22}$$

where: Δ—the separation exergy changes due to overlap, it can be derived:

$$\Delta = -RT_0\sum_{j=1}^{n} N_j ln(\beta^{\Omega j}) \tag{4.23}$$

$$EX_{TS}^{'} = -RT_0\sum_{j=1}^{n} N_j ln(\beta^{\Omega j}.N_j) \tag{4.24}$$

In the formula, β is the degree of overlap of the components of the j fraction, which depends on the degree of emptiness or overlap between this fraction and the adjacent fraction of En's distillation or real boiling point distillation at 5 or 95% of the point. N_j is the molecular fraction of fraction j in the petroleum fraction.

In the absence of associated value data, especially for the purpose of energy analysis, it is recommended to take $\beta = 1.4$, and the error will not exceed 10%, which can meet the engineering error requirements.

4.3.2 Thermodynamic Analysis of Actual Separation Process

Separated exergy or mixed exergy is only the minimum work consumed or the maximum work made when the process is completely reversible, but due to the irreversibility of the actual process, the exergy loss in the mass transfer process is much greater than the separated exergy, the difference is called Process exergy loss, its size varies with the process and operating conditions. Separation exergy is the minimum exergy that must be consumed and is effective.

4.3.2.1 Simple Distillation

Taking a simple distillation column of a binary ideal solution as an example, the exergy loss of the actual separation process is discussed. As shown in Fig. 4.3, the feed is 1 kmol, the components are A, B, and the component fractions are X_A, X_B, the top product A is $X_A$kmol, and the bottom product B is $X_B$kmol.

The energy consumed is the heat supply Q_R from the bottom reboiler. At the same time, the top condenser must discharge the heat Q_D. If the temperature of the feed and the outlet product are the same and are liquid, $Q_R \approx Q_D$, which reflects the relationship of energy quantity. In fact, the energy of reboiling heating duty is discharged in the same amount at the top of the tower, but the temperature is lower than the heating temperature at the bottom of the tower, so there must be heat transfer exergy loss. The essence of the fractionation process is because of the temperature

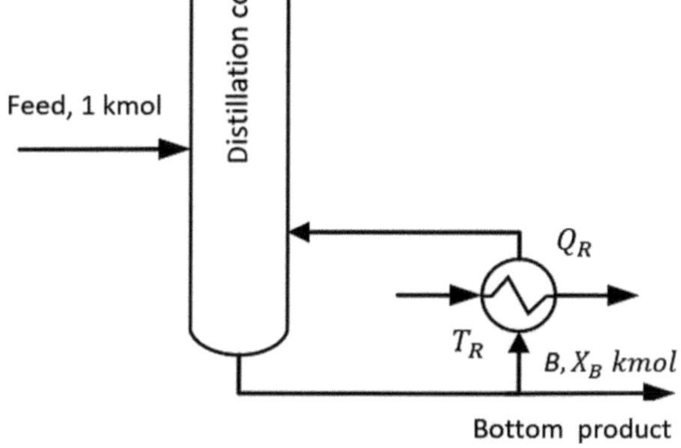

Fig. 4.3 Binary distillation

difference, the concentration difference is formed and the mass transfer is carried out. The exergy loss of the rectification is:

$$D_{KL} = QT_0\left(\frac{1}{T_D} - \frac{1}{T_R}\right) = QT_0[(T_R - T_D)/(T_R.T_D)] \tag{4.25}$$

where:

T_D—Condenser inlet temperature, K;

T_R—Reboiler outlet temperature, K;

Q—Reboiler heat load, kW;

D_{KL}—Distillation exergy loss, kW.

It can be seen from the formula (4.25):

(a) Exergy loss is related to the heat load, that is, it is related to the reflux ratio. Reducing the reflux ratio can reduce Q and thus reduce the exergy loss.
(b) The exergy loss is related to the temperature difference between the heat supplying the tower and the heat leaving the tower. The smaller the difference between the two, the smaller the exergy loss. The exergy loss is also related to the reboiler outlet temperature. The lower the the temperature, the smaller the exergy loss.

4.3.2.2 Actual Distillation Process

The above discussion is about simple binary distillation or distillation without sideline. The actual rectification process is complicated, and there are more inlet streams and also sidelines. At this time, the actual process exergy loss is much more complicated than binary distillation. According to the heat balance and exergy balance, the heat entering the tower must be equal to the heat output from the tower. Separation work and process exergy loss:

$$D_{KL} = T_0(\sum \frac{Q_P}{T_{pM}} - \sum \frac{Q_I}{T_{IM}}) \tag{4.26}$$

The derivation is as follows:

$$\begin{aligned} D_{KL} &= \sum Q_I(1 - \frac{T_0}{T_{IM}}) - \sum Q_p\left(1 - \frac{T_0}{T_{pM}}\right) \\ &= \sum Q_I - \sum Q_P - \sum Q_I\left(\frac{T_0}{T_{IM}}\right) + \sum Q_p\left(\frac{T_0}{T_{pM}}\right) \end{aligned}$$

Because of the energy balance of the tower, the energy leaving from the tower is equal to the energy entering the tower $\sum Q_I = \sum Q_p$, so there is formula (4.26):

$$D_{KL} = T_0(\sum\left(\frac{Q_P}{T_{pM}}\right) - \sum\left(\frac{Q_I}{T_{IM}}\right))$$

Where: T_{PM} Average temperature of outlet product, reflux and heat dissipation, K;

T_{IM}—The average temperature of various raw materials entering the tower and reboiling heating, K;

Q_I—Energy supplied to the tower, kW;

Q_p—Energy off the tower, kW.

The process exergy loss caused by the irreversibility of the process should be subtracted from the separation work:

$$D'_{KL} = T_0\left(\sum\left(\frac{Q_P}{T_{pM}}\right) - \sum\left(\frac{Q_I}{T_{IM}}\right)\right) - EX_{TS} \tag{4.27}$$

It can be seen from the formula (4.27):

(a) The process exergy loss is related to the total heat entering the tower (called the total process energy). The smaller $\sum Q_I$, process exergy loss is small. To save energy, attention should be paid to improving the process flow and conditions to reduce the total process energy consumption.
(b) The process exergy loss is related to the average temperature of the supply and leaving the tower products. The temperature of the energy supplied is high, and the temperature of the energy leaving the tower is low, the exergy loss is large, and vice versa. Therefore, to reduce the exergy loss of the process, the temperature of the energy supplied should be reduced as much as possible, and the temperature of the leaving streams from the tower should be increased as much possible. For example, improving the ratio of reflux heat extraction in each part of the tower, increasing the ratio of high temperature heat extraction in the lower part, reducing the temperature difference between the leaving and returning to the tower and increasing the average temperature of the stream, etc., can reduce the process exergy loss.

Exergy efficiency η_x:

$$\eta_x = (\sum Q_p\left(1 - \frac{T_0}{T_{pM}}\right) + EX_{TS})/\sum Q_I\left(1 - \frac{T_0}{T_{IM}}\right) \tag{4.28}$$

4.3.2.3 Exergy Loss on Absorption Process

The absorption process is governed by Henry's law, and it is mostly operated at low temperature and high pressure. The energy consumption of the absorption process is reflected in the regeneration of the absorbent. In order to restore the absorption capacity of the absorbent and restore the concentration to the concentration before absorption, energy must be consumed to separate the absorbed components from the absorbent. The minimum work consumed by reversible regeneration is separation work. The absorption process is the mixing process of the absorbed component and the absorbent, and the driving force of the absorption process is the pressure difference. The exergy loss in the absorption process is [2]:

$$D_{KS} = RT_0 ln\left(\frac{p_i}{p_{eq}}\right)n_i \tag{4.29}$$

where: n_i—kilomoles of absorption components;

p_i—The partial pressure of the gas phase of the absorption component on a certain section of the tower;

p_{eq}—The partial pressure of the gas phase at which the concentration of the i component in a certain cross-section liquid absorbent becomes equilibrium.

The greater the difference between the component partial pressure and its equilibrium partial pressure, the greater the process rate, the greater the exergy loss, and the loss work is proportional to the logarithmic value of the pressure ratio.

4.4 Thermodynamic Analysis of Chemical Reaction Process

Chemical reaction process is another common unit operation process in petrochemical process, such as catalytic cracking, catalytic reforming, conversion and synthesis in fertilizer production and so on. The chemical reaction process often constitutes the main body of the process. In addition, the established three-link analysis method of energy consumption analysis focuses on analyzing the changes in fuel and power consumption during the process. The chemical energy changes of raw materials and products are reflected in the form of reaction heat (exergy), in the form of thermodynamic energy consumption (exergy difference). Therefore, it is particularly important to accurately calculate the reaction exergy.

4.4.1 Calculation of Chemical Reaction Exergy

According to the basic definition formula, the chemical reaction exergy:

$$EX_R = \Delta H - T_0 \Delta S \tag{4.30}$$

Ignoring the difference between the reference state and the standard state (25 °C, 1 atm), the reaction exergy of the isothermal reaction at the reference temperature is equal to the free enthalpy change of the reaction in the standard state

$$EX_R = -\Delta G \tag{4.31}$$

For chemical plant with known reactant and product composition, the free enthalpy data of the standard state can be directly calculated from the manual, which is the reference reaction exergy. When the reaction is not at the reference temperature, the isothermal chemical reaction with reaction temperature T is defined by the free enthalpy

$$\Delta G = \Delta H - T\Delta S \tag{4.32}$$

$$\Delta S = (\Delta H - \Delta G)/T \tag{4.33}$$

Therefore

$$\begin{aligned} EX_R^{'} &= \Delta H - T_0((\Delta H - \Delta G)/T) \\ &= (T_0/T)\Delta G + \Delta H(1 - T_0/T) \end{aligned} \tag{4.34}$$

Generally speaking, the function relationship between free enthalpy and equilibrium constant:

$$\Delta G = RTln(\frac{K_p}{J_p})$$

$$EX_R^{'} = RT_0ln(\frac{K_p}{J_p}) + \Delta H(1 - T_0/T) \tag{4.35}$$

Equation (4.35) is the calculation formula of reaction exergy under reaction conditions, and ΔH is the reaction heat effect. It can be expressed by the thermal effect of a unit reactant, which is actually a comprehensive expression of the thermal effect of a complete reaction multiplied by the progress of the reaction (conversion rate).

When considering that there are still inert components that do not participate in the reaction, the following formula is used when entering and leaving the reactor along with the reactants and products:

$$EX_R^{'} = RT_0ln\left(\frac{K_p}{J_p}\right) + \Delta H\left(1 - \frac{T_0}{T}\right) + RT_0\sum n_i ln((p_i)_0/(p_i)_I) \tag{4.36}$$

where: $(p_i)_0$, $(p_i)_I$—Are the partial pressures of components i leaving and entering the reactor;

n_i—the number of kilomoles of the inert component relative to 1 kmol of the reactant.

For Isomolecular reaction, that is, when the stoichiometric coefficients of reactants and products are the same, and the pressure drop of the reactor itself can be ignored,

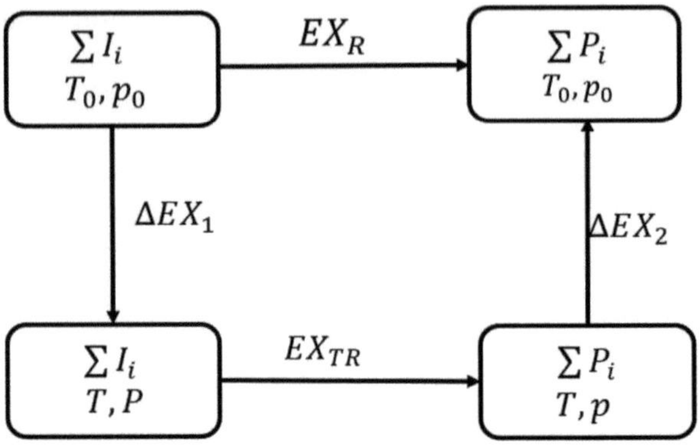

Fig. 4.4 Reaction exergy calculation graph

the influence of inert components can be ignored, and the reaction exergy calculation can be calculated according to formula (4.35).

For the adiabatic reaction process, the enthalpy is equal before and after the reaction, and the temperature is not equal. It can be divided into two steps to find EX. First, calculate the required heat ΔH from outside under the constant temperature T_i. In the second step, the reaction product releases heat, and the temperature changes from T_i toT_e, so (Fig. 4.4):

$$EX_{TR}^{'} = RT_0 ln\left(\frac{K_{pi}}{J_p}\right) + T_0\Delta H\left(\frac{1}{T_i} - \frac{1}{T_m}\right) + RT_0\sum n_i \ln((p_i)_0/(p_i)_I) \tag{4.37}$$

where: T_m—Thermodynamic average temperature of inlet and outlet temperature;

$$T_m = (T_i - T_e)/ln(T_i/T_e)$$

K_{pi}—Equilibrium constant at inlet temperature.

The thermodynamic exergy difference of the reaction (reference temperature reaction) can be corrected by the change value of physical exergy obtained from the change in operating temperature and pressure.

$$EX_R = EX_{TR}^{'} + \Delta EX_1 - \Delta EX_2 \tag{4.38}$$

4.4.2 Reaction Exergy Loss and Reaction Exergy Calculation on Complex Reaction

The actual reaction process exergy loss consists of two parts: one is the reaction exergy loss caused by the irreversibility of the reaction itself. The second is the exergy loss caused by the irreversibility of the heat transfer process in the reactor.

The irreversibility of the reaction process is due to the existence of the reaction driving force $ln(K_p/J_p)$, that is, the difference between the equilibrium constant and the reaction pressure quotient promotes the reaction to proceed. $J_p = K_p$, the chemical equilibrium is reached, it becomes a reversible process. In chemical thermodynamics, the free enthalpy change $-\Delta G$ is called chemical affinity and represented by A.

$$A = -\Delta G = RTln(K_p/J_p) \tag{4.39}$$

It is the driving force of the reaction process. And loss work:

$$D = (T_0/\mathrm{T})A \tag{4.40}$$

$$D = RT_0 ln(K_p/J_p) \tag{4.41}$$

The chemical reaction exergy loss is generally not large. It can usually be ignored, especially for complex petroleum processing processes, it is difficult to know K_p and J_p accurately. For a rough calculation, the reaction heat effect ΔH can be directly taken as the reaction exergy. This can be explained from the following derivation process.

In engineering, the concept of equilibrium temperature moment is often used, that is, the difference between the equilibrium temperature T_{eq} equivalent to the composition of the material system and the actual reaction temperature. The equilibrium constant corresponding to T_{eq} is also K_p. According to the functional relationship between the equilibrium constant and temperature:

$$ln\left(\frac{K_{p2}}{K_{p1}}\right) = (\frac{\Delta H}{R})(\frac{1}{T_1} - \frac{1}{T_2}) \tag{4.42}$$

Therefore

$$ln\left(\frac{K_p}{J_p}\right) = (\frac{\Delta H}{R})(\frac{1}{T_{eq}} - \frac{1}{T}) \tag{4.43}$$

Lost work (exergy Loss):

$$T_0\Delta S = T_0\Delta H(\frac{1}{T_{eq}} - \frac{1}{T}) \tag{4.44}$$

Table 4.1 Equilibrium temperature moment versus β

Reactor	Reaction equilibrium temperature (°C)	Measured temperature (°C)	Equilibrium temperature moment	β
Stage one reformer	787	808	21	0.005
High temperature shift	955	979	24	0.0045
High temperature shift	442	435	−7	0.004
Low temperature shift	250	235	−15	0.0163

Substitute the definition formula:

$$EX_R = \Delta H - T_0 \Delta S$$
$$= \Delta H - T_0 \Delta H\left(\frac{1}{T_{eq}} - \frac{1}{T}\right) = \Delta H\left(1 - \frac{T_0\left(T - T_{eq}\right)}{T_{eq}.T}\right)$$
$$= \Delta H(1 - \beta)$$

The ratio of the exergy loss of the chemical reaction to the heat of reaction is:

$$\beta = \frac{T_0\left(T - T_{eq}\right)}{T_{eq}.T}$$

The equilibrium temperature moment and β value of a certain synthetic ammonia plant are shown in Table 4.1.

Analyzing the above formula, for the petroleum processing, the reaction temperature is above 250 °C (573 K), the equilibrium temperature moment is generally less than about 30 °C, and the β is about 0.01. From Table 4.1, β is slightly larger at low temperature, and the low temperature shift is 1.63%. The higher the temperature, the smaller the effect. At high temperature, the exergy loss accounts for less than 0.5% of the reaction heat; therefore, for the reaction of petroleum and organic hydrocarbons of unknown composition, the operating condition reaction exergy is approximately taken as the heat of reaction, which will not cause too much error and can meet the requirements of engineering calculations.

4.5 Thermodynamic Analysis of the Combustion Process [5, 6]

Combustion is a strong oxidation reaction. Through combustion, the fuel releases chemical energy. The combustion process is a highly irreversible process, which will inevitably cause a large amount of exergy loss (Fig. 4.5).

In the combustion process of petrochemical process, such as fired furnace and boiler, combustion and heat transfer are often carried out at the same time. Therefore, the temperature in the furnace is not the combustion product temperature, but the combustion product temperature after partial heat transfer. The total exergy loss from the chemical energy of the fuel to the heated fluid is essentially the combined exergy loss of the combustion and heat transfer process. Although the combustion process of a gas turbine is a complete combustion process, the amount of excess air is large (excess air coefficient $\alpha = 3 \sim 4$), which essentially includes the exergy loss of the mixed heat transfer of combustion products and cold air.

Therefore, for the general petrochemical process, the combustion exergy loss refers to the combustion exergy loss at the adiabatic combustion temperature. The actual air volume is also slightly more than the theoretical air volume. Different from general chemical reactions, the combustion process heats the substances involved in the reaction, or even inert components, to the adiabatic combustion temperature. The temperature of the combustion products gradually decreases during the heat transfer process. To facilitate the analysis of the combustion process, we assume that both air and fuel are combusted adiabatically in the furnace. The combustion heat heats the combustion products and inert components to the adiabatic combustion temperature reached by the combustion before transferring heat to the heated fluid, and finally discharged from the furnace at the exhaust gas temperature. The combustion and heat transfer process model in the furnace is shown in Fig. 4.6.

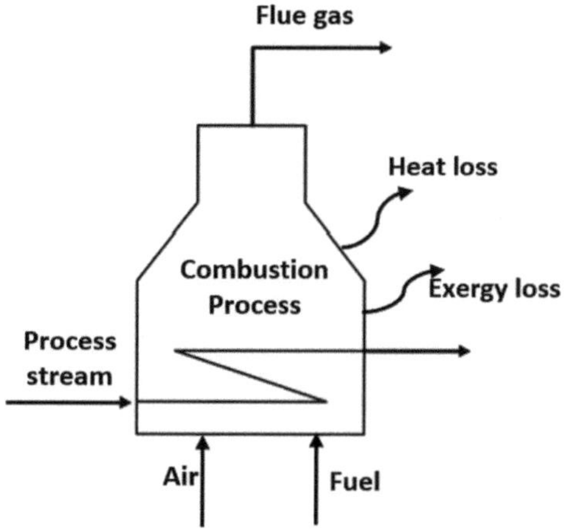

Fig. 4.5 Actual combustion process

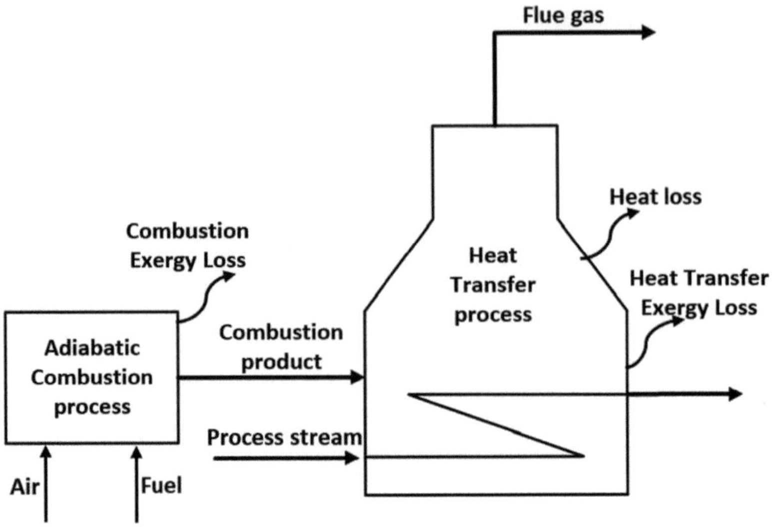

Fig. 4.6 Combustion exergy loss calculation model

It can be seen from Fig. 4.6 that the actual combustion process can be divided into a two-step calculation of the adiabatic combustion process and the heat transfer process in the furnace, which also conforms to the characteristics of the thermodynamic state function.

4.5.1 *Adiabatic Combustion Process*

The adiabatic combustion process in Fig. 4.6 assumes that air and fuel are combusted under adiabatic conditions. At this time, the energy of fuel and the energy of combustion air will all be converted into the energy of combustion gas, reaching the adiabatic combustion temperature, and its exergy loss can be calculated that the exergy of air and fuel is subtracted from the exergy of combustion products. It can also be calculated according to the derived following formula:

$$D_{KC} = T_0 \Delta S + Q_p \left(\frac{T_0}{T_{ad} - T_0} \right) \ln\left(\frac{T_{ad}}{T_0}\right) \tag{4.46}$$

where: T_{ad}—Adiabatic combustion temperature, K;

$$T_{ad} = T^0 + Q_p / (L_0 \alpha C_p) \tag{4.47}$$

Q_p—Fuel LHV, kJ/kg;

D_{KC}—Adiabatic combustion exergy loss, kJ/kg;

L_0—Theoretical combustion air volume, kg/kg;

α—Excess air coefficient,$\alpha < 1.3$, When $\alpha > 1.3$, the exergy loss of mixed heat transfer with cold air should be considered.

C_p—Average specific heat capacity of combustion products, kJ/kg K;

T^0—Combustion air temperature, K.

In many cases, the entropy changes ΔS before and after combustion is difficult to obtain accurately. As mentioned in the previous section, it can be ignored without large errors. In this way, the Eq. (4.46) becomes:

$$D_{KC} = Q_p\left(\frac{T_0}{T_{ad} - T_0}\right)\ln(\frac{T_{ad}}{T_0}) \tag{4.48}$$

The above combustion exergy loss is the combustion exergy loss under the actual air volume, which includes the excess air used to ensure complete combustion. Generally, the excess air coefficient of petrochemical fired furnaces, boilers and reformers is below 1.3, so there is no need to distinguish exergy loss caused by excess air; For the combustion process of gas turbines, due to the temperature resistance of the construction material, the gas temperature cannot be too high. A lot of air is often added to dilute and cool the combustion gas, and α reaches 3–4. The mixed heat transfer of a large amount of cold air and combustion gas cannot be ignored and should be evaluated. At this time, combustion exergy loss should be divided into theoretical combustion exergy loss ($\alpha = 1.0$) and cold air dilution cooling exergy loss. At this time, the theoretical combustion exergy loss is still calculated according to the formula (4.46). Only T_{ad} is the adiabatic combustion temperature under the theoretical air volume.

$$T_{ad} = T^0 + Q_p/(L_0C_p) \tag{4.49}$$

Assuming that the actual exhaust gas temperature of the gas turbine is T_3, the dilution cooling exergy loss is:

$$D_{KA} = Q_pT_0\frac{T_{ad} - T_3}{ln(T_{ad}/T_3)} \tag{4.50}$$

4.5.2 *Heat Transfer Process*

In Fig. 4.6, it is assumed that the combustion products after adiabatic combustion transfer heat to the heated fluid in the furnace, and the heat transfer exergy loss can be obtained from the furnace exergy balance, that is, the exergy loss combined combustion and heat transfer is calculated first, and then deduct the adiabatic combustion Exergy loss to determine the exergy loss of heat transfer. It can also be calculated by the heat transfer exergy loss equation (4.3):

$$D_{KH} = QT_0(T_h - T_c)/(T_h.T_c)$$

where: T_h—the average temperature of the flue gas, which can be the average of *Tad* and the exhaust temperature, K;

T_c—Is the average temperature of the heated stream, K;

Q—Furnace heat transferred duty, kW.

For multiple heated streams, the average temperature of each stream can be obtained first, and then the temperature of the cold stream can be obtained from the heat load of each stream according to the weighted average of the heat load.

4.5.3 Approaches to Reduce Combustion Exergy Loss

The approaches to reduce exergy loss in the combustion process are as follows.

1. **Choose the right grade of energy instead of fuel combustion**

   In the petrochemical process, fired furnaces and low-pressure steam boilers are limited by technological conditions, so their exergy efficiency is very low, and combustion is a highly irreversible process. On the other hand, many process units only require heat sources at medium and low temperatures. At this time, looking for a suitable low-temperature heat source, and even using the back pressure steam from the steam turbines instead of fuel combustion heating, can avoid the combustion exergy loss and reduce the exhaust flue gas and heat dissipation loss.
2. **Pay attention to the energy-saving effect of air preheating**

   At present, in actual industrial processes, air is used as the oxide that participates in combustion. During the combustion process, air and inert components are heated from ambient temperature to combustion temperature to absorb a large amount of heat. The air preheating energy is directly reflected in increasing the adiabatic combustion temperature and saving fuel. Generally speaking, when the process conditions in the furnace do not change, the heat load of the heated fluid is relatively fixed. It is also required that the total heat of the combustion products remain unchanged. The preheated air reduces the heat from the ambient temperature to the furnace inlet temperature of the fired heater, which can reduce the fuel that is roughly the same as the preheat load. On the other hand, when the fuel combustion heat load decreases, the fuel consumption decreases. When the excess air coefficient remains unchanged, the total air volume decreases, and the adiabatic combustion temperature rise is the same, so that the adiabatic combustion temperature increases, reduces the process exergy loss. Therefore, air preheating is an effective measure to reduce the stew loss during the combustion process and should be used first.

3. **Reduce excess air consumption**

   In order to ensure that the fuel can be completely burned, it is necessary to maintain a certain amount of excess air, but a large amount of excess air will not only increase the total exhaust flue gas rate and increase the heat loss, but will also reduce the adiabatic combustion temperature and increase the process exergy loss, the mixed cold air heat transfer exergy loss is produced. Generally, a burner with good performance has an excess air coefficient between 1.1 and 1.2 that can meet the requirements of complete combustion.

References

1. H. Ben, *Analysis and Synthesis of Energy Consumption in Process* (Hydrocarbon Processing Press, 1989), p. 1
2. Y. Yi, H. Desheng, *Chemical Process Thermodynamic Analysis* (Chemical Industry Press, 1985)
3. C. Junwu, H. Xiren, Analysis of energy consumption process of catalytic cracking unit-discussion on energy saving direction. Refinery Design **4**, 4–5 (1982)
4. H. Ben, Calculation of exergy difference in the separation process of petroleum and its fractions. Acta Petrolei Sinica **3** (1986)
5. C. Anmin, L. Kun, Thermodynamic analysis of gas turbine-heating furnace combined system. Oil Refinery Design **4** (1988)
6. Z. Mingshan, *Exergy Analysis of Energy System* (Tsinghua University Press, 1988), p. 4

Part II
Three-Link Energy Analysis Approach and Energy/Exergy Balance

Chapter 5
Energy Utilization Three-Link Analysis Method

Abstract For years, people have been accustomed to analyzing and researching processes from the perspective of material change clues and unit operations. However, according to the energy change clues to study the process, there are three characteristics: (1) The main forms of process energy use are heat, fluid work, and steam; and heat, work, and steam are converted from fuel and electricity; (2) Heat, work, steam provided by the conversion equipment enters the core process together with the recovered energy to promote the completion of the process; (3) the stream leaving core process has high temperature and pressure, its energy can be recovered to reduce the energy input to the plant. Based on the three energy utilization characteristics, this chapter proposed updated 3-link energy and exergy analysis models and their detail items; it divides the energy utilization system into 4 levels: equipment, energy utilization links, process units, auxiliary system, and plant. The reader can evaluate the energy efficiency for every level and identify energy utilization issues, put forward solutions for energy-saving and carbon reduction.

Keywords Energy characteristics · Energy utilization · 3-link energy model · Energy analysis · Carbon reduction · Energy solutions

Abbreviation and Symbology for Energy Balance

Energy link	Energy symbol		Definition	Exergy symbol
Energy conversion	Energy supply	E_P	Total energy supply	EX_P
		E_{PF}	Furnace fuel energy supply	EX_{PF}
		E_{PB}	Boiler fuel energy supply	EX_{PB}
		E_{PBF}	Power plant boiler fuel energy supply	EX_{PBF}
		E_{PG}	Coke energy supply	EX_{PG}
		E_{PS}	Steam energy supply	EX_{PS}
		E_{PE}	Electricity energy supply	EX_{PE}

(continued)

T. A. Chen, *Energy Saving and Carbon Reduction*,
https://doi.org/10.1007/978-981-19-5295-1_5

(continued)

Energy link	Energy symbol		Definition	Exergy symbol
		E_{PH}	External heat energy supply	EX_{PH}
	Output	E_B	Conversion link output energy	EX_B
		E_{BS}	Conversion link output Steam	EX_{BS}
		E_{BH}	Conversion link output heat	EX_{BH}
		E_{BE}	Conversion link output electricity energy	E_{XBE}
	Direct loss	E_W	Direct loss of energy	D_{JU}
		E_{WC}	Incomplete combustion	EX_{WC}
		E_{WX}	Flue gas energy loss	EX_{WX}
		E_{WD}	Heat loss	EX_{WD}
		E_{WP}	Ineffective power	EX_{WP}
			Process exergy loss	D_{KU}
	E_U		Effective energy supply	EX_U
	E_{UO}		Effective power of non-process fluids	EX_{UO}
Process energy utilization	E_{AR}		Endothermic reaction heat	EX_{AR}
	E_{RE}		Exothermic reaction heat	EX_{RE}
	E_N		total process energy used	EX_N
	E_{TT}		Thermodynamic Physical Energy Difference	EX_{TT}
	E_T		Thermodynamic energy consumption	D_T
	E_{UD}		Dissipate heat of Energy utilization link	EX_{UD}
			Process exergy loss	D_{KP}
Energy recovery and recycling	E_O		Energy to be recovered	EX_O
	E_R		Recovery and Recycling Energy	EX_R
	E_{RH}		Heat exchanger recovery and recycling energy	EX_R
	E_{RM}		Streams Recovery and Recycling Energy	EX_R

(continued)

(continued)

Energy link	Energy symbol	Definition	Exergy symbol
	E_{UD}	Dissipate heat of Energy utilization link	EX_{UD}
	E_E	Energy Recovered for output	EX_E
	E_J	Rejected energy	D_{JR}
	E_{JC}	Cooling rejected energy	D_{JC}
	E_{JD}	Heat dissipation rejected energy	D_{JD}
	E_{JM}	Streams rejected energy	D_{JM}
	E_{JO}	Other rejected energy	
	E_N	Total process energy used	EX_N
		Process exergy loss	D_{KR}
Summary indicators	E_A	Net energy (exergy) consumption	D_A
	E_C	Converted to primary energy (exergy) consumption	EX_C
	η_U	Energy conversion efficiency %	η_{XU}
	η_P	Energy process utilization efficiency %	η_{XP}
	η_R	Energy recovery efficiency %	η_{XR}
		Total rejection energy	D_J
		Total process exergy loss	D_K

The petrochemical process is a complex process. It not only includes the unit operation process of "three transfers and one reaction", but it is not a simple superimpose. It often crosses and penetrates each other. For such a complex petrochemical process, the thermodynamic analysis simply by unit operations is monotonous and insufficient. Moreover, each unit operation is interrelated and restricted with other unit operations. Therefore, it is necessary to find an analysis method that can anatomy the energy utilization law in complex processes.

Most petrochemical processes are thermal processing processes. In order to achieve the conditions for process operation, pressure and temperature must be increased. The processed products need to be cooled and transported. In order to reduce the heating duty, people have turned to recover energy from the product stream that needs to be cooled. With the development of modern science and technology and the strengthening of awareness of energy conservation and CO_2 mitigation, the degree of energy recovery is also increasing year by year. Figure 5.1 indicates a typical segmentation of the chemical process system, some people call the thermal processing process visually as "heating up, processing, and cooling down" [1].

As energy prices fluctuate and rise, global warming, people have begun to realize the importance of energy conservation and CO_2 mitigation, and take the physical energy consumption such as water, electricity, steam, and Instrument air etc. as an evaluation index and count them. However, whether the process energy consumption is reasonable and whether the energy quality is matched with energy demand, these questions cannot to be answered with plant energy consumption indicators, it also

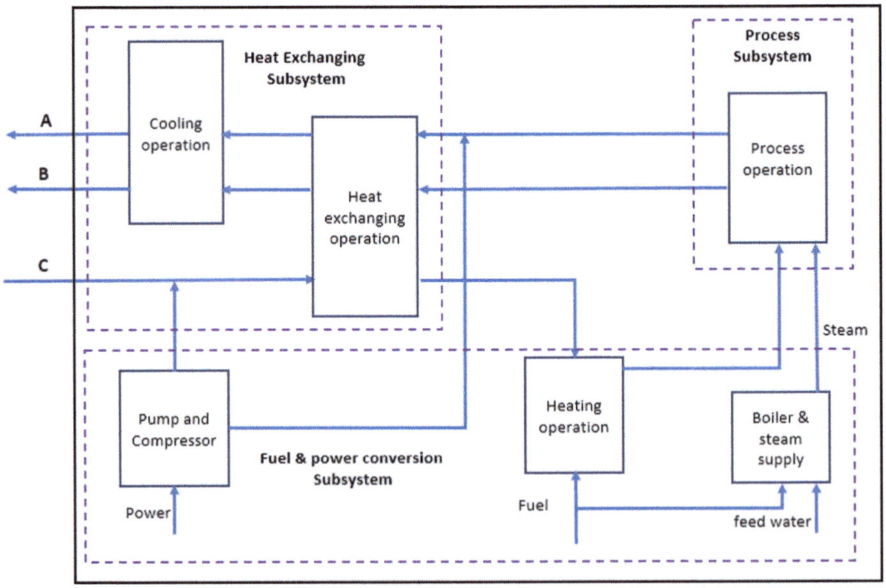

Fig. 5.1 Segmentation of the chemical process system

requires people to propose a set of energy analysis methods that scientifically reflect the law of energy changes and are practical in engineering.

5.1 Energy Consumption Characteristics of Petrochemical Industry

For many years, people have been accustomed to analyzing and researching processes from the perspective of material change clues and unit operations. However, according to the energy change clues to study the process, it is found that the energy consumption of petrochemical process has the following three characteristics:

1. The main forms of process energy use are heat, fluid work and steam; and heat, work and steam are converted from fuel and electricity. Energy conversion is generally achieved through conversion equipment, such as furnaces, pumps and compressors etc.
2. The energy in the form of heat, work, steam, etc. provided by the conversion equipment enters the core process (tower, reactor, etc.) together with the recovered recycling energy to promote the completion of the process, except for part of the energy that is transferred to the product, the rest enter the energy recovery system.
3. After the energy has completed its mission in the core process section, the energy quality will drop, but it still has a higher temperature and pressure; to a large extent, it can be recovered and recycled through heat exchange equipment, power exchange equipment (such as hydraulic turbines), etc., so that it can be recycled and utilized to reduce the energy input to the conversion equipment. However, due to engineering and economic constraints, the recycling/recovered cannot be done to the end, and it is finally discharged into the environment through channels such as cooling and heat dissipation. Together with the energy loss of the energy conversion, it constitutes the energy consumption in quantity for process plant.

Therefore, there are three major links in the energy use of petrochemical processes, that is, the energy conversion and transmission link that converts primary energy source into energy that can be directly used in the process, the energy utilization link that completes the core process of the process, and the Energy recovery link which recovers the energy discharged from the process energy utilization link and supply it to energy utilization link again.

From the above characteristics, it can be seen that the energy-consuming object is the core equipment of the process, but its quantity remains unchanged (no consumption), and the energy conversion and transmission link and energy recovery link serving to process energy utilization link with huge consumption of energy in quantity. Obviously, energy use analysis should examine changes in the quantity and quality of energy use from an entire system perspective.

Based on the above-mentioned characteristics of energy use, Hua Ben professor [2, 4] proposed three-link model of energy use analysis systematically solves the

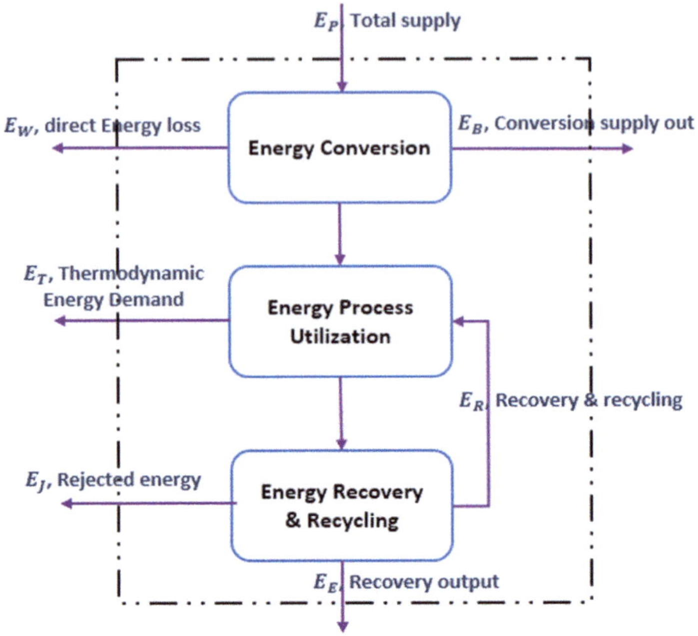

Fig. 5.2 The three-link model of process energy utilization

method problems of energy balance and exergy balance of oil refining equipment [2, 3]. The three-link model of energy utilization analysis is shown in Figs. 5.2 and 5.3. However, since the model was developed in the analysis of energy use in the oil refining process, considering that the petrochemical industry characteristics is not enough, it needs to be improved appropriately.

5.2 Improvement of the Three-Link Model of Energy Consumption Analysis [3]

The outstanding feature of the energy analysis three-Link model is that the process energy utilization is analyzed and studied according to the law of energy change rather than the law of material change; the chemical energy of raw materials is not counted, and the difference between the chemical energy of raw materials and products is treated in thermodynamic energy consumption; The process is divided into three links of energy conversion and transmission, process utilization and energy recovery, and the quantity and quality changes of energy use are investigated respectively. This is a scientific and practical simplification method for the refinery industry, and is proven by the practice of energy balance in the refinery.

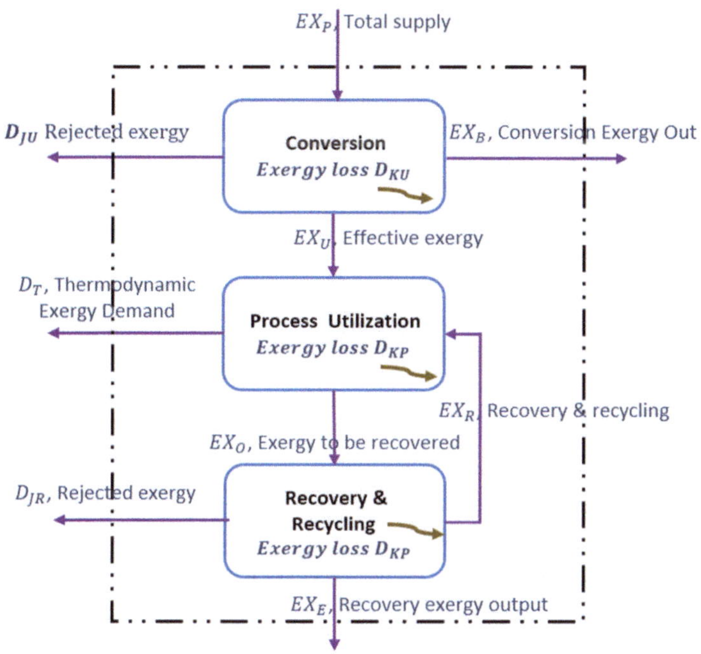

Fig. 5.3 Exergy balance diagram of three-link model

For petrochemical processes, especially when raw materials and fuels can be separated, it is often feasible to ignore the chemical energy of raw materials. The raw materials and products are still macromolecular hydrocarbons, but they are different in structure and type, and they all contain high chemical energy, all of which are reflected in the chemical reaction of the process. Therefore, the difference between the chemical energy of the raw material and the product is included in the thermodynamic energy consumption through the heat of reaction. Even if it is difficult to separate the energy consumption of raw materials and fuels, the above method can still be used to analyze the energy consumption of the process. At this time, the three-link model mainly analyzes the energy consumption of fuels and power energy, the chemical energy of raw materials is only summarized and counted as the energy consumption of plant.

Therefore, combined with the characteristics of the petrochemical process, the following improvements are proposed to the original energy analysis three-link model, so that it can be applied to the petrochemical process for energy analysis. The improved three-link model of energy consumption analysis is shown in Figs. 5.4 and 5.5.

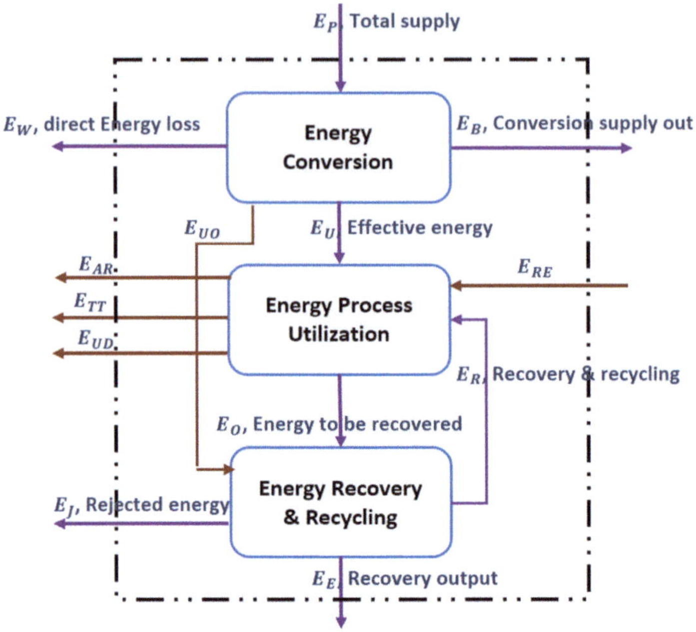

Fig. 5.4 Improved energy balance three-link model. E_{UO}—effective power of non-process fluids; E_{TT}—thermodynamic physical energy Difference; E_{UD}—dissipate heat of Energy utilization link; E_{RE}—exothermic reaction heat; E_{AR}—endothermic reaction heat

5.3 The Consideration of Using Energy Utilization Three-Link Model

5.3.1 Energy Consumption Analysis and Calculation Benchmark

Based on the common practice of the oil refining industry, the early three-Link energy analysis model used the unit raw material as the calculation basis to calculate the energy balance and the exergy balance. The reason is that the main raw materials of the refinery are single and the products are widely distributed. It is more convenient to use unit raw materials as the calculation basis.

The petrochemical industry is often made of multiple raw materials, and the products are relatively single and clear. It is customary to use the target product as the basis for calculation. For multiple raw materials and multiple products, it is more convenient to use unit time (hour) as the basis for calculation. Therefore, the calculation basis of energy consumption analysis should be based on the specific characteristics of different industries. The principle is: it is more practical and convenient to select single raw materials or products, combined with the local regulations and authority's

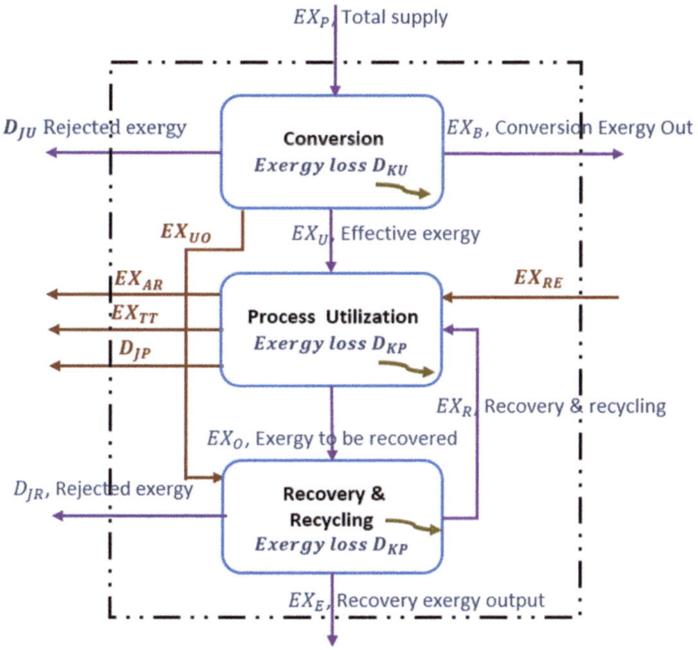

Fig. 5.5 Improved exergy balance three-link model. D_{Jp}—dissipate heat exergy of energy utilization link; EX_{TT}—thermodynamic physical exergy difference; EX_{UO}—effective power of non-process fluids; EX_{RE}—exothermic reaction exergy; EX_{AR}—endothermic reaction exergy

requirement, and respect the habits of the industry to make it comparable in the same industry.

5.3.2 *Effective Power of Pumps for Non-Process Fluids*

Non-process fluid pumps and compressors refer to pumps and compressors that boost water, wind, and other non-process fluids pressure for plant use. Some of these are located inside the plant, but most of them are located in the utility system and shared by several plants. This part of energy is actually difficult to enter the energy balance system. In order to deal with the problem conveniently, the early analysis model regards this part of the pump and compressors energy consumption as the power consumption of the plant, and converts the consumption index of its unit product (water, steam instrument air, nitrogen etc.) into the energy consumption of the plant, and then calculates this part of the consumption according to its consumption divide them into ineffective power and effective power per conversion efficiency.

The ineffective power is discharged from the conversion link together with the ineffective power of other process pumps, and the effective power is regarded as effective energy supply and enters the process utilization link.

Since the effective power of the non-process fluid pump and compressor does not actually enter the utilization link, in order to truly reflect the relationship between process energy use, this part of the effective power from conversion link should be directly led to the energy recovery link for disposal, bypassing the process utilization link. And this part of the effective power is recorded as E_{UO}; in this way, the non-process fluid pump and compressors are also regarded as the conversion link equipment, and its effective power is included in the link conversion efficiency, but does not enter the utilization link, avoiding impact on the effective energy supply E_U, total process use energy E_N and energy to be recovered E_O.

5.3.3 Equipment Heat Dissipation in the Energy Use Link

The process energy utilization link is the core link of the plant energy consumption, and the heat dissipation of the equipment is inevitable, and it accounts for a considerable proportion. The energy consumption analysis model shown in Figs. 5.2 puts the heat dissipation of the equipment in the process energy utilization link into the energy to be recovered link, introduces to the energy recovery link, and discharges to outside together with the heat dissipation of the energy recovery link equipment. As a result, the energy to be recovered is increased. The energy recovery rate is low. Due to the irreversibility of heat dissipation, the heat dissipation of process utilization equipment will not and cannot go to the energy recovery link to be recovered/recycled. It can only be reduced by strengthening heat preservation in the energy utilization link. Therefore, the heat dissipation of the equipment in the process utilization link should be listed as a single item in the energy consumption analysis model, and it shall be directly discharged from the utilization link and recorded as E_{UD}.

5.3.4 Feed Raw Materials Chemical Energy

The petrochemical process is a processing process of hydrocarbon substances, and its raw materials and products contain high chemical energy. The energy consumption of the processing process is relatively small compared with the chemical energy of the raw materials, and the purpose of energy consumption analysis is to reveal the law and rationality of energy use during processing. The difference between the chemical energy of raw materials and products is mostly reflected in the chemical reaction heat of the process, which is reflected in the thermodynamic energy consumption of the plant.

For the chemical fertilizer industry that is accustomed to calculating the chemical energy of raw materials, the energy consumption of the plant can be regarded as the two parts of the fuel-power energy consumption and the raw material energy consumption. The fuel-power energy consumption is the energy used and consumed in the process, and the above-mentioned energy consumption analysis model should be adopted, and carry out analysis to examine the characteristics and laws of its use process. The chemical energy of raw materials is only reflected in the primary energy consumption of the entire plant. That is, the fuel-power energy consumption and the physical energy of the raw materials enter the balance system, and the chemical energy of the raw materials is placed in the energy consumption of the plant.

5.3.5 The Reaction Exotherm Should Be Included in the Total Process Energy Consumption

The total process energy consumption of the process is a main analysis and evaluation index of the energy consumption of the plant. It refers to the total amount of energy used by the process equipment of the plant, which reflects the design and operation level of the plant. However, because the energy analysis three-link model does not count the chemical energy of the raw materials, only the physical energy of the raw materials is included in the balance system.

In the case of chemical reaction, it should be carefully analyzed. For endothermic reactions, the heat of reaction comes from external energy supply and recycling the recovered energy, which has been included in the total process energy consumption of the process. For exothermic reactions, most of the reaction heat is derived from the chemical energy of the raw materials, in addition to the effective energy supply and recycling recovered energy. The chemical energy of raw materials does not enter the energy balance system according to regulations. In fact, the exothermic reaction heat often plays a role in increasing the temperature of the reactants to promote the process. If the exothermic reaction heat is used well, the energy supply can be directly reduced. Therefore, in an analysis system with exothermic reaction, the heat of reaction should be regarded as the energy of the supply system, and it should be counted as the input energy in the process energy utilization link, recorded as E_{RE}, and the heat of reaction should be regarded as the part of total process energy used in the process.

The total process energy used consists of three parts:

$$E_N = E_U + E_R + E_{RE} \tag{5.1}$$

When the process is an endothermic reaction, take $E_{RE} = 0$; where E_R is recovered and recycled energy.

5.3.6 Principles for Handling Some Special Equipment

The core of the energy use analysis method is to divide the process energy use into three links, and each link has different characteristics. According to the characteristics and laws of the three-link model, some special equipment in the plant may have one kind of equipment belonging to two or even three links.

The hydrogen production reformer is an example. This can be divided according to the different functions of each component of the equipment: fuel combustion and heat transfer belong to the conversion link, the reformer tube of the raw material conversion can be regarded as the process energy use link of the reaction equipment, and the waste heat boiler is the energy recovery link that takes the heat from the reformed gas and cools and recovers energy, which is regarded as the recovery and utilization of the energy to be recovered. The recovery and utilization of flue gas heat is still a conversion link. Therefore, when dealing with this kind of equipment, it should be classified into the corresponding link depending on its functions.

5.4 The Detail Items of the Improved Three-Link Model

The improved three-link model of energy use analysis points out the balance and restriction of energy use from a macro perspective. The energy used by the plant runs through the entire process, and the energy grade is reduced, and finally it is lost when it can no longer be recycled in the plant, resulting in a consumption of energy. From the exergy balance model, we see that energy devaluation and loss occur gradually in three different links.

The following describes the composition and content of the energy and exergy parameters of the energy analysis model.

5.4.1 Energy Balance Parameters

(1) Total energy supply E_P

The total energy supply E_P is summarized by the equipment energy supply items of the conversion link, including the following content.

(1) Fuel supply;
(2) Supply coke (including catalytic cracking coke);
(3) Steam supplied from outside;
(4) Electricity supply (including electricity consumption of the plant and electricity consumed proportional on water and compressed air);
(5) External heating: refers to the hot feed heat input of the plant (the part above the specified temperature) and the energy used in the conversion

process after the recovery link (air for the fired furnace, steam generation for the conversion process, etc.);

(6) The raw materials bring chemical energy (used when calculating the energy consumption of the plant in the fertilizer industry).

(2) Conversion output energy E_B

The conversion output energy refers to the energy output from the conversion link to the outside of the process unit, mainly:

(1) Steam turbine extraction or back pressure steam;
(2) Fired furnace heating external plant materials heat;
(3) The cogeneration of heat and power in the conversion link supplies the electricity outside the plant.

(3) Direct loss of energy E_W

(1) Flue gas heat loss includes: furnace combustion flue gas heat loss; chemical incomplete combustion loss; mechanical incomplete combustion loss; furnace atomization steam (calculated based on the enthalpy at the exhaust flue gas temperature); furnace electricity (fan) Effective power and so on.
(2) Waste heat boiler blow down loss, etc., steam transmission leakage loss.
(3) Heat dissipation, including heat dissipation from conversion equipment, hot feed outlet; heat dissipation loss from external heat supply and steam transmission, and heat dissipation loss from the transfer of heated fluid from conversion equipment to process equipment, etc.
(4) Ineffective power energy refers to the ineffective power of the process pump and compressors, Ineffective power for water, compressed air and Nitrogen etc. and the ineffective power of the steam-driven pump (steam reciprocating pump, steam jet pump) (calculated by subtracting the effective power from the total supplied steam energy).

(4) Effective power of non-process fluids E_{UO}

Refers to the effective power such as the pump/compressors that does not enter the process fluid, mainly the effective power of water, compressed air, nitrogen and air-cooled fans, and the effective power of other non-process fluids. This part of the energy has not entered the process utilization link and should be calculated separately.

(5) Effective energy supply E_U

Refers to the effective energy supplied by the conversion and transmission link in the total input energy E_P, including the energy and exergy of the fired heaters, regenerators, pumps, compressors and steam transmission.

(6) Total process energy used E_N

The total process energy used E_N refers to the actual amount of energy used by the system's process equipment (excluding the chemical energy of raw materials) under the operating conditions of the test period, and generally consists of the following three parts:

$$E_N = E_U + E_R + E_{RE}$$

In the formula, E_{RE} is the reaction exotherm. For the endothermic reaction, the heat of reaction has been derived from E_U and E_R, therefore, it is regarded as $E_{RE} = 0$.

The total process exergy used also consists of the above three parts.

(7) Thermodynamic energy consumption E_T

Refers to the energy that is converted into the product from the energy consumed in the process. It is equal to the difference between the energy brought by the product and the energy (including physical energy and chemical energy) brought by the raw material. It is called thermodynamic physical energy difference E_{TT}. In an energy balance system that does not account for the chemical energy of the raw materials, the thermodynamic energy consumption is calculated from the difference in physical energy (exergy) between the product and the raw materials plus the reaction heat (exergy) at the reference temperature of the system, namely: $E_T = E_{TT} + E_{AR} - E_{RE}$, E_{AR} is the endothermic reaction heat. The same with thermodynamic exergy difference D_T.

(8) Energy to be recovered E_O

Except for the thermodynamic energy consumption and heat dissipation in the energy utilization link, the total energy used in the process enters the recycling and recovery link and is regarded as the energy to be recovered.

The energy to be recovered does not include the chemical energy of the output product of the plant.

(9) Energy recovered and recycled E_R

Energy recovered and recycled is the energy recovered from the energy to be recovered E_O and used in the process utilization link of the plant, including the following four items:

(1) The energy in the heat exchange recovery energy used to improve the energy quality of the streams of the process equipment, and it will increase and decrease with the effective energy supply;
(2) The part of the steam generated in the energy recovery link and used in the process utilization link (note that the steam used in the conversion link and the external transmission in the steam generation should be separated);
(3) The hydraulic turbine recovers the power from the high-pressure process stream to drive the energy of the process pump/compressor;
(4) Bring the energy in the form of streams to energy utilization equipment.

(10) Recovery output energy E_E

The energy recovered from the energy to be recovered E_O and used outside supply except the process utilization link of this plant includes:

(1) The hot feed output energy of the plant (the part above the specified temperature);
(2) Exchange heat output energy with other plants;
(3) Preheating furnace air, the energy for steam generated and used for the energy conversion link.

(11) Rejection energy E_J

Refers to the energy to be rejected into the environment in various ways from the energy to be recovered. Including heat dissipation, cooling, streams (condensed water, exhaust steam, etc.) and other waste energy four items.

5.4.2 *Exergy Balance Parameters*

The parameters in the Exergy balance model are also similar to the energy balance, except that the exergy balance takes into account the energy quality factor, and increases the internal process exergy loss in each link.

(1) Exergy loss during the energy conversion process D_{KU}

Including the irreversible loss of the conversion equipment, mainly the irreversible combustion and heat transfer exergy loss of the combustion equipment; the ineffective friction loss of the pump and compressors.

(2) Exergy loss in energy utilization Link D_{KP}

The exergy loss of the process energy utilization link is caused by the irreversible operation of the reaction, fractionation and absorption, and extraction equipment.

(3) Exergy loss in the energy recovery and recycling process D_{KR}

Refers to the irreversible exergy loss of heat exchangers, coolers, hydraulic turbines and heat transfer equipment in the energy recovery link.

5.5 Balance Relationship and Evaluation Index of the Three-Link Model

From the energy balance and exergy balance relationship shown in Figs. 5.4 and 5.5, the balance and restriction relationship can be found. Then, according to the equipment parameter content and the equipment balance result, the plant energy balance and exergy balance results can be easily summarized, which becomes the solid foundation for diagnosing energy use problems.

5.5.1 Energy Balance Relationship and Evaluation Index

(1) The system energy balance relationship is shown in Eq. (5.1), note that $E_T = E_{TT} + E_{AR} - E_{RE}$, then:

$$E_P = (E_W + E_T + E_{UD} + E_J) + (E_E + E_B) \tag{5.2}$$

The second term on the right side of the formula (5.1) is the energy supplied outside from the system, which does not belong to the energy used and consumed by the system. The items in the first brackets on the right side of the formula (5.1) are the energy that enters the product or is lost to the environment in four forms in the three links of the system in the energy use process, which is called net energy consumption, namely:

$$E_A = E_W + E_T + E_{UD} + E_J \tag{5.3}$$

The balance between supplying energy to the system and supplying energy out from the system is as follows:

$$E_A = E_P + E_{RE} - E_B - E_E \tag{5.4}$$

The results of formula (5.2) and formula (5.3) should be the same.

(2) Energy balance relationship of energy conversion and transmission link:

$$E_P = E_B + E_U + E_{UO} + E_W \tag{5.5}$$

The effective supply of energy from the conversion and transmission link is:

$$E_U = E_P - E_B - E_{UO} - E_W \tag{5.6}$$

(3) The energy balance relationship of the energy process utilization link:

$$E_U + E_{RE} + E_R = E_{TT} + E_{UO} + E_O + E_{AR} \tag{5.7}$$

The total process energy consumption of the system is:

$$E_N = E_U + E_R + E_{RE} \tag{5.8}$$

The energy to be recovered can be:

$$E_O = E_N - E_{TT} - E_{UD} - E_{AR} \tag{5.9a}$$

The above formula can also be written as:

$$E_O = E_N - E_T - E_{UD} - E_{RE} \tag{5.9b}$$

(4) The energy balance relationship of the energy recovery link

$$E_O + E_{UO} = E_J + E_E + E_R \tag{5.10}$$

The energy to be recovered can be:

$$E_O = E_J + E_E + E_R - E_{UO} \tag{5.11}$$

The calculation results of Eqs. (5.9) and (5.11) should be the same. Otherwise, the reason should be found and the results need be corrected.

The energy rejected from the energy recovery and recycling link can be balanced according to formula (5.12):

$$E_J = E_O - E_R - E_E + E_{UO} \tag{5.12}$$

Effective power such as water, compressed air, and air-cooled fans are rejected as other rejected energy from energy recovery and the recycling process. The effective power of the pump and compressors can be regarded as other rejected energy items.

The rejected energy from energy recovery and the recycling link is composed of cooling rejected energy E_{JC}, heat radiation rejected energy E_{JD}, stream rejected energy E_{JM} and other rejected energy E_{JO}. Therefore:

$$E_J = E_{JC} + E_{JD} + E_{JM} + E_{JO} \tag{5.13}$$

Generally, the balanced heat dissipation data by the following formula can be used to verify the measured heat dissipation rejected energy from energy recovery and the recycling link:

$$E_{JD} = E_J - E_{JC} - E_{JO} - E_{JM} \tag{5.14}$$

(5) Plant energy consumption E_C

Plant energy consumption can generally be calculated directly from the physical consumption according to its unified conversion index, or it can be summarized and converted according to the energy balance data by the following formula:

$$E_C = E_A + \Delta E_M + E_{IC} \tag{5.15}$$

$$\Delta E_M = \sum (\mp E_i \xi_i) \tag{5.16}$$

ξ_i = Primary energy consumption when producing i energy/(actual enthalpy value) − 1.

(6) Evaluation index

Energy conversion and transmission efficiency

$$\eta_U = (E_U + E_B + E_{UO})/E_P \times 100\% \quad (5.17)$$

Plant total process energy used E_N, ref to Eq. (5.8).
Energy efficiency in process energy utilization link

$$\eta_P = (1 - E_{UD}/E_N) \times 100\% \quad (5.18)$$

Energy recovery efficiency of energy recovery and recycling link

$$\eta_R = (E_R + E_E)/(E_O + E_{UO}) \times 100\% \quad (5.19)$$

Plant energy efficiency

$$\eta_C = (E_U + E_{UO} + E_B)/E_{PC} \times 100\% \quad (5.20)$$

where: E_{PC}—Primary energy consumption for the plant (without deduction of plant output energy), kW or MJ/t;

$$E_{PC} = E_P + \Delta E_M$$

ΔE_M—Energy consumption correction item, kW or MJ/t;

E_{IC}—Chemical energy of raw materials at reference temperature, kW or MJ/t;

$E_i i$—Energy and energy-carrying working medium supply energy, kW or MJ/t;

ξ_i—Conversion factor for i types of energy and energy-carrying working fluid.

5.5.2 Exergy Balance Relationship and Its Evaluation Index

(1) System exergy balance relationship, noting $D_T = EX_{TT} + EX_{AR} - EX_{RE}$ then:

$$EX_P = D_T + (EX_E + EX_B) + (D_{JU} + D_{JP} + D_{JR}) + (D_{KU} + D_{KP} + D_{KR}) \quad (5.21)$$

On the right side of the formula (5.21), the first term is the thermodynamic exergy consumption, which is the theoretical minimum exergy consumption of the system; the second term is the exergy output and transfer out of the system, which does not belong to the exergy consumption of the system; the third term is the three links rejecting exergy lost to the environment, it is called rejection exergy

$$D_J = (D_{JU} + D_{JP} + D_{JR}) \tag{5.22}$$

The fourth term on the right side of the formula (5.21) is the process exergy loss caused by the irreversibility of the three links, which is called the process exergy loss of the system:

$$D_K = D_{KU} + D_{KP} + D_{KR} \tag{5.23}$$

The system process exergy loss, rejecting exergy and thermodynamic exergy consumption can be grouped together, called the net exergy consumption of the process:

$$D_A = D_K + D_J + D_T \tag{5.24a}$$

or

$$D_A = EX_P - EX_E - EX_B \tag{5.24b}$$

(2) Exergy balance relationship of energy conversion link

$$EX_P = EX_B + EX_U + EX_{UO} + D_{JU} + D_{KU} \tag{5.25}$$

Provided effective Exergy is:

$$EX_U = EX_P - EX_B - EX_{UO} - D_{JU} - D_{KU} \tag{5.26}$$

(3) Exergy balance relationship of process energy utilization link

$$EX_U + EX_R + EX_{RE} = D_{JP} + EX_{TT} + EX_{AR} + EX_O + D_{KP} \tag{5.27}$$

The total exergy value supplied to process energy utilization link (that is, the total exergy used in the process)

$$EX_N = EX_U + EX_R + EX_{RE} \tag{5.28}$$

The Exergy to be recovered can be calculated from the positive sum, or can be obtained from the Eq. (5.27) balance:

$$EX_O = EX_N - EX_{TT} - EX_{AR} - D_{KP} - D_{JP} \tag{5.29a}$$

The following formula can also be used to determine the exergy to be recycled

$$EX_O = EX_N - D_T - D_{KP} - D_{JP} - EX_{RE} \tag{5.29b}$$

(4) Energy recovery link exergy balance relationship

$$EX_O + EX_{UO} = EX_E + D_{JR} + EX_R - D_{KR} \tag{5.30}$$

The exergy to be recovered can be obtained from the positive sum of the exporter, and match the result of formula (5.29a, 5.29b).

$$EX_O = D_{JR} + EX_E + EX_R + D_{KR} - EX_{UO} \tag{5.31}$$

Exergy rejected in the energy recovery and recycling link is:

$$D_{JR} = EX_O - EX_E - EX_R - D_{KR} + EX_{UO} \tag{5.32}$$

In exergy balance calculation, due to the calculation accuracy and calculation error of the data, it is impossible to achieve a complete "balance". Therefore, under the premise of reasonable data, a small amount of imbalance is allowed. Generally, it does not exceed 2% of the net exergy consumption of the plant, otherwise it should be further reviewed and corrected until it meets the requirements. The unbalanced item can appear alone, or it can be put together in rejected exergy.

(5) Plant exergy consumption, which reflects the quality of the plant energy consumption

$$D_C = D_A + \Delta EX_M + EX_{IC} \tag{5.33}$$

$$\Delta EX_M = \sum (\mp EX_i \xi_{xi}) \tag{5.34}$$

where: ξ_{xi}—premiary exergy consumption when producing i type of energy/actual exergy value − 1.

EX_{IC}—For industries where the chemical energy of raw materials is counted as the energy consumption of the plant (such as the chemical exergy of the chemical fertilizer industry), if the chemical energy of the raw materials is not counted, $EX_{IC} = 0$.

EX_i—i energy supplied exergy;

ΔEX_M—Correction factor for primary exergy consumption of the plant;

$\xi_{xi} i$— energy exergy consumption conversion factor.

(6) Evaluation index

Exergy efficiency of conversion and transmission link

$$\begin{aligned}\eta_{XU} &= \frac{EX_U + EX_B + EX_{UO}}{EX_P} \times 100\% \\ &= \left(1 - \frac{D_{JU} + D_{KU}}{EX_P}\right) \times 100\%\end{aligned} \tag{5.35}$$

Exergy efficiency of process energy utilization link

$$\begin{aligned}\eta_{XP} &= \frac{EX_O + D_T}{EX_N} \times 100\% \\ &= \left(1 - \frac{D_{KP} + D_{JP}}{EX_N}\right) \times 100\%\end{aligned} \tag{5.36}$$

The exergy efficiency of the energy process utilization link is an important evaluation index. The quantity of energy in the process utilization link remains conserved, but the quality is reduced. The exergy efficiency reflects the degree of energy degradation and external dissipation during the energy utilization process. In addition, the total process energy used in the process utilization link EX_N is also a very important evaluation index. It is the amount of exergy used in the process energy utilization link and reflects the grade and quality of the total energy used.

Energy recovery efficiency in Energy recovery and recycling link

$$\eta_{XR} = \frac{EX_R + EX_E}{EX_O + EX_{UO}} \times 100\% \tag{5.37}$$

5.5.3 *Energy Balance and Exergy Balance Result (Diagram)*

(1) The energy balance results can be expressed in the form of energy flow diagram, so as to visually, intuitively and clearly show the ins and outs of energy changes, which is convenient for evaluation, analysis and improvement. The drawing method of energy flow diagram is as follows:

1. According to the result of energy balance and the changing law of energy in the three links
2. Determine the energy flow width according to the numerical value;
3. According to the top-down representation of the input, discharge and circulation of energy flow, etc.;
4. Mark the symbols and values of each energy flow in the energy flow diagram, and indicate the unit of energy under the diagram;
5. Indicate energy evaluation indicators under the energy flow diagram to give a complete and clear concept.

   The energy flow diagram mode is shown in Fig. 5.6.

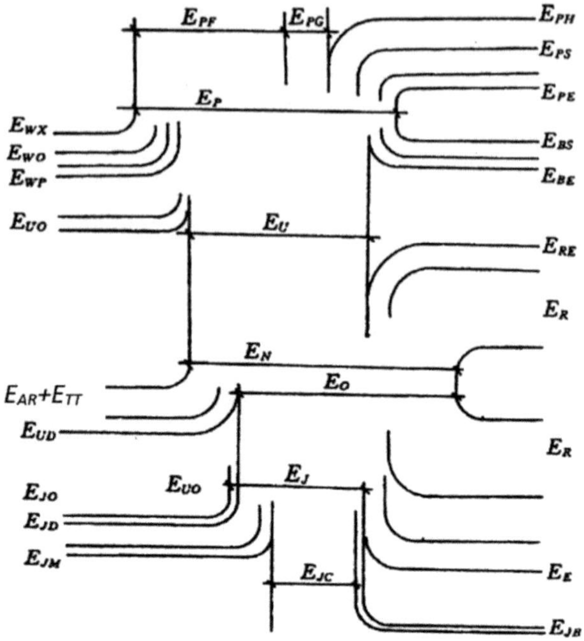

Fig. 5.6 Plant/enterprise energy flow diagram. $\eta_U = \%$; $\eta_R = \%$; $\eta_C = \%$; $E_N = MJ/t$; $E_c = MJ/t$

(2) Like the energy balance, the exergy balance result of the production plant can also be represented by the exergy flow diagram. The drawing method of the exergy flow diagram is the same as that of the energy flow diagram, except that the internal irreversible exergy loss item is added. The basic mode of the exergy flow diagram is shown in Fig. 5.7.

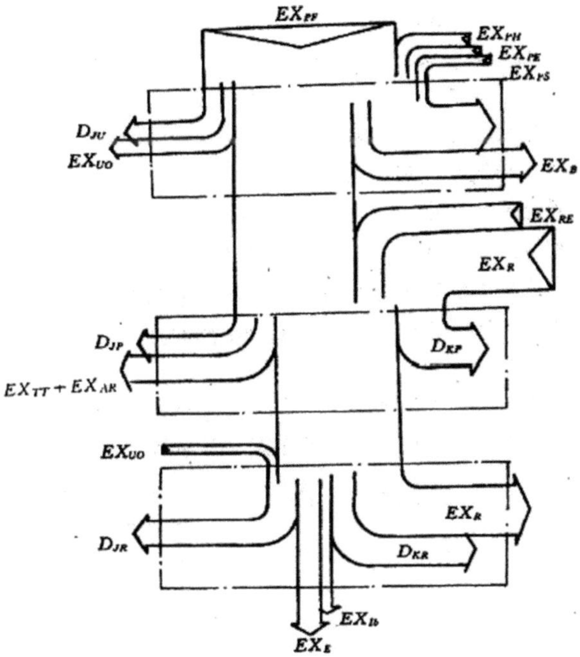

Fig. 5.7 Plant exergy flow diagram. $\eta_{XU} = \%$; $\eta_{XP} = \%$; $\eta_{XR} = \%$; $D_c = MJ/t$

References

1. H. Yaowen, *Thermodynamic Analysis of Chemical Process* (Education Committee of China Chemical Industry Society, New Chemical Technology, 1982)
2. H. Ben, *Analysis and Synthesis of Energy Consumption in Process* (Hydrocarbon Processing Press, 1989), p. 1
3. C. Anmin, Several improvements of the three-link model of energy consumption analysis in petrochemical process. Petrochemical (2) (1990)
4. H. Ben, Theoretical basis and practical analysis of energy consumption in oil refining process. Petrol. Refining (2) (1982)

Chapter 6
Equipment Energy Balance and Exergy Balance

Abstract Equipment is the basis of the petrochemical production plant. The energy balance and exergy balance testing and analysis of the plant usually start from the equipment. This chapter discusses the Equipment Energy Balance and Exergy Balance for the main equipment used in chemical plants, petrochemical plants, and oil refineries, including the content and requirements of the plant energy analysis test trial, energy and exergy balance for pumps, compressors, furnace and fired heaters, FCCU regenerator, column and reactor process equipment, heat exchanger and networks, power recovery equipment, etc.

Keywords Energy test-trial · Pump and compressors · Fired heaters · FCCU regenerator · Process equipment · Energy recovery

Equipment is the basis of petrochemical production plant. The energy balance and exergy balance testing and analysis of the plant usually start from the equipment. On the basis of equipment energy balance and exergy balance test, the three-link model of energy consumption analysis proposed in Chap. 5 is used to summarize, analyze and synthesize from the local to the overall (plant), and then look at the local energy consumption reasonability from the overall perspective. It is the main purpose of energy consumption analysis to make a practical evaluation, identify the energy utilization issues and opportunities, and point out the direction and measures to be taken for energy saving and tapping potential. So first introduce the energy balance and exergy balance of the equipment.

6.1 The Content and Requirements of the Plant Energy Analysis Test

6.1.1 Determination of Test Conditions and Test Scope

The investigation, testing, statistics, and accounting of energy consumption data for process equipment and utility systems for process energy balance and exergy balance

T. A. Chen, *Energy Saving and Carbon Reduction*,
https://doi.org/10.1007/978-981-19-5295-1_6

is called energy consumption analysis and verification. It contains two levels of content: energy use data measurement and data verification processing. The main means of checking is energy balance. Therefore, the data verification of energy-using equipment is completed, and the energy balance of energy-using equipment is basically completed.

To perform energy consumption analysis, first determine the working conditions of the test. Different petrochemical plants have different production load rates in different periods due to different transportation and production plans. Even if it is operated under a stable load rate, the operation plan could be different. Therefore, a production condition used for energy consumption analysis and evaluation should be selected, which is called a test condition. The selection of test conditions should meet the following conditions

(1) The processing capacity of the plant should meet the material balance of the whole plant throughout the year or on a regular basis as much as possible;
(2) The operation plan should be representative, that is, the production plan with the longest application throughout the year should be selected;
(3) The operation requires stable operation, it can represent the operation level of the plant, and has a certain degree of advancement.

For the oil refining industry, due to the large number of production units, it is difficult to conduct plant tests at the same time. The material balance may also be inconsistent with the material balance of the entire plant. On the other hand, there is also a difference between the energy consumption of the plant test and the statistical energy consumption of the testing period.

Generally speaking, energy consumption statistics are the final and statistical average results of cumulative consumption (year, quarter, month) within a period. It is a macroscopic statistical result, which does not reflect the energy use problems of internal equipment and among equipment under specific conditions during the energy use process. It is difficult to "prescribe the right medicine" to find the potential for improvement. The plant energy consumption analysis and verification are a concrete reflection of the instantaneous energy consumption process under a certain test condition during the test period. The energy balance and exergy balance carried out on this basis can reflect the specific problems in the energy use of the process and equipment, reveal the specific parts and reasons for the unreasonable energy use, and point out the potential for energy-saving improvement.

The energy consumption results of the plant proposed by the energy balance should be exactly the same as the statistical results of the test period, but the energy balance can analyze the changing law and results of the energy use of the process in a deeper and more detailed manner. It is obviously different from the average statistical result of a certain period. This is mainly due to the difference between the raw material variety, production plan, load rate and other factors and the test period. It is usually called additional energy consumption. The specific value can be estimated through analysis.

The test scope should be all equipment involved in energy use and change in the plant, that is, the equipment in the three links of energy utilization should be measured and verified. It could be classified as pumps, compressors, heating furnaces, regenerators, turbines, towers, reactors and heat exchange equipment, etc. This is because the purpose of energy analysis and testing is to dissect the law of energy change, energy loss and dissipated parts in the process. Therefore, the content of the test is not limited to energy conversion and heat exchange equipment. All three-link equipment involved in energy analysis should be tested. For example, process equipment (towers, reactors), because the process focuses mostly on energy balance, when heat dissipation is neglected, the energy quantity of inlets and outlets is balanced with energy efficiency of 100%. In the past, testing and analysis were not usually done, but the process equipment must be tested to dissect the causes and locations of the exergy loss during the process.

After establishing the concept of exergy analysis, it is not difficult to find that the core process of energy use is process equipment. Although the energy quantity is conserved, the energy quality is different. This can also be seen from the energy parameters (T, p) entering and leaving the equipment. For physical processes and endothermic reaction processes, the temperature and pressure of the product leaving the equipment are mostly lower than the temperature and pressure of the entering equipment. Even for an exothermic reaction, although the exit temperature of the product may be higher than the temperature of the reactant at the inlet, the molecular structure and quantity of the raw material and product have been changed, and the exergy of the outlet product is less than the exergy of the supplied exergy. Therefore, it is necessary to test and verify the energy consumption of the process equipment in order to obtain a comprehensive energy consumption analysis result, and make a practical diagnosis and evaluation of the plant.

6.1.2 Testing Trial Requirements [1]

The test trial period—the company's energy balance test trial is required to be completed within two months. During the test period, the trial tests each unit are required to be carried out simultaneously, and thc oil refinery requires that the production load of each unit during the test period be stable and as close to the same time as possible.

The test data should include the temperature, flow, pressure and corresponding physical properties and composition analysis data of each part of the equipment. The number of test record groups should not be less than four. Try to use online instruments for testing and company operating database, and at the same time, install necessary various energy measuring instruments as needed, and be equipped with some portable test instruments. All test instruments must be calibrated to indicate the error range, and the measurement accuracy shall meet the specified requirements, and the liquid stream flow rates measurement data shall be checked against the tank gauge data and meet the material balance of the plant.

The laboratory analysis data required for fuel, raw materials, products, flue gas and other energy balance should be complete and accurate, and it should be carried out simultaneously with energy balance. Each analytical test item shall not be less than two sets of data, and the average value can generally be taken.

The test trail data should be analyzed and verified to meet the thermal balance of the test equipment and system. The data should be selected reasonably, and the average value should be generally used. The original data should be fully and accurately recorded and loaded into the test trail file.

6.2 Pump and Compressor Equipment

The Pump and Compressor are important process equipment to realize the transportation of petrochemical fluids. Its function is to increase the pressure of the fluid to enable the fluid to obtain flow work, so as to drive the fluid to flow or to ensure that the fluid is operated at the required pressure level. Pump and Compressor can be divided into two categories: pump (liquid), fan and compressor (gas) according to the state of the medium. According to the driving mode of the prime mover, they can be divided into two categories: electric and steam. In addition, there is a class of power-consuming equipment that does not drive process fluids, such as electrical desalination, high-voltage static electric field power consumption in the electro-refining process, electrical dust removal, rotating disk extraction towers, and lighting are called process power demand. In this regard, a specific analysis can be made, which is generally included in the energy balance of the process equipment, and the power consumption is treated as a supply.

6.2.1 Centrifugal Pump

In petrochemical plants, feed materials, products, intermediate products, the reflux of tower equipment for heat removal, solvents and utilities (cooling water) and other liquid materials are transported and boosted by pumps. Although there are many types of pumps, the most commonly used is the centrifugal pump, which constitutes more than 80% of the pumps used in the plant, and most of them are driven by electric motors. For large pumps, it is also driven by steam turbines.

Before calculating the pump, the energy parameters of the pump should be tested, mainly: stream flow rates, properties (relative density, viscosity, etc.), the pressure entering and exiting the pump, and the Current, Voltage, power factor of the motor, and the model and rated parameters of the pump and motor should be collected.

6.2.1.1 Pump's Energy Supply

The pump and the motor should be considered as a unit, so the supplied energy (power consumption) can be calculated from the measured current, voltage and the verified power factor:

$$N_{cons} = \sqrt{3}.U.I.cos\varphi \times 10^{-3} \quad (6.1)$$

where: U—Motor test voltage, V;

I—Motor test current, A;

$cos\varphi$—Measured power factor of the motor;

N_{cons}—Power consumption of the motor, kW.

The power factor of the motor is a key parameter in the calculation of the motor pump, and the rated power factor of the motor should not be blindly applied. Otherwise, the sum of the power consumption of the pumps may be greater than the total power consumption of the plant. Taking into account the transmission line loss of the plant's transformation and distribution, the reading of the plant's electricity meter should be slightly greater than the sum of the power consumption of all electrical equipment.

In actual operation, most of the pumps are not operated under the rated load, and some motors have a very low load rate, and the motor $cos\varphi$ is greatly reduced as the motor load rate decreases, and the rated parameters (load rate of 1) cannot be applied (Table 6.1). The efficiency of the motor does not change much, even when the load rate is as low as 50%, the efficiency of the motor is only reduced by about 1% (see Table 6.2). The relationship between motor load rate and motor power factor is shown in Table 6.1. When performing energy analysis and calculation, it can also be checked by load factor according to the rated power factor of different motors [2]. Many companies and energy monitoring units have electrical power testers that can directly measure current, voltage, power factor and power.

6.2.1.2 Effective Work of the Pump

The function of pumps (including compressors) is to increase the flow work (pressure energy) of the fluid. The magnitude of the flow work is marked by the increase in pressure after the fluid passes through the pump and is obtained by the fluid. The purpose of the pump energy balance is to investigate its efficiency and determine the effective work and ineffective losses. Therefore, the effective work of the pump refers to the energy consumed by the pump unit and transferred to the fluid through the pump. The pump efficiency is usually calculated with the pump outlet (before the regulating valve) as the boundary, which is reasonable for the conversion equipment from electric energy to flow work. The large amount of throttling loss that we often encounter in the pump outlet regulating valve is a problem in the process energy use

Table 6.1 Motor loading versus power factor

Loading	0.25	0.5	0.75	1	Loading	0.25	0.5	0.75	1
Power factor, **cosφ**	0.62	0.85	0.90	0.92	Power factor, **cosφ**	0.55	0.67	0 77	0 82
	0.61	0.83	0.89	0.91		0.54	0.66	0.76	0.81
	0.60	0.82	0.88	0.90		0.53	0.65	0.75	0.80
	0.59	0.78	0.85	0.88		0.53	0.63	0.74	0.79
	0.58	0.77	0.84	0.87		0.52	0.61	0.73	0.78
	0.57	0.75	0 83	0.85		0.51	0.59	0.72	0.77
	0.57	0.74	0.82	0.85		0.50	0.57	0.71	0.76
	0.56	0.71	0.81	0.84		0.50	0.56	0.70	0.75
	0.55	0.70	0.80	0.83					

Table 6.2 Motor loading versus motor efficiency

loading	0.25	0.5	0.75	1	loading	0.25	0.5	0.75	1
Efficiency	0.87	0.935	0.945	0.95	Efficiency	0.78	0.85	0.86	0.86
	0.86	0.93	0.94	0.94		0.775	0.84	0.85	0.85
	0.85	0.92	0.93	0.93		0.77	0.83	0.84	0.84
	0.84	0.91	0.92	0.92		0.76	0.82	0.83	0.83
	0.83	0.90	0.91	0.91		0.75	0.81	0.82	0.82
	0.82	0.89	0.90	0.90		0.74	0.80	0.81	0.81
	0.81	0.88	0.89	0.89		0.73	0.79	0.80	0.80
	0.80	0.87	0.88	0.88		0.72	0.78	0.79	0.79
	0.79	0.86	0.87	0.87		0.71	0.77	0.78	0.78

link. Together with the pressure drop of the pipeline, equipment, regulating valve, and metering instrument, it constitutes the power process exergy loss, which is an improvement in the power process (fluid transportation), the potential of energy use is not part of the energy conversion link.

Effective work of pump:

$$N_e = Q\Delta p/3.67 \tag{6.2}$$

where: N_e—Effective work of pump, kW;

Q—Pump media flow, m^3/h;

Δp—Pump inlet and outlet pressure difference, MPa.

6.2.1.3 Total Pump Efficiency

For the calculation of pump energy balance for the purpose of energy consumption analysis, it is not necessary to distinguish the efficiency of the pump and the motor. The ratio of the effective work of the pump to the power consumption of the motor is often called the total efficiency (unit). In fact, this is the product of pump efficiency and motor efficiency, which not only meets the needs of energy balance, but also avoids the complicated calculation of the pump performance curve.

$$\eta_{total} = \frac{N_e}{N_{Cons}} \times 100\% \tag{6.3}$$

where: η_{total}—pump efficiency;

N_{Cons}—pump power consumption, kW.

6.2.1.4 Ineffective Power of the Pump

Except for the effective work transferred to the fluid, the electricity consumed by the motor is lost due to friction becoming the heat energy during the conversion process, which is called the ineffective power.

There are three causes of ineffective power: the mechanical friction loss consumed by the motor, coupling, and gearbox is finally discharged into the environment by means of heat dissipation and cooling; the internal eddy current impact friction loss of the pump, part of which is lost or converted into heat energy at fluid temperature heat energy; the energy of the low-pressure steam discharged from the steam-driven pump, condensing turbine, and Steam Jet pump is lost in the condenser and coolers (sometimes the condensate and its low-temperature heat can be recovered).

For liquids, the frictional heat energy is either lost from the casing of the pump or enters the streams, and is transformed into heat energy at the fluid temperature. However, because the amount of frictional heat is small, it is not enough to cause a significant change in the temperature of the flow, which is much smaller than the error of the engineering measurement of the flow temperature; In addition, the mechanical pump is based on the flow energy (work) as the object for balance analysis, so it is omitted. Ineffective power is:

$$N_W = N_{Cons} - N_e \tag{6.4}$$

It is worth noting that for electric centrifugal pumps, electricity and effective work belong to exergy, and its energy grade is 1. Therefore, the results of energy balance and exergy balance are the same, but ineffective power is used as exergy loss during the conversion process in exergy balance.

The ineffective power of the centrifugal pump system can be divided into the ineffective power of the pump and the motor. The efficiency of the motor is related to the load rate of the motor. When the load rate is above 0.5, the motor efficiency changes less than 1%. When the load rate is lower than 0.4, the motor efficiency decreases greatly.

The ineffective power of the pump is related to the pump efficiency. The pump efficiency is not only dependent on the type and model of the pump, but also related to the load rate of the pump. According to the characteristic curve of the pump, there is a high-efficiency load zone in the operation of the pump, and when the load deviates from the high-efficiency zone (generally low load), the efficiency drops. The way to improve pump efficiency is that the load rate of the pump reaches the rated flow rate, and the pump is required to run in the high-efficiency area. Therefore, the key to reducing the ineffective power of the pump is to avoid lower loading as much possible, thereby increasing the load rate of the motor and the load rate of the pump.

It is required that in the design and selection, the margin of the pump should be reasonable, and the motor should be matched accordingly. It should be calculated that the normal load rate of the motor is above 70%, and the pump and motor that have been operating at low load for a long time should be replaced.

For pumps whose load rate varies with the production operating conditions, production plan, and load change frequently, consider adopting a speed control device. At present, motor frequency conversion speed control, hydraulic coupler and cascade speed control devices have all been applied examples. According to the relationship between the characteristics of the pump and the speed, the speed control device is adopted to reduce the head of the pump with reducing flow rate, so as to avoid the pump head increasing caused by the decrease of the flow rate, and mitigating the throttling loss.

6.2.1.5 Energy Balance Result of Centrifugal Pump

The energy balance results of the centrifugal pump are summarized in Table 6.3.

Example 5.1 For a atmospheric and vacuum distillation unit, the crude oil pump model is 250YSI-150B, the relative density of the fluid is 0.8861, the flow rate of the pump is 289.6 m^3/h, and the pressure difference of the pump in and out is 13.6 kg/cm^2, the motor model is JK132-2, the current measured is 19.6 A, the voltage is 6000 V, after correction, the power factor is 0.876. Find the effective power, power consumption, ineffective power and total efficiency of the pump.

Solution Motor power consumption:

$$N_{Cons} = \sqrt{3} \times 19.6 \times 6000 \times 0.876 \times 10^{-3} = 178.4 \text{ (kW)}$$

Effective Work: $N_e = 13.6 \times \frac{289.6}{36.7} = 107.3 \text{ (kW)}$

Table 6.3 Pump and compressor efficiency summary table

Name	Pump or Compressor									Prime mover									Unit Efficiency
	Model	Medium	flow rate	Temp.	Pressure, MPa		Effective Work	Efficiency	Shaft work	Model	Voltage	Current	Power factor	Steam Pressure MPa		Flow Rate	Consumed Power	Efficiency	
			t/h	℃	In	Out	kW	%	kW		V	A	$cos\varphi$	In	Outt	/h	kW	%	%

Ineffective power: $N_W = 178.4 - 107.3 = 71.1$ (kW)

Total unit efficiency: $\eta_{total} = \left(\frac{107.3}{178.4}\right) \times 100\% = 60.15\%$

6.2.2 *Compressor*

Compressors are widely used in chemical and petrochemical plants, such as catalytic main fans, hydrogen compressors in ammonia synthesis plants, hydrogen compressors in reforming hydrogenation plants, and air compressors. There are many types of compressors, which can be divided into two categories: volumetric type and speed type according to their working principles. Usually, reciprocating compressors are of volumetric type, while centrifugal compressors are of speed type. These two categories are also most commonly used in petrochemical plants.

6.2.2.1 Electric Motor Driven Compressor

(1) Data collection

Before calculating the compressor, the energy parameters of the compressor should be calibrated, mainly the inlet flow rate, medium properties, atmospheric pressure, compressor inlet and outlet pressure, inlet and outlet temperature, motor current, voltage and power factor, motor and compressor models, specifications and rated parameters.

(2) Power consumption (electricity) of electric compressors

Same as the calculation of electric power consumption of centrifugal pump motor.

(3) Effective energy output of the compressor

Due to the smaller specific heat capacity of the gas, all the shaft work of the gas compression enters the compressed gas stream, which increases the gas pressure and the temperature. Therefore, from the energy balance of the first law, the shaft work of the adiabatic compressor is all considered effective, the mechanical and transmission losses are comparative small.

There are three scenarios in the compressor compression process, isothermal compression, adiabatic compression and variable compression [3]. Isothermal and adiabatic compression is the limit of the actual compression process. The general compression process can be regarded as the variable compression process of an ideal gas. The shaft power of the centrifugal compressor is:

$$N_s = (16.34 p_1 V_1 \left(\frac{m}{m-1}\right)(\varepsilon^{(m-1)/m} - 1))/(\eta_p.\eta_g.\eta_c) \tag{6.5}$$

Where: p_1—Inlet pressure, MPa;

V_1—Volume flow under suction condition, m^3/min;

ε—Compression ratio, $\varepsilon = p_2/p_1$;

p_2—Compressor discharge pressure, MPa;

η_p—Variable efficiency;

η_g—Compression machinery efficiency;

$N > 1000$ kW; $\eta_g = 96\% \sim 98\%$; $N < 1000$ kW, $\eta_g = 94\% \sim 96\%$.

η_c—Transmission efficiency, in direct transmission, $\eta_c = 1.0$, When using a gear box for transmission, $\eta_c = 0.93$–0.98.

The compression process of a multi-stage reciprocating compressor can be regarded as multi-stage variable compression, and its shaft power is:

$$N_S = 16.34 p_1 V_1 \left(\frac{Sm}{Sm - 1} \right) (\varepsilon^{(m-1)/Sm} - 1)/\eta_g . \eta_c \tag{6.6}$$

In the formula, S is the number of compression stages, and other symbols are the same as before.

Theoretically, between the given initial pressure and final pressure, the more compression stages and the closer to isothermal, the lower the power consumption of the compressor. However, the number of stages increases, the structure is complicated, and the equipment cost also increases, so it is generally around three. The mechanical efficiency and transmission efficiency of the reciprocating compressor vary with the size of the compressor and the transmission mode. Generally, large and medium compressors $\eta_g = 0.9 - 0.95$, small compressors $\eta_g = 0.85 - 0.90$; transmission efficiency of belt drive: $\eta_c = 0.96 - 0.99$, gear transmission $\eta_c = 0.97 - 0.99$, direct-coupled $\eta_c = 1.0$.

The above discussion is the calculation of shaft power of ideal gas. For actual gas that cannot be regarded as ideal gas, the shaft power can be multiplied by the correction coefficient of the actual gas [4].

$$\varphi = (Z_1 + Z_2)/2Z_1$$

In the formula, Z_1, Z_2 are the compressibility factors of the compressor in the inlet and outlet states respectively, which can be checked from the manual according to the reduced parameters of the working fluid in the inlet and outlet states. It can also be estimated as follows [5].

$$Z = 1 + \frac{9p_r}{128T_r} - \frac{54p_r}{128T_r^3} \tag{6.7}$$

where: T_r—Reduced temperature, $T_r = T/T_c$;

p_r—Reduced pressure, $p_r = p/p_c$.

Therefore, the critical parameters of the streams can be calculated first, and then the import and export reduced parameter values can be determined from the import and export parameter values of the compressor working fluid, and the Z_1, Z_2 value can be calculated according to the above formula.

(4) Physical exergy supplied by the compressor

For the exergy balance calculation of the second law of thermodynamics, the exergy entering the compressed fluid is different from energy, mainly physical exergy, namely heat exergy and pressure exergy. The reason is that part of the shaft power (the actual power consumption of the compressor) is converted into heat, which increases the stream temperature. Although the energy is not lost, the capability of performing work is obviously reduced. The physical exergy of the compressed stream is the sum of the pressure exergy and the heat exergy.

Pressure exergy

$$N_{e1} = V_1/(60 \times 22.4) R T_0 \ln(\frac{p_2}{p_1}) \tag{6.8}$$

Thermal exergy

$$N_{e2} = \int_{T_1}^{T_2} C_p(1 - \frac{T_0}{T}) V_1 m/(60 \times 22.4) dT$$

When the specific heat capacity can be regarded as a constant

$$N_{e2} = \left(C_p \Delta T - C_p T_0 \ln\frac{T_2}{T_1}\right). V_1 m/(60 \times 22.4) \tag{6.9}$$

where: C_p—Specific heat capacity of streams, kJ/(kg °C);

V_1—compressed stream flow rate, kg/h, m^3/min;

R—Gas constant, 8.314 kJ/(°C kmol).

The total physical exergy is $N_e = N_{e1} + N_{e2}$

(5) Total compressor efficiency

$$\eta = N_s / N_{cons} \times 100\% \tag{6.10}$$

It can be seen that the total energy efficiency of the compressor is approximately the efficiency of the motor. Exergy efficiency:

$$\eta_x = N_e/N_{cons} \times 100\% \tag{6.11}$$

Exergy efficiency and energy efficiency values are different.

(6) Ineffective power

$$N_W = N_{cons} - N_s \tag{6.12}$$

Process exergy loss

$$D_K = N_{cons} - N_e \tag{6.13}$$

The calculation results of compressor energy balance are summarized in Table 6.3.

6.2.2.2 Steam Turbine Driven Compressor

The steam-driven compressor is driven by a back pressure or condensing steam turbine. The calculation methods of compressor shaft power, effective energy and effective exergy are completely the same as those of electric compressors. It's just that the calculation method of energy supply is different. In addition to the relevant content of the compressor, the calibration data also calibrate the inlet and outlet temperature and pressure of the steam turbine to determine the sum of the supplied energy.

(1) Condensing Turbine

The supplied energy and exergy are the energy and exergy supplied by the steam, which can be calculated from the enthalpy and entropy at the corresponding temperature and pressure.

$$E_{in} = (H - H_{15})G_s/3600 \tag{6.14}$$

$$EX_{in} = ((H - H_{15}) - T_0(S - S_{15}))G_s/3600 \tag{6.15}$$

where: G_s—Steam usage, t/h;

H, H_{15}—Respectively are the enthalpy values of the steam turbine inlet state and 15 °C saturated water, kJ/kg;

S, S_{15}—Respectively are the enthalpy values of the steam turbine inlet state and 15 °C saturated water, kJ/(kg K).

Noted that the exhaust steam from steam condensing turbine contains energy, but it will be condensed by cooler, the condensate temperature is usually close to reference temperature. Therefore, this energy is regarded as rejected energy.

The output power of the condensing turbine (when directly connected) is the shaft power of the driven compressor, and the calculation is the same as before. The output power of a condensing turbine is much smaller than the supplied energy (exergy), especially for a low-pressure condensing turbine, and its energy utilization efficiency is low. With the deepening of energy-saving improvements, companies have gradually reduced the use of low-pressure condensing turbines, and replaced them with high-pressure condensing turbines or back pressure turbines. Even high-pressure condensing turbines have far lower efficiency then power generation efficiency of the power plant.

(2) Back pressure steam turbine

The net energy supplied is the energy consumption of the enthalpy difference between the steam entering and exiting the back pressure turbine.

$$E_{in} = (H_1 - H_2)G_s/3600 \tag{6.16}$$

where: H_1, H_2—the enthalpy of the steam entering and exiting the back pressure turbine respectively, kJ/kg;

E_{in}—Net energy consumption of back pressure turbine, kW.

When calculating the energy consumption of petrochemical plants, it is often calculated based on the specified energy consumption or the actual energy consumption of unit high pressure steam or back pressure steam:

$$E'_{in} = (E_1 - E_2)/3600 \tag{6.17}$$

The input exergy value is:

$$EX_{in} = \frac{((H_1 - H_2) - T_0(S_1 - S_2))G_s}{3600} \tag{6.18}$$

where: S_1, S_2—Respectively are the entropy values of the steam entering and exiting the back pressure turbine, kJ/(kg K);

E_1, E_2—Respectively, specified or the actual energy consumption of steam entering and exiting the back pressure turbine, kW;

E'_{in}—The actual energy consumption of the turbine, kW;

EX_{in}—Exergy consumption of turbine, kW.

Back pressure turbine shaft power

According to the first law of thermodynamics, for the adiabatic expansion process ignoring the position head and velocity head, the expansion shaft work of the back pressure turbine is equal to the enthalpy drop of the back pressure steam, that is, the shaft work is equal to the net supply energy, which is calculated by formula (6.16).

The steam entropy value decreases as the steam pressure increases, and the steam entropy value increases with the superheat temperature increases; there is often a feature that the entropy of the inlet steam is smaller than the entropy value of the back pressure steam. The exergy difference is greater than the energy difference of the back pressure turbine. This means that the energy efficiency of the back pressure turbine can be close to 1, while the exergy efficiency is less than 1. The energy consumed by the back pressure turbine is the high-energy part of the steam.

The physical properties (enthalpy, entropy) of the above-mentioned steam can be checked and calculated from the relevant graphs, and can also be calculated according to the calculation method of the physical properties of the steam stream in Chap. 2.

The balance results of the compressor and pump system are summarized in Table 5.3, and the electric motor driven and steam turbines driven compressors and pumps should be filled in separately.

The steam Jet pump can be regarded as a steam driven compressor, which compresses the stream from a vacuum state to atmospheric pressure. The calculation of energy supply is the same as that of a condensing turbine. The effective work can be considered according to the adiabatic compression process, but the efficiency of the compression process is all taken as 1.

Example 5.2 The centrifugal main air compressor of a catalytic cracking unit is

responsible for supplying the coke-burning air to the regenerator. The air volume is 833 m^3/min, the air compressor inlet pressure is atmospheric pressure, the outlet pressure is 0.255 MPa, the outlet temperature is 185 °C, and the motor current is 330 A, the voltage is 6000 V, and the power factor of the motor is 0.88. Find the power consumption, effective work, energy and exergy efficiency of the main air compressor. The average specific heat capacity of air is 1.005 kJ/ (kg K).

Solution

(1) Compressor power consumption

$$\begin{aligned} N_{cons} &= \sqrt{3}.U.I.cos\varphi \times 10^{-3} \\ &= \sqrt{3} \times 330 \times 6000 \times 0.88 \times 10^{-3} \end{aligned}$$

$$= 3017.93\ (\text{kW})$$

(2) The effective output energy of the compressor is the shaft work of the compressor:

$$N_s = (16.34 p_1 V_1 (\frac{m}{m-1})(\varepsilon^{(m-1)/m} - 1))/\eta_p$$

Taking the mechanical efficiency and transmission efficiency as 1, the variable efficiency η_p of the centrifuge at this volume flow rate is 0.78. The variable index is 1.4.

$N_s = (16.34\times 0.1 \times 833_1(\frac{1.4}{1.4-1})(2.55^{(1.4-1)/1.4} - 1))/0.78 = 1872.7$ kW

(3) Compressor output physical exergy

Pressure exergy

$$\begin{aligned} N_{e1} &= \frac{V_1}{(60 \times 22.4)} RT_0 \ln\frac{p_2}{p_1} \\ &= \frac{833}{(60 \times 22.4)} \times 8.314 \times 288.ln2.55 = 1389.2\ (\text{kW}) \end{aligned}$$

Thermal exergy

$$\begin{aligned} N_{e2} &= \frac{V_1 M}{(60 \times 22.4)}\left(C_p \Delta T - C_p T_0 \ln\frac{T_2}{T_1}\right) \\ &= \frac{833 \times 29}{(60 \times 22.4)}\left(1.005 \times (185 - 15) - 1.005 \times 288 \times \ln\frac{458}{288}\right) \\ &= 657.4\ (\text{kW}) \end{aligned}$$

Total physical exergy: $N_e = 1389.2 + 657.4 = 2046.6$ kW

(4) Unit efficiency

Energy efficiency:

$$\eta = \frac{N_s}{N_{cons}} \times 100\% = \frac{1872.7}{3017.93} \times 100\% = 62.5\%$$

Exergy efficiency:

$$\eta_x = \frac{N_e}{N_{in}} \times 100\% = \frac{2046.6}{3017.93} \times 100\% = 67.81\%$$

(5) Process exergy loss

$$D_k = 3017.93 - 2046.6 = 971.33\ (\text{kW})$$

6.3 Industrial Furnace Equipment

Industrial furnace is another basic unit operation process of petrochemical and chemical process. It is the process equipment that transforms chemical energy of fuel into heat energy and transmits it to process streams, including fired heaters, reforming furnaces, cracking furnaces, etc. in petrochemical and chemical process.

6.3.1 Process Fired Heater Energy and Exergy Balance Calculation

The fired heater is a device that provides heat energy to process raw materials in petrochemical plants and oil refining plants. Due to the difference in the process, the parameters (T, p, phase state) of the fired heater are different. Generally speaking, the temperature of the process stream at the furnace outlet is above the bubble point, so the outlet is often in a gas-liquid two-phase state.

There are two methods for determining the energy balance and efficiency of the fired heater: "Forward balance" and "Reverse balance". The calculation of Forward balance requires the calculation of the vaporization rate of the streams based on the principle of phase equilibrium and the calculation of the vaporization rate of the streams to determine the heating load of the gas and liquid phases. The complexity and accuracy of the calculation will also cause large errors due to measurement and chart errors. On the other hand, an important principle of furnace balance is to meet the heat balance of downstream equipment. If the effective heating load of the atm pressure furnace balance in the atmospheric and vacuum equipment cannot meet the heat balance of the atmospheric pressure tower, the calculation results of the furnace must be corrected. Otherwise, the system will be unbalanced, and the excess or insufficient heat will be attributed to heat dissipation, and the heat dissipation has a reasonable range; too large or too small will be inappropriate.

It is recommended that for each heating item that has no phase change or the exact vaporization rate is known, a forward balance calculation can be carried out, and it will be better to cross check with reverse balance. Without knowing the vaporization rate, the total effective load can be calculated from the reverse balance, and other heating loads that are easy to calculate for without phase change are subtracted to determine the heating load with phase change. Finally, it should be included in a furnace-process equipment system to meet its heat balance results.

According to the function of the heating furnace, it can be divided into three sections: fuel combustion, radiation and convective heat transfer. The calculation with energy consumption analysis as the main purpose does not need to go deep into the three sections process calculation, but only calculate the energy and exergy supplied or left the furnace (including loss) from a macro perspective.

1. Energy and exergy of fuel burning

Fired heater fuels are generally divided into liquid and gaseous fuels. The heating furnaces of petrochemical plants rarely use solid fuels. The heat of combustion is the heat released when a substance molecule is oxidized to the final oxidation product, also known as the Heat value. According to the condensation heat of water vapor generated in the combustion products, it can be divided into high heat value and low heat value. The so-called low heat value refers to the complete combustion of the fuel, and the heat released when the water vapor in the combustion product still exists in a gaseous state is called low cheat value. Compared with the high calorific value, the difference is the condensation heat of the generated water vapor, because this part of the condensation heat is the same diffusion, it is difficult to actually use, it is suggested to use lower Heat value for process energy analysis calculation.

The low heat value of general petroleum distillate fuel is:

$$Q_p = \{4725.2 + 470.5K + 63.84API + 0.0667API^2 - 5.5283K.API\} \times 4.1868 \quad (6.19)$$

For heavy oil fuel low heat value:

$$Q_p = (26.37API + 9272.2) \times 4.1868 \quad (6.20)$$

where: API—Oil API degree, It can be calculated by d_4^{20}:

$$API = \frac{141.5}{0.9942 \times d_4^{20} + 0.009181} - 131.5$$

K—Oil UOP characteristic factor;

Q_p—Low calorific value of fuel oil, kJ/kg.

The heat value of gas fuel can also be obtained from the heat value of each single component according to the sum rule:

$$Q_p = \left(\sum_{i=1}^{n} X_i Z_i Q_i\right) / Z_m \quad (6.21)$$

where: X_i—i component mole fraction;

Q_i—i component low calorific value at 15.6 °C at 1 atm, kJ/kg;

Z_i, Z_m—i component and gas mixture compressibility factor. For gaseous fuel that can be regarded as an ideal gas, Z_i and Z_m are regarded as 1.

Correspondingly, the energy supplied to the furnace is:

$$E_1 = FQ_p/3600 \quad (6.22)$$

where: F—Fuel rate, kg/h;

E_1—Fuel energy, Kw.

The chemical exergy of the fuel is the chemical reaction exergy of the fuel, and it is the free enthalpy change before and after the reaction when the diffusion exergy is neglected under isothermal and isostatic conditions at 25 °C.

$$ex = Q_p + T_0 \Delta S \tag{6.23}$$

where: ex—Fuel exergy value, kJ/kg.

In most cases, the entropy change ΔS of fuel combustion is difficult to actually measure, and the fraction of $T_0 \Delta S$ is also very small. Usually, the low heat value of the fuel can be taken as fuel exergy.

For liquid fuels with known elemental composition, when the fuel chemical exergy needs to be accurately calculated, it can be calculated by the following formula:

$$ex = Q_p\left(1.0038 + \frac{0.1365H}{C} + 0.0308\frac{O}{C} + 0.104\frac{\mathrm{S}}{\mathrm{C}}\right) \tag{6.24}$$

And fuel exergy supplied to furnace is:

$$EX_1 = F.ex/3600 \tag{6.25}$$

where: EX_1—fuel exergy supplied to furnace, kW;

ex—Fuel exergy value, kJ/kg;

ΔS—Fuel combustion entropy increase, kJ/(kg K).

2. Sensible heat of fuel

For liquid fuels, when the temperature is higher, it can be calculated according to the fuel oil temperature T, characteristic factor K and relative density, according to the physical energy and exergy of petroleum distillate calculation method. The sensible heat of the fuel close to room temperature is not counted.

$$\begin{aligned} E_2 = {} & F\{1.556 + 8.256 \times 10^{-3} API(t - 15) + 1.99 \\ & \times 10^{-3}(t^2 - 15^2)\} \times (0.0533K + 0.3604)/3600 \end{aligned} \tag{6.26}$$

$$\begin{aligned} EX_2 = {} & F\left\{A_0(T - T_0) + 1.99 \times 10^{-3} API\left(T^2 - T_0^2\right) - A_1 \ln(T/T_0)\right\} \\ & \times (0.0533K + 0.3604)/3600 \end{aligned} \tag{6.27}$$

where: A_0, A_1—constant

$A_0 = 8.256 \times 10^{-3} API - 0.6767$

$A_1 = 135.3 + 2.3778API$

t—Fuel oil temperature, °C;

T_0—Reference temperature, K (Generally take 288 K).

For gaseous fuels, the sensible heat energy E_2 of the fuel can be obtained by summing the enthalpy difference of a single component between the enthalpy at operating temperature and the enthalpy at the reference temperature, and the physical exergy can be obtained from the sensible heat energy of the fuel according to the fuel temperature:

$$EX_2 = E_2\left(1 - \frac{2T_0}{T + T_0}\right) \quad (6.28)$$

3. Atomizing steam brings energy E_3 and exergy EX_3

When the furnace uses liquid fuel, in order to improve the atomization quality of the burner, steam atomization is often required. The energy and exergy of the atomized steam can be calculated by the calculation method of the steam stream energy and exergy in Chap. 2 according to the parameters of the steam.

4. Energy and exergy of combustion air

For the un-preheated combustion air temperature is the ambient temperature, and its physical energy and exergy are close to zero. The diffusion exergy caused by the composition concentration difference between the air and the outlet flue gas is not considered. With the deepening of energy-saving and the energy-saving improvement of furnaces, the combustion air of many large and medium-sized furnaces have been preheated (preheated by external heat sources or preheated by its own flue gas) and contains a certain amount of physical energy and exergy (mainly thermal energy and exergy).

From the specific heat capacity of air and the temperature correlation formula

$$C_p = 0.9956 + 0.000093(T - 273) \quad (6.29)$$

Then the physical energy from air temperature T_0 to T:

$$\begin{aligned} E_4 &= W\int_{T_0}^{T}\{0.9956 + 0.000093(T - 273)\}dT \\ &= W\{0.9702(t - 15) + 4.65 \times 10^{-5}(T^2 - T_0^2)\}/3600 \end{aligned} \quad (6.30)$$

Then the physical exergy from air temperature T_0 to T:

$$EX_4 = W\int_{T_0}^{T} C_p(1 - T_0/T)dT \tag{6.31}$$

$$EX_4 = W\left\{0.9434(T - T_0) + 4.65 \times 10^{-5}\left(T^2 - T_0^2\right) - 279.4.ln(\frac{T}{T_0})\right\}/3600 \tag{6.32}$$

where: W—Air flow rate, kg/h.

5. Electricity consumption of heating furnace $E_5(EX_5)$

In order to improve the efficiency of the heating furnace, many large and medium-sized furnaces are equipped with flue gas waste heat recovery systems, which are mostly used for air preheating, and need to be equipped with daft fans and blowers. The energy consumption can be determined by the calibration data of the furnace fan motor. The energy grade of electricity is 1, so the energy is the same as the exergy value, but it should be noted that the unit should be consistent with other supplied energy.

Total supplied energy

$$E_P = E_1 + E_2 + E_3 + E_4 + E_5 \tag{6.33}$$

Total supplied exergy

$$EX_P = EX_1 + EX_2 + EX_3 + EX_4 + EX_5 \tag{6.34}$$

6. Flue gas energy E_6 and exergy EX_6

Flue gas heat loss should generally include two parts: heat of flue gas and heat carried by atomized steam. The atomized steam is depressurized and heated in the furnace. After the atomization is completed, it is still discharged in the form of steam. Its heat is the steam energy under the exhaust parameters. It should be noted that the energy of the atomized steam is the enthalpy (including Latent heat), then the latent heat should still be calculated in the exhaust flue gas, because the latent heat of this part of the steam is not released in the furnace, and the amount of combustion flue gas is the sum of the amount of air in the combustion chamber and the amount of fuel.

Under the general exhaust composition, the correlation formula of the specific heat capacity of the flue gas is

$$C_{pm} = 1.05 + 1.235 \times 10^{-4}t \tag{6.35}$$

Exhaust flue gas energy and exergy

$$E_6 = \frac{W + F}{3600}\int_{T_0}^{T}\left(1.05 + 1.235 \times 10^{-4}\right)(T - 273)dT + G_s(H - H_{15})$$

$$= \frac{W+F}{3600}\left\{1.05(T-T_0)+0.618\times 10^{-4}\left(T^2-T_0^2\right)\right\}+\frac{G_s}{3600}(H-H_{15}) \tag{6.36}$$

where: H, H_{15}—steam enthalpy value at exhaust gas temperature and the enthalpy value of saturated water at 15 °C

C_{pm}—Specific heat capacity of flue gas, kJ/(kg K);

G_s—Atomized steam consumption, kg/h.

The exhaust flue gas exergy can be calculated according to the simplified formula of heat exergy at variable temperature:

$$EX_6 = E_6\left(1-\frac{2T_0}{T+T_0}\right) \tag{6.37}$$

7. Furnace heat dissipation energy E_7 and exergy EX_7

The surface heat dissipation loss of the heating furnace should be measured and calculated according to the surface temperature and surface area [8]. For actual measurement is not practical, it should be based on the test data in the past or taking 2–4% of the nominal heating load of the furnace. Using the ratio of the heating load under low load rate is not acceptable. The heat dissipation exergy loss is calculated and determined according to the flue gas temperature in the radiation and convection sections. It is also possible to simply take the flue gas temperature at the outlet of the radiant section as the average temperature of the flue gas in the furnace, and calculate the heat dissipation exergy loss from the following formula:

$$EX_7 = E_7\left(1-\frac{T_0}{t_p+273}\right) \tag{6.38}$$

where: t_p—Average temperature of flue gas in furnace, °C.

Direct loss energy:

$$E_W = E_6 + E_7 \tag{6.39}$$

Direct loss exergy:

$$EX_W = EX_6 + EX_7 \tag{6.40}$$

The effective heating load of the furnace can be determined by reverse balance calculation: $E_u = E_P - E_W$. However, the effective heating exergy cannot be directly determined. The total effective heating load can be obtained through the inverse balance and checked with each sub-item heating load to finally determine the heating item with phase change or chemical change. After determining the load of each heating item, the exergy value of each heating item can be determined according to the heat exergy calculation method, thereby determining the total effective heating exergy, and then determining the total process exergy loss and furnace efficiency in the furnace.

In addition to heating the main process streams, the heating furnace generally also superheats steam, preheats raw materials (cold feed), and generates steam. The gas preheating air by its own flue gas is not counted at the entrance and exit.

8. Steam superheating energy E_8 and exergy EX_8

$$E_8 = (2.004 + 0.176p_s)(T_2 - T_1) \times GS/3600 \tag{6.41}$$

$$EX_8 = (2.004 + 0.176p_s)\left(T_2 - T_1 - T_0 ln\frac{T_2}{T_1}\right) \times GS/3600 \tag{6.42}$$

where: p_s—Pressure of superheated steam, MPa;

T_1, T_2—The temperature before and after the steam is superheated, K.

9. Energy E_9 and exergy EX_9 for preheating raw material in convection section

The preheating raw materials (cold feed) are generally petroleum fractions, which can be calculated according to the calculation method of petroleum fraction energy and exergy, and calculated by relative density, characteristic factor and temperature of entering and exiting the furnace. The preheating of raw materials is generally only to improve the thermal efficiency of the furnace, without phase change, only a liquid temperature change process.

$$E_9 = \frac{G_{CF}}{3600} \times \big((1.5566 + 8.256 \times 10^{-3} API\big)(t_2 - t_1) \\ + 1.99 \times 10^{-3}(t_2^2 - t_1^2))(0.0533K + 0.3604) \tag{6.43}$$

$$EX_9 = \frac{G_{CF}}{3600}(A_0(T_2 - T_1) + 1.99 \times 10^{-3}\big(T_2^2 - T_1^2\big) \\ - A_1 ln\frac{T_2}{T_1})(0.0533K + 0.3604) \tag{6.44}$$

where: t_1, t_2—Preheat the temperature of raw materials in and out of the furnace, °C

G_{CF}—Preheating raw material flow, kg/h;

T, T_2—Preheat the temperature of the feed material in and out of the furnace, K.

For preheating other streams, calculate streams energy and exergy according to Chap. 2.

10. Steam generation energy E_{10} and exergy EX_{10}

Some heating furnaces use waste heat boilers to generate steam in order to improve efficiency. At this time, the steam energy and exergy are calculated by the following formula (the steam pressure $\leq$1.0 MPa).

$$E_{10} = \left(2310.4 - 613.4p_s + 511.6p_s^2 - 187p_s^3\right) \times GS/3600 \tag{6.45}$$

$$EX_{10} = E_{10}\left(1 - \frac{T_0}{t_s + 273}\right) \tag{6.46}$$

where: t_s—Saturation temperature of steam, °C;

p_s—Saturated steam pressure of steam, MPa.

When the steam pressure is greater than 1.0 MPa, it can be calculated from the formula (2.98 in Chap. 2).

11. Main heating energy of heating furnace E_{11} and exergy EX_{11}

When the main heating load has a phase change or a chemical change, first obtain the effective energy output E_u from the furnace inverse balance:

$$E_U = E_P - E_W \tag{6.47}$$

The main heating heat load is E_U minus other heating load items:

$$E_{11} = E_U - E_8 - E_9 - E_{10} \tag{6.48}$$

If there are items such as preheating water, subtract the corresponding items;

The heat exergy of the main heating item can be obtained from the arithmetic mean temperature T_m entering and exiting the furnace.

$$EX_{11} = E_{11}\left(1 - \frac{T_0}{T_m}\right) \tag{6.49}$$

At this time, the furnace has heat energy and heat exergy can be calculated by adding and summing

$$E_U = E_8 + E_9 + E_{10} + E_{11} \tag{6.50}$$

$$EX_U = EX_8 + EX_9 + EX_{10} + EX_{11} \tag{6.51}$$

12. Process exergy loss in the heating furnace

After obtaining the effective output energy output and output exergy of the heating furnace from the forward balance, the process exergy loss in the heating furnace can be calculated. From the energy conservation of the first law, the energy supplied is equal to the energy loss and the output energy effectively. From the balanced point of view of the second law, the supplied exergy is greater than the sum of effective exergy gained by streams, flue gas exergy loss and heat dissipation exergy loss. The difference is called exergy loss in the furnace process, which is mainly caused by the irreversible loss of combustion and heat transfer in the heating furnace.

$$EX_L = EX_P - EX_W - EX_U \tag{6.52}$$

13. Furnace efficiency

Furnace efficiency is an important evaluation index that reflects the energy utilization level of heating furnaces. From the results of energy balance and exergy balance, two furnace efficiencies can be proposed- energy efficiency and exergy efficiency. Energy efficiency reflects the loss and utilization relationship in energy quantity, and exergy efficiency reflects the result of energy quality changes, taking into account irreversible factors of the process.

$$\eta = \frac{E_U}{E_P} \times 100\% \tag{6.53}$$

$$\eta_x = \frac{EX_U}{EX_P} \times 100\% = (1 - (EX_L + EX_W)/EX_P \times 100\% \tag{6.54}$$

14. The relationship between energy efficiency and exergy efficiency [6]

The energy supplied to the heating furnace is mainly the fuel energy, and its exergy value is roughly the same as the energy value. The temperature of the fluid to be heated in the process has been determined by the process operating conditions, and the corresponding exergy efficiency is generally determined, but it varies slightly with the difference in thermal efficiency. The exergy efficiency of the heating furnace under different process conditions is different due to the temperature of the heated fluid. When the heat efficiency η of the heating furnace is known and the fuel exergy value is approximately the low heating value of the fuel, the exergy efficiency η_x of the heating furnace can be estimated according to the temperature range change of each heated fluid in the furnace (ε_i is the energy grade of a certain heated fluid).

$$\eta_x = \sum_{i=1}^{n} \frac{Q_i \varepsilon_i}{Q_P} \tag{6.55}$$

If there is only one heated fluid, then:

$$\eta_x = \eta\left(1 - \frac{2T_0}{T_2 + T_1}\right) \tag{6.56}$$

In the process of converting chemical energy into heat energy and transferring it to the heated fluid, exergy loss includes two aspects: one is the flue gas energy loss and heat dissipation loss; the other is the irreversible loss of combustion and heat transfer.

15. Method of calibrating furnace energy balance

Due to the difference in the accuracy of the measured data of the heating furnace, the furnace efficiency and the main heating load obtained from the reverse balance or even the forward balance often have large errors, and it is difficult to reflect the true situation. The main indicator of the correct result is whether it can meet the heat balance between the heating furnace and the process equipment. At this time, the heat balance of the process equipment should be met when the heat dissipation data of the process equipment (associated with the heating furnace) has been tested according to the heat balance of the heating furnace-process equipment system, so as to determine the main heating load of the heating furnace. Analyze and correct factors such as the amount of fuel and the excess air coefficient until the thermal balance of the system is met.

16. Determination of burning exergy loss

We don't calculate the energy and exergy balance of the heating furnace according to the three sections of combustion, radiation, and convection, we only take the heating furnace as a device, and calculated from the supply and output according to the "black box" method, but it is still very necessary to determine the combustion exergy loss, so that we could analyze the changes in the energy consumption process in the furnace, it is necessary to distinguish the exergy loss of combustion and the exergy loss of the heat transfer process in the furnace. Combustion exergy loss can be calculated by formula (6.57). Therefore

$$D_{KC} = \frac{F}{3600} Q_P \frac{T_0}{T_{ad} - T_0} ln \frac{T_{ad}}{T_0}, \text{kW} \tag{6.57}$$

where: T_{ad}—Adiabatic (no heat transfer) combustion temperature, K;

$$T_{ad} = T^0 + \frac{Q_P}{L_0 \alpha C_p} \tag{6.58}$$

where: T^0—Combustion air temperature, K;

In this way, the irreversible heat transfer process exergy loss in the furnace is:

$$D_{KU} = EX_L - D_{KC} \tag{6.59}$$

By distinguishing the process exergy loss in the heating furnace, it is possible to accurately point out the cause of the exergy loss and improvement approaches, identify the energy saving opportunities. For example, the temperature of the preheated air can be increased to increase T_{ad} or decrease α, which can save heating furnace fuel and improve combustion temperature, reduce combustion exergy loss; avoid heating low-temperature streams, and replace heating low-temperature heat sources with other heat sources or energy-carrying working fluids (steam).

Example 5.3 Energy balance and exergy balance calculation example for process heating furnace.

An atmospheric furnace of crude distillation unit of an oil refinery heats the Initial distillation column bottom oil. The initial distillation column bottom oil flow rate is 324.91 t/h, the relative density is 0.9122, temperature is heated from 303 to 366 °C; the furnace fuel is 4th sideline of vacuum column, the fuel rate is 1.97 t/h, its relative density is 0.9256, the temperature is 248 °C; the amount of atomizing steam is 0.608 t /h with the steam pressure of 1.0 MPa and the temperature of 218 °C; the combustion air is preheated to 235 °C with the third sideline of crude tower, and the actual air volume has been verified to be 31.173 t/h; The exhaust flue gas temperature is 144 °C, and the exhaust flue gas volume is 37.8 t/h (minor air leakage identified in the convection chamber). In order to improve the efficiency of the furnace, cold feed, steam superheating and preheating the demineralized water are used. The cold feed is crude oil (the crude oil is preheated by the vacuum furnace), the relative density is $d_4^{20} = 0.9068$, the cold feed volume is 30.121 t/h, and it is heated from 180 to 224 °C. The superheated 1.0 MPa steam is 10.21 t/h, which is superheated from 185 to 218 °C. There are two streams for demineralized water preheating: preheating 4.87 t/h demineralized water heated from 57 to 120 °C, and the other is preheating demineralized water 10.92 t/h from 60 to 155 °C. Try to calculate the energy balance and exergy balance of the heating furnace.

Solution Supply energy

1. Supply energy E_P and exergy EX_P

Combustion heat energy E_1 and exergy EX_1

Furnace fuel is 4th sideline of vacuum column of this plant, $d_4^{20} = 0.9256$, the low heat value of heavy oil can be calculated per Eq. (6.20):

$$Q_p = (26.37API + 9272.2) \times 4.1868$$

Among them

$$API = \frac{141.5}{0.9942d_4^{20} + 0.009181} - 131.5$$
$$= \frac{141.5}{0.9942 \times 0.9256 + 0.009181} - 131.5 = 20.75$$

Then, $Q_p = (26.37 \times 20.75 + 9272.2) \times 4.1868 = 41.111(\mathrm{MJ/kg})$
So, fuel combustion heat energy

$$E_1 = \frac{FQ}{3600} = 1.97 \times 10^{-3} \times 41.11\frac{1}{3600} = 22.497\ (\text{MW})$$

Its exergy $EX_1 = E_1 = 22.497$ (MW)

Sensible heat of fuel energy E_2 and exergy EX_2

$$E_2 = F\{C_3(T - 288) + 1.99 \times 10^{-3}((T - 273)^2 - 225)\} \times C_1/3600$$

among them, $C_3 = 1.556 + 8.256 \times 10^{-3}API = 1.556 + 8.256 \times 10^{-3} \times 20.75 = 1.729$

The characteristic factor correction factor is regarded as 1, namely $C_1 \approx 1$ $E_2 = 1.97 \times 10^3\left\{1.729(248 - 15) + 1.99 \times 10^{-3}\left(248^2 - 225\right)\right\} \times \frac{1}{3600} = 287.04$ (kW)

$$EX_2 = F\{A_0(T - 288) + 1.99 \times 10^{-3}API\left(T^2 - 82944\right) - A_1\ln\left(\frac{T}{288}\right)\}$$
$$\times C_1/3600$$

among them: $A_0 = 8.256 \times 10^{-3}API - 0.6767 = 8.256 \times 10^{-3} \times 20.75 - 0.6767 = -0.5054$

$$A_1 = 135.3 + 2.3778API = 135.3 + 2.3778 \times 20.75 = 184.64$$

Then,

$$EX_2 = 1.97 \times 10^3\left\{-0.5054 \times (248 - 15) + 1.99 \times 10^{-3}\left(521^2 - 82944\right) -184.64\ln\left(\frac{521}{288}\right)\right\} \times \frac{1}{3600} = 80.93\ \text{kW}$$

Atomizing steam brings energy E_3 and exergy EX_3

We can check the steam graph and calculate, here is calculated by the empirical formula.

$$\begin{aligned} E_3 &= \frac{G_s\Delta H}{3600} = \frac{G_s(2417.7 - 87.9 \times p_s + (2.01 + 0.188p_s)t)}{3600} \\ &= 0.608 \times 10^3\frac{(2417.7 - 87.9 \times 1.0 + (2.01 + 0.188 \times 1) \times 218)}{3600} \\ &= 473.9\ (\text{kW}) \end{aligned}$$

$$EX_3 = G_s(\Delta H - T_0\Delta S)/3600$$

Among them:

$$\Delta S = 4.1868ln\{(t_s + 273)/T_0\} + H_R/)(t_s + 273) \\ + (2.004 + 0.176p_s)ln\{t + 273)/(t_s + 273)\}$$

Where the saturation temperature of steam

$$t_s = 179.5p_s^{0.254} = 179.5 \times 1.0^{0.254} = 179.5\ °C$$

Vaporization heat of steam

$$H_R = 2310.4 - 613.4p_s + 511.6p_s^2 - 187p_s^2 = 2011.67\ \frac{kJ}{kg}$$

Then:

$$\Delta S = 4.1868ln\{(179.5 + 273)/288\} + 2011.67/)(179.5 + 273) \\ + (2.004 + 0.176 \times 1)ln\{218 + 273)/(179.5 + 273)\} = 6.5013\ kJ/(kg\ °C)$$

$$EX_3 = \frac{G_s(\Delta H - T_0\Delta S)}{3600} = E_3 - G_s\frac{T_0\Delta S}{3600} \\ = 473.9 - 608 \times 288 \times \frac{6.5013}{3600} = 157.7\ kW$$

Preheating air energy E_4 and exergy EX_4

$$E_4 = \frac{W\Delta H}{3600} = \frac{W\{0.9702(t - 15) + 4.65 \times 10^{-5}(T^2 - 288^2)\}}{3600} \\ = \frac{31173\{0.9702(235 - 15) + 4.65 \times 10^{-5}(508^2 - 288^2)\}}{3600} = 1918.8\ kW$$

$$EX_4 = \mathrm{W}\frac{\{0.9434(\mathrm{T} - \mathrm{T_0}) + 4.65 \times 10^{-5}(T^2 - T_0^2) - 279.4\ln\left(\frac{T}{T_0}\right)\}}{3600} \\ = 31173\frac{\{0.9434(508 - 288) + 4.65 \times 10^{-5}(508^2 - 288^2) - 279.4\ln\left(\frac{508}{288}\right)\}}{3600} \\ = 494.6\ kW$$

Total energy supplied:

$$E_P = E_1 + E_2 + E_3 + E_4 = 25176.7\ kW$$

Total exergy supplied

$$EX_P = EX_1 + EX_2 + EX_3 + EX_4 = 23230.2\ kW$$

2. Direct energy loss E_W and exergy loss EX_W

Direct energy loss of flue gas E_6 及 exergy loss EX_6

$$E_6 = \frac{W + F}{3600}\left\{1.05(T - T_0) + 0.618 \times 10^{-4}\left(T^2 - T_0^2\right)\right\} + \frac{G_s}{3600}(H - H_{15})$$

Among them: fuel rate $F = 1970$ kg/h
Atomized steam rate $G_s = 608$ kg/h
Look up the steam table to know that the steam enthalpy value at the exhaust gas temperature is 2739 (kJ/kg).
15 °C saturated water enthalpy $H_{15} = 62.97$ (kJ/kg)
Then

$$E_6 = \frac{37{,}800 + 1970}{3600}\left(\left(1.05(417 - 288) + 0.618 \times 10^{-4}\left(417^2 - 288^2\right)\right)\right) + \frac{608}{3600}(2739.0 - 62.97) = 2010.4 \text{ kW}$$

$$EX_6 = E_6\left(1 - \frac{2T_0}{T + T_0}\right) = 2010.4 \times \left(1 - \frac{2 \times 288}{417 + 288}\right) = 367.9 \text{ kW}$$

Heat loss energy E_7 及 exergy EX_7

Furnace body heat loss energy $E'_7$and exergy EX'_7, the average temperature of flue gas in the furnace is 422 °C.
Measured: $E'_7 = 824.7$ (kW)

$$EX_7' = E_7' \times \left(1 - \frac{T_0}{t_p + 273}\right) = 824.7 \times \left(1 - \frac{288}{422 + 273}\right) = 438.6 \text{ kW}$$

Oil transfer line Heat dissipation loss energy E_7'' and exergy EX_7''
Measured $E_7'' = 150.56$ (kW)

$$EX_7' = E_7' \times \left(1 - \frac{T_0}{t + 273}\right) = 150.56 \times \left(1 - \frac{288}{366 + 273}\right) = 82.7 \text{ kW}$$

Heat loss energy $E_7 = E'_7 + E"_7 = 842.7 + 150.56 = 975.28$ (kW)
Heat loss exergy $EX_7 = EX_7' + EX_7'' = 438.6 + 82.7 = 521.3$ (kW)
Direct loss of energy $E_W = E_6 + E_7 = 2010.4 + 975.28 = 2985.68$ (kW)
Direct loss of exergy $EX_W = EX_6 + EX_7 = 367.9 + 521.3 = 889.2$ (kW)

3. Effective energy supply and effective supply exergy

Superheated steam energy E_8 and exergy EX_8

$$E_8 = \frac{(2.004 + 0.176p_s)(T_2 - T_1)G_S}{3600}$$
$$= \frac{(2.004 + 0.176 \times 1.0)(218 - 185) \times 10210}{3600} = 205.8 \text{ kW}$$

$$EX_8 = \frac{(2.004 + 0.176p_s)(T_2 - T_1 - T_0 \ln(T_2/T_1)G_S}{3600}$$
$$= \frac{(2.004 + 0.176 \times 1.0)(218 - 185 \ln(491/458) \times 10210}{3600} = 80.56 \text{ kW}$$

Cold feed energy E_9 and exergy EX_9

Cold feed (crude oil) rate 30,121 kg /h

$$E_9 = \frac{G_{CF}}{3600} \times C_3(t_2 - t_1) + 1.99 \times 10^{-3}(t_2^2 - t_1^2))C_1$$

Among them: $C_3 = 1.5566 + 8.256 \times 10^3 API$, $C_1 \approx 1$

$$API = \frac{141.5}{0.9942 \times 0.9068 + 0.009181} - 131.5 = 23.87$$

$$C_3 = 1.5566 + 8.256 \times 10^{-3} \times 23.87 = 1.7537$$

$$E_9 = \frac{30121}{3600}(1.7537(224180) + 1.99 \times 10^{-3}(224^2 180^2) \times 1$$
$$= 941.59 \text{ (kW)}$$

$$EX_9 = \frac{G_{CF}}{3600}\left(A_0(T_2 - T_1) + 1.99 \times 10^{-3}(T_2^2 - T_1^2) - A_1 ln\frac{T_2}{T_1}\right)C_1$$

$$A_0 = 8.256 \times 10^{-3} API - 0.6767 = -0.4796$$

$$A_1 = 135.3 + 2.3778 API = 192.06$$

$$EX_9 = \frac{30121}{3600}\left\{-0.4796(224 - 180) + 1.99 \times 10^{-3}(497^2 453^2) - 192.06 \ln\frac{497}{453}\right\}$$
$$\times 1 = 370.46 \text{ (kW)}$$

Demin. water heating energy E_{10} and exergy EX_{10}

Demin. water heating 1 (E_{10}' 及 exergy EX_{10}')

$$E_{10}^{'} = \frac{WC_p(T_2 - T_1)}{3600} = \frac{4870}{3600} \times 4.1868 \times (120 - 57) = 356.82\ (\text{kW})$$

$$EX_{10}' = E_{10}'(1 - \frac{\mathrm{T}_0}{T_m})$$

Among them: $T_m = \frac{(T_2 - T_1)}{ln\frac{T_2}{T_1}} = (120 - 57)/\ln(393/330) = 360.58\ (\text{K})$

$$EX_{10}' = E_{10}'\left(1 - \frac{T_0}{T_m}\right) = 356.82\left(1 - \frac{288}{360.58}\right) = 71.83\ (\text{kW})$$

Demin. water heating 2 ($E_{10}^{''}$ 及 exergy $\mathrm{EX}_{10}^{''}$)

$$E_{10}^{''} = \frac{WC_p(T_2 - T_1)}{3600} = \frac{10920}{3600} \times 4.1868 \times (155 - 60) = 1206.5\ (\text{kW})$$

$$\begin{aligned} EX_{10} = E_{10}\left(1 - \frac{T_0}{T_m}\right) &= 1206.5\left(1 - \frac{288}{(155 - 60)/ln\left(\frac{428}{333}\right)}\right) \\ &= 288.51\ (\text{kW}) \end{aligned}$$

Demin. water heating energy: $E_{10} = E_{10}' + E_{10}^{''} = 1563.32\ (\text{kW})$
Demin. water heating exergy: $EX_{10} = EX_{10}' + EX_{10}^{''} = 360.34\ (\text{kW})$

Main heating E_{11} and exergy EX_{11}

Here, the main heating load is calculated by using the reverse balance method.

According to energy balance:

$$\begin{aligned} E_{11} = E_P—E_W - (E_8 + E_9 + E_{10}) &= 25176.7 - 2985.68 \\ &- (205.08 + 941.59 + 1563.32) = 19481\ (\text{kW}) \end{aligned}$$

$$EX_{11} = E_{11}\left(1 - \frac{T_0}{T_m}\right) = 19481\left(1 - \frac{288}{(366 - 303)/ln(639/576)}\right) = 10{,}237.3\ (\text{kW})$$

Subtotal: $E_U = E_P - E_W = 22190.8\ (\text{kW})$

$$EX_U = EX_8 + EX_9 + EX_{10} + EX_{11} = 11{,}048.7\ (\text{kW})$$

4. Process exergy loss EX_L.

The energy in and out of the process is equal, so there is $E_P = E_U + E_W$

However, the process exergy varies in and out, and there is process exergy loss, namely

$$EX_P = EX_U + EX_W + EX_L$$

$$EX_L = EX_P - EX_U - EX_W = 23230.2 - 11{,}048.7 - 889.2$$
$$= 11{,}292.3\ \text{(kW)}$$

5. Furnace efficiency

Furnace energy efficiency

$$\eta = \frac{E_U}{E_P} \times 100\% = \frac{22{,}190.8}{25{,}176.7} \times 100\% = 88.14\%$$

Furnace exergy efficiency

$$\eta_x = \frac{EX_U}{EX_P} \times 100\% = \frac{11{,}048.7}{23{,}230.2} \times 100\% = 47.56\%$$

The energy and exergy balance results of industrial furnaces can be summarized in Table 6.4.

6.3.2 Reactor and Tail Gas Incinerator

Reaction furnaces such as hydrogen production reformers, cracking furnaces, etc., in principle, can use the same method as the heating furnace for energy and exergy balance. The difference is that these furnaces not only transfer heat from the combustion gas to the process stream, but also conduct chemical reactions in the furnace tubes while transferring heat. At this time, the heating and reaction heat of the process stream in the furnace tube should be determined in detail according to the relevant process calculation method. Please note that the flue gas waste heat recovery should be in the furnace system. However, the thermal energy recovery (conversion gas, etc.) of the process reactants is classified as the energy recovery link according to the three-link analysis method of energy use, so this is not considered.

In order to eliminate the impact of harmful components in the tail gas discharged from the plant, many environmental protection devices are equipped with incinerators (such as sulfur tail gas incinerators, asphalt exhaust incinerators).

The purpose of the incinerator is not to supply heat, but for environmental protection purposes, so that the furnace can reach a given incineration condition and burn off harmful components in the exhaust gas. The tail gas after incineration is discharged out of the furnace together with the flue gas. As far as heating is concerned, it seems that the thermal efficiency is zero. But the purpose of the incinerator is incineration. All the energy required to reach the required incineration conditions should be considered effective energy. According to this statement, except the heat loss of the furnace body, the flue gas energy is effective. Because the incineration gas heat rate is generally small, the recovery is restricted by the technology and economy, and generally not recovered. For large-scale incinerators, incineration gas can still be

Table 6.4 Industrial furnace energy balance sheet

Description		Flow rate	Temperature	Energy	Exergy	Note
		t/h	°C	kW	kW	
Supplied energy	Fuel chemical energy					
	Sensible heat of fuel					
	Electricity					
	Steam					
	Combustion air					
	Reaction exotherm					
	Total					
Effective provided energy	Reaction endotherm					
	Total					
Direct energy loss	Flue gas exhaust					
	Boiler Blowdown					
	Heat dissipation					
	Ineffective power					
	Discarded slagging					
	Process exergy loss (combustion heat transfer)					
Efficiency, %						

(1) The effective part of electric energy generally enters the flue gas
(2) Atomization and soot blowing steam should be calculated based on its enthalpy value at the flue gas temperature

regarded as the effective part, entering the total energy used in the process (for environmental protection purposes), and the heat recovery is considered in the energy to be recovered link for the recovery of flue gas waste heat. Preheating the combustion air by its own flue gas or changing operating conditions to reduce fuel consumption are measures to reduce the total energy consumption of the process.

6.4 Catalytic Cracking Regenerator

Although the catalytic cracking regenerator is divided into energy conversion equipment according to its function, it has the same functions as the heating furnace, but its energy consumption and characteristics are different from those of the heating furnace, and the energy and exergy balance calculations should be carried out separately.

1. Regenerator coke combustion chemical energy E_1 and exergy EX_1
The fuel of the catalytic cracking regenerator is coke attached to the surface of the catalyst, which is difficult to measure by measuring instruments. The coke burning amount of the regenerator can only be obtained by the calibration data, which is composed of the main air volume and the regeneration flue gas. The amount of coke burning can be calculated by the following simplified formula, and the error is less than 2% [7] ·

$$C_h = V_2\{0.125(CO_2 + CO) + 0.1\}/23 \tag{6.60}$$

where: CO_2, CO—Respectively, the percentages of CO_2 and CO in the regenerated flue gas (dry);

V_2—Air volume for regeneration (including main air, pressurized air, etc.), Nm^3/h;

C_h—Regenerator coke burning rate, kg/h.

The low heat value of coke is the complete combustion heat of coke. The combustion heat released at CO_2 and H_2O (steam) as the final product can be determined by the element composition ratio H/C in the coke.

$$Q_p = \frac{32{,}790 + 120{,}060 \times H/C}{1 + H/C} \tag{6.61}$$

where: $\frac{H}{C}$—Hydrogen to carbon ratio in coke (mass);

Q_p—Coke low heat value (complete combustion), kJ/kg.

The ratio of hydrogen to carbon in coke can also be calculated from flue gas composition data

$$H/C = \frac{8.93 - 0.425(CO_2 + CO) - 0.257CO}{CO_2 + CO} \tag{6.62}$$

Therefore, the coke burning energy of the regenerator is

$$E_1 = C_h.Q_p/3600 \tag{6.63}$$

The chemical exergy of coke is approximately taken as its heat value, so there are:

$$EX_1 \cong E_1 \tag{6.64}$$

where: E_1, EX_1—Respectively, the chemical energy and chemical exergy coke burnt in the regenerator, kW.

2. Regenerative combustion oil energy E_2 and exergy EX_2
   For some light oil feed catalytic cracking units, the heat released from regenerator is insufficient to meet reactor heat demand, and the regenerator is often sprayed with fuel oil. The calculation of fuel injection energy and exergy is the same as that of heating furnace fuel.
3. Regeneration air brought to Regenerator energy E_3 and exergy EX_3

   The catalytic cracking reaction-regeneration system is mostly operated under 0.2–0.5 MPa (absolute) pressure. The coke burning air is provided by the air compressor (main fan). Because of the almost adiabatic compression process of the air, not only the air stream pressure rises, but the temperature of the air also rises, bringing a certain amount of heat energy into the regenerator. The calculation method of the heat energy and heat exergy into the regenerator is the same as the preheated air of the heating furnace).

   Although the main air pressure is low, after the regenerator burns and the exhaust gas temperature reaches 600 °C or more, the pressure energy of the regeneration flue gas can be expanded to recovery the power using flue gas turbines, which is equivalent to a combustion gas cycle process: low-temperature compression, combustion increases temperature, and high-temperature expansion to do work. Moreover, in recent years, flue gas energy recovery units have been rapidly developed, and they have been adopted by many refineries, and have exerted greater energy-saving benefits. Therefore, the pressure exergy should be calculated for the flue gas stream in and out of the regenerator. Because the main air and flue gas can be regarded as ideal gas, the influence of pressure on the enthalpy difference (energy) can be ignored.

   Since the main air is mostly expressed by volume flow, substituting into the air volume specific heat correlation, integrate the heat energy and heat exergy.

   Thermal energy and exergy:

$$E_3 = \frac{V_2}{3600}\{1.254(T - T_0) + 6.0 \times 10^{-5}(T^2 - T_0^2)\} \tag{6.65}$$

$$EX_3 = \frac{V_2}{3600}\{1.2194(T - T_0) + 6.0 \times 10^{-5}\left(T^2 - T_0^2\right) - 361.1ln(\frac{T}{T_0})\} \tag{6.66}$$

   Pressure exergy is:

$$EX_{p3} = \frac{V_2}{3600 \times 22.4} RT_0 ln(\frac{p}{p_0}) \tag{6.67}$$

   where: R—Gas constant, $R = 8.314kJ/(\text{kmolK})$.

   The exergy supplied to the regenerator is the physical exergy of the main air, which is the sum of heat exergy and pressure exergy.
4. Coke brings sensible heat energy E_4 and exergy EX_4

The sensible heat of coke is that the coke is adsorbed on the surface of the catalyst during the cracking process of the catalytic reactor and is brought to the regenerator through the circulation of the catalyst. It is equivalent to the sensible heat of the fuel and can be calculated from the amount of coke burned, the reaction temperature and the specific heat of the coke;

$$E_4 = C_h \times 0.3047 \times 10^{-3} \times (t_R - 15) \tag{6.68}$$

where: t_R—Reactor operating temperature, °C.

Coke sensible heat exergy can be calculated according to formula (2.16).

5. Regenerator supplies steam energy E_5 and exergy EX_5

The energy and exergy calculation of regenerator steam supplied refer to the energy and exergy calculation method of steam stream in Chap. 2.

The supplied energy is the sum of the above 5 items.

$$E_P = E_1 + E_2 + E_3 + E_4 + E_5 \tag{6.69}$$

The supplied exergy is the sum of the above 5 items.

$$EX_P = EX_1 + EX_2 + EX_3 + EX_4 + EX_5 \tag{6.70}$$

6. Flue gas energy loss $E_6$and exergy EX_6

Like the flue gas of the heating furnace, the flue gas loss energy of regenerator should include the energy loss of the regeneration steam used under the flue gas parameters. The flue gas exergy should also include the flue gas pressure energy loss. The flue gas volume is determined by the main air volume V_2 and the flue gas-air ratio β:

$$\beta = \left(\frac{121 - 2(CO_2 + CO + O_2) + 0.79CO}{100 - (CO_2 + CO + O_2)} + e\right)/(1 + e) \tag{6.71}$$

where: β—flue gas to air ratio; CO_2, CO and O_2 are volume percent, %;

e—The molar ratio of water vapor to dry air in the air is checked by temperature and relative humidity. In most cases, it can be 0.02.

The flue gas rate is

$$V_{flue} = V_2.\beta \tag{6.72}$$

Flue gas thermal energy loss:

$$E_6 = \frac{V_{flue}}{3600} \times 1.295\left\{1.05(T - T_0) + 0.618 \times 10^{-4}\left(T^2 - T_0^2\right)\right\}$$

$$+\frac{G_{sg}}{3.6}\left(H_g - H_{15}\right) \tag{6.73}$$

where: G_{sg}—Regenerator steam used, t/h;

H_g—Enthalpy of regeneration steam, kJ/kg.

The exhaust flue gas exergy is the sum of heat exergy and pressure exergy.

The exhaust flue gas heat exergy:

$$EX_6 = E_6\left(1 - \frac{2T_0}{T + T_0}\right) \tag{6.74}$$

The flue gas pressure exergy:

$$EX_{pg} = \frac{V_{flue}}{3600 \times 22.4} RT_0 ln(\frac{p}{p_0}) \tag{6.75}$$

From the flue gas pressure exergy loss, we can see the importance of recovering the flue gas pressure energy, but the flue gas expansion work is much larger than the pressure exergy value. This is because the expansion process can not only recover the pressure exergy, but also part of the enthalpy of the flue gas is converted to work also.

7. Incomplete combustion loss in exhaust flue gas

For the process of incomplete regeneration, every Nm^3 CO in the flue gas oxidizing to CO_2 will release 12. 635 MJ of heat, therefore

$$E_7 = V_2\beta \times CO \times 12.635/3.6 \tag{6.76}$$

In the formula, the concentration of flue gas CO is m^3/m^3.

The CO combustion process in the flue gas is equivalent to fuel combustion. Therefore, the CO chemical energy can be regarded as chemical exergy, $EX_7 \cong E_7$.

8. Heat dissipation loss

Calculate the combined radiation and convection heat dissipation coefficient of the regenerator and its system according to the method in Chap. 2, and calculate the heat dissipation from the heat dissipation surface area and surface temperature of equipment and system:

$$E_8 = F\alpha_T(T_W - T_f)/1000 \tag{6.77}$$

$$\alpha_T = \frac{4.33(T_W^4 - T_f^4) \times 10^{-6}}{T_W - T_f} + 1.47(T_W - T_f)^{1/3} + 4.36\frac{W^{0.6}}{D^{0.4}} \quad (6.78)$$

where: α_T—Combined heat transfer coefficient of radiation and convection, $W/(m^2K)$;

T_W—Surface wall temperature of regenerator system, K;

T_f—Ambient temperature, K;

F—heat dissipation area, m^2;

D—Outer diameter of heat dissipation of equipment, m;

W—Ambient wind speed, m/s.

Heat loss exergy is:

$$EX_8 = E_8(1 - \frac{T_0}{T_g}) \quad (6.79)$$

where: T_g—Operating temperature of regenerator bed, K.

Direct energy loss from the regenerator:

$$E_W = E_6 + E_7 + E_8 \quad (6.80)$$

Direct exergy loss from the regenerator:

$$EX_W = EX_6 + EX_7 + EX_8 \quad (6.81)$$

9. Coke desorption heat.

The coke desorption of the regenerator and the coke adsorption of the reactor are a pair of reverse processes. Generally, the heat of desorption is not considered when the reverse system is used as the heat balance for reaction-regeneration system. In particular, it is customary not to consider this item when calculating the catalyst circulation amount based on the coke energy utilization rate.

However, when the regenerator is used as the system, the heat of desorption is not only the leaving heat, but also considered to be effective energy. This is because the coke is desorbed and burned in the regenerator to ensure the conversion activity of the catalyst in the reactor and make the coke to be adsorbed on the catalyst and releases the heat of adsorption, which contributes to the supply of heat to the reactor.

In the process calculation, it is customary to take the desorption heat as 11.5% of the coke combustion heat, which is suitable for the incomplete combustion process (combustion heat is about 7000 kcal/kg). At present, the regeneration process tends to be completely regenerated, and the combustion heat also varies. Depending on the regeneration temperature, it is more reasonable to take the desorption heat per unit of coke. According to the experimental results [13], the desorption heat of coke is 2220 kJ/kg:

$$E_9 = C_h \times 2220/3600 \tag{6.82}$$

$$EX_9 = E_9\left(1 - \frac{T_0}{T_g}\right) \tag{6.83}$$

10. System generates steam

    There are generally two aspects to steam generation in the regenerator. One is that with the application of complete regeneration technology and the progress of residual oil cracking technology, the coke yield has increased, and a large amount of excess heat removal facilities in the regenerator-reactor system needs to be installed inside or outside the regenerator, these heat removal facilities to take out the excess heat to generate steam;

    The second is the recovery and utilization of flue gas waste heat to generate steam, and in the future, incomplete regeneration could be used to balance the excess heat, and CO will be burned in a CO incinerator on the flue gas duct, and the waste heat boiler will be further recovered to generate steam.

    For calculation of steam energy E_{10} and exergy EX_{10}, please refer to Chap. 2 steam energy and exergy calculation.
11. Steam superheat energy E_{11} and exergy EX_{11}

    For the installation of steam superheating facilities, the calculation method of steam superheating energy and exergy is the same as the calculation method of heating furnace superheating steam energy and exergy.
12. Circulating catalyst brings out net heat E_{12} and exergy EX_{12}

    The effective energy output of the regenerator can be obtained from this balance first:

    $$E_U = E_P - E_W \tag{6.84}$$

    Then subtract several other known loads from the effective energy supply to obtain the energy used for heating the circulating catalyst

    $$E_{12} = E_U - E_9 - E_{10} - E_{11} \tag{6.85}$$

    The exergy carried by the catalyst can be calculated based on

$$EX_{12} = E_{12}\left(1 - \frac{T_0}{T_m}\right) \tag{6.86}$$

$$T_m = \frac{T_g - T_R}{ln(T_g/T_R)}$$

Regenerator's effective heat exergy output:

$$EX_U = EX_9 + EX_{10} + EX_{11} + EX_{12} \tag{6.87}$$

13. Regenerator efficiency

    Like the furnace efficiency, there are the energy efficiency and exergy efficiency of the regenerator.

    The energy efficiency of the regenerator is the ratio of the total effective energy output by the regenerator to the total energy supplied:

$$\eta = \frac{E_U}{E_P} \times 100\% = \left(1 - \frac{E_W}{E_P}\right) \times 100\% \tag{6.88}$$

    The exergy efficiency of the regenerator is the ratio of the total effective exergy supplied by the regenerator to the total exergy supplied:

$$\eta_x = \frac{EX_U}{EX_P} \times 100\% \tag{6.89}$$

    The difference with energy efficiency is that the sum of the effective exergy from regenerator and the direct loss of exergy from regenerator is less than the total exergy supplied. The difference is the process exergy loss in the regenerator.

14. Process exergy loss of regenerator

    The total process exergy loss of the regenerator can be calculated by the following equation:

$$D_{kg} = EX_P - EX_W + EX_U \tag{6.90}$$

    Like the heating furnace, process exergy loss can be further divided into coke combustion exergy loss and heat transfer process exergy loss. Although the coke burning process of the spent catalyst in the regenerator is carried out on the surface of the catalyst, the combustion heat has been transferred to the circulating catalyst at the same time as the coke burning. Because the circulating flow rate and specific heat capacity of the catalyst are both large, the temperature rise of the catalyst (temperature differences between Reactor and regenerator) is not too large.

Combustion exergy loss of regenerator:

$$D_{kgC} = C_h Q_p \left(\frac{T_0}{T_{ad} - T_0} \right) ln(\frac{T_{ad}}{T_0}) \tag{6.91}$$

$$T_{ad} = T^0 + Q_p/(b\alpha C_{pV})$$

where: T_{ad}—Adiabatic combustion temperature of regenerator, K;

T_0—Main air temperature, K;

α—Excess air coefficient

b—main air consumption index, Nm^3/kg;

$$b = \frac{V_1}{C_h} = V_2/(1.02C_h) \tag{6.92}$$

where: V_1—Dry air volume of regenerator, Nm^3/h.

$$\alpha = \frac{100 - (CO_2 + CO + O_2)}{100 - (CO_2 + CO + O_2) - 3.62O_2} \tag{6.93}$$

The corresponding exergy loss in the heat transfer process of regeneration can be obtained by the following equation balance

$$D_{KH} = D_{kg} - D_{kgC} \tag{6.94}$$

Generally speaking, the exergy efficiency of the regenerator is higher than that of the heating furnace due to the higher catalyst recirculation temperature. Ways to reduce exergy loss during regeneration are: reducing coke generation rate to avoid combustion exergy loss, and the heat released is constrained by the heat balance of the rector-regenerator, and the energy of excessive coke generation must be taken out, and the energy grade of recovered heat is low. Under the condition of a certain coke generation rate, increase the energy grade of effective heat output, and the heat utilization is developing toward high temperature. Many catalytic cracking units in oil refinery are using regenerators to generate medium-pressure steam to increase the energy grade of recovered energy; improve the reactor-regenerator system temperature can also reduce exergy loss.

Regenerator energy balance and exergy balance results can be summarized in Table 6.5.

Example 5.4 For a catalytic cracking unit in an oil refinery, the regenerator operating data is: the main air and booster air volume are 62,226 Nm^3/h, and the flue gas composition is CO_2, 14–6%, CO, 1.4%, O_2, 2.0%. The main air temperature

Table 6.5 FCCU regenerator and energy recovery system energy balance sheet

Item		Flow rate	Temperature °C	Pressure M Pa	Energy kW	Exergy kW
Energy supplied	Coke chemical energy					
	Sensible heat of coke					
	Steam					
	Air brought in					
	Burning fuel oil					
	Total					
Effective energy	Coke desorption heat					
	power recovered from flue gas turbine					
	Steam generation					
	Dense bed heat removal					
	Catalyst brings out					
	Total					
Energy loss	Sensible heat of flue gas					
	CO chemical energy					
	Regenerator heat dissipation					
	Recovery system heat dissipation					
	Total					
Process exergy loss (combustion, heat transfer)						
Efficiency, %						

is 185 °C, and the regenerator injects steam 2.6 t/h (0.5 MPa, 300 °C). After the flue gas passes through the waste heat boiler, the exhaust flue gas temperature is 248 °C; and the waste heat boiler produces 13.86 t/h steam with steam temperature of 180 °C, and the steam pressure of 1.0 MPa, the boiler feed water temperature is 114 °C, the steam is superheated to 310 °C, and the reaction temperature is 460 °C, the regeneration temperature is 640 °C, and the regeneration operating pressure is

0.137 MPa (1.37 kg/cm^2). Try to conduct energy balance and exergy balance of the regenerator to determine the energy and exergy efficiency of the regenerator.

Solution

Supplied energy regenerator

1. Coke burning amount of regenerator

$$C_h = V_2 \frac{\{0.125(CO_2 + CO) + 0.1\}}{23} = 62{,}226\{0.125 \times (14.6 + 1.4) + 0.1\}$$
$$= 5.681 \text{ (t/h)}$$

2. Coke combustion heat

(1) Coke H/C ratio

$$H/C = \frac{8.93 - 0.425(CO_2 + O_2) - 0.257CO}{CO_2 + O_2} = 0.0947$$

(2) Coke low heat value

$$Q_p = \frac{32{,}790 + 120{,}060 \times \mathrm{H/C}}{1 + \mathrm{H/C}} = 40{,}339.5 \text{ (kJ/kg)}$$

(3) Coke combustion heat

$$E_1 = \frac{C_h Q_p}{3600} = \frac{5681 \times 40{,}339.5}{3600} = 63{,}658 \text{ (kW)}$$

The chemical exergy of coke is similar to the chemical energy of coke, so there is:

$$EX_1 \cong E_1 = 63{,}658 \text{(kW)}$$

3. No fuel injection $E_2 = EX_2 = O$
4. Main air brings in energy and exergy

$$E_3 = \frac{V_2}{3600}\left\{1.254(T - T_0) + 6.0 \times 10^{-5}\left(T^2 - T_0^2\right)\right\}$$
$$= \frac{62{,}226}{3600}\left\{1.254(185 - 15) + 6.0 \times 10^{-5}\left(458^2 - 288^2\right)\right\} = 3816 \text{ (kW)}$$

The regeneration system does not have a flue gas turbine, but the inlet pressure exergy is (deemed as an ideal gas):

$$EX_{p3} = \frac{V_2}{3600 \times 22.4} RT_0 \ln\left(\frac{p}{p_0}\right) = 1593 \text{ (kW)}$$

5. Sensible heat of coke

$$E_4 = C_h \times 0.3047 \times 10^{-3} \times (t_R - 15)$$
$$= 5681 \times 0.3047 \times 10^{-3} \times (460 - 15) = 770.3\ (\text{kW})$$

$$EX_4 = E_4\left(1 - \frac{T_0}{T_m}\right) = 770.3 \times \left(1 - \frac{288}{476.35}\right) = 304.58\ (\text{kW})$$

$$T_m = \frac{733 - 288}{\ln(\frac{733}{288})} = 476.35$$

6. Steam for regenerator

Check the manual, the enthalpy of 0.5 MPa, 300 °C superheated steam is 3064.2 kJ/kg, the entropy is 7.4606 kJ/(kg °C), and the water entropy at 15 °C is 0.2244 kJ/(kg °C).

$$E_5 = 2600 \times (3064.2 - 62.8)/3600 = 2168\ (\text{kW})$$

$$EX_5 = 2600 \times \{(3064.2 - 62.8) - 288(7.4606 - 0.2244)\}/3600 = 663\ (\text{kW})$$

Total Energy to Regenerator

$$E_P = 63{,}658 + 3816 + 770.3 + 2168 = 70{,}412.3$$

$$EX_P = 63{,}658 + 819.1 + 1593 + 304.58 + 663 = 67{,}037.7$$

7. Exhaust flue gas energy loss

$$\begin{aligned}E_6 &= \frac{V_{\text{flue}}}{3600} \times 1.295\left\{1.05(TT_0) + 0.618 \times 10^{-4}\left(T^2 - T_0^2\right)\right\} \\ &\quad + \frac{G_{sg}}{3.6}\left(H_g - H_{15}\right) = \frac{65{,}341.9}{3600} \\ &\quad \times 1.295\left\{1.05(24{,}815) + 0.618 \times 10^{-4}\left(521^2 - 288^2\right)\right\} \\ &\quad + \frac{2.6}{3.6}(885 - 15) = 8649.7\ \text{kW}\end{aligned}$$

$$EX_6 = E_6\left(1 - \frac{2T_0}{T + T_0}\right) = 8649.7\left(1 - \frac{2 \times 288}{521 + 288}\right) = 2491.2\ (\text{kW})$$

The exhaust flue gas pressure exergy loss, ignoring the regenerator pressure drop, the flue gas pressure exergy loss is the same as the main air pressure exergy:

$$EX_{p6} = 1593\ (\text{kW})$$

8. Waste boiler blowdown loss energy (Blowdown 0.7 t/h)

$$E_7 = 0.7 \times (180 - 15) \times 1.163 = 134.32\ (\text{kW})$$

$$EX_7 = 134.32\left(1 - \frac{2 \times 288}{453 + 288}\right) = 29.91\ (\text{kW})$$

9. Emission of CO chemical energy

$$E_8 = V_{flue} \times \text{CO} \times \frac{12.635}{3.6} = 65{,}341.9 \times 0.014 \times \frac{12.635}{3.6}$$
$$= 3210.64\ (\text{kW})$$

$$EX_8 \cong E_8 = 3210.64\ (\text{kW})$$

10. Heat dissipation loss energy

The heat dissipation loss of the regeneration system has been measured to be 1419.15 (kW)

$$E_9 = 1419.15\ (\text{kW})$$

$$EX_9 = 1419.15 \times \left(1 - \frac{288}{913}\right) = 971.5\ (\text{kW})$$

Regenerator energy loss

$$E_W = 8649.7 + 134.32 + 3210.64 + 1419.5 = 13413.81$$

$$EX_W = 2491.2 + 29.91 + 1593 + 3210.64 + 971.5 = 8566.25$$

11. Coke desorption heat

$$E_{10} = C_h \times \frac{2200}{3600} = 0.617C_h = 0.617 \times 5681 = 3505.2\ (\text{kW})$$

$$EX_{10} = E_{10} \times \left(1 - \frac{288}{913}\right) = 2399.5\ (\text{kW})$$

12. Steam generation and superheating of the waste boiler: 13.86 t/h of steam, all of which are superheated to 310 °C at pressure of 1.0 MPa in the waste heat boiler

$$E_{11} = 13.86 \times (H - H_{15})$$

Check steam table, get the steam enthalpy of 3070.6 kJ/kg, and the steam entropy is 7.1238 kJ/(kg °C).

Water enthalpy and entropy at 15 °C are 62.8 kJ/kg and 0.2244 kJ/(kg °C) separately.

$$E_{11} = 13860 \times (3070.6 - 62.8)/3600 = 11580\ (\text{kW})$$

$$EX_{11} = 13{,}860 \times \{(3070.6 - 62.85) - 288(7.1238 - 0.2244)\}/3600 = 3929\ (\text{kW})$$

13. Circulating catalyst brings out net heat

Effective energy output:

$$E_U = E_P - E_W = 70{,}410.3 - 13{,}413.8 = 56{,}996.5\ (\text{kW})$$

$$E_{12} = E_U - E_{10} - E_{11} = 56996.5 - 3505.2 - 11580 = 41911.3$$

14. Reactor and regenerator average temperature:

$$T_m = \frac{640 + 460}{2} + 273 = 823\ (\text{K})$$

$$EX_{12} = 41911.3 \times \left(1 - \frac{288}{823}\right) = 27{,}244.9\ (\text{kW})$$

Effective heat supply exergy by regenerator:

$$EX_U = 2399.5 + 3929 + 27244.9 = 33{,}573.4\ (\text{kW})$$

Regenerator energy efficiency and exergy efficiency

$$\eta = \frac{E_U}{E_P} \times 100\% = \frac{56{,}996.5}{70{,}410.3} \times 100\% = 80.95\%$$

$$\eta_x = \frac{EX_U}{EX_P} \times 100\% = \frac{33{,}573.4}{67051.7} \times 100\% = 50.07\%$$

15. Process exergy loss

$$D_{kg} = EX_P - EX_W - EX_U = 67{,}051.7 - 8566.25 - 33{,}573.4 = 24{,}912.05 \text{ (kW)}$$

6.5 Process Energy Utilization Equipment

Process energy utilization equipment includes towers (fractionation towers, absorption towers, extraction towers, etc.), reactors, and vessels. Process energy utilization equipment link is the core process that constitutes a process unit or plant. The energy usage of the plant is carried out around this link. In this link, the general energy loss is not large, only the energy loss caused by heat dissipation, so the quantity of supplied energy and discharged energy does not change much, but the quality is greatly reduced. Carrying out the energy balance and exergy balance of this link, it can be found that the real cause of the rear energy loss lies in the energy use link, which can be used to find out the existing energy utilization issues and ways to improve.

According to the energy analysis method proposed in Chap. 5, the energy balance of equipment does not consider the calculation of streams chemical energy and chemical exergy, but only calculates the chemical energy and chemical exergy changes of the materials before and after the process through the heat of reaction. Although the operating pressures of process equipment are different, the pressure difference between entering and leaving the equipment is generally not large. For process equipment dominated by heat energy, pressure energy accounts for a small proportion of total energy, so detailed calculation of pressure energy and exergy changes is meaningless. Therefore, the effect of pressure is not considered in the energy and exergy balance of equipment. The flow energy is considered uniformly in the plant. For the output stream that does have the value of power recovery, the pressure energy and pressure exergy should be calculated and analyzed.

6.5.1 *Column Equipment*

6.5.1.1 Features of Column Equipment

Tower equipment is a kind of mixture separation equipment widely used in petrochemical plants, and its process belongs to the process of physical change, such as fractionation, absorption, extraction, etc., but in terms of its separation principle, it has its own characteristics.

The fractionation process is based on the different boiling points of the mixture components. The light components are continuously heated and vaporized on the fractionation tray, and the heavy components are continuously condensed and cooled to achieve the purpose of separating pure (purer) components.

There are two states of vapor phase and liquid phase on the fractionating tower plate, and the vapor phase temperature under the plate is higher than the liquid phase temperature on the plate. When the vapor and liquid phases contact on the tray, heat transfer occurs. The lower temperature liquid mixture is heated by the higher temperature vapor phase mixture to vaporize the light components in the liquid phase; at the same time, the higher temperature gas mixture after a part of the heat is given, the temperature drops and part of the heavy components condense. Fractionation is performed on the trays of the fractionation tower, and the method of partial vaporization and partial condensation is used many times to carry out the separation operation. The concentration difference and temperature difference between the phases are the driving force and prerequisites for the fractionation.

Gas absorption is another important separation operation, which uses the different solubility of each component in the gas mixture in the solvent to separate the gas mixture. That is, when the mixed gas comes into contact with an appropriate liquid, one or more components in the gas are dissolved in the liquid to form a solution, and the insoluble components remain in the gas phase, so the mixed feed gas enters the two phases of the liquid and gas, as a result, the gas mixture has been being separated. The components that can be dissolved are usually called solutes, the components that are not absorbed are called inert components or carriers, the solvent used in the absorption operation is called the absorbent, and the solution obtained by the absorption operation is called the absorption liquid. In order to enable the absorbent to be reused and to ensure absorption efficiency, it is often necessary to regenerate the solution. According to the solubility curve of the gas, generally speaking, pressurization and cooling are beneficial to the absorption operation and can improve the solubility of the gas. Conversely, heating and decompression are beneficial to the desorption process. Therefore, unlike the fractionation operation, the energy consumption of the absorption operation is mainly the energy consumption of the desorption (regeneration) operation.

The absorption process is divided into two categories: physical absorption and chemical absorption. The so-called chemical absorption is an absorption operation accompanied by a chemical reaction in the absorption process. For example, in a gas desulfurization device in a refinery, ethanolamine absorbs H_2S to generate ammonium sulfide, which is a chemical absorption operation.

The separation of rich gas in a catalytic cracking unit belongs to multi-component physical absorption. The absorbent used is stabilized gasoline and raw gasoline. The process in which the heavier components in the rich gas are dissolved by gasoline is the absorption process. In the absorption and desorption operation of catalytic cracking unit, the desorption operation is to separate the C_2 components that are brought to the bottom oil of the desorption tower due to excessive absorption. Use reboiling heating to desorb C_2, leaving the de-ethanized gasoline. The absorbent does not need to be regenerated. The purpose of the absorption process is to recover gasoline and liquefied gas components in the compressed rich gas. Desorption is only to desorb the excessive absorption part in the absorption operation, and the regeneration degree is low, unlike the general solvent regeneration that must release all the solutes. After absorption and desorption, the stabilization tower is a fractional distillation operation under pressure to fractionate the liquefied gas components from the de-ethanized gasoline.

Liquid-liquid extraction uses the difference in solubility of the components in the raw material liquid in the extractant to achieve the purpose of separation. Liquid-liquid extraction is often used when the boiling points of the components to be separated are very close, or the concentration of the components to be separated is very low, or the mixture contains highly heat-sensitive components. Solvent recovery generally uses distillation, so the solvent recovery process in the extraction operation is the reason for the higher energy consumption.

6.5.1.2 Minimum Separation Work

The separation process in the tower equipment is to separate the mixture into pure or relatively pure substances. To complete such a process, energy is consumed. The work (energy) required for the separation operation under reversible conditions is called the separation work, which is the theoretical limit value of the minimum energy consumption under the conditions that require the separation accuracy. Obviously, the actual energy consumption will be much greater than the value of the separation work. However, as a benchmark for comparison, it still has a certain meaning. It is possible to find the exergy loss caused by the irreversibility of the process and the final energy price paid.

1. **Separation work and mixing work**

The process of separating the mixture into pure or relatively pure substances requires external separation work:

$$I_W = RT_0 \sum X_i lnX_i \tag{6.95}$$

For the process of mixing pure or relatively pure substances into a mixture, the reversible mixing work can be released to the outside world, the value is the same as the separation work, and the sign is opposite:

$$I_W = -RT_0 \sum X_i lnX_i \tag{6.96}$$

where X_i—mole fraction of substance i in the mixture;

I_W—Separation work of mixture.

Both gas absorption and desorption belong to the separation process. The absorption process is the separation process of the gas mixture, and for the solute, it is also the mixing process of the absorbent and the solute. The reverse process of the absorption process belongs to the separation process of solute and absorbent.

2. Separation exergy (work) of petroleum fractions in the fractionation process [9]

$$I_W = -RT_0 \sum N_j ln(1.33N_j) \tag{6.97}$$

where: N_j—Number of moles per unit petroleum fraction.

Separation work is the minimum work consumed in theory, and it is the real effective part in thermodynamic analysis. Strictly speaking, it should be calculated. However, the amount is generally small, and many cases where the requirements are not strict are not calculated. At this time, the separation exergy is placed in the process exergy loss.

Example 5.5 The atmospheric pressure tower data of an atmospheric and vacuum unit of a refinery is shown in the table below. Try to calculate the separation exergy from topping crude oil to products.

Solution According to the formula (6.97), the calculation is as follows (using 1 t crude oil as the calculation basis).

Fraction	Yield kg/t	Molecular weight	No. moles n	Mole fraction N_j	1.33 N_j	$N_j ln(1.33N_j)$
Atm top oil	52.3	126	0.4151	0.1497	0.1991	−0.2416
1st Atm draw oil	46	165	0.2788	0.1006	0.1337	−0.2024
2nd Atm draw oil	103	210	0.4905	0.1769	0.2353	−0.256
3rd Atm draw oil	70	300	0.2333	0.0842	0.112	−0.843
Atm bottom	677.8	500	1.3546	0.4886	0.6948	−0.2106
Total	948.6		2.7723	1.00		−1.0949

Calculate the separation exergy of atmospheric tower by formula (6.97):

$$\begin{aligned} I_W &= -RT_0 \sum N_j \ln(1.33N_j) \\ &= 8.314 \times 2.7723 \times 288 \times (-1.0949) = 7268 \text{ (kJ/t)} \end{aligned}$$

6.5.1.3 Calculation of Energy Balance and Exergy Balance of Tower Equipment

The following takes the fractionation tower as an example to illustrate the energy balance and exergy balance calculation of tower equipment.

The operating conditions of the atmospheric tower of an atmospheric and vacuum unit are shown in Fig. 6.1.

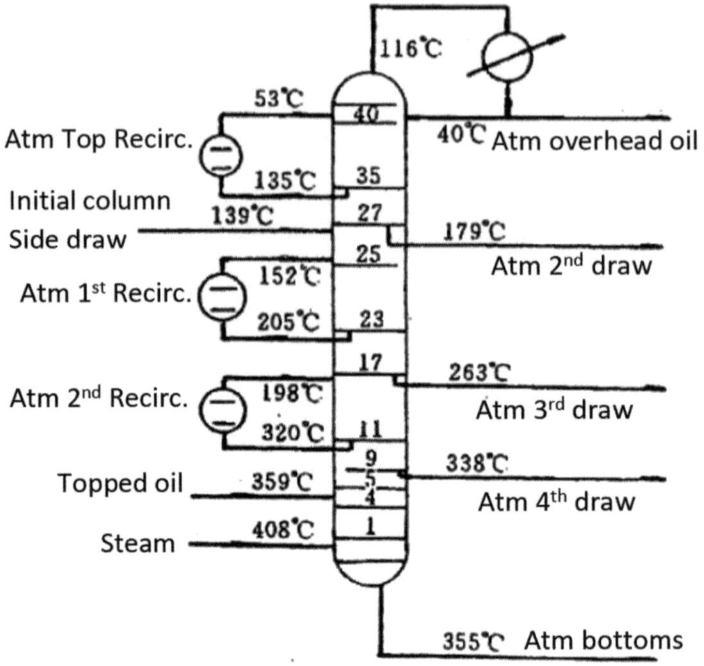

Fig. 6.1 Test trial parameters of an atmospheric tower

Most of the rectification towers in petrochemical plants are complex towers. Its characteristic is that it is dealing with a complex mixture of petroleum fractions, and the product is also a mixture, but the fraction is narrower than the raw material, and the number of mixtures is reduced. In addition to tower top products, there are often sideline products. Often more than one raw material is supplied. The energy difference between the heated raw materials and the products (excess heat) is not only taken out by reflux at the top of the tower, but also taken out by the side line reflux and recycled for use.

Traditionally, the energy balance of the tower is divided into the input energy (raw material brought in, steam brought in, heating or heating from the bottom reboiler), products brought out energy, reflux heat removal energy and equipment heat loss, etc.; for exergy balances, the calculation of separation exergy and process exergy loss should also be considered. The separation exergy is not considered in the energy balance. The minimum work of the separation process has been included in the total energy consumption and converted into heat during the separation process, which is carried out by the product and the reflux.

1. Supply energy of tower equipment

The input energy is generally two types of energy supplies: raw materials brought in (including heating furnace heating) and tower bottom reboiler heating. For many petroleum fractionation towers with large heat load, most of the input energy is heated by the heating furnace to reach the required operating temperature ensuring the required vaporization rate in the flash section of the tower.

The calculation basis of the supplied energy is usually 15 °C, 1 atm, and its phase state is also the phase state at that temperature and pressure. For tower equipment using heating furnace heating, the raw material heat and heating load can be calculated separately.

(1) Energy in feed materials E_{P1} and EX_{P1}

According to the phase state of the raw materials entering the heating furnace (usually liquid phase), calculate the energy and exergy based on the calculation method of energy and exergy of the streams (petroleum fraction, light hydrocarbons and others) in Chap. 2.

(2) Effective heating load provided by heating furnace E_{P2} and EX_{P2}

It can be obtained according to the calculation result of the heating furnace and incorporated into the energy balance of the tower equipment for verification.

(3) Heat supply from the reboiler E_{P3} and EX_{P3}

For the reboiler, because the calculation of the circulating oil of the tower is more complicated, it is difficult to directly obtain the energy supplied to the tower by the reboiler, and indirect calculation is usually used. First, determine the heat dissipation loss of the reboiler, test or estimate the heat dissipation of the reboiler, and then take the heat from the heat source of the reboiler, which is generally a heat carrier or the reflux of other tower equipment. Its characteristics are pure liquid phase. After measuring flow rate and the temperature in and out of the reboiler, the heating load can be determined, and the heat dissipation is deducted, which is the amount of heat supplied to the tower. For the steam heating source, a similar method can also be used to determine the energy supplied into the tower.

(4) Steam brought energy E_{P4} and EX_{P4}

According to the steam energy and exergy calculation method.

Supplied energy:

$$E_P = E_{P1} + E_{P2} + E_{P3} + E_{P4} \tag{6.98}$$

Supplied exergy

$$EX_P = EX_{P1} + EX_{P2} + EX_{P3} + EX_{P4} \tag{6.99}$$

2. Products bring out energy and exergy E_O and EX_O

The output energy of the tower is the energy released by the product under the conditions of (T, p) to the reference state. Except for the product at the top of the tower and the cold reflux in the gas phase, the products at the side and bottom oil leaving the tower are all liquid. The calculation of the energy and exergy of each product and the vapor from the top of the tower can be carried out in accordance with the relevant methods in Chap. 2.

3. Process heat removal energy and exergy E_C and EX_C

For the reflux heat removal operation of the tower, and the flow from the tower and the return tower remain unchanged. After the energy recovery system, the temperature of the reflux returning tower is reduced. Therefore, it does not need to be reflected in the incoming energy calculation in the balance calculation. It is only used as the outgoing party to calculate the reflux energy difference between leaving tower and returning tower. The calculation of its energy and exergy is also carried out in accordance with the calculation method of streams energy and exergy in Chap. 2. Note that the return temperature should be used to replace the original reference temperature at this time to obtain the energy taken out of the reflux leaving and returning tower.

4. Heat loss energy E_D and EX_D

The heat dissipation loss of the tower equipment should also be determined by testing. The calculation method is the same as the heat dissipation item of the heating furnace. The surface area, surface temperature, ambient temperature and wind speed measured will be used to calculate the heat dissipation loss and exergy according to the method in Sect. 2.5 in Chap. 2. As the tower equipment is bottom-up, the temperature gradually decreases. Therefore, the test and calculation should be performed in sections, and attention should be paid to the estimation of the surface area of the equipment components.

The heat dissipation test results must satisfy the heat balance of the tower.

5. Separation exergy EX_{TS}

The separation exergy of the fractionation process can be carried out according to the method of column equipment in the section of 6.5.1.2. For simple binary or multiple distillation systems, it can be calculated according to formula (6.95), and the separation exergy of petroleum fraction separation process can be calculated according to formula (6.97).

6. Process exergy loss

The irreversible exergy loss in the fractionation process can be obtained from the exergy balance:

$$DK_L = EX_P - EX_C - EX_O - EX_D - EX_{TS} \tag{6.100}$$

The exergy loss in the fractionation process is generally composed of two parts. One is the excessive heating of the feed, which causes the exergy loss in the mixing heat transfer in the tower. For example, the exergy loss caused by the mixing of reaction oil and gas at about 460 °C in the catalytic fractionation tower with the bottom oil in the desuperheating section.

Second, in order to maintain the fractionation process, the mass transfer driving force (concentration difference) are established on each plate in the tower. The formation of the concentration difference depends on the temperature difference. Therefore, the irreversible losses on the trays in the tower are mainly caused by the heat and mass transfer in the tower. In addition, heat transfer and mass transfer are related to each other. If there is no heat transfer temperature difference, there will be no concentration difference between the phases, and mass transfer separation is impossible. Therefore, the heat and mass transfer in the tower are all attributed to the exergy loss of mass transfer. Generally, the fractionation tower feeds at the boiling point, and there is rarely any exergy loss in the tower mixing. The mixed exergy loss of catalytic reaction oil and gas is determined by the process route and conditions, and people have begun to work on improvements to increase the exergy utilization efficiency.

7. Tower equipment efficiency

The energy efficiency of the tower equipment is relatively high, and the energy carried by the leaving streams can generally be recovered and used, and the loss part is mainly heat loss. This is due to the conservation of energy in the first law of thermodynamics. Energy efficiency is always less than 1, that is to say, for equipment with exothermic reaction, although the chemical energy of raw materials is not counted, the exothermic reaction heat should be considered as the energy supplier at this time.

$$\eta = \frac{E_O + E_C}{E_P} \times 100\% = \left(1 - \frac{E_D}{E_P}\right) \times 100\% \tag{6.101}$$

The exergy efficiency of the tower equipment is much lower than the energy efficiency, mainly because the irreversible process in the tower causes the exergy loss of the process. As a result, the exergy value of the leaving stream and heat dissipation is much lower than the value of the exergy supplied to the equipment.

$$\eta_x = \frac{EX_O + EX_C}{E_P} \times 100\% = \left(1 - \frac{D_{KL} + EX_D}{E_P}\right) \times 100\% \tag{6.102}$$

Tower equipment energy balance and exergy balance summary see Table 6.6.

The energy balance and exergy balance calculation of the absorption and extraction equipment are in principle the same as that of the fractionation tower, except that there is generally no mid reflux heat removal. The separation exergy can generally be ignored and placed in the process of exergy loss.

Example 6.6 The basic data and calculation results of the calibration test of an atmospheric column of atmospheric and vacuum distillation unit are [10] shown in Table 6.7.

The heat supply of the tower is composed of five items: the initial distillation bottom oil brings heat, heat exchange and furnace heating, stripping steam brings

Table 6.6 Tower equipment energy balance and exergy balance

Description			Flow t/h	Temperature °C	Phase	Pressure MPa	Energy kW	Exergy kW
Energy supply	Stream Bring In							
		Subtotal						
	Steam Bring In							
	Process Heating							
	Total							
Energy Output	Leaving Stream							
		Subtotal						
	Steam Leaving energy							
	Process Heat Removal							
		Subtotal						
	Process exergy loss							
	Heat loss							
	Total							
	Efficiency, %							

heat, and the initial distillation tower over vaporized oil brings heat. The output heat is the heat taken by the product, the heat taken by the reflux, the heat dissipation, and the water vapor at the top of the tower is taken out. According to the calculation method of streams energy and exergy, the calculation results are also listed in Table 6.7.

Table 6.7 Energy (exergy) balance result of an atmospheric tower

Items		Basic Data						Calculation results	
		Flow rate t/h	Relative density	Temperature, °C		Characteristic factor	MW	Energy, kW	Exergy kW
				in	out				
Energy supply	Initial column bottom oil							38,811.27	9767.46
	Over-vaporized oil							918.87	185.86
	Stripping steam							1831.22	733.06
	Feed oil heat exchange							21,899.84	10,033.82
	Feed oil heating							1964.44	10,327.69
	Total							83,105.67	31,047.89
Energy output	Atm. tower Steam	2.01			108.00			1467.29	311.65
	Atm. tower overhead oil	12.30	0.74		108.00	11.98	116.00	1718.14	238.84
	Atm. tower 1st draw	17.00	0.79		182.00	11.64	172.00	1801.51	396.04
	Atm. tower 2nd draw	42.65	0.84		268.00	12.83	280.00	7580.78	2258.20
	Atm. tower 3rd draw	7.63	0.87		343.00		327.32	1751.95	620.40
	Atm. tower bottom	254.99	0.94		355.00	11.71	500.00	57,985.22	21,056.61
	Sum of products							72,304.92	24,881.75
	Separation exergy								633.05
	Cold reflux	12.96	0.74	38.00	108.00	11.98		1632.18	273.60
	1st mid-stage reflux	45.00	0.77		134.00	11.80		1938· 83	444.35
	2nd mid-stage reflux	77.00	0.81	144.00	202.00	11.80		3175.14	1123.86
	3rd mid-stage reflux	38.40	0.85	198.00	320.00	11. 6		3690.17	1690.10

(continued)

Table 6.7 (continued)

Items	Basic Data						Calculation results	
	Flow rate t/h	Relative density	Temperature, °C		Characteristic factor	MW	Energy, kW	Exergy kW
			in	out				
Sum of heat removal							10,436.32	3531.92
Heat loss							364.43	156.39
Process exergy Loss								1844.78
Total							83,105.67	31,047.89

6.5.2 Reaction Equipment

Reaction equipment is another important process equipment in the process energy utilization link. In the reaction equipment, the raw materials undergo qualitative changes, and the raw materials are transformed into products. Reaction equipment is the core process of petrochemical plants. Carrying out the energy balance and exergy balance calculation of the reaction equipment is helpful to find the locations of energy and exergy loss, so as to improve the operation and modification of the plant. In the calculation for the purpose of energy analysis, we only calculate and analyze the quantity and quality of energy and energy (exergy) loss of the equipment to determine the energy efficiency of the equipment without involving complicated process calculations.

The reaction process can be divided into two categories: endothermic reaction and exothermic reaction. Although some reactions are endothermic and other reactions are exothermic in some reaction processes, they are nothing more than two categories of endothermic and exothermic reactions.

6.5.2.1 Endothermic Reaction Equipment

The endothermic reaction process of petrochemical equipment includes catalytic cracking reaction, naphtha platinum reforming reaction, hydrocarbon conversion and ethylene cracking, etc., all of which are endothermic reactions. The outstanding feature of this type of reaction is that not only the operating conditions must be achieved by increasing the temperature and pressure, but also the continuous supply energy during the reaction process. The raw material absorbs the reaction heat to change its structure and transform it into a product. Therefore, the raw materials need to bring in a large amount of heat. For energy analysis methods that do not consider the chemical energy of raw materials and products, the difference in chemical energy between raw materials and products can be reflected through the change of reaction heat.

1. Supply energy and exergy: E_P and EX_P

The energy and exergy supplied to the reaction equipment are generally as follows.

(1) Feed materials bring in

The raw materials are nothing more than the streams commonly used in petrochemical plants (oil, chemical raw materials, air, water and steam, etc.). Most of these streams enter the reaction equipment in the gas or liquid state. Therefore, its energy and exergy calculation of can be referred to the method in Chap. 2.

(2) Steam Bring in

In addition to the raw materials, the reaction equipment often blows steam for the purpose of dilution and loosening. The energy and exergy calculation of steam stream is also carried out with reference to the related methods in Chap. 2.

(3) Chill raw material

In the reaction equipment, in order to stop the further reaction of the leaving products and prevent the increase in the yield of the by-product's reaction, in many cases, the chilled stream is injected to make the reaction products quickly out of the reaction conditions and terminate the ongoing reaction. The calculation of stream energy and exergy can also be carried out by referring to the method in Chap. 2.

(4) Energy provided by heating furnace

The supplied energy and exergy are determined by the calculation result of the furnace balance.

2. Leaving products bring out energy and exergy

The calculation is carried out by the relevant method in Chap. 2, but it should be calculated and determined in detail based on the physical properties of the product and the component fractions, and is consistent with the calculation results of the subsequent separation equipment. For example, the catalytic cracking reaction oil and gas should not only be calculated in the gas phase, but also be calculated and summed based on the physical properties of the narrow fractions separated separately in the fractionation tower. If the chilled raw materials are counted, then the export product should be the product flow rate and parameters after chilled.

3. Reaction heat energy and exergy E_R and EX_R

The endothermic reaction uses the role of a catalyst to convert a part of the energy provided by the raw material and the heating furnace into the product under the reaction conditions. At this time, the structure of the product has changed, and the chemical energy of the product has been significantly improved. The increase value is considered to be approximately the same as the heat of reaction value. For the reaction process with known raw material and product composition, the heat of reaction can be calculated according to the relevant graphs, according to Guess's law, and corrected by Kirchhoff's formula. For complex reaction processes with unknown reaction composition, especially the reaction heat of petroleum refining process, it can be determined by empirical formula or heat balance method. The reaction energy under operating conditions is the heat of reaction. The calculation of exergy can refer to Chap. 2. Generally, the energy (reaction heat) entering the product can be approximated as the exergy value.

Therefore, the endothermic reaction process is the process of increasing the exergy value, which converts the disordered thermal energy (energy grade less than 1) brought by the furnace or raw materials into the ordered chemical energy of the product (energy grade close to 1), but only the reactor itself, due to the irreversibility of heat transfer inside the equipment, there is still a large amount of process exergy loss, and the overall result is that the process exergy loss is significantly reduced compared to the exothermic reaction process.

4. Steam carries energy E_S and EX_S

For steam that is not used as a raw material to participate in the reaction, the sensible heat and latent heat of the steam taken out at the outlet state should be taken as the energy brought out by the outgoing stream. The steam quantity is the same as that of the supply. The energy and exergy are calculated by the method in Chap. 2 according to the parameters at outlet state.

Effective brought out energy:

$$E_U = E_O + E_S + E_R + E_C \tag{6.103}$$

$$EX_U = EX_O + EX_S + EX_R + EX_C \tag{6.104}$$

The above formulas E_C and EX_C are heat removal, and E_O and EX_O are products that are brought out.

For the endothermic reactions, energy is generally not taken from the reactor, but unconverted raw materials continue to be recycled into the reactor, because of the energy entering and leaving the reactor with difference in parameters is equivalent to the heat taken by the process (such as the refining of the catalytic plant), but the input and output are carried through the raw materials and products, which can be calculated together with the raw materials and the products.

5. Surface heat dissipation of reaction equipment E_D, EX_D

The surface heat dissipation of the equipment is the same as that of the tower equipment. The surface area of the equipment should be calculated per dimension, the surface temperature, ambient temperature, humidity, wind speed, etc. should be tested and recorded, and then the heat dissipation loss can be calculated. The heat flow meter can also be used to measure the heat dissipation per unit surface area, and according to the internal temperature (catalyst and reaction streams of the equipment) determines the heat dissipation exergy loss. The heat dissipation test result should meet the heat balance of the reactor.

6. Reactor equipment process exergy loss

According to the first law of thermodynamics of energy conservation, the energy entering and leaving the reaction equipment is conserved, but the energy quality is not conserved. The energy quality difference between entering and leaving the reaction equipment is called the process exergy loss of the reaction equipment.

$$D_{KR} = EX_P - EX_U - EX_D \tag{6.105}$$

7. Reactor efficiency

Among the energy output from the reactor, except for the heat dissipation loss that has been eventually lost to the environment and cannot be recovered, the remaining energy is effectively used (reaction heat enters the product) and can be recovered and recycled, which is considered to be effective, so its energy efficiency is:

$$\eta = \frac{E_U}{E_P} \times 100\% = \left(1 - \frac{E_D}{E_P}\right) \times 100\% \tag{6.106}$$

In the outgoing exergy flow, except the heat dissipation exergy loss, the process exergy loss is also the energy cost of the irreversible process, which cannot be recovered. Therefore, the exergy efficiency is much lower than the energy efficiency, which is calculated by the following formula:

$$\eta_x = \frac{EX_U}{EX_P} \times 100\% = \left(1 - \frac{EX_D + D_{KR}}{EX_P}\right) \times 100\% \tag{6.107}$$

6.5.2.2 Exothermic Reaction Equipment

Exothermic reactions are another type of reaction process, and the energy use characteristics are also different from endothermic reactions. The exothermic reaction requires the outside to provide a suitable operating condition (temperature, pressure, catalyst). Providing this operating condition often requires energy consumption. Once this condition is reached, the reaction starts, and no external energy is required. In fact, as the reaction progresses, the raw materials continue to react and release heat, which increases the temperature of the system. This reaction heat does not come from external energy, but come from the chemical energy of raw materials, similar to the combustion process. In view of this characteristic, in the calculation of energy balance and exergy balance, the method is in principle the same as the endothermic reaction, except for the following points.

1. The exothermic heat of the reaction should be regarded as the input energy.

Because of the energy analysis method used, the chemical energy of the raw materials and products are not counted, and only chemical energy differences are reflected in the form of reaction heat. The exothermic reaction makes the physical energy of the product greater than that of the raw material, and the chemical energy is less than that of the raw material. As a result, the energy efficiency of the exothermic reaction equipment may be caused greater than 100%. For a strict count for chemical energy system, this situation does not occur, but the above phenomenon occurs when only physical energy is considered. However, the reaction exotherm is regarded as part of the raw material chemical energy converted into heat energy supply, which can avoid the above phenomenon and conform to the analysis energy method proposed by us.

2. Chilling and heat removal items of Exothermic equipment should be counted as output.

Due to the nature of the exothermic reaction, the temperature of the system will increase, and the equilibrium conversion rate of some reactions will decrease as the temperature increases. In order to improve the equilibrium conversion rate, heat is often taken out in the middle of the reactor, such as the medium temperature carbon monoxide conversion in fertilizer production, which is divided into multiple conversion sections. When the temperature of the shift gas in the first stage reaches the requirements of the second stage after being cooled by the heat exchange equipment, it enters the second stage. In this case, it can be considered as the heat removal item of the reactor. The calculation can be carried out according to the relevant method in Chap. 2 based on the flow rate of the withdrawn stream and the energy parameters leaving and returning to the reactor.

The corresponding process exergy loss is generally greater than that of the endothermic reaction equipment. This is because the chemical energy of the raw materials is released during the reaction process, which changes from order to disorder, and the energy grade decreases.

The energy and exergy balance results of the reaction equipment can also be summarized in Table 6.8.

6.6 Energy Recovery and Utilization Equipment

Energy recovery equipment is a device set up purely to recover energy to reduce the energy provided to the plant. If the energy is not recovered after use, it is hard to imagine how amazingly high energy consumption will be.

Energy recovery and recycling is an important aspect of energy saving. Therefore, we must first do a good job of the energy and exergy balancing of energy recovery link in order to discover the potential and make improvements.

The energy recovery link includes heat exchange (recovery and recycling, recovery output heat exchange), cooling and power recovery equipment.

6.6.1 Cooling and Heat Exchange Equipment

Heat exchange equipment is a type of equipment widely used in petrochemical plants, including heat exchangers, coolers, and condensers. The purpose of using this type of equipment is different from process equipment, it is not to complete the processing process but to recover and reuse the energy discharged from the process equipment; It is mostly used for heating materials. There are also many types of heat exchange equipment. Commonly used are shell and tube type, double pipe type, and various plate heat exchangers. The most commonly used is the shell and tube heat exchanger.

Table 6.8 Reactor equipment energy and exergy balance sheet

Description			Flow rate, t/h	Temperature °C	Phase	Pressure MPa	Energy, kW	Exergy, kW
Energy incoming	Stream in							
		Subtotal						
	Steam in							
	Reaction exotherm							
	Process heating							
	Quench feed							
	Total							
Energy outgoing	Stream out							
		Subtotal						
	Steam out							
	Process heat removal							
		Subtotal						
	Reaction endotherm							
	Equipment heat loss							
	Total							
	Process exergy loss							
	Efficiency, %							

6.6.1.1 Calculation of Energy and Exergy Balance of a Single Heat Exchange Equipment

1. Energy released from hot fluid E_h, EX_h; the energy absorbed by the cold flow E_C, EX_C

The materials handled by the heat exchanger are mostly raw materials, processed products and utility streams. After testing to determine the flow rate and inlet and outlet temperatures of the cold and hot streams, it can be carried out by the related streams calculation method in Chap. 2.

2. Heat exchanger heat dissipation energy E_D and exergy EX_D

The heat dissipation of heat exchange equipment requires to be measured and calculated. There are two test methods: one is to calculate the heat dissipation from the surface area and temperature of the equipment according to the method in Sect. 2.5 in Chap. 2, which requires the area of the equipment components such as the head flange Estimate; and calculate the heat dissipation of the exposed part that is not insulated. The second is through the calculation of heat balance, the difference between the heat released by the hot flow and the heat absorbed by the cold flow is the heat dissipation ($E_D = E_h - E_c$). These two methods should be checked for consistency.

$$EX_D = E_D(1 - \frac{T_0}{T}) \tag{6.108}$$

where: T—Take shell side fluid temperature, K.

3. Exergy loss in heat exchange process EX_L

It can be obtained by the following balance:

$$EX_L = EX_h - EX_C - EX_D \tag{6.109}$$

Since heat dissipation is dissipated through the outermost surface area of the device, the heat source for heat dissipation is mainly shell-side fluid. In the case of a heat exchanger with a hot flow in the shell side, the heat load that is actually transferred from the hot flow to the cold flow is the cold flow heat load. Therefore, the heat transfer exergy loss:

$$EX_L = E_C T_0 (T_H - T_C)/(T_H . T_C) \tag{6.110}$$

For a heat exchanger with cold flow in the shell side, the heat dissipation load is the heat transfer from the tube side hot flow to the shell side cold flow, and then the heat is dissipated from the shell side. At this time, the process exergy loss should be based on the hot flow released heat:

$$EX_L = E_h T_0 (T_H - T_C)/(T_H . T_C) \tag{6.111}$$

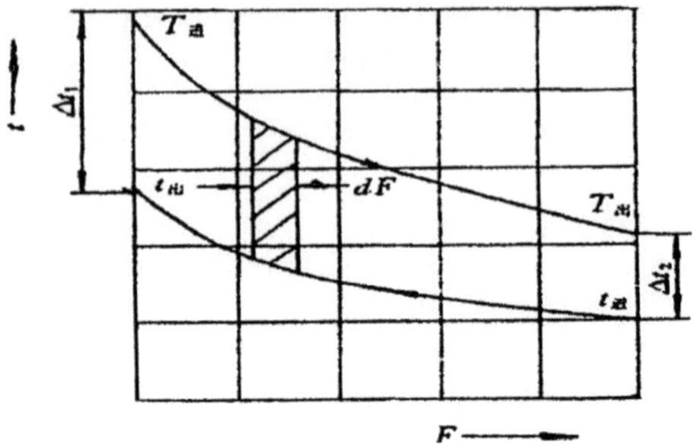

Fig. 6.2 Countercurrent operating temperature distribution

The calculation results the exergy loss of the process using the two methods should be the same.

4. Heat transfer temperature difference and heat transfer coefficient

For the countercurrent heat transfer operation shown in Fig. 6.2, the logarithmic average heat transfer temperature difference:

$$\Delta t_{lm} = (\Delta t_1 - \Delta t_2)/\ln(\frac{\Delta t_1}{\Delta t_2}) \tag{6.112}$$

In the case of cross-flow and baffle, the actual heat transfer temperature difference of the multi-pass and multi-tube-pass shell-&-tube heat exchangers used in practice is multiplied by the logarithmic average temperature difference by the temperature difference correction coefficient F_t.

$$\Delta t_m = \Delta t_{lm} \times F_t \tag{6.113}$$

F_t can be found in the relevant manuals, but for the convenience of computer applications, the following direct calculation methods are cited.

For m shell pass, $2m.n$ tube pass (such as 1-2, 1-4, 2-4, 2-8, 3-6, 3-12, ...) heat exchanger temperature difference correction coefficient F_t can be calculated using following two parameters P and R by the computer model [11] according to formula (6.114).

$$R = \frac{T_{in} - T_{out}}{t_{out} - t_{in}},\ P = \frac{t_{out} - t_{in}}{T_{in} - t_{in}} \tag{6.114}$$

For m shell pass, $2m.n$ tube pass, heat transfer temperature difference correction coefficient:

$$R \neq 1,\ F_t = \frac{\sqrt{R^2+1}/(R-1)ln((1-Y)/(1-YR)}{ln\frac{2/Y-1-R+\sqrt{R^2+1}}{2/Y-1-R-\sqrt{R^2+1}}} \tag{6.115}$$

Among them, $Y = ((\frac{PR-1}{P-1})^{\frac{1}{m}} - 1)/((\frac{PR-1}{P-1})^{\frac{1}{m}} - R)$

$$when \quad R = 1,\ F_t = \frac{\sqrt{2}(\frac{Y}{1-Y})}{ln\frac{2/Y-2+\sqrt{2}}{2/Y-2-\sqrt{2}}} \tag{6.116}$$

Among them, $Y = P/(m - mP + P)$

The total heat transfer coefficient of a single heat exchanger can be calculated from the heat transfer equation:

$$K = Q/(F\Delta t_m) \tag{6.117}$$

where: F_t—Heat transfer temperature difference correction coefficient;

Δt_{lm}—Log mean temperature difference, K;

Δt_m—Corrected actual heat transfer temperature difference, K;

F—Heat transfer area of heat exchange equipment, m^2;

Q—The heat load of the heat exchanger adopts cold flow absorbed heat or hot flow released heat according to whether the shell side fluid is hot or cold, kW

K—Total heat transfer coefficient, kW/(m^2 K).

***Example** 5.7* In a catalytic cracking unit, one heat exchanger heat exchanged between feedstock oil and 2nd mid. reflux of fractionation. The raw material (wax oil) travels to the shell side, the flow rate is 119.34 t/h, the relative density is 0.85, the characteristic factor is 12, and the temperature rises from 121.5 to 168 °C; The 2rd mid. reflux travels to the tube side, the flow is 92.96 t/h, the relative density is 0.86, the characteristic factor is 12, and the stream temperature is reduced from 240 to 183 °C. The heat exchanger is a single shell-pass heat exchanger with a heat transfer area of 260 m^2. Carry out the energy balance and exergy balance of the heat exchanger and calculate the heat transfer temperature difference and the total heat transfer coefficient.

Solution The cold and hot streams of the heat exchanger are both liquid petroleum fractions, energy and exergy of cold and hot streams can be calculated by formula (2.50) and formula (2.51) in Chap. 2

1. Cold flow heat absorption and exergy

$$E = \Delta H = \{C_3(t_2 - t_1) + 1.99 \times 10^{-3}(t_2^2 - t_1^2\}C_1$$

$$EX = \left\{A_0(T_2 - T_1) + 1.99 \times 10^{-3}(T_2^2 - T_1^2) - A_1 ln(\frac{T_2}{T_1})\right\}C_1$$

Among them: $API = \frac{141.5}{0.9942 \times 0{\cdot}85 + 0.009181} - 131.5 = 34.14$

$C_1 = 0.0533 \times 12 + 0.3604$

$C_3 = 1.5566 + 8.256 \times 10^{-3} API$

$A_0 = 8.256 \times 10^{-3} API - 0.6767$

$A_1 = 135.3 + 2.3778 API$

For cold flow, substitute relevant physical property data to get: $API = 34.14$, $C_1 = 1$, $C_3 = 1.8385$, $A_0 = -0.3948$, $A_1 = 216.478$, therefore:

$$\begin{aligned} E_c &= \{1.8385(168 - 121.5) + 1.99 \times 10^{-3}(168^2 - 121.5^2)\} \times \frac{119.34}{3.6} \\ &= 3722.5\ (\text{kW}) \end{aligned}$$

$$\begin{aligned} EX_c = &\{-0.3948(168 - 121.5) + 1.99 \times 10^{-3}(441^2 394.5^2)\} \\ &- 216.478 ln\left(\frac{441}{394.5}\right) \times \frac{119.34}{3.6} = 1154.74\ (\text{kW}) \end{aligned}$$

2. Hot flow released energy and exergy

For hot flow, substitute relevant physical property data and get: $API = 32.24$, $C_1 = 1$, $C_3 = 1.8228$, $A_0 = -0.4105$, $A_1 = 211.96$, therefore,

$$\begin{aligned} E_h &= \{1.8228(240 - 183) + 1.99 \times 10^{-3}(240^2 - 183^2)\} \times \frac{92.96}{3.6} \\ &= 4144.1\ (\text{kW}) \end{aligned}$$

$$\begin{aligned} EX_h = &\left\{-0.4105(240 - 183) + 1.99 \times 10^{-3}(513^2 - 456^2) - 211.96 \ln\left(\frac{513}{456}\right)\right\} \\ &\times \frac{92.96}{3.6} = 1682.3\ (\text{kW}) \end{aligned}$$

3. Heat dissipation and heat dissipation exergy

The amount of heat dissipation can be determined by the balance of the difference between hot flow heat release and cold flow absorption:

$$E_D = 4144.1 - 3722.5 = 421.6\ (\text{kW})$$

As the hot stream flows through the tube pass, the temperature of the heat dissipation fluid is taken as the average temperature of the shell pass, and the amount of exergy dissipated.

$$T_m = \frac{121.5 + 168}{2} + 273 = 417.75\ ^\circ\mathrm{C}$$

$$EX_D = 421.6 \times \left(1 - \frac{T_0}{T_m}\right) = 130.9\ (\mathrm{kW})$$

4. Heat transfer Exergy loss

Determined by exergy balance:

$$EX_L = EX_h - EX_c - EX_D = 1682.3 - 1154.74 - 130.9 = 369.9\ (\mathrm{kW})$$

5. Average heat transfer temperature difference

Log mean temperature difference:

$$\Delta t_{lm} = \frac{\Delta t_1 - \Delta t_2}{ln\left(\frac{\Delta t_1}{\Delta t_2}\right)} = (72 - 61.5)/ln(72/61.5) = 66.61\ ^\circ\mathrm{C}$$

Check the temperature difference correction coefficient or calculated by formula (6.115), $F_t = 0.897$

Average heat transfer temperature difference: $\Delta t_m = 0.897 \times 66.61 = 59.73\ ^\circ\mathrm{C}$

6. Average heat transfer coefficient

$$K = \frac{Q}{F\Delta t_m} = 4144.1 \times \frac{10^3}{260 \times 59.73} = 266.8\left(W/\left(\mathrm{m}^2\mathrm{K}\right)\right)$$

6.6.1.2 Calculation and Analysis of Energy and Exergy Balance of Heat Exchange Network

In petrochemical plants, depending on the process, there are many heat exchangers to form a heat exchange network. The analysis and improvement of the heat exchange network is a very important aspect of the energy saving of the plant.

The energy balance and exergy balance results of a single heat exchanger is summarized in Table 6.9.

In view of the convenience of energy consumption analysis and summary, the energy balance of the heat exchange system can be divided into six categories.

Table 6.9 Energy and exergy balance summary table heat exchanging equipment

Order	Name	Model	Fluid name	Shell or Tube pass	Flow kg/h	temperature ℃		energy kW	exergy kW	heat loss kW	heat loss exergy kW	Exergy loss kW	Heat transfer Approach ℃	Heat transfer coefficient W/ (m2. K)	Remarks
						inlet	oulet								
			Cold												
			Hot												
			Cold												
			Hot												
			Cold												
			Hot												
			Cold												
			Hot												
Total			Cold												
			Hot												

(1) Heat exchange of raw materials

Raw material heat exchange is the recovery of energy from the energy recovery link, and the energy is brought to the process utilization link through the raw material, which belongs to the heat exchange and recycling energy.

(2) Steam generation

Using the excess heat of the plant to generate steam (including preheating demineralized water), depending on its use, can be classified as heat exchange recycle recovery or heat exchange output, which should be specifically analyzed.

(3) Heat exchange output

For plants that use heat integration, the excess heat of the plant is often used to exchange heat with raw materials or heat carriers of external plant, or as the heat source of the reboiler of the external plant to save the fuel consumption of external plants. This recovered energy belongs to the recovery output energy. It should be summarized separately.

(4) The preheated air is recovered from the energy to be recovered and used to preheat the furnace air can save the furnace fuel with the same preheating load.

(5) As the heat source of the reboiler of this plant, it can also be used as the heat exchange recycle energy, but it can be classified in the summary.

(6) Coolers

Including water-cooled and air-cooled coolers, energy balance and exergy balance calculations are the same as heat exchangers. The cooler is generally not insulated, therefore, when considering the heat absorbed by the cold flow and calculating the amount of cooling water from this, the heat dissipation factor should be deducted.

The summary of the heat exchange network can be obtained by the sum of the energy balance and exergy balance results of each individual heat exchanger, and summarize the hot flow energy (exergy), cold flow absorption energy (exergy), heat dissipation loss (exergy) and process exergy loss summary. However, the average heat transfer temperature difference and average heat transfer coefficient of various types of heat exchange systems cannot be simply added up and should be calculated according to the heat load average and area average methods respectively.

Heat load Average method to calculate network heat transfer temperature difference:

$$\Delta T_{ms} = Q_{total}/\sum\left(\frac{Q_i}{\Delta T_{mi}}\right) \tag{6.118}$$

Area average method to calculate network heat transfer coefficient:

$$K_S = \sum_{i=1}^{n}\left(K_i.\frac{F_i}{F_{total}}\right) \tag{6.119}$$

where: ΔT_{mi}—The heat transfer temperature difference of the *i* heat exchanger, °C;

F_i—The heat exchange area of the *i* heat exchanger, m^2;

Q_i—The heat transfer load of the *i* heat exchanger, kW;

K_i—Heat transfer coefficient of the *i* heat exchanger, kW/ (m^2 °C);

K_S—Network average total heat transfer coefficient, kW/ (m^2 °C);

ΔT_{ms}—Network average heat transfer temperature difference, °C;

F_{total}—Total heat transfer area of the network, m^2;

Q_{total}—Total heat transfer load of the network, kW.

A deeper analysis of the heat exchange network is the analysis of thermal resistance and pressure drop. According to the stream's parameters, physical properties and structural parameters of the equipment, the film heat transfer coefficient and pressure drop on both sides of the tube side and the shell side can be calculated to determine the key thermal resistance.

Running heat exchanger design and rating program such as HTRI or Aspen EDR can easily conduct this type of deeper analysis. When necessary, it can also be combined with the monitoring and analysis of the thermal resistance of the equipment structure, so as to fully grasp and evaluate the working conditions of the entire heat exchange equipment.

6.6.1.3 Analysis and Calculation of Pinch Technology in Heat Exchange Network

The above-mentioned balance and summary of the heat exchange network are only for the analysis and evaluation of the problems in the equipment and the network, and to propose directions and ways for improvement. It has not yet fundamentally solved the phenomenon of how to avoid the degraded energy of the heat exchange network and eventually be discarded, thereby increasing the consumption of cooling and heating utilities. The advantageous tool to solve this problem is the pinch technology principle proposed by Mr. Linhoff. It is based on the second law of thermodynamics and concepts, but does not do quantitative exergy analysis and calculations, uses the most optimized concept, but does not require sophisticated mathematical methods, so it is easy to popularize [12]. The principle of pinch design will be introduced in the following chapters.

6.6.2 Power Recovery Equipment

Power recovery equipment is a device that recovers streams power from the process discharge stream. It mainly includes hydraulic turbines and gas expansion turbines. There are two types of power recovery. One is to directly drive the process pump or compressors replacing the motor with a turbine. The other type is to expand high pressure stream using gas expansion turbines to generate power and connect it to the power grid. No matter what the situation is, the turbine and connecting equipment (pump or generator) can be treated as a unit.

1. Energy supplied to the unit

That is the work that the high-pressure stream can do through the turbine. For hydraulic turbines, it is the same as the effective work calculation formula of centrifugal pump (6.2)

$$E_{in} = Q\Delta p/3.67 \tag{6.120}$$

where: Q—High-pressure stream flow, m^3/h;

Δp—The pressure difference between the high-pressure stream entering and leaving the hydraulic turbine, MPa.

For a gas expansion turbine, it can be regarded as an ideal gas stream

$$E_{in} = \frac{V}{3600 \times 22.4} RT_m ln(\frac{p_1}{p_2}) \tag{6.121}$$

where: p_1, p_2—Pressure of the high-pressure gas stream entering and leaving the turbine, MPa;

T_m—The arithmetic mean temperature of the high-pressure gas stream entering and leaving the turbine, K;

R—Gas constant, 8.314 kJ / (kmol.°C);

V—Gas flow rate, m^3/h.

When it cannot be regarded as an ideal gas, it should be treated as a real gas and multiplied by the correction factor φ

$$E_{in} = \frac{V}{3600 \times 22.4} RT_m ln\left(\frac{p_1}{p_2}\right) \times \varphi \tag{6.122}$$

$$\varphi = (Z_1 + Z_2)/2Z_1 \quad (6.123)$$

where: Z_1, Z_2—The compression factor of the gas stream at the inlet and outlet of the turbine, it can be calculated by Eq. (6.7).

2. Output work E_e

It is the effective work driving the process pump or compressors, and the calculation method is shown in the Sect. 6.2 of this chapter.

When the turbine is used to generate electricity, the net output power is the amount of electricity generated.

3. Total process exergy loss

The process exergy loss of the turbine unit is the sum of the ineffective power of the expansion turbine and the ineffective power of the pump (generator), which can be found by the balance:

$$D_{ex} = E_{in} - E_e \quad (6.124)$$

4. Unit efficiency

It is the product of the efficiency of the expansion turbine and the efficiency of the pump (generator set), which can be calculated by the following formula:

$$\eta = \frac{E_e}{E_{in}} \times 100\% \quad (6.125)$$

The energy balance and exergy balance results of the power recovery equipment are summarized in Table 6.10.

For low-temperature heat upgrade utilization systems like heat pump systems, heat pumps can be considered together with process equipment. The application of heat pumps reduces low-temperature heat reinjections, reduces the external supply heating load at the bottom of the tower, and consumes more electricity. At this time, just put the compressor etc. into the conversion equipment, and use the compressor power supply energy and exergy as the input of the tower. For particular case, you can also refer to the relevant literature for detailed thermodynamic analysis.

Table 6.10 Summary of work recovery equipment

Number	Expander turbines									
	Model	Rated parameters				Actual parameters				Shaft power kW
		Flow rate m^3/h	Pressure MPa	Rotating speed rpm	Power kW	Flow rate m^3/h	Temperature °C outlet/Inlet	pressure Mpa outlet/Inlet	Effective work kW	
Number	Pump or Compressors									
	Model	Rated parameters				Actual parameters				Unit power recovery %
		Flow rate m^3/h	Pressure MPa	Rotating speed rpm	Power kW	Flow rate m^3/h	Temperature °C outlet/Inlet	pressure Mpa outlet/Inlet	Effective work kW	

References

1. China Petrochemical Corporation Standard SH2600-92 "Energy Balance Method for Petrochemical Enterprises" (1992)
2. Z. Xipeng, Calculation and analysis of oil refining process. Maoming Petrochem. Ind. **4** (1988)
3. *Process Design Data of Oil Refining Equipment, Compressor Process Calculation* (Petrochemical Press, 1978), p. 4
4. H. Ben, C. Anmin, Analysis and estimation of the impact of load rate on energy consumption of refinery plants. Pet. Refining **1** (1987)
5. Jiangsu Normal University, etc., Physical Chemistry (People's Education Press, 1980), p. 12
6. C. Anmin, L. Kun, Thermodynamic analysis of gas turbine-heating furnace combined system. Oil Refinery Des. **4** (1988)
7. C. Anmin, Calculation of coking amount and thermal efficiency of catalytic cracking regenerator. Oil Refining Des. **2** (1982)
8. H. Ben, C. Anmin, Practical calculation and economic analysis of heat dissipation in refineries. Pet. Refining **2** (1984)
9. H. Ben, Calculation of the exergy difference in the separation process of petroleum and its fractions. Acta Petrolei Sinica **3** (1986)
10. C. Anmin, Z. Junzheng et al., Maoming No. 1 distillation unit energy consumption analysis report. Oil Refining Des. **4** (1989)
11. Shanghai Institute of Chemical Technology, *Basic Chemical Engineering* (Shanghai Science and Technology Press, 1987)
12. S. Meisheng, Principles of pinch point design. Oil Refining Des. **3** (1989)
13. Z. Lixin, Overview of thermal balance of catalytic cracking unit. Catalytic Cracking **5** (1991)

Chapter 7
Energy Balance and Exergy Balance of Petrochemical Plants

Abstract After completing the analysis and calculation of the energy utilization equipment of the plant, according to the energy balance and exergy balance data of the equipment, it can be summarized as the energy balance and exergy balance of the plant. This Chapter verifies the system energy consumption data in the plant, including process plant material balance, the calculation of thermodynamic energy and exergy consumption, calculation of energy and exergy of recycling and output streams, stream rejection energy and exergy; plant heat loss verification and summary; Balance of supply and consumption of steam, electricity, and water; finally, summarize the energy and exergy balance of energy conversion link, energy process utilization link and energy recovery link, and integrated into the plant energy balance and exergy balance.

Keywords Petrochemical plant · Energy verification · Heat dissipation · Utility verification · Energy balance · Exergy balance

Abbreviation and Symbology for Energy Balance

Energy link	Energy symbol		Definition	Exergy symbol
Energy conversion	Energy supply	E_P	Total energy supply	EX_P
		E_{PF}	Furnace fuel energy supply	EX_{PF}
		E_{PB}	Boiler fuel energy supply	EX_{PB}
		E_{PBF}	Power plant boiler fuel energy supply	EX_{PBF}
		E_{PG}	Coke energy supply	EX_{PG}
		E_{PS}	Steam energy supply	EX_{PS}
		E_{PE}	Electricity energy supply	EX_{PE}
		E_{PH}	External heat energy supply	EX_{PH}
	Output	E_B	Conversion link output energy	EX_B

(continued)

T. A. Chen, *Energy Saving and Carbon Reduction*,
https://doi.org/10.1007/978-981-19-5295-1_7

(continued)

		E_{BS}	Conversion link output Steam	EX_{BS}
		E_{BH}	Conversion link output heat	EX_{BH}
		E_{BE}	Conversion link output electricity energy	E_{XBE}
	Direct loss	E_W	Direct loss of energy	D_{JU}
		E_{WC}	Incomplete combustion	EX_{WC}
		E_{WX}	Flue gas energy loss	EX_{WX}
		E_{WD}	Heat loss	EX_{WD}
		E_{WP}	Ineffective power	EX_{WP}
			Process exergy loss	D_{KU}
	E_U		Effective energy supply	EX_U
	E_{UO}		Effective power of non-process fluids	EX_{UO}
Process energy utilization	E_{AR}		Endothermic reaction heat	EX_{AR}
	E_{RE}		Exothermic reaction heat	EX_{RE}
	E_N		total process energy used	EX_N
	E_{TT}		Thermodynamic physical energy difference	EX_{TT}
	E_T		Thermodynamic energy consumption	D_T
	E_{UD}		Dissipate heat of Energy utilization link	EX_{UD}
			Process exergy loss	D_{KP}
Energy recovery and recycling	E_O		Energy to be recovered	EX_O
	E_R		Recovery and recycling energy	EX_R
	E_{RH}		Heat exchanger recovery and recycling energy	EX_R
	E_{RM}		Streams recovery and recycling energy	EX_R
	E_{UD}		Dissipate heat of energy utilization link	EX_{UD}
	E_E		Energy recovered for output	EX_E
	E_J		Rejected energy	D_{JR}

(continued)

(continued)

	E_{JC}	Cooling rejected energy	D_{JC}
	E_{JD}	Heat dissipation rejected energy	D_{JD}
	E_{JM}	Streams rejected energy	D_{JM}
	E_{JO}	Other rejected energy	
	E_N	Total process energy used	EX_N
		Process exergy loss	D_{KR}
Summary Indicators	E_A	Net energy (exergy) consumption	D_A
	E_C	Converted to primary energy (exergy) consumption	EX_C
	η_U	Energy conversion efficiency %	η_{XU}
	η_P	Energy process utilization efficiency%	η_{XP}
	η_R	Energy recovery efficiency %	η_{XR}
		Total rejection energy	D_J
		Total process exergy loss	D_K

The energy balance of the plant is a preliminary form of energy analysis, which is the balance of energy quantity. Using energy balance can find equipment and positions with large energy loss, and many energy-saving potentials and measures are proposed based on this. However, it is inevitable that only the phenomenon of energy loss, it is difficult using energy balance to find the root of causes for the real high energy consumption. Therefore, it is necessary to use the concept of energy quality to guide. For example, the cooler discharges a large amount of energy, but this energy couldn't be recovered easily; efforts must be spent on the heat exchange network to increase the heat transfer coefficient and reduce the heat transfer temperature difference, thereby increasing the heat recovery rate and reducing the cooling waste energy.

If the exergy balance calculation of the plant is performed in conjunction with the energy balance, it will be found that the reason for the low energy grade of cooling and rejection energy is that the energy is degraded in the process equipment, and the energy quality is gradually degraded during the recovery process, and finally has to be cooled and discarded. The potential for improvement lies in reducing the degradation of energy used in the process equipment, in reducing the process exergy loss of the recovery system, and avoiding a large amount of energy being reinjected by cooling. Improve the recovery efficiency and reduce the energy supply of the plant from the outside.

It should be pointed out that the concept of exergy (energy quality) has long been used by most engineering and technical personnel (perhaps subconsciously). For example, in the tower design, the proportion of reflux heat from the bottom of the tower is increased to recover more heat energy. The heat exchange network synthesis is designed according to the temperature of the cold and hot flow, it is just that there is no systematic and comprehensive consideration.

The exergy balance analysis is a comprehensive and systematic evaluation of energy quality. Indicate the magnitude of exergy loss in each energy-using device. Combined with the process operating parameters, we can point out the potential and measures for improvement. For example, through exergy analysis and calculation, it can be pointed out that the different heat removal ratios in each part of the tower will cause the exergy loss in the process of the tower and the impact on the energy recovery system. Therefore, the energy utilization analysis of the plant (energy balance combined with exergy balance) can find out the amount of energy loss and energy degradation locations, so as to propose improvement directions and specific improvement measures to achieve the purpose of reducing the energy consumption and CO_2 mitigation of the plant.

The previous chapter focused on the specific calculation methods of energy balance and exergy balance of energy-using equipment. This chapter discusses the energy data balance method of the plant and its system, and combines the balance of the equipment to summarize the energy balance and exergy balance of the plant.

7.1 Balance Verification of System Energy Consumption Data in the Plant

According to the three-link model of energy use and its balance relationship established in Chap. 5, in order to complete the energy balance and exergy balance of the plant, in addition to energy-using equipment, it is necessary to verify and account for the energy use data of other utility systems of the plant [1]. This section and the Sects. 7.2 and 7.3 are the verification of the system energy consumption data in the plant.

7.1.1 Process Plant Material Balance

The material balance of the plant is the basis for energy consumption analysis and calculation, so it is necessary to accurately reflect the material balance of the device during the test period.

Table 7.1 Material balance of production plant

Inlet			Outlet		
Description	Feed rate, kg/h	Yield, %	Description	Products, kg/h	Yield, %
Total			Total		

(1) The flow data should be checked with the gauge data of the container (oil tank, gas tank, etc.) on the basis of meter measurement. The test instrument should be checked beforehand, and its accuracy should comply with the relevant standards to ensure that the material balance data objectively reflects the actual situation during the test period.
(2) Raw materials and products must be mass balanced, processing losses are included in the output, and carefully checked.
(3) Some gas at the top of the tower is usually not metered under normal conditions. Temporary measures can be used for measurement. Such as measuring with anemometer. When it is used as a furnace fuel, it can be determined by deactivating the gas as fuel, observing the change in the amount of liquid or gas fuel in the furnace (the same working condition), and combining with the analysis of the gas composition.

The material balance results of the device can be filled in Table 7.1.

7.1.2 Thermodynamic Energy Consumption

As mentioned earlier, thermodynamic energy consumption is the difference between the total energy of feed raw materials and products of the plant, including physical energy and chemical energy, which is the theoretical minimum energy consumption of the plant. That is, the difference between the energy of the raw material at the temperature and pressure of entering the plant and the energy of the product at the temperature and pressure of the output plant. Here we mainly discuss how to calculate and determine.

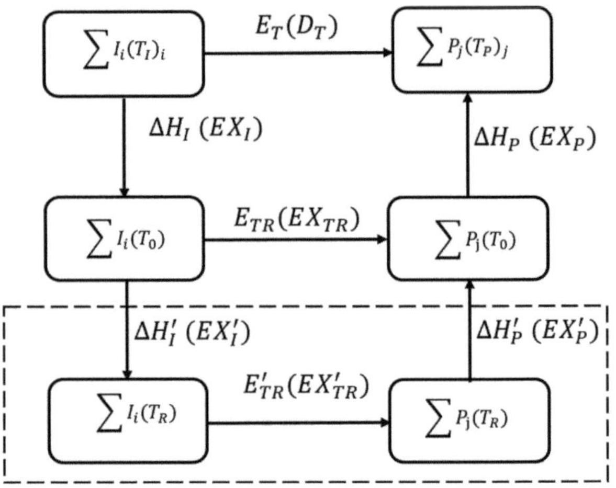

Fig. 7.1 Thermodynamic energy consumption calculation approach

Regardless of the chemical energy of raw materials, thermodynamic energy consumption consists of two contents: (1) the difference in physical energy between the raw materials entering the plant and the products leaving the plant, it is not only due to the difference in temperature and pressure of the raw materials and products, but also due to the products composition different as well as the phases different. (2) Reaction heat energy consumption at reference temperature. Thermodynamic energy consumption can be calculated by designing a reversible path as shown in Fig. 7.1.

The thermodynamic energy consumption E_T is calculated by the following formula:

$$E_T = E_{TT} - E_{TR} \tag{7.1}$$

$$E_{TT} = \Delta H_I + \Delta H_P, E_{TR} = E_{AR} - E_{RE} \tag{7.2}$$

where: E_{TR}—Reaction heat energy consumption at reference temperature;

E_{TT}—The physical energy difference between raw materials and products, ref to Table 7.2;

E_{AR}—Endothermic reaction heat;

E_{RE}—Exothermic reaction heat;ΔH_I, ΔH_P is the enthalpy difference of the raw material or product, specified as the final state minus the initial state.

It can be seen from Fig. 7.1 that the reaction heat energy consumption for calculating the thermodynamic energy consumption refers to the reaction heat at the reference temperature. For a chemical process reaction with a known composition, the heat of formation of the reactants and products can be easily found from the relevant manuals, and the reaction heat can be obtained from Guess's law. Most of the manuals contain physical property data at 25 °C. Therefore, the calculated heat of

Table 7.2 Product and feed physical energy (exergy) difference summary table

Description		flow rate kg/h	Temperature °C	Pressure MPa	Pressure exergy kJ/kg	Thermal Energy kW	Thermal Exergy kW
Feed							
	Total						
Product							
	Total						
Physical energy (exergy) difference							

reaction is basically the reaction heat at the reference temperature and does not need to be corrected. However, for a large number of complex petrochemical reactions, it is difficult to accurately know the molecular structure and composition of the raw materials, products and reaction types. Therefore, it is difficult to calculate the reaction heat at the reference temperature from the manual data. In engineering design and operation, the reaction heat is mostly estimated through heat balance or empirical formulas. This reaction heat is mostly the reaction heat under operating conditions. It can be converted into the reaction heat at the reference temperature through the calculation approach of the part of the dashed frame in the lower part of Fig. 7.1

$$E_{TR} = E'_{TR} - (\Delta H'_P + \Delta H'_I) \tag{7.3}$$

For those are difficult to check the reaction heat at 25 °C from the manual data, but the reaction heat at the operating temperature can be directly estimated from the path shown in Fig. 7.2, and it is not necessary to return to the reference state for calculation.

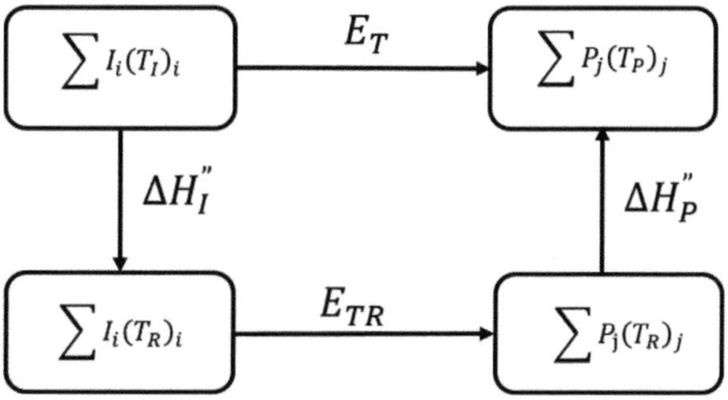

Fig. 7.2 Thermodynamic energy consumption calculation diagram

Thermodynamic energy consumption

$$E_T = E_{TR} + (\Delta H''_P + \Delta H''_I) \quad (7.4)$$

This kind of estimation is more convenient. However, Fig. 7.1 is of practical significance for clarifying the concept and solving the calculation of the known composition reaction. According to Fig. 7.2, it should be noted that $(\Delta H''_P + \Delta H''_I)$ is not the usual physical energy difference.

In actual operation, since the reaction heat effect is mostly not isothermal reaction, but has a certain temperature gradient, the temperature gradient is usually not large at this time, or when the temperature difference between the inlet and outlet is not large due to the heat removal (supply), it can be regarded as an isothermal reaction process. And take the average temperature of temperature in and temperature out. For catalytic cracking, the average reaction operating temperature can be used. The reformer can calculate the average reaction temperature and reaction heat of each reactor in stages.

Like the difference in physical energy, the energy consumption of reaction heat can be positive or negative. It is stipulated that the endothermic reaction heat is positive and the exothermic reaction heat is negative. This is because the endothermic reaction transfers the energy supplied to the reaction to the product, which changes the molecular structure, and is an effective part of energy consumption; the released heat from exothermic reaction caused by the change in the molecular structure and composition of the raw material, and it does not require energy supplied from the outside, the energy consumption is negative, which means that the heat effect of the exothermic reaction can be fully utilized and the energy consumption of the plant can be reduced. The good utilization of exothermic reaction heat is an important aspect of energy saving, which is worthy of study and attention. The low energy consumption of large-scale ammonia plants is mainly due to the better utilization of reaction heat, high and medium pressure steam is generated to drive the back pressure turbine, which not only saves power consumption, but exhaust steam can also be used further in processes. For the synthetic ammonia of small chemical fertilizers, intermediate chilling is used in the medium CO transformation, and the exothermic reaction is not fully used, and the energy consumption per ton of ammonia is very high.

7.1.3 Thermodynamic Exergy Consumption D_T

Corresponding to thermodynamic energy consumption, thermodynamic exergy difference indicates the total exergy difference between the device product and the raw material input to the plant. For the energy analysis system excluding chemical energy (exergy), it can be calculated by the exergy consumption of the reaction heat exergy and the physical exergy difference between the product and the raw material. The calculation approach is the same as that shown in Fig. 6.1.

$$D_T = EX_{TT} + EX_{TR} + EX_{TS} \tag{7.5}$$

$$EX_{TT} = EX_I + EX_P \tag{7.6}$$

where: EX_{TT}—physical exergy difference;

EX_{TR}—Reference reaction exergy. EX_{TR} consists of two parts: endothermic reaction heat EX_{AR}, exothermic reaction heat EX_{RE}, according to regulations, we have $EX_{TR} = EX_{AR} - EX_{RE}$.

The physical exergy difference is calculated and determined based on the material balance of the plant and the parameters such as the yield of each stream, temperature and pressure. Physical exergy difference includes two parts: heat exergy and pressure exergy. For calculation of heat exergy and pressure exergy, please refer to the related content in Chap. 2. Under normal circumstances, the pressure exergy is very small compared to the heat exergy, so the calculation of pressure exergy is not considered; for high-pressure operation devices, when the pressure exergy cannot be ignored, it can be calculated. The calculation result of physical exergy difference can also be filled in Table 7.2.

The chemical exergy difference refers to the difference between the chemical exergy of the product and the raw material, which corresponds to the calculation of the chemical exergy, and also includes the chemical reaction exergy and the concentration change exergy. Therefore, in the case of both a chemical reaction process and a separation process, such a chemical exergy difference includes two parts: the chemical reaction exergy and the separation exergy.

For the calculation of separated exergy, please refer to the Sect. 6.5 of Chap. 6.

Chemical reaction exergy difference means that a chemical reaction occurs at the reference temperature and pressure T_0, p_0, which makes the raw materials and products have different chemical exergy. The difference can also be determined by the thermal effect of the reaction.

(1) When the chemical composition of the raw materials and products are known, and the corresponding thermodynamic data can be found, the following formula can be used to calculate the reaction exergy at T_0, p_0.

$$EX_{TR} = G\left(RT_0 ln\frac{K_P}{J_P} + RT_0 \sum n_i ln\frac{(p_i)_O}{(p_i)_I}\right) \tag{7.7}$$

where: K_P—the equilibrium constant under T_0, p_0;

G—Raw material or product rate;

J_P—Product and reactant quotient;

$(p_i)_O$, $(p_i)_I$—Inert reactant pressure entering and leaving reactor.

Find out the equilibrium constant K_P under the condition of T_0, p_0, the pressure quotient J_P of product and reactant, and the pressure of inert reactant entering and leaving the reaction equipment $(p_i)_O$, $(p_i)_I$, then the reaction exergy can be easily calculated.

(2) For actual complex reaction processes such as petroleum cracking, the reaction exergy EX'_{TR} under operating conditions can be obtained first, and then the reference reaction exergy EX_{TR} can be obtained from the idea of Fig. 7.1.

$$EX_{TR} = EX'_{TR} + (EX'_I + EX'_P) \quad (7.8)$$

Example 7.1 Calculate the thermodynamic energy consumption and thermodynamic exergy consumption of an atmospheric and vacuum distillation unit. The temperature of raw material and product, yield data are shown in Table 7.3. The separation exergy of the initial distillation tower, atmospheric tower and vacuum tower of the plant has been calculated according to formula (6.97).

Solution Thermodynamic energy consumption and thermodynamic exergy consumption consist of two parts: physical energy (exergy) difference and chemical energy (exergy) difference. There is no chemical reaction process in the atmospheric and vacuum distillation unit, and the chemical exergy difference only includes the separation exergy of the concentration change. The calculation of physical energy difference and physical exergy difference are carried out in accordance with the method in Chap. 2. The calculation result is also shown in Table 7.3, indicated that the thermodynamic energy consumption is 122.4 MJ/t, and the thermodynamic exergy difference is 38.7 MJ/t.

7.1.4 Calculation of Energy and Exergy of Recycling and Output Streams

(1) Recycling streams energy and exergy

Recycled streams refer to the streams that does not pass through the heat exchange equipment, but directly carries energy back to the upstream process equipment by the downstream equipment. For example, the re-processing oil of the catalytic cracking unit is pumped from the fractionation tower and returned to the reactor. The corresponding recycle streams energy (exergy) is the energy (exergy) carried by stream of the downstream equipment directly back to the upstream equipment without heat exchange. The characteristic of this type of energy (exergy) is that there is no exergy loss in the indirect heat exchange process, only the energy and exergy loss caused by the heat dissipation of the transmission pipeline.

Calculation of recycling Stream Energy and exergy refer to the calculation method of corresponding streams energy and exergy introduced in Chap. 2, and the results are summarized in Table 7.4. The energy (exergy) in the table refers to the energy

Table 7.3 Thermodynamic energy consumption (exergy) consumption summary table

Description		Temperature °C	Specific density	Yield, %	Energy MJ/t	Exergy MJ/t
Crude oil		38	0.907	100	42.3002	1.617
Products	Initial tower overhead	40	0.7105	1.205	0.670	0.028
	Atm overhead gas	38	0.7357	3.632	1.797	0.069
	2nd Atm draw	40	0.7923	5.020	2.519	0.104
	3rd Atm draw	61	0.8375	12.594	12.160	0.891
	4th Atm draw	54	0.8699	2.253	1.709	0.108
	Vacuum overhead gas	38	0.8332	0.354	0.156	0.006
	1st Vacuum draw	42	0.8597	0.966	0.501	0.022
	2nd Vacuum draw	66	0.8823	14.983	14.823	1.195
	3rd Vacuum draw	90	0.9152	13.896	20.075	2.286
	4th Vacuum draw	90	0.9256	2.507	3.615	0.412
	5th Vacuum draw	142	0.9614	42.590	106.708	18.987
	Product Total			100.000	164.732	24.107
Total energy (exergy) difference					122.432	22.490
Separation exergy	Initial distillation tower					5.573
	Atmospheric tower					6.730
	Vacuum tower					3.942
Total separation exergy						16.245
Thermodynamic energy consumption (exergy consumption)					122.432	38.734

(exergy) from the outlet temperature of the downstream equipment to the reference temperature, and the energy (exergy) after deducting the heat loss energy (exergy) is called the net recycling energy (exergy).

The heat loss exergy is calculated by the average temperature of the fluid in the pipe according to formula (2.105).

The energy and exergy of each stream are balanced and checked according to the following formula.

Table 7.4 Recycle streams energy (exergy) recovery summary table

Item	Stream name	Flow rate kg/h	Relative density	Pressure MPa	Temperature °C	Bring out energy kW	Bring out exergy kW	Heat loss kW	Heat loss exergy, kW	Net Energy kW	Net exergy kW	Remarks
Total												

$$\text{Brought out energy (exergy)} = \text{Net recycling energy(exergy)} + \text{Heat dissipation energy (exergy)}.$$

(2) Output streams refers to the streams used in the plant product or intermediate product to carry energy outside the plant (downstream plant or system) or in the conversion link of the plant. Usually called hot discharge stream. In order to simplify the energy consumption analysis model, the energy recycled from the recovery link back to the conversion link is classified as heat output, and the supply side is also included this heat output. The hot discharge stream to the outside of the plant will gradually increase as the heat integration deepens. For example, the hot wax oil from the atmospheric and vacuum unit is output to the catalytic cracking unit, and the hot vacuum residue is transported to the coking unit etc.

Since this kind of stream has a certain temperature in the case of cold discharge, it is difficult to use by both the supplier and the buyer. Even if the plant is not hot discharged, the energy is difficult to use to the reference temperature. Therefore, in order to benefit both the supplier and the buyer and be accepted by both parties, according to the different properties of the streams (oil products), a specified temperature is proposed for the output streams. The energy and exergy of the output stream are calculated to the specified temperature. The energy and exergy from the specified temperature to the reference temperature are included in the thermodynamic energy consumption (exergy difference). For example, the specified temperature of the output streams of the oil refining industry is 60 °C for gasoline, 80 °C for diesel and wax oil, and 100 °C for heavy oil.

The calculation of output stream energy and exergy can be performed by the temperature, pressure and related physical properties of the stream leaving the plant according to the method introduced in Chap. 2, and the results are summarized in Table 7.5. The energy of the output stream should be the energy after deducting the heat loss. However, when the device is calibrated, it is difficult to measure the inlet temperature of the downstream device, and it is difficult to accurately estimate the heat loss of the system. The device output temperature can be used in the energy balance (at this time, the specified temperature can be adjusted appropriately), which should be deducted when the energy consumption is counted.

Table 7.5 Summary table of recovery output streams energy

Item	Stream Name	Flow rate kg/h	Relative density	Plant output stream		Specified Temp. °C	Energy kW	Exergy kW	Stream work kW
				°C	MPa				
Total									

In the output streams, if there is a higher pressure, and installing a power recovery equipment can make the downstream device or system power saving or reducing, the effective work of high pressure streams shall also be counted, and other circumstances shall not be counted. For effective work calculation, see Chap. 4 Power Recovery Equipment.

7.1.5 *Streams Rejection Waste Energy and Exergy*

Stream rejection energy (Exergy) is the energy (exergy) that is carried into the environment in the form of stream in the energy recycling and recovery link. Including process steam condensate, plant heating and tracing condensate, waste heat steam generator blowdowns, electric desalination water, various washing water and coking cold coke water, etc. When the plant is not well energy managed, the amount of non-process steam used increases, and most of them are rejected in a state of exhausted steam, which greatly increases the stream reinjection energy. The streams reinjection energy and exergy should be calculated to the reference temperature. You can also refer to the aforementioned calculation method of streams energy and exergy. The results are summarized in Table 7.6.

However, due to the streams in the environment that are discharged in the conversion process, such as heating furnace flue gas, incomplete combustion products, waste boiler blowdown, etc. have been included in the loss of the conversion process, which will not be included here.

The reinjection energy of streams is mainly heat energy, and individual high-pressure stream rejected into the environment, and the pressure energy is also accounted for, such as the venting of synthetic ammonia gas. Pressure energy can be filled in the remark's column in Table 7.6.

Table 7.6 Summary table of plant stream discharge energy

Item	Stream name	Flow rate kg/h	Temp. °C	Energy, kW	Exergy, kW	Remarks
Total						

7.2 Plant Heat Loss Verification and Summary

The heat dissipation loss of the plant is distributed in the three links of the energy use process. The heat dissipation of the energy conversion and energy use links has been determined by calculating the energy balance and exergy balance of energy-using equipment. The heat dissipation of the energy recovery and recycling link is mainly composed of equipment and pipelines. The energy balance and exergy balance calculations of the heat exchange equipment and power recovery equipment have also been determined. It is still necessary to determine the heat dissipation loss of the pipeline (fittings).

7.2.1 Analysis of the Characteristics of Heat Dissipation

Except for the thermodynamic energy consumption that enters the product or is brought out by the product, the energy consumption of the plant is rejected to the environment in the form of heat dissipation, cooling, and stream rejection. Moreover, the heat dissipation energy consumption occupies a relatively high proportion, and the heat dissipation comes from the high temperature fluid, and the exergy value is also high. Due to the heat dissipation, the same or even more energy (fuel) consumption will be supplied from the outside world.

The heat dissipation in the energy conversion and energy use links reduces the temperature of the process stream, in order to compensate for the heat dissipation, the conversion link must supply equivalent effective energy. Considering the furnace efficiency factor, more fuel will be added.

The heat dissipation in the recovery and recycling link reduces the temperature of the streams to be recovered, the energy recovery rate decreases, and ultimately fuel consumption increase to make up for the decline in the recycling and recovery efficiency caused by heat dissipation, but because the energy recovery has a recovery efficiency and is close to the conversion efficiency, the general recovery and recycling process heat dissipation, Will increase the energy equivalent to the heat loss. The cooling rejection energy are rejected after being recovered and degraded to the extent that it can no longer be used, and its energy grade is very low. Therefore, reducing heat dissipation is more practical than reducing cooling.

Table 7.7 shows the energy consumption analysis of heat dissipation and cooling for a certain integrated Crude distillation-catalysis unit.

It can be seen from Table 7.7 that although the heat dissipation energy only accounts for 28% of the total rejection energy, exergy accounts for 60.4%; on the contrary, the cooling energy accounts for 56.1% of the total rejection energy, and the rejection exergy only accounts for 20.47%.

In order to reduce heat dissipation, we must first find out the current situation and determine the location and value of heat dissipation. This requires heat dissipation testing and verification work.

Table 7.7 Heat dissipation and cooling waste energy table

Description	Energy		Exergy	
	MJ/t	%	MJ/t	%
Total exhaust energy	189.77	100.0	41.26	100.0
Cooling	106.45	56.1	8.83	20.47
Heat dissipation	53.08	27.97	24.93	60.43

7.2.2 Calculation Summary of Heat Loss of Equipment and Pipeline

The heat dissipation loss should be determined by actual measurement and calculation. There are two methods for heat dissipation calculation: one is to calculate using the surface area of the equipment, surface temperature and environmental conditions (wind speed, temperature), etc., first obtain the combined radiation and convection heat dissipation coefficient, and then calculate the heat dissipation; the other uses streams calibration to calculate heat loss by the enthalpy drop method from the T, p at beginning and the end of the pipeline. However, due to the large heat capacity of the fluid, the temperature drop due to heat dissipation is not large; while the temperature measurement has errors, and the accuracy is low for short distances. However, the two methods should be cross checked against each other. After calculated the amount of heat dissipation; the heat dissipation exergy can be calculated by the internal average fluid temperature according to the formula $EX_D = E_D\left(1 - \frac{T_0}{T_b}\right)$.

Pipeline (pipe fittings) heat loss can be summarized in Table 7.8. This table only summarizes the heat dissipation of the pipeline (fittings) for the recycling and recovery link, and the heat dissipation of the equipment has been summarized in the corresponding equipment balance sheet. The difficulty of the pipeline heat loss test lies in the determination of the exposed surface area of the pipe. Valves and flanges are irregular surfaces, and it is difficult to calculate and determine their surface area with a simple method.

For 350 °C oil products, the heat dissipation loss of one Dg250 bare valve is equivalent to the heat dissipation loss of nearly 100 m of well-insulated pipelines. Literature given the surface area and equivalent pipe length of the valves and flanges of the petrochemical plants in Japan [2]. A certain water pump factory in China has performed calculations on the surface areas of the valves and flanges, and the results are shown in Tables 7.9.

Table 7.8 Pipeline (fittings) heat loss summary table

Item	Pipe (fittings) Name	Nominal Diameter	Pipe length m	Insulation thickness mm	Insulation material	Surface Temperature °C	Surface area, m2	Heat flux kW/m2	Heat Loss coefficient W/ (m2.K)	Heat loss kW	Heat loss exergy kW	Remarks
Total												

Table 7.9 Outer surface area of valve and flange

Nominal diameter	Dg50	Dg80	Dg100	Dg150	Dg200	D 250
Surface area (m^2)	0.31	0.46	0.58	1.02	1.41	1.75
Equivalent length (m)	1.73	1.645	1.62	2.04	2.05	2.27

7.2.3 Pipeline Heat Dissipation Verification and Summary of Plant Heat Dissipation

Due to the accuracy of the heat dissipation test and the estimation accuracy of the surface area of the pipeline (pipe fitting), and the gaps between the insulation shells are not estimated, there are errors in the test results and should be verified. In addition to the verification by the enthalpy drop method mentioned above, it should be able to meet the total energy balance of the pipeline streams. That is, the energy carried by each stream of the energy to be recovered from the process equipment should be the sum of the following four items:

The total heat released by the stream in the heat exchanger (maybe more than one); the energy released by the cooler; the energy carried by the plant output streams; Pipeline heat dissipation. If it is unbalanced, the cause should be found and corrected.

As mentioned earlier, the heat dissipation of petrochemical plants is distributed in the three links of the energy use process. After completing the equipment balance calculation and pipeline (fittings) heat dissipation calculation of the energy use process, the heat dissipation of the entire plant can be summarized by link. Thereby there is a general result for the heat dissipation of the plant, and the proportion of heat dissipation to total energy consumption and the focus of further improvement are grasped. The heat dissipation of the plant can be summarized in Table 7.10. When filling in the form, the heat dissipation of the conversion link and process energy use link should be filled out for each equipment.

7.3 Balance of Supply and Consumption of Steam, Electricity and Water

7.3.1 Steam

Steam is a relatively common energy-carrying medium used in process plant. Energy analysis requires not only the amount of steam supplied, but also the usage details, so as to provide a basis for the improvement potential of steam in combination with the process. With the deepening of energy saving of plants, many plants are using waste heat to generate steam. Therefore, the steam balance should include the balance self-produced and self-consumed items. In order to facilitate the summarization of energy consumption analysis, the steam used is divided into energy conversion link, process

Table 7.10 Plant heat dissipation summary

<table>
<tr><th colspan="2" rowspan="2">Description</th><th>Energy</th><th>Exergy</th><th rowspan="2">Remarks</th></tr>
<tr><th>kW</th><th>kW</th></tr>
<tr><td rowspan="5">Energy conversion</td><td></td><td></td><td></td><td></td></tr>
<tr><td></td><td></td><td></td><td></td></tr>
<tr><td></td><td></td><td></td><td></td></tr>
<tr><td>Steam line heat dissipation</td><td></td><td></td><td></td></tr>
<tr><td>Sum</td><td></td><td></td><td></td></tr>
<tr><td rowspan="5">Process equipment</td><td></td><td></td><td></td><td></td></tr>
<tr><td></td><td></td><td></td><td></td></tr>
<tr><td></td><td></td><td></td><td></td></tr>
<tr><td></td><td></td><td></td><td></td></tr>
<tr><td>Sum</td><td></td><td></td><td></td></tr>
<tr><td rowspan="5">Energy recycling & recovery</td><td>Heat exchanger</td><td></td><td></td><td></td></tr>
<tr><td>Cooler</td><td></td><td></td><td></td></tr>
<tr><td>Process pipeline</td><td></td><td></td><td></td></tr>
<tr><td>Other</td><td></td><td></td><td></td></tr>
<tr><td>Sum</td><td></td><td></td><td></td></tr>
<tr><td colspan="2">Total</td><td></td><td></td><td></td></tr>
</table>

energy use link and energy recovery link (heating and heat tracing). The balance of steam quantity, energy and exergy balance are summarized in Table 7.11 described as follows:

(1) The plant uses different pressure levels, and the steam temperature and pressure are indicated in the table.
(2) The intermittent use of steam can be converted into continuous use of steam based on the number of hours in use, and it should be noted in the remarks:

$$\text{Converted flow} = \text{intermittent flow} \times \text{conversion coefficient}$$
$$\text{Conversion factor} = \text{daily use time (hours)}/24$$

(3) The energy and exergy of the steam are carried out according to the calculation method of the steam stream in Chap. 2. But it should be calculated based on the enthalpy of water at 15 °C. When converting to primary energy consumption, consider converting to actual (comparative) energy consumption.
(4) When the steam is superheated in the energy conversion link, list this item separately, do not write the quantity, only indicate the superheating energy and exergy value.

Table 7.11 Plant steam balance sheet

Description			Flow rate kg/h	Steam parameters		Energy, kW	Exergy kW	Remarks
				°C	MPa			
Total steam supply	Supply in							
	output							
	Total							
Steam usage details	Energy conversion link							
		Sum						
	Process energy use Link							
		Sum						
	Heating & heat tracing							
		Sum						
	Steam line loss							
	Total							

(5) Pay attention to the verification of steam flow. The amount of steam supplied by the system should be in line with the steam balance of the whole plant, and the steam flowmeter parameters should be compared with the design parameters to correct the flow rate. The amount of steam generated should be consistent with the amount of deoxidized water and the amount of blowdown discharged.

(6) The heat and water converted into electricity calculated by the deoxidized water will not be put into the steam balance. In the conversion of primary energy, it can be converted back according to the actual (comparative) energy consumption of deaeration water.

7.3.2 Water

Plant water includes cooling water (fresh water, cooling water and chilled water), flushing and purging water, and deoxidized water for steam generation is also summarized here. The water consumption summary table is Table 7.12, and the description is as follows.

(1) For various types of water, the water supply should be balanced with each water usage item. For cooling water, there is generally no metering instrument for a single cooler. The heat output of the hot flow can be used to subtract the

Table 7.12 Summary table of plant water consumption (t/h)

Order	Description	Fresh water	Cooling water	Demin water or deoxidized water	Chilled water
Total					

heat loss, and the water volume can be obtained from the heat balance, or the ultrasonic flowmeter can be used to measure. But in the end, supply and use must be balanced, otherwise they should be revised one by one.

(2) The deoxidized water is balanced according to supply and consumption. In the electricity balance below, the water is converted into electricity, and the heat carried is included in the heat brought in the plant, and converted into actual (comparative) energy consumption in calculating the energy consumption of the plant.

(3) Intermittent water consumption should also be converted into continuous consumption based on the daily usage time.

In view of the purpose of water use, water is only used as a working fluid to carry energy (cooling) in the energy use of the plant, and generally does not bring energy to the plant. But in the process of producing all kinds of water, water pumps and cooling tower fans consume a lot of electricity. Therefore, in order to simplify the energy consumption analysis model, the water consumption and energy consumption are truthfully and objectively reflected. Water does not appear in the energy consumption, and the energy consumption of water is converted into electricity consumption according to the energy consumed. This can be understood as treating the water production system as a part of the plant, and the power consumption of the pump should be counted as the power consumption of the plant.

Method of water discount to electricity:

(1) Actual consumption method. The consumption of water production is mainly electricity consumption. Others, such as steam and wind, are small and vary from season to season (such as heating period) and are put into the system for unified consideration.

$$A = W/G \tag{7.9}$$

where: A—Unit fresh water (cooling water) discount index, kW h/t;

W—power consumption for water production, kW h/h;

G—Water production, t/h.

Demineralized water can also be calculated based on the actual electricity consumed in water treatment and combined with the water consumption per unit of demineralized water:

$$A^{'} = \varphi.A + W^{'}/G^{'} \tag{7.10}$$

where: $A^{'}$—demineralized water discount index, kW h/t;

φ—Water consumption of unit demineralized water, t/t;

$W^{'}$—– Electricity consumption for production of demineralized water kW h/h;

$G^{'}$— Production of demineralized water, t/h.

(2) Relative energy consumption method. According to the calculation method of energy consumption in the refinery [3], the relative energy consumption of cooling water is 4.19 MJ/t (1000 kcal/t), the fresh water is 7.54 MJ/t (1800 kcal/t).

Therefore, the relative power conversion index is: Fresh water, 0.60 kWh/t; cooling water, 0.33 kWh/t; demineralized water and deoxidized water, according to its power consumption and water consumption, take 1 kWh/t.

7.3.3 Power Consumption

The power consumption of the Plant can be divided into the following parts.

(1) The power consumption of the motor driving process pump has been calculated in the balance of the pump equipment and summarized in Table 6.3.
(2) Electricity for plant process, such as electric heating, electric refining, electric desalination, electric dust removal, electric mist removal, etc.
(3) Other electricity consumption refers to lighting and non-process fluids driving electricity. Such as the electricity consumed for water and instrument air production, and this plant uses electricity for furnaces fans and air-cooled fans. The characteristic of this kind of electricity is that it does not drive the process fluid, and its flow work does not enter the process energy use link.
(4) Plant power transmission and transformation losses

The power supplied to the plant is the sum of three parts: the power supplied by the power supply system, the self-generation of the heat and power combined system, and the plant water and air converted to electricity. In some cases, the electricity transferred from the substation of this plant to other units should be deducted. Generally, the proportion of air energy consumption is relatively small, and it is

Table 7.13 Plant power balance

Description		Electricity kW	Energy MJ/t	Remarks
Supply	Power supply			
	Self-power generation			
	Discounted power of water and air			
	Output to other units			
	Total			
Pump and compressors	Effective power			
	Ineffective power			
	Total			
Process equipment	Effective power			
	Ineffective power			
	Total			
Other	Effective power			
	Ineffective power			
	Total			
Transmission and transformation loss				
Plant total	Effective power			
	Ineffective power			
	Total			

ignored in statistical energy consumption. However, when the plant uses a large amount of wind, the unit should be considered for electricity consumption.

The difference between the total power consumption of the plant and the converted power of the water and air should be the same as the difference between the total power input to the substation and the power output to other plant. Otherwise, the reason should be analyzed, the results should be corrected.

The power balance of the device is summarized in Table 7.13.

7.4 Plant Energy Balance and Exergy Balance Calculation and Summary

After completing the energy balance (exergy balance) of the plant energy-consuming equipment and the utility system balance test calculation of other energy data of the system in the plant, it can be summarized according to the energy consumption analysis mode proposed in Chap. 5 to complete the transfer from a single equipment to the entire plant.

Table 7.14 Summary table of energy conversion links

Description			Energy		Exergy		Remarks
			kW	MJ/t	kW	MJ/t	
Total supply	Coke						
	Fuel						
	Steam						
	Electricity						
	supply Heat						
		Subtotal					
	Total supply						
	Conversion output						
Direct energy loss	Flue gas						
	heat loss						
	CO chemical energy						
	Ineffective power						
	Total						
	Process exergy loss						
Non-process effective power							
Effective utilization	Cold feed						
	Main heating						
	Effective power						
	Steam generation						
	Preheating water						
	Steam superheating						
	Total						
Efficiency (%)							

7.4.1 Energy Conversion Link

7.4.1.1 Total Energy Supply and Net Supply Energy (Exergy) $E_P(EX_P)$

The difference between the two is whether there is a conversion link output energy and exergy. The supply energy is composed of the following components.

(1) Fuel energy and exergy of the heating furnace: calculated from the sum of fuels balanced by the equipment in Chap. 6. Note that the regenerator burning oil and CO boiler supplementary fuel should be included in this item.

(2) Catalytic cracking coke burning $E_{PG}(EX_{PG})$: It is only useful for the summary of the catalytic cracking unit and can be filled in by the calculation result of the catalytic cracking regenerator.
(3) Steam supply: Resulted from the steam balance results in Sect. 7.3 of this chapter. Energy and exergy are calculated based on actual steam parameters.
(4) Power supply: derive from the results of the power supply balance table in Sect. 7.3 of this chapter. Including the actual power consumption of the plant, discounted power for water use, industry and instrument air.
(5) Supply heat: The heat supplied to plant, which consists of the following parts.

(1) Hot feed. From the specified temperature to the temperature of the raw materials entering the plant, the energy and exergy can be calculated in accordance with the stream calculation method in Chap. 2.

The energy from the specified temperature to the reference temperature is the raw material input heat, which is reflected in the thermodynamic energy consumption. When the energy conversion equipment of this plant burns the hot oil of the external plant, the sensible heat above the reference temperature should also be included.

(2) Heat exchange between plants. As the heat integration deepens, it has become common to use excess heat between plants to heat raw materials. At this time, the energy obtained by the raw material from the external plant should be included in the heat supply item. For example, the top crude oil of the atmospheric and vacuum unit of the refinery is heated by the circulating oil slurry of the catalytic cracking unit to increase the final heat exchange temperature.
(3) The energy brought in by the utility. Such as the thermal energy brought in by deoxdized water.
(4) In order to simplify the energy consumption analysis model, the energy recovered and recycled to the conversion link of the plant is also called input heat; it is also counted as output heat on the output side. In this way, the large loop that exists objectively in the energy use process is eliminated, and the energy use analysis model is simplified. For example, the preheating of furnace air and the medium and low-pressure steam generated in the recovery link are used in the equipment of the conversion link.

When the selected reference temperature is different from the ambient air temperature, the energy obtained during the preheating process of the air should be taken.

Total energy supply:

$$E_P = E_{pF} + E_{pE} + E_{pG} + E_{pS} + E_{pH} \tag{7.11}$$

Total exergy supply:

$$EX_P = EX_{pF} + EX_{pE} + EX_{pG} + EX_{pS} + EX_{pH} \tag{7.12}$$

(6) Energy output from the conversion link E_B, EX_B

The energy supplied from the conversion link is mainly: ① the back pressure steam output from plant high and medium pressure steam turbine in the conversion link. At this time, the incoming party should also account for the amount of high and medium pressure steam supplied in; ②When the furnace load rate is low, in order to improve furnace efficiency and energy saving, the part of the convection section and even the radiant section of the furnace may be used to heat the external plant stream, superheat steam, and heat carrier, it will provide output heat energy; ③ in order to improve the exergy utilization efficiency of fuel, heat and power combined generation provides power to external plant etc. are included in the conversion and outgoing energy E_B and Exergy EX_B

(7) Net supplied energy and net supplied exergy

$$E_P^{'} = E_P - E_B \tag{7.13}$$

$$EX_P^{'} = EX_P - EX_B \tag{7.14}$$

If there is no conversion of output energy, the supplied energy and the net supplied energy are the same.

7.4.1.2 Direct Loss of Energy $E_W(EX_W)$

In the energy supplied, part of the energy is directly lost to the environment during the conversion and transmission process. This part of the energy loss is lost before the process energy being used, and has nothing to do with the process energy use. In general, it includes three aspects:

(1) Stream rejection energy loss $E_{WX}(EX_{WX})$

Mainly include: (1) the sensible heat loss of exhaust flue gas; (2) waste boiler blowdown; (3) fuel atomization steam used for the furnace, the exhausted steam energy discharged with the flue gas; (4) leakage of steam transmission; (5) carbon particles and combustible gas (CO, etc.) produced by incomplete combustion. All above can be summarized in the calculation results of the conversion link equipment.

(2) Direct heat dissipation loss $E_{WD}(EX_{WD})$

Refers to the heat loss on the surface of equipment and pipelines during the energy conversion process. Including: (1) heating furnace, regenerator, etc. surface heat dissipation; (2) external heat output transmission process heat dissipation, such as hot feed intermediate storage tank and pipeline heat dissipation, steam pipeline heat dissipation (sometimes small, can be ignored); (3) The heat dissipation of the intermediate transmission pipeline of main heated stream, this part of heat dissipation is generally not reflected in the furnace efficiency, but because this heat dissipation will reduce the energy entering the process equipment, it should be verified in the conversion link.

(3) Ineffective power E_{WP}

Refers to the difference between the input energy and the output effective work during the power conversion process. It is divided into two parts: (1) the ineffective power of the motor driven pump or compressor, including the efficiency loss of the pump or compressor and the efficiency loss of the motor, is also equal to the process exergy loss of the pump; (2) the ineffective power of the steam-driven pump or compressor is the difference between supply of energy and effective power. The supplied energy should also be calculated based on the actual enthalpy value (difference) of the steam.

Ineffective power is reflected the process exergy loss in the exergy balance. The value is not equivalent to the ineffective power exergy, that is, the ineffective power of the steam-driven pump should be converted into the exergy value. The ineffective power and exergy loss of the pumps have been summarized in Table 6.3. When summarized as the plant balance, it can be directly checked from Table 6.3. The explanation is as follows: (1) If the energy of the condensing turbine regenerator is used, it should be deducted from it; (2) the loss of power transmission and transformation is also included in the ineffective power; (3) the ineffective power of converted electricity of water, and air is also included in the ineffective power. In the balance, it can be carried out according to the verification results of the actual water plant (air compressor station).

Direct loss of energy:

$$E_W = E_{WX} + E_{WD} + E_{WP} \tag{7.15}$$

Direct loss of exergy:

$$EX_W = EX_{WX} + EX_{WD} \tag{7.16}$$

7.4.1.3 Effective Power of Non-Process Fluids E_{UO}

Refers to the effective power that has not entered the process fluid and therefore cannot be brought to the process equipment. Mainly the effective power of water, air converted electricity, the effective power of air-cooled fans and furnace fans, and electricity for lighting. The details are summarized by the other electricity columns in the plant electricity balance sheet.

The effective power of non-process fluids is also a conversion link effectively supplies exergy EX_{UO}.

7.4.1.4 The Conversion Link Provides Effective Energy E_U, Effective Exergy EX_U

Refers to the energy and exergy that can be supplied to the process energy use link after deducting the of the conversion supply energy to external plant and the effective power of the non-process fluid after the external loss and internal irreversible loss in the conversion and transmission link.

The energy (exergy) supplied by the conversion link mainly includes: (1) the effective energy (exergy) of main heated stream of the heating furnace or the catalytic cracking regenerator; (2) the effective power output by various pumps (excluding the effective power of non-process fluids); (3) Energy and exergy carried by steam entering the process energy use link; (4) Energy and exergy carried by hot feed or external heating. Generally, it is required to sum up from forward balance. The above four items have been balanced in the corresponding equipment and system and the results can be summarized. However, the summary result must be consistent with the result obtained by the balance, and check:

$$E_U = E_P - E_B - E_{UO} - E_W \tag{7.17}$$

The process exergy loss of the conversion link is D_{KU}, which can be obtained by adding up the process exergy loss of each conversion equipment. Mainly the combustion and heat transfer irreversible exergy loss of the furnace where the chemical energy of the fuel (coke) is converted into the heat energy of the process stream, the ineffective power of the motor driven pump and compressor and the exergy loss in the ineffective power of the steam-driven pump or compressor. And pay attention to the exergy loss outside the conversion equipment, such as the mixed exergy loss of different temperature streams, the throttling exergy loss (desuperheater for steam letdown) and the power transmission loss.

$$EX_U = EX_P - EX_B - EX_{UO} - EX_W - D_{KU} \tag{7.18}$$

7.4.1.5 Energy Conversion Efficiency

$$\eta = \frac{(E_U + E_{UO} + E_B)}{E_P} \times 100\% \tag{7.19}$$

Exergy conversion efficiency

$$\eta_x = \frac{(EX_U + EX_{UO} + EX_B)}{EX_P} \times 100\% \tag{7.20}$$

The energy balance results of the conversion link can be summarized in Table 7.16.

7.4.2 Energy Process Use Link

In order to determine the total process energy E_N and the total process exergy EX_N, we must first summarize the recycled energy (exergy) from the energy to be recovered (exergy).

1. Recycling and recovery energy(exergy)

It refers to the energy (exergy) recovered from the energy (exergy) to be recovered and recycled to the process energy use link of the plant. It includes three parts:

(1) Heat exchange recovery and recycling energy (exergy). The hot streams that have the energy to be recovered enters the process utilization link through heat exchange equipment, preheating raw materials or generated steam is called heat exchange recovery and recycling energy (exergy), which can be summarized in Tables 6.9. In the case of steam generation, the quantity entering the process utilization link shall be checked, and the quantity shall not be recounted.
(2) Recycling streams recovery energy (exergy). It is the Stream Energy (exergy) that is directly brought back to the upstream equipment from the downstream equipment. For example, the reprocessing oil of the catalytic cracking unit returns to the reactor from the fractionation tower. Although this method increases the energy (exergy) recovery, because the circulating streams brings more energy in the process equipment, generally speaking, the total energy consumption of the process is increased, which is not good for energy saving, so the catalytic plant are reducing the reprocessing oil ratio in order to achieve more energy-saving, but according to the destination and classification of energy balance, it should be classified here. The recycling stream recovery energy can be summarized in Table 7.4.

2. Total process energy E_N and total process exergy EX_N

The total process energy E_N reflects the amount of energy used by the plant process equipment under its equipment operation conditions, and reflects the design and operation level. According to the form of energy use, it can be divided into three categories: total process energy used in steam E_{NS}, total process energy used in heat E_{NH}, and total process energy used in power E_{NP}. Generally, in the balance calculation, the total energy consumption E_N can be calculated from the total effective energy output, the recovery and recycling energy and the heat release during the reaction process.

$$E_N = E_U + E_R + E_{RE} \tag{7.21}$$

Then the total process steam energy E_{NS} and the total process power energy E_{NP} determined by the aforementioned steam and electricity balance, and then the total process heat energy E_{NH} is obtained inversely.

$$E_{NH} = E_N - E_{NS} - E_{NP} \tag{7.22}$$

More in-depth, the total process heat E_{NH} should be obtained from each sub-item of the process equipment for verification.

Total process exergy used not only reflects the energy consumption of process equipment from the quantity of energy, but also the quality of energy. Total process exergy used can also be divided into three categories: steam, heat and power. The summarization and calculation methods are also similar to the total process energy consumption of the process, first find out EX_N:

$$EX_N = EX_U + EX_R + EX_{RE} \tag{7.23}$$

where EX_{RE} is the exergy of the exothermic reaction heat under its operating conditions. Then determine the EX_{NS} and EX_{NP} from the steam balance and the electric balance, and finally find the EX_{NH}.

$$EX_{NH} = EX_N - EX_{NS} - EX_{NP} \tag{7.24}$$

3. Thermodynamic energy consumption and exergy consumption

Refer to the first section of this chapter for the calculation method.

When the heat carried by the raw materials is greater than the heat carried by the products, the physical energy difference E_{TT} is negative. The opposite is positive.

When the reaction process is exothermic, the thermal effect is negative, and the endothermic is positive. Therefore, when the raw material temperature is high, the product temperature of the output plant is low, and the process is exothermic, the thermodynamic energy consumption is negative, and the corresponding thermodynamic exergy consumption is also negative vice versa.

4. Heat dissipation E_D, EX_D and process exergy loss D_{KP} in the process utilization link

The energy loss in the process utilization link is the heat dissipation loss of the equipment. It should not be attributed to the heat dissipation in the energy recycling and recovery link, otherwise it will increase the amount of (energy) exergy to be recovered [3]. It can be summarized by the balance calculation results of the process equipment in Chap. 6.

The irreversible exergy loss of the actual process can be divided into the heat and mass transfer exergy loss of the tower equipment and the heat transfer and reaction process exergy loss of the reaction equipment according to its type. A large amount exergy is lost to the irreversibility of the process, thereby reducing the amount of exergy to be recycled. The process exergy loss in the process energy use link can be calculated by the exergy loss sum of the process equipment that forms the core process of the plant. And pay attention to the exergy loss of the fluid flow process of the pressure difference entering and exiting the process equipment. It can be summarized from the calculation results of the balance of process equipment in Sect. 6.5 of Chap. 6.

5. Energy to be recovered E_O and exergy to be recovered EX_O

The energy to be recovered reflects that the energy that leaves the process energy use link and enters the energy recovery link "waiting" for further recovery and recycling, it can be obtained by summarizing the energy (heat, power, steam) carried by each product leaving the process equipment and the heat removal stream. The stream carries the energy calculated to the temperature of the product out of the plant. It can also be calculated from the energy parameter balance of the process energy utilization link:

$$E_O = E_N - E_T - E_{UD} - E_{RE} \tag{7.25}$$

$$EX_O = EX_N - D_T - D_{JP} - D_{KP} - EX_{RE} \tag{7.26}$$

6. Process energy use link efficiency

The energy efficiency of the energy use link is only caused by heat dissipation. Compared with the total process energy consumption of the process, the proportion of heat dissipation is small, and the energy efficiency is close to 1. For the exothermic reaction heat $E_{RE}(EX_{RE})$, it has been included in E_N as the supply energy. Therefore, E_T and E_O are both effective energies, and the energy efficiency is:

$$\eta = \frac{(E_O + E_T)}{E_N} \times 100\% \tag{7.27}$$

Exergy efficiency is an important indicator that reflects the characteristics of energy use in the energy use link more than energy efficiency. First of all, the process energy use link is the core process of the plant's energy use; serving for it is the two links of energy conversion and energy recovery and recycling. Although almost all the energy provided by the two links is effectively output in terms of quantity, the quality and grade of the energy are greatly reduced, which is fully reflected in the exergy loss in the exergy balance process. Therefore, the exergy efficiency is much less than 1. Objectively reflects the attribute of energy degradation in the process energy use link.

$$\eta_x = \frac{(EX_O + D_T)}{EX_N} \times 100\% = \left(1 - \frac{D_{KP} + EX_D}{EX_N}\right) \times 100\% \tag{7.28}$$

The balance results of the process energy utilization link can be summarized in Table 7.15.

Table 7.15 Summary of energy utilization link

Description		Energy		Exergy		Remarks
		kW	MJ/t	kW	MJ/t	
Effective supply						
Recovery and recycling						
Reaction exotherm						0 for endothermic reaction
Total process energy	Steam					
	Flow energy					
	Process heat					
	Total					
Thermodynamic energy consumption	Thermal energy (exergy) difference					
	Reaction heat					
	Total					
Process exergy loss	Reactor					
	Fractionator					
	Other equipment					
	Flow exergy loss					
	Total					
Heat dissipation						
Energy to be recovered						
Efficiency, %						

7.4.3 Energy Recovery and Utilization Link

The energy (exergy) discharged from the process use link enters the energy recovery and recycling link, and is generally recovered and recycled and rejected in four forms: recovery and recycling, recovery output, rejecting losses to the environment, and irreversible exergy loss during the recovery and recycling process.

1. Recovery and recycling energy $E_R(EX_R)$

The results have been summarized before.

2. Recovery output energy $E_E(EX_E)$

Refers to the energy (exergy) recovered from the energy (exergy) to be recovered and used outside of the plant energy process utilization link. As mentioned above, for the simplified model, the energy (exergy) recovered for the conversion of this plant is counted as the supplied heat, and accordingly should be counted as the recovery output item here.

Table 7.16 Summary of Energy Recovery Link

Description		Energy		Exergy		Remarks
		kW	MJ/t	kW	MJ/t	
Recovery & recycling	Feed heat exchange					
	Reboiler heating					
	Steam generation recycling					
	Recycling Stream					
	Total					
Recovery Output	Heat exchange output					
	Hot feed output					
	Total					
Rejected Energy	Heat loss					
	Stream rejected					
	Other					
	Cooling					
	Total					
Process exergy loss	Heat exchangers					
	Other					
	Total					
Energy to be recovered						
Effective power of non-process fluids						
Recovery efficiency, %						

Recovery output energy (exergy) also includes two types: heat exchange output and hot feed output.

The heat exchange output is used to heat materials outside the plant process energy utilization link, such as the preheating air used in the furnace, the preheating boiler feed water, and the process streams of the external plant. They can also be summarized by the heat exchange equipment summary table (recovery category). Recovering heat to generate steam and sending it to the outside of the plant or conversion link also belongs to the heat exchange recovery output category.

Hot feed discharging includes the delivery of hot raw materials above the specified temperature to the outside plant, and furnace fuel. Summarized by Table 7.5 of this chapter.

3. Rejection energy E_J rejection exergy D_J

Refers to the energy (exergy) that is rejected into the environment in various ways from the energy (exergy) to be recovered, In the energy to-be-recovered, all the energy except the recovery cycle and the recovery output will enter the rejection energy. And for the rejection exergy, in addition to the recycling cycle and recovery output in the exergy to be recycled, the exergy loss in the recovery and recycling process link should also be deducted.

Rejection energy (exergy) is rejected into the environment in four (three) forms.

(1) Cooling rejection energy

In order to carry energy (exergy) through the cooling medium (water or air), the energy that is indirectly or directly rejected into the environment can be summarized in Table 6.9. The flow exergy loss of the cooling medium is not reflected in the equipment, count it as other exergy losses.

(2) Heat dissipation rejection energy

Refers to the energy to be recovered carried by the streams leaving the process energy use link, and the energy lost to the environment through the surface of the equipment and pipeline during the transportation and recovery process, as summarized in Tables 7.9 and 7.12. The heat dissipation rejection energy here refers to the heat dissipation in the recovery and recycling link, and does not include the heat dissipation energy in the conversion and process use links.

(3) Output stream rejection energy

Refers to the energy carried into the environment by streams, which can be summarized in Table 7.6. Be careful not to omit items in the summary.

(4) Other rejection energy

After the effective power of the fluid enters the system, it flows in the energy process utilization and energy recovery links to cause exergy loss, and the energy balance is attributed to other rejection energy. It can be summarized by other electricity consumption categories of electricity balance Table 7.15. The effective power of the pump is reflected the process exergy loss in the exergy balance. Therefore, there are no other rejection exergy item in the exergy balance.

4. Energy recovery and recycling exergy loss

It is the irreversible exergy loss of the recovery equipment in the energy recovery link. Including the following two:

(1) Exergy loss of heat transfer and flow process of heat exchange equipment. Including the process exergy loss of all heat exchange equipment, that is, the process exergy loss including the recovery cycle and the recovery output heat exchange. The cooler only counts the exergy loss in the heat transfer process, and the flow exergy loss is included in the other process exergy losses.
(2) Other process Exergy loss. It is mainly the effective power of the non-process fluids. In the process of recovering energy, allowing the cooling medium to flow to take away the cooling heat belongs to the category of other process exergy loss, and the lighting electricity is also attributed to this. The other process exergy loss is numerically equal to other effective power in the energy balance.

5. Energy recovery efficiency

The above four calculations for the completion of the recovery link should be balanced and checked. The energy (exergy) of the entering recovery and recycling link is the energy to be recovered and the effective power of the non-process fluid (from energy conversion link). The energy discharged is the recovery cycle, recovery output and rejection energy loss. The discharged exergy still has exergy loss during the energy recycling process. The entry and exit must be balanced, and the effective power of non-process fluids must not be missed.

The recovery cycle and recovery output in the energy recovery link are both the targeted process and the effective part of the process. Its energy recovery efficiency is:

$$\eta = (E_R + E_E)/(E_O + E_{UO}) \times 100\% \tag{7.29}$$

Exergy recovery efficiency is:

$$\eta_x = (EX_R + EX_E)/(EX_O + E_{UO}) \times 100\% \tag{7.30}$$

The balance results of the energy recovery and recycling link are summarized in Table 7.16.

7.4.4 *Entire Plant Summary of the Balance*

1. Regarding the plant unbalanced item

In the balance of energy and exergy, due to the error of measurement and calculation, if the energy consumption parameters are calculated strictly according to the forward balance. There may be an imbalance. The balance calculation should allow for the existence of unbalanced terms, but not too large, and should be less than 2% of energy consumption or exergy consumption.

The energy imbalance item is combined with other rejection energy, and the exergy unbalance item is combined into the rejection exergy. If the reverse balance method is used, there is no such item.

2. Regarding entire plant energy consumption

According to the method of energy analysis, because in the balance process, it is absolutely calculated according to the enthalpy balance. Therefore, the calculated energy consumption is less than the statistical energy consumption under the test conditions, and the balance energy consumption is called the net energy (exergy) consumption. The difference lies in the energy loss during the conversion to the energy medium (steam, electricity) of the plant, which is reflected in the statistical energy consumption; secondly, for industries that are accustomed to measuring the chemical energy of raw materials, there is a big difference in chemical energy. In order to be consistent with the statistically customary energy consumption in this industry, the balanced net energy consumption can be converted into primary energy consumption. The conversion method is carried out according to formulas (5.33) and (5.34). The conversion result should be completely consistent with the statistical energy consumption during the test period.

3. Plant energy (exergy) balance results and (exergy) energy flow diagram

For the convenience of analysis and evaluation, the balance results of the above links can be summarized as the balance results of the entire plant, as shown in Table 7.17, and the energy (exergy) flow diagram can be drawn according to the method in Chap. 5.

The energy consumption analysis summary table and exergy flow diagram of a catalytic cracking unit are shown in Table 7.18 and Figs. 7.3, and 7.4, respectively.

Table 7.17 Process plant energy (exergy) balance analysis summary

Link	Description		Energy		Exergy		Remarks
			Symbol	MJ/t	Symbol	MJ/t	
Energy conversion	Total supply	Furnace fuel	E_{PF}		EX_{PF}		
		Coke	E_{PG}		EX_{PG}		
		Steam supply	E_{PS}		EX_{PS}		
		Electricity supply	E_{PE}		EX_{PE}		
		Heat supply	E_{PH}		EX_{PH}		
		Total	E_P		EX_P		
	Conversion supply out		E_B		EX_B		
	Direct loss	Incomplete combustion	E_{WC}		EX_{WC}		
		Flue gas	E_{WX}		EX_{WX}		
		Heat loss	E_{WD}		EX_{WD}		
		Ineffective power	E_{WP}		EX_{WP}		
		Conversion rejection	E_W		EX_W		
	Process exergy loss				D_{KU}		
	Effective energy		E_U		EX_U		
	Effective power of non-process fluids		E_{UO}		E_{XUO}		
Energy utilization	Reaction exotherm		E_{RE}		EX_{RE}		
	Recovery and recycling		E_R		EX_R		

(continued)

Table 7.17 (continued)

Link	Description		Energy		Exergy		Remarks
			Symbol	MJ/t	Symbol	MJ/t	
	Process total energy used		E_N		EX_N		
	Thermodynamic energy (exergy) consumption		E_T		D_T		
	Equipment heat dissipation		E_{UD}		EX_{UD}		
	Process exergy loss				D_{KP}		
Energy recovery and recycling	Energy to be recovered		E_O		EX_O		
	Effective power of non-process fluids		E_{UO}		EX_{UO}		
	Recovery and recycling		E_R		EX_R		
	Recovery output		E_E		EX_E		
	Rejection	Cooling	E_{JC}		D_{JC}		
		Heat dissipation	E_{JD}		D_{JD}		
		Streams	E_{JM}		D_{JM}		
		Other	E_{JO}		D_{JO}.		
		Subtotal	E_J		D_J		
	Process exergy loss				D_{KR}		
Summary indicators	Net energy(exergy) consumption		E_A		D_A		
	Primary energy (exergy) consumption		E_C		EX_C		
	Energy conversion efficiency %		η_U		η_{XU}		
	Energy process utilization efficiency%		η_P		η_{XP}		
	Energy recovery efficiency %		η_R		η_{XR}		
	Total rejection		E_J		D_J		
	Total process exergy loss				D_K		

Table 7.18 FCCU Process plant exergy analysis summary

Link	Description		Energy		Exergy	
			kW	MJ/t	kW	MJ/t
Energy conversion	Total supply	Coke + Fuel	64,756.19	1953.48	64,756.19	1953.48
		Steam supply	40,959.01	1235.60	16,375.34	493.99
		Electricity supply	5953.49	179.60	5953.49	179.60
		Heat supply	4560.341	137.57	1194.03	36.02
		Total	116,229.00	3506.24	88,279.06	2663.09
	Conversion rejection		17,572.22	530.10	6893.01	207.94
	Process exergy loss				29,877.76	901.29
	Non-process effective power		555.98	16.77	555.98	16.77
	Output steam		43,778.89	1320.66	14,969.96	451.59
Energy utilization	Effective energy		54,321.96	1638.71	35,982.35	1085.44
	Recovery and recycling		32,467.92	979.45	10,418.82	314.30
	Process total energy used		86,789.88	2618.16	46,401.17	1399.74
	Thermodynamic energy (exergy) consumption		8700.95	262.47	9727.00	293.42
	Process exergy loss				11,113.72	335.26
	Heat dissipation		3774.57	113.87	1039.83	31.37
Energy recovery and recycling	Energy to be recovered		74,314.36	2241.76	24,529.62	739.69
	Recovery output		8398.24	253.35	1577.82	47.60
	Rejection		33,448.20	1009.00	4265.03	128.66
	Process exergy loss				8258.95	249.13
Summary indicators	Net energy(exergy) consumption		64,051.91	1932.23	71,731.27	2163.89
	Primary energy (exergy) consumption		79,620.09	2401.87	91,669.10	2765.35
	Energy conversion efficiency %		84.88		58.35	
	Energy process utilization efficiency%		95.65		75.00	
	Energy recovery efficiency %		54.99		49.28	
	Total rejection		54,241.28	1636.29	12,197.90	367.97
	Total process exergy loss				40,250.43	1485.68

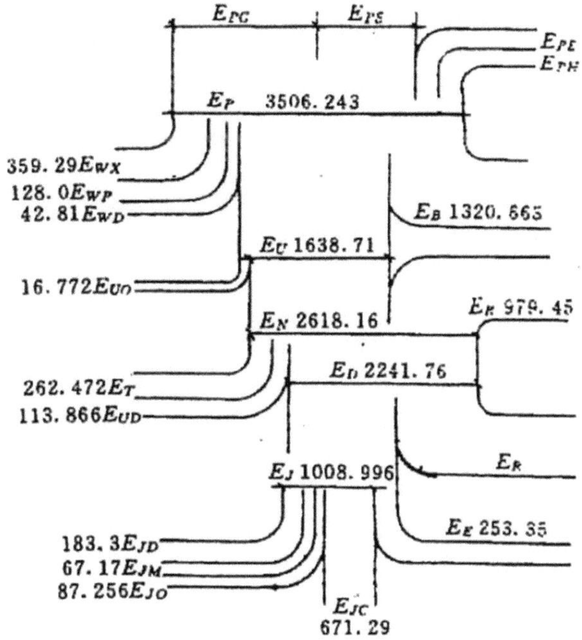

Fig. 7.3 Energy flow diagram of a catalytic cracking unit in a refinery

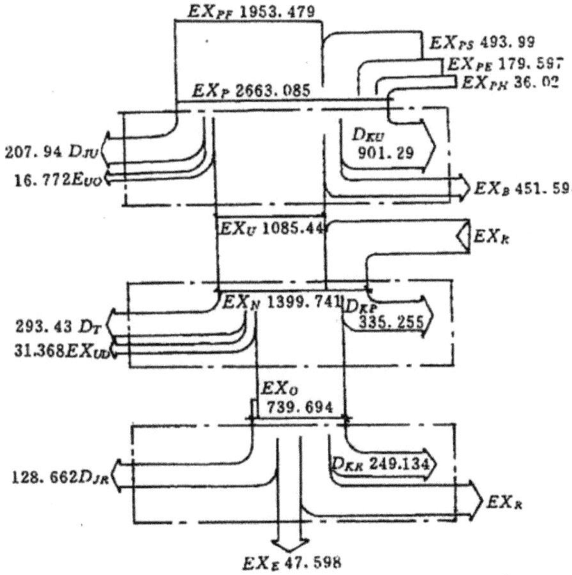

Fig. 7.4 Exergy flow diagram of a catalytic cracking unit in a refinery

References

1. China Petrochemical Corporation Standard SH2600-92 "Energy Balance Method for Petrochemical Enterprises (1992)
2. H. Ben, C. Anmin, Practical calculation and economic analysis of heat dissipation in refineries. Pet. Refining **2** (1984)
3. C. Anmin, Several improvements to the three-link model of energy consumption analysis in petrochemical process. Petrochem. Ind. **2** (1990)

Chapter 8
Utility/Auxiliary System and Energy Balance of the Whole Plant

Abstract In addition to the plant for the production of products, petrochemical plants also have many auxiliary systems and utility systems that provide services and support for plant stable production, such as storage and transportation systems, steam, water supply, air supply systems, power generation and transformation systems, and wastewater treatment systems. This Chapter verifies the energy balance of the utility system including the energy balance of steam, power, water, air, and nitrogen system, and auxiliary system including Storage and transportation system, Wastewater treatment system verification of auxiliary systems for indirect production, etc.; finally incorporates all process plant energy balance results into entire plant energy balance, proposes the energy utilization index.

Keywords Utility system · Auxiliary system · Process units · Energy balance · Energy index

Abbreviation and Symbology for Energy Balance

Energy link	Energy symbol		Definition	Exergy symbol
Energy conversion	Energy supply	E_P	Total energy supply	EX_P
		E_{PF}	Furnace fuel energy supply	EX_{PF}
		E_{PB}	Boiler fuel energy supply	EX_{PB}
		E_{PBE}	Power plant boiler fuel energy supply	EX_{PBE}
		E_{PG}	Coke energy supply	EX_{PG}
		E_{PS}	Steam energy supply	EX_{PS}
		E_{PE}	Electricity energy supply	EX_{PE}
		E_{PH}	External heat energy supply	EX_{PH}
	Output	E_B	Conversion link output energy	EX_B

(continued)

T. A. Chen, *Energy Saving and Carbon Reduction*,
https://doi.org/10.1007/978-981-19-5295-1_8

(continued)

Energy link	Energy symbol		Definition	Exergy symbol
		E_{BS}	Conversion link output Steam	EX_{BS}
		E_{BH}	Conversion link output heat	EX_{BH}
		E_{BE}	Conversion link output electricity energy	E_{XBE}
	Direct loss	E_W	Direct loss of energy	D_{JU}
		E_{WC}	Incomplete combustion	EX_{WC}
		E_{WX}	Flue gas energy loss	EX_{WX}
		E_{WD}	Heat loss	EX_{WD}
		E_{WP}	Ineffective power	EX_{WP}
			Process exergy loss	D_{KU}
	E_U		Effective energy supply	EX_U
	E_{UO}		Effective power of non-process fluids	EX_{UO}
Process energy utilization	E_{AR}		Endothermic reaction heat	EX_{AR}
	E_{RE}		Exothermic reaction heat	EX_{RE}
	E_N		Total process energy used	EX_N
	E_{TT}		Thermodynamic physical energy difference	EX_{TT}
	E_T		Thermodynamic energy consumption	D_T
	E_{UD}		Dissipate heat of energy utilization link	EX_{UD}
			Process exergy loss	D_{KP}
Energy recovery & recycling	E_O		Energy to be recovered	EX_O
	E_R		Recovery and Recycling Energy	EX_R
	E_{RH}		Heat exchanger recovery and recycling energy	EX_R
	E_{RM}		Streams recovery and recycling energy	EX_R
	E_{UD}		Dissipate heat of energy utilization link	EX_{UD}
	E_E		Energy recovered for output	EX_E

(continued)

(continued)

Energy link	Energy symbol	Definition	Exergy symbol
	E_J	Rejected energy	D_{JR}
	E_{JC}	Cooling rejected energy	D_{JC}
	E_{JD}	Heat dissipation rejected energy	D_{JD}
	E_{JM}	Streams rejected energy	D_{JM}
	E_{JO}	Other rejected energy	
		Process exergy loss	D_{KR}
Summary indicators	E_A	Net energy(exergy) consumption	D_A
	E_C	Converted to primary energy (exergy) consumption	EX_C
	η_U	Energy conversion efficiency %	η_{XU}
	η_P	Energy process utilization efficiency%	η_{XP}
	η_R	Energy recovery efficiency %	η_{XR}
		Total rejection exergy	D_J
		Total process exergy loss	D_K

The characteristics of these utility and auxiliary systems are that the process is generally intuitive and simple, and the energy balance relationship is not as complicated as that of the production plant. The focus of energy analysis in petrochemical processes lies in the analysis and improvement of production plant, and the only condition for improvement is to find problems through energy balance and exergy balance. The auxiliary system is not the case, because the process is simple, relying on energy balance can find problems, and propose improvement directions and measures. Of course, for more in-depth improvement, the concept of energy quality must be integrated into the energy balance, and overall optimization and overall consideration of improvement must be made from the perspective of the entire plant.

The energy balance of the whole plant is a summary expression of the production plant and utility/auxiliary systems, which is only the macroscopically linking the plant and the utility/auxiliary system together and expressing it in a unified form. The real energy-saving improvement also lies in the improvement and global optimization of plants and systems. Therefore, for the utility/auxiliary system and the whole plant summary, only the energy balance needs to be performed, and the exergy balance calculation is not necessary.

Different industries have different scopes and systems for enterprise energy balance, and the relationship between energy consumption of various parts is also different. Take a refinery as example to indicate the scope of energy balance.

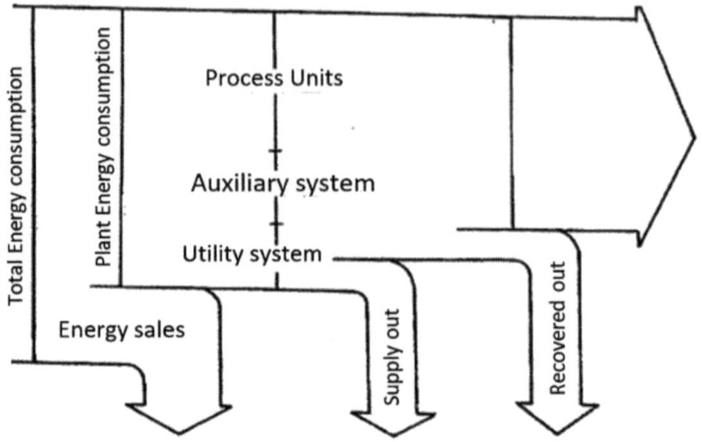

Fig. 8.1 Enterprise energy balance system

The energy balance scope of the entire refinery and the relationship between the energy consumption of each units and systems are shown in Fig. 8.1 [1]. Its purpose is to distinguish the energy consumption of oil refining from the energy consumption of non-refining production equipment, such as chemical equipment, large-scale mechanical instrument manufacturing system, etc., although it belongs to the energy consumption of the whole plant.

Generally, for petrochemical plants, unless there is a special need, it is not necessary to distinguish between oil refining and chemical plants, and the two are classified as energy consumption of production plants. However, for large-scale mechanical instrument manufacturing plants whose functions far exceed the normal maintenance needs of petrochemical plants, the energy balance system of the enterprise is not included, and it is included the energy export scope.

Generally, the energy balance system of petrochemical enterprises includes production plant, utility system (heating supply, steam supply, power supply, water supply, air/nitrogen supply) and auxiliary systems (storage and transportation and sewage treatment plants and in-plant laboratory, and maintenance) within the plant area [2].

For scientific research, laboratory tests, maintenance (mechanical, electrical, instrument), administrative offices, warehouses, fire protection, etc., although it may be included in the energy balance of the enterprise, it belongs to indirect production due to the intermittent and irregular energy use patterns. The auxiliary system only counts the verified consumption. The instrument manufacturer does not enter the energy balance system. But for the catalyst manufacturing plant, it can be treated as a production plant and enter the energy balance system.

Therefore, the energy balance range of the enterprise is actually the process production equipment, the utility and auxiliary systems that provide services and guarantees for it, eliminating the auxiliary systems of indirect production, which is convenient for grasping the key points reducing the energy consumption of plants and enterprises through analysis and improvement. Since the energy consumption characteristics of utility and auxiliary systems are different from those of production plant, the focus of energy balance is to verify consumption, determine the energy consumption per unit of product, summarize it to the whole plant, and understand the ins and outs of energy changes, and energy flow diagrams are generally not drawn.

8.1 Energy Balance of Utility System

The utility system is generally embodied as an energy conversion link in the energy balance of the whole plant. In the task of energy balance, in addition to determining the loss and efficiency of the process, the energy consumption per unit product should mainly be verified, so that it can be used when the plants are summarized and converted into primary energy consumption.

8.1.1 Energy Balance of Heating System

The heating system includes industrial boilers for generating steam (or hot water), steam turbine generator sets, steam (including condensate) and hot water delivery pipe networks.

8.1.1.1 Energy Balance of Industrial Boilers

The energy balance of industrial boilers includes water treatment systems. However, in the case of not only supplying boilers but also sending them to various plants for waste heat boiler to generate steam, and even to provide demineralized water required by the process, the unit energy consumption of the demineralized water treatment system should be verified.

1. Energy verification of water treatment station

It is required that the balance results of the demineralized water (deoxidized water) and the demineralized water demand of whole plant during the test period are consistent. This is not only the difference in the load rate, but also the slightly different unit energy consumption. More importantly, it reflects the difference in water consumption per unit of raw material or product (demineralized water, deoxidized water). The unit energy consumption verification of the water treatment station is summarized in Table 8.1 described as follows.

(1) Supply energy

It is mainly composed of four items: water, electricity, steam and heat supply.

Table 8.1 Summary table of energy approval for water treatment station

Description		Quantity		Parameters		Energy	Remarks
		Unit	Value	Temperature, °C	Pressure, MPa	kW	
Supply energy	Water						
	Electricity						
	Steam						
	Heat supply						
	Total						
Unit consumption	Demineralized water						
	Deoxidized water						

(1) Water consumption: due to the amount of water consumed in the treatment process, 1 ton of demineralized water (or deoxidized water) is produced, and the water consumption is between 1.1 and 1.5 t/t. The water consumption is calculated according to the actual unit consumption (or relative unit consumption) of the water supply unit consumption.
(2) Power consumption: Mainly used in the water treatment process, the power consumption of the water flow and booster pumps, and the vacuum pump power consumption for vacuum deoxygenation. The actual or unified conversion index of the whole plant is used to calculate the power consumption.
(3) Steam consumption: includes two types of steam for heating and deaeration, and steam for vacuum power, calculated according to the actual energy consumption of steam produced by the plant.
(4) Supply heat: includes the low-temperature heat provided by the plant and the energy carried by the condensate water. For the flash evaporation steam of industrial boiler blowdown and used for heating demineralized water, this item is also included.

The total energy supply is the sum of the above four items.

(2) Energy consumption per unit of water production

For a water treatment device that produces only one type of water, energy consumption can be easily calculated by statistics. For the case where the water treatment station produces both demineralized water and deoxidized water, Energy consumption per ton of demineralized water production can be calculated based on the consumption of water and electricity. According to the energy consumption of heating and steam supply and the output of deoxidized water, the energy consumption of steam and heat for deoxidized water is calculated, the energy consumption of steam and heat plus the energy consumption of demineralized water is the energy consumption of deoxidized water.

(3) Loss of water treatment station

The ineffective power of the pump and the energy brought out by the streams will be not verified here, but it is verified in the energy balance of the industrial boilers.

2. Industrial boiler energy balance

For the energy balance of industrial boilers, it is necessary to verify energy loss and efficiency, and verify the primary energy consumption per unit of steam production.

The verified steam output of the boiler must be in line with the steam balance results of the whole plant during the test period. Do not use artificially sufficient load for calibrating the so-called high-load thermal efficiency.

The energy balance results of industrial boilers (including water treatment systems) can be entered in summary Table 8.2. For the test methods of related items, please refer to the relevant standards of the thermal industry. Pay attention to when filling in the form:

(1) The energy balance of industrial boilers is calculated based on steam production. Therefore, the part of the self-consumed steam in the system should be included in the consumption item and also used as the energy supply;
(2) The low temperature heat of the production plant (system) sent to the water treatment station or preheating boiler feed water is counted as the heat supply. In the case of a high degree of integration, the use of the heat of the plant or system to preheat the combustion air or feed water of the boiler is also included in this item;
(3) The reference temperature for energy balance calculation must be consistent with the whole plant, and the equivalent heat value under the streams parameters participates in the balance calculation;
(4) The economizer and its own flue gas preheating the combustion air are not shown in the table. The effect has been reflected in fuel saving and loss reduction. However, it should be noted that the exhaust flue gas temperature at this time should be the exhaust gas temperature of the last recovery equipment (component);
(5) Verification of unit consumption of steam production is another important part of energy balance, which is used to determine the actual energy consumption conversion index of each production plant or the whole plant. At this time, all consumption items of supplied energy should be calculated based on actual energy consumption (or relative energy consumption), and fill in the column of energy consumption for steam production in Table 8.2, and then calculate the energy consumption per unit of steam production based on the steam production. For the case where demineralized water (or deoxidized water) is supplied to the system, the total energy of the water supplied to the system can be calculated from the unit water production energy consumption and the water output volume of the water treatment station, then calculate the energy consumption of steam production;

Table 8.2 Industrial boiler energy balance sheet

Description			Parameters		Energy		Energy consumption for steam production
			Temperature (°C)	Pressure (MPa)	kW	MJ/t	
Energy supply in	Fuel	Fuel oil					
		Fuel gas					
		Coke					
		Subtotal					
	Steam	Heating steam					
		dioxygen steam					
		Steam for steam pump					
		Subtotal					
	Power	Pump power					
		Water (converted electricity)					
		Subtotal					
	Heat supply						
	Feed water						
	Total						
Energy loss	Flue gas						
	Chemical incomplete combustion						
	Mechanical incomplete combustion						
	Blowdown, exhaust steam						
	Ineffective power						
	Total						
Supply out energy (hot water)							Flow rate:
Unit energy consumption	Supply deoxidized water						
	Supply demineralized water						
	Steam production						
Thermal efficiency, %							

Table 8.3 Plant-wide steam balance sheet (t/h)

Description		P= MPa		P= MPa		P= MPa		Deoxidized water
		t= °C		t= °C		t= °C		
		Generation	Usage	Generation	Usage	Generation	Usage	
Purchase								
Output (sale)								
Utility								
	Subtotal							
Auxiliary								
	Subtotal							
Process plant								
	Subtotal							
Total								

(6) The thermal efficiency should be based on the energy balance data, and the supplied out energy is determined by the supplied in energy minus the loss energy.

$$\eta = \frac{\text{Supplied out energy}}{\text{Supplied in energy}} \times 100\% \tag{8.1}$$

3. Plant-wide steam balance

During the test period, the quantity of steam of different pressure levels in the whole plant should be balanced, and it is required to fill in according to different levels. Both self-produced and self-used should be filled out. Deoxidized water (demineralized water) should also be balanced and filled out item by item. The balance is filled out according to the three categories of utility system, auxiliary systems, and process plants. The self-use part of the boiler is included in the utility system category. The deoxidized water is balanced according to the quantity of the supply out system, excluding the part used by the boiler to generate steam.

The balance of steam is summarized in Table 8.3.

8.1.1.2 Power Station Generator Set

Many petrochemical plants have their own power stations. The energy balance of the boiler part of the self-supplied power station is the same as that of the steam supply boiler. The energy balance of the steam turbine generator unit is summarized

Table 8.4 Energy balance sheet for steam turbine generator set

Description			Quantity	Parameters			Remarks
				Pressure (MPa)	Temperature (°C)	Energy kW	
Energy supply in	Steam						
	Power	Auxiliary power					
		Cooling water (Converted Power)					
		Subtotal					
	Total						
Energy loss	Turbine heat loss						
	Steam used for turbine						
	Transmission equipment						
	Generator						
	Ineffective power						
	Cooling rejection						
	Total						
Energy supplied out	Power generated						
	Extraction steam or back pressure steam						
	Recycle condensate						
	Total						
Unit efficiency, %							

in Table 8.4. For the test calculation method of the relevant parameters, please refer to the relevant standards of the power industry system. In the balance, it should be noted that the self-consumption of the power station should be included in the energy consumed, and at the same time as the output power, the two should not offset each other, that is, the total power generation is used as the calculation basis for balance.

The efficiency of the generator set is the overall efficiency of the supplied in energy, that is, the ratio of the supplied-out energy to the supplied in energy.

8.1.1.3 Verification of the Whole plant's Thermal Pipe Network

The plant-wide thermal pipeline network refers to the thermal system pipeline and the hot water system transmission pipeline that supply steam to each steam-using unit from the steam produced by the boiler or the steam purchased from outside the plant. The demineralized water and condensate pipelines transported at room temperature are not included. The system boundary is generally determined at the branch valve of the unit or plant with temperature and pressure measurement and flowmeter.

The energy balance of the thermal network focuses on verification of the heat dissipation loss of pipe network and the steam condensation caused by the heat dissipation; find the locations with serious heat dissipation, and then improve the insulation to reduce the heat dissipation, and improve the efficiency of steam transmission. The energy of the thermal pipe network of the whole plant is summarized in Table 8.5, which is explained as follows.

(1) The energy loss of the steam pipe network is mainly heat loss. In some cases, there are also steam condensation losses (such as long-distance transportation pipelines) and leakage losses. There are usually two test methods for heat loss: surface heat dissipation test and enthalpy drop method. The enthalpy drop method determines the energy loss according to the flow rate and enthalpy changes at the beginning and the end, it includes the heat dissipation, condensation, and leakage loss; if there is no condensation and leakage loss, it can also be checked with the surface heat dissipation test data. Pay attention to the surface heat dissipation of the test pipe fittings and exposed parts;
(2) For superheated steam transportation, when both ends are in a superheated state, the steam flow rate remains unchanged, and the heat dissipation only manifests as a decrease in the degree of superheating of the steam, resulting in energy loss. Material loss is only the leakage loss;
(3) For superheated at the beginning and saturation at the end, there may be condensation loss, which reduces the amount of steam, which can be estimated from the difference in flow rate. If there is no flow meter, it can be calculated from the heat dissipation test results:

$$\text{Condensed steam} = \frac{\text{heat loss} - \text{End flow} \times \text{Enthalpy difference (start, end)}}{\text{Heat of condensation}} \tag{8.2}$$

(4) For saturated steam transportation, the heat dissipation may all become the condensation loss of steam, and the pipe network loss is shown as leakage loss in the balance of steam rate, that is, the steam rate decreases during the transportation process.

8.1.2 Test and Balance of Power Supply System

The energy balance of the power supply system refers to the electrical balance of the power supply network, including the measurement of various power losses from external and self-generated transformers, power distribution, transmission lines, etc., the power balance of the power supply network and the calculation of the enterprise's power utilization efficiency.

1. Principle of energy balance of power supply system [2]

(1) The power balance of the whole plant shall comply with

$$\sum W_{zw} = \left(\sum W_{zi} + \sum W_{zh}\right) - \sum W_{yg} \tag{8.3}$$

Table 8.5 Thermal pipe network energy loss summary

Order	Description	Diameter Dg	Length, m	Beginning				End				Energy Loss, kW	Remarks
				Flow kg/h	Pressure MPa	Temp. ℃	Enthalpy kJ/kg	Flow kg/h	Pressure MPa	Temp. ℃	Enthalpy kJ/kg		
	Exhaust loss and condensati on												
	Total												

where: $\sum W_{zw}$—Total electricity consumed by the whole plant, kW h;

$\sum W_{zi}$—Total power supply of each supply line, kW h (Subject to watt-hour meter measurement);

$\sum W_{zh}$—Self-generation, kW h (Subject to steam turbine power generation);

$\sum W_{yg}$—Electricity transferred out from enterprises, kW h.

(2) The total power loss of the whole plant should meet

$$\sum W_s = \sum P_x - \sum P_b - \sum P_s \tag{8.4}$$

where: $\sum W_s$—Total power loss of the whole plant, kW h;

$\sum P_x$—Sum of active power loss of total line, kW h;

$\sum P_b$—Sum of active power loss of each transformer, kW h;

$\sum P_s$—Sum of motor power loss, kW h.

2. Test summary of energy balance of power supply system

(1) The power loss test results of distribution lines are summarized in Table 8.6;

(2) The load factor and transformation loss of the transformer are summarized in Table 8.7;

(3) The power load balance of the whole plant is summarized in Table 8.8.

In the utility, water and air have been converted electricity into the production unit, but the self-use and loss are also included in Table 8.8. The difference between the actual and the converted index should also be included for the use of converted electricity.

3. Enterprise Electricity Utilization Efficiency

In the total electricity supplied to enterprise, after deducting various losses, the ratio of effectively used by the enterprise is called the enterprise electric energy utilization efficiency.

$$\eta = \left(1 - \sum W_s / \sum W_{zw}\right) \times 100\% \tag{8.5}$$

Table 8.6 Summary of distribution line loss test

Description	Line name	Length km	Measured current A	Measured voltage V	Power loss during test period kW h/d	Remarks
Total						

Table 8.7 Transformer load rate and transformer loss summary

Order	Transformer location	Equipment name	Rated data					Measured current or monthly average, A		Load factor %	Actual substation loss		Remarks
			Capacity kVA	Voltage , kV		Current, A							
				High side	low side	High side	low side	High side	low side		Active power loss	Reactive power loss	
	Total												

Table 8.8 Transformer load rate and transformer loss summary

Description		Supply (Self-generation)	Usage (output)	Remark
Purchase				
Output				
Utility units				
Process plants and auxiliary units				
Total				

8.1.3 Energy Balance of Water Supply and Air Supply System

8.1.3.1 Water Supply System

The water supply system consists of two parts, a fresh water station and a cooling water system. The demineralized water treatment has been merged into the heating system, it is not considered here. The energy consumption of the water supply system is mainly the power consumed by the pump and the cooling tower fans, which is generally driven by a motor. In the optimization and improvement of energy use, some people have begun to consider using back pressure turbines to drive cooling water pumps, and even seasonally adopting condensing turbines as a means of large-scale system steam balance. In addition to electricity consumption, there are also a small amount of makeup water for cooling water system, heating steam in winter, etc. The energy balance of the water supply system consists in verifying the total amount of energy consumption and determining the energy consumption per unit of water supply. The energy balance of the water supply system is summarized in Table 8.9 described as follows.

(1) As the cooling water rate and the turning on of fans are affected by seasonality, the cooling water rate in summer is higher, the power consumption of the cooling tower fans and the loss of water evaporation are also large; the water consumption is small in winter, and the fans are not turned on, so the unit water supply energy consumption may vary greatly. Pay attention to the season of the test period and coordinate with the plant test, that is, the calibration working conditions are consistent with the calibration working conditions of the whole plant. When necessary, the difference between winter and summer of energy consumption per unit of water supply is determined by statistical data.

Table 8.9 Energy balance sheet for water supply system

Description		Fresh water		Cooling water		Chilled water	
		t /h	kW	t /h	kW	t /h	kW
Water production							
Power	Effective power						
	Ineffective power						
	Total						
Steam							
Heat supply							
Fresh water							
Energy consumption (MJ/t)							
Electricity per ton water							

Effective power and ineffective power are calculated by the efficiency of pumps, including water pumps and fans

Electricity per ton water index is adopted in the plant and the energy balance of the whole plant

Table 8.10 The whole plant water consumption balance sheet

Order	Description	Fresh water t/h	Cooling water t/h	Demineralized water t/h	Chilled water t/h	Remarks
	Total					

(2) The water produced should meet the requirements of the water balance of the whole plant.
(3) Each consumption should be calculated based on actual energy consumption or relative energy consumption, and determine the unit water supply energy consumption and convert it into electricity per a ton of water (converted electricity index) for process plant energy balance use.

The water balance of the whole plant is summarized in Table 8.10.

According to the water production data, the water consumption of the whole plant can be balanced. It can be balanced by units and system, and the loss and self-use part of the water system should also be reflected.

8.1.3.2 Air Supply System

Many metering and regulating instruments and equipment in petrochemical plants require industrial air and instrument air. In order to produce instrument air, a certain desiccant (such as silica gel, molecular sieve, etc.) needs to be used for dehydration

Table 8.11 Air compressor station energy balance sheet

Description		Parameters		Quantity		Energy	
		Temperature °C	Pressure MPa	Unit	Quantity	kW	MJ/Nm³
Air production							
Supply energy	Electricity						
	Steam						
	Heating						
	Water (Converted power)						
	Total						
Energy loss	Ineffective power						
	Cooling water out						
	Total						
Unit consumption of air supply							
Effective energy supply							

and de-oiling, and the dehydration and regeneration are intermittently switched. The purpose of the energy balance of the air supply system is mainly to verify the unit consumption of the air supply to provide the converted electricity index of the air usage units, and then to verify the total energy consumption of the air supply and the efficiency of the conversion equipment. The energy consumption is mainly electricity, steam, cooling water, and in particular, it is also used for regeneration of the desiccant with a suitable heat source. The energy balance of the air compressor station is summarized in Table 8.11. When calculating the air supply unit consumption, each consumption item should be calculated according to the actual energy consumption of the plant (relative energy consumption).

There are often nitrogen and oxygen systems in petrochemical plants. The energy balance of the nitrogen and oxygen production system can be carried out in accordance with Table 8.11 of the air supply system. The air separation unit can be processed as a production unit.

The balance results of compressed air, nitrogen and oxygen for the whole plant are listed in Table 8.12. The amount of wind (N_2, O_2) produced by the whole plant must be consistent with the balance result.

8.2 Energy Balance of Auxiliary System

According to the division of the energy balance system of the enterprise, the auxiliary system includes the storage and transportation system of raw materials and products and the sewage treatment system. Here, in addition to the energy balance of storage

Table 8.12 Summary of compressed air, nitrogen and oxygen (Nm3/h)

Order	User name	Compressed air	Nitrogen	Oxygen	Remark
	Total				

and transportation and wastewater treatment, the energy consumption of systems such as laboratory testing, scientific research, and maintenance is also verified to be excluded from the summary of the whole plant.

8.2.1 Storage and Transportation System

The scope of the storage and transportation system can be determined by itself according to the regulations of the industry. For the oil refining industry, it is stipulated that the unloading, storage, loading and transportation of raw materials (crude oil), products, and intermediate products belong to the storage and transportation system.

Because the unloading, storage, loading, and transportation of raw materials and intermediate products require power and heat, with this large system, the storage and transportation system accounts for a large proportion of the energy consumption of the entire plant, reaching 5–7% in the refining industry.

The energy consumption of each part of the storage and transportation system can be attributed to the consumption of the storage and transportation unit, and the energy balance of the storage and transportation unit can be conducted. The energy consumed by the storage and transportation system is mainly steam, electricity, water and heating supply. Except that a small part increases the product temperature, enters the thermodynamic energy E_T and part of the steam condensate can be used for recovery, most of the rest are discarded and lost to the environment. The energy balance of the storage and transportation system is similar to that of the production plant. The results can be summarized in Table 8.13, as described below.

1. Supply energy

The supply of energy is the sum of the consumption of each part in the unloading, storage, loading and transportation process of the storage and transportation system. The fuel consumption during the transportation of individual motorcars shall be counted as fuel consumption.

(1) Water can be converted into electricity into the balance system, and the method of converting electricity is the same as that of the production plant.

(2) If the pump runs intermittently, it can be converted into continuous operation power consumption according to the time in use.

Table 8.13 Summary table of energy balance of storage and transportation system

Description			Temperature °C	Pressure MPa	Quantity		Energy		Remarks
					Value	Unit	kW	MJ/t	
Supply energy	Steam								
	Electricity								Including water and air power
	External supply heating								
	Total								
Energy Loss	Ineffective power								
	Transmission heat loss								Pipeline
Effective use of energy									
Streams (In/out) enthalpy difference									
Rejected Energy	Stream rejection energy	Condensate							
		Exhaust steam							
		Water							
		小计							
	Heat dissipation								
	Total								
Recovery output condensate									

(3) The input heat includes the energy of hot water heat tracing and the energy of the product or intermediate product above the specified temperature when the hot storage process is used. The energy below the specified temperature is counted as the energy of the product or intermediate product.

(4) Steam is obtained by adding up the consumption of each part in the unloading, storage, loading, and transportation process of the storage and transportation system, and there should be a detailed list of steam used in each part.

2. Direct loss of energy

(1) For the ineffective power of the pump, the efficiency and ineffective power of the pump are calculated and summarized in accordance with the section on pump equipment in Chap. 5.

The effective energy and efficiency of the solid material conveying equipment (conveyor belt) shall be carried out according to industry regulations, and the ineffective power shall be determined according to it, or determined by itself according to the amount of work obtained or performed by the material in the conveying process.

(1) The surface heat dissipation loss of the heating and steam supply equipment and pipeline entering the storage and transportation system can be determined by the surface heat dissipation test of the equipment and pipeline

(2) In particular, when there is steam leakage in the storage and transportation system (before use), it is counted as a stream loss.

3. Effective use of energy

Effective utilization of energy is also the total energy used by the process, which can be obtained by subtracting the direct loss energy from the total supply energy.

4. Thermodynamic energy consumption

The thermodynamic energy consumption of the storage and transportation system is mainly due to the physical energy difference of the materials entering and leaving the storage and transportation system, and there is no reaction heat, that is, the energy carried difference between the materials (raw materials, products and intermediate products) leaving the storage and transportation system and the materials(raw materials, products and intermediate products) entering the storage and transportation system. Raw materials, products and intermediate products may all pass through (in and out) the storage and transportation system, which is the characteristic of the storage and transportation system. The thermodynamic energy consumption can be calculated item by item of the energy change value of the raw materials, products, and intermediate products through the storage and transportation system, and summed up. It can be calculated according to the method in Table 8.14.

5. Rejection of energy

The energy to be recovered can be obtained by subtracting the thermodynamic energy consumption from the effective energy supply. Except for a small amount of recovered energy (recovered condensate), most of the energy to be recovered in the storage and transportation system is rejected.

(1) Stream rejection energy: the part of energy that is not recovered from steam condensate or exhaust steam used for heating tracing.
(2) Heat dissipation rejection energy: heat loss for the surface of the equipment and pipeline. For the sake of simplification, the energy carried away by the cooling

Table 8.14 Physical energy difference calculation for storage and transportation system

Description		Flow rate t/h	Specific heat capacity or relative density	System temperature, °C		Energy difference kW	Remarks
				Inlet	Outlet		
Raw material							
Intermediate product							
product							

water (such as the spray cooling of liquefied petroleum gas) and the ineffective work of water converted electricity are included in this item.

(3) The effective power of the pump can be included in this item as other rejection energy.

6. Recovery output energy

Recovered condensate from more concentrated condensate system and used in utility system accounted as recovery output energy.

7. Calculation basis of storage and transportation system

The hourly quantity can be used as the calculation basis, and the energy consumption per unit of raw material (or product) can be verified. Due to the overhaul and start-up arrangements of the production plant, the operating hours are less than that of the storage and transportation system, and the difference comes down to additional energy consumption.

8.2.2 Wastewater Treatment System

The stripping water and washing water discharged from various plants in petrochemical plants contain oil, phenol and other harmful substances and need to be treated to avoid harm to the soil and the environment. The wastewater treatment plant is set up to accomplish this task.

The process and equipment of the wastewater treatment plant are generally relatively simple, and the main forms of energy use are steam and electricity (some s wastewater plants also use the industrial air in the plant for surface aeration, at this time the use of air can be converted into electricity consumption). For the energy balance of the equipment (pumps and fans, etc.), please refer to Chap. 6. For advanced treatment devices (such as wastewater stripping plant and activated carbon adsorption, etc.), it can be treated as a production plant.

During the test period, the wastewater treatment load should be consistent with the discharge of all production plants of the whole plant, and should not be carried out at the peak or low valley of wastewater treatment.

The energy balance of the wastewater treatment plant is summarized in Table 8.15, and the description is as follows.

1. The calculation of the supplied energy and the ineffective power and effective power of pumps is the same as the calculation of production plant and storage and transportation unit.
2. The wastewater treated by the wastewater treatment plant is not a process stream, and its energy has been included in the process equipment separately as the stream rejected energy item, and it flows to the wastewater treatment plant mostly at room temperature, so thermodynamic energy consumption is not calculated.

Table 8.15 Wastewater treatment plant energy balance sheet (t/h)

Description			Quantity		Energy kW	Energy consumption MJ/t
			Unit	Value		
Energy supply	Supply electricity	Electricity for pumps				
		Water (converted to power)				
		Compressed air (converted to power)				
		Total				
	Steam supply					
	Total					
	Ineffective power of pump					
Effective energy	Effective power of pump					
	Effective load of thermal equipment					
	Total					
Recovery output energy	Recovery output condensate					
	Total					
Rejection energy	Heat dissipation of equipment and pipelines					
	Discharge condensate					
	Total					
Unit consumption						

3. The heat dissipation of the wastewater treatment unit is mainly through the heat dissipation on the surface of the pool, and the heat of water evaporation under the convection and radiation heat transfer between the water surface and the atmosphere is all attributed to the surface heat dissipation. The effective power of the pump is also converted into heat by friction into the water, and finally rejected through heat dissipation.
4. In the effective energy supply, except for the condensate recovered output to the condensate station, the rest is rejection energy, and the energy is lost to the environment in the form of heat dissipation and streams rejection. The rejection energy of streams mainly refers to the energy carried by the streams such as the discharge of condensate and exhaust steam.
5. In the energy balance, each energy consumption is calculated based on the equivalent heat value. When calculating the energy consumption per unit of wastewater treatment, it is converted into a primary energy consumption.

8.2.3 Energy Consumption Verification of Auxiliary Systems for Indirect Production

1. Energy use form and characteristics

Indirect production departments such as maintenance, laboratory, and firefighting mainly use electricity, water, steam, and a small amount of fuel in some cases. Its energy use characteristics are different from process plant and direct production auxiliary units. It is not continuous and constant, but discontinuous and unstable. For example, electricity for machine tools and electric welding is mostly used during the day; electricity for lighting is mostly used at night. The use of steam is also very complicated, and the use time is often unruly.

2. Energy consumption statistics and its coordination with energy verification

Since their above-mentioned energy use characteristics are completely different from the process plant, the period (year or month) cumulative energy consumption can best summarize their energy use level. However, in the energy balance process of a certain test condition of the aforementioned production plant, it is necessary to verify the amount of non-production energy used by these systems during the enterprise energy balance test period, and clarify their relationship in the system energy balance.

For electricity and water, as the proportion of the consumption in the total load is small and the load is not continuous, it can generally be deducted from the total after verifying the amount of external energy consumption.

For steam, it is based on verified usage and energy using characteristics, If it is low-pressure steam and connected to the steam generator outlet pipe network, it can be counted as the recovery output item E_E of the entire plant production energy analysis system. If it is medium-pressure steam, or low-pressure steam, but it is directly supplied by the boiler, the conversion output energy E_B item can be counted, and its intermittent load can be converted into an average continuous load, so as to balance the steam production and usage of the whole plant.

8.3 Summary of Plant-Wide Energy Balance

The petrochemical process includes oil refining, chemical, fertilizer, chemical fiber and other industries. Many companies have two or more industries, and some of the companies generally are with a single industry. Enterprise (whole plant) energy balance summary refers to enterprises in a single industry. For joint enterprises, the single industry can be treated as second-level enterprises, and the entire plant energy balance can be summarized from second-level enterprises. According to the balance results of the secondary enterprises, the joint enterprise can summarize the physical balance of the energy consumption of the whole plant, and analyze the energy utilization status of the enterprise as a whole to study the direction and measures for improvement. There is no need to balance the energy of the joint enterprise.

When the energy balance of plants, utility system and auxiliary systems is completed, it can be summarized through the energy consumption analysis model established in Chap. 5 to point out the improvement direction and potential of the enterprise's energy use.

8.3.1 Energy Balance Summary Method and Calculation Basis

In a petrochemical plant, there are as few as a few plants and as many as dozens of plants (refining industry). After the plants have completed the energy balance work, they can be summarized as the energy balance of the entire plant, so that people have a general view and evaluation of the energy consumption of the entire plant.

The summary method is still the energy consumption analysis mode discussed in Chap. 5, that is, the entire petrochemical plant is regarded as a system, and the unit process equipment that constitutes each plant and system is divided into three links of energy conversion, energy process utilization and energy recovery and recycling. But we have completed the energy balance of the plant and the system the energy balance of the whole plant can be summarized based on energy consumption parameters of the plant or system, from the perspective of the whole plant, analyze the law of enterprise energy consumption and propose evaluation indicators.

For different production plants and systems, because their products and raw materials are different, the summary calculation basis used is also different. Generally, the unit raw material or product of the plant or system is used as the basis of energy balance, which is different from the basis used by the whole plant. Therefore, the energy consumption parameters proposed by the plant should be converted before they can be summarized.

Generally, as described in Chap. 5, the determination of the calculation basis of the enterprise energy balance should be based on the relatively single main raw material or main product of the enterprise. For example, in the oil refining industry, the raw materials processed are different for each plant, but for the whole plant, it is mainly crude oil, so it is based on the unit crude oil processing volume; for the synthetic ammonia plant, it can be based on ton of ammonia; fertilizer plants sometimes use the unit fertilizer (Urea) as the basis, or the nitrogen content of the ex-factory product as the basis, that is, the ton of nitrogen product shall be used the basis. When the products and raw materials are not single and it is difficult to find a suitable basis, the hourly quantity shall be used as basis.

When the summary basis of the plant is different from the summary basis of the whole plant, it shall be revised.

(1) When the energy balance results of production plant, utility and auxiliary systems are summarized as the energy balance of the whole plant, they must be multiplied by the processing capacity or product volume conversion coefficient δ_i:

$$\delta_i = G_i / G \tag{8.6}$$

where: G_i—During the test period, the processing volume or product volume of a plant or system as a calculation basis, t/h;

G—The processing volume or product volume used as the basis for calculation of the whole plant (enterprise) during the test, t/h.

For the plant and the whole plant are based on the hourly quantity, there is no need to carry out the conversion factor correction.

(2) The energy balance test of the whole plant requires that the product volume (or processing volume) of each plant should be consistent with the material balance of the whole plant. At this time, the sum of the water, electricity, steam, and fuel quantities of each unit should also meet the total consumption of the whole plant during the test period. Under special circumstances, such as the oil refining industry, the test period is too long to meet the above requirements. When there are differences, the test consumption and energy balance parameters of each unit of water, electricity, gas, and fuel should also be corrected.

$$\alpha_i = \frac{G_i}{G_{i0}} \tag{8.7}$$

where: $\alpha_i i$—plant water, electricity, gas, fuel consumption correction coefficient;

G_{i0}—Test period i plant plant-wide balance processing capacity or product capacity, t/h.

8.3.2 Summary of Energy Balance of Test Conditions

8.3.2.1 Summary of Basic Data

1. The energy balance project parameters of the whole plant are summarized in Table 8.16.

According to the energy balance results of each production plant, utility system, and auxiliary system, multiply its energy use parameters by the processing volume (or product volume) correction coefficient and fill in Table 8.16. Pay attention to when filling out the form:

(1) The production plant can be filled in one by one, and the auxiliary system can also be filled in by categories. The items that are not available are not filled in.
(2) For utility system, since most of the water, electricity, steam, and fuel have been included in the energy consumption and energy balance of the production plant, it should not be double-calculated, but should include the difference caused by various reasons (including self-use) between the actual consumption of the whole plant and the sum of plants consumption:

Table 8.16 Summary table of main energy balance parameters of the enterprise

Plant name	Feed or product,	conversion coefficient	Fuel	electricity	Coke	Conversion streams Rejection	Conversion Heat loss	Ineffective power	Use link Heat loss	Reaction exotherm	Recycling in plant	Output among the plants	cooling Rejection	Heat loss rejection	Stream rejection	other rejection
symbol	t/h	δ_i	E_{PF}	E_{PE}	E_{PG}	E_{WX}	E_{WD}	E_{WP}	E_{UD}	E_{RE}	E_R	E_E	E_{JC}	E_{JD}	E_{JM}	E_{JO}

(1) Steam supply part. Power self-consumption steam, system losses and part of the energy of auxiliary systems and the corresponding energy consumption parameters;
(2) water supply system. For self-use and loss; convert electricity to find the corresponding energy consumption parameters (effective power and ineffective power); pay attention to deducting the consumption of indirect production systems;
(3) Air supply system. Most of the small air consumption is not included in the energy consumption, and it is not included in the energy balance system of each unit. You can subtract the converted electricity part from the total power consumption of the air compressor station;
(4) Power supply system. It is the loss of power transmission and transformation and the consumption of indirect production auxiliary system. Transmission and transformation losses are counted as ineffective power and are direct losses.

(3) The total consumption of electricity and fuel (or purchased steam) of the whole plant should be balanced.

2. Under the test conditions, the balance of steam, electricity and water of the whole plant are shown in Tables 8.3, 8.8 and 8.10 respectively, and the fuel balance of the whole plant is summarized in Table 8.17. The conversion output in the table refers to the amount of output outside the system.

3. The recovered output energy of the whole plant is summarized in Table 8.18. described as follows:
(1) All the waste heat of the process streams within the energy balance system, whether directly (heat exchange) or indirect (feed water preheating, etc.) used outside the system are recovery output energy;
(2) The low-pressure steam generated from process stream has merged with the pipe network and the back-pressure turbine steam, which is difficult to distinguish strictly when it is used by the system or exported. The treatment principle is: first think that the output is back pressure steam. When the output is large, it can be considered as part of the generated steam output. For generated low-pressure steam directly sent out or the generated medium-pressure steam entering the

Table 8.17 Plant-wide fuel balance

Order	Item	Flow rate t/h	Temperature °C	Pressure MPa	Energy kW	Unit Energy MJ/t	Remarks
Total							

Table 8.18 Enterprise (entire plant) output energy summary table

Order	Item	flow rate t/h	Temperature °C	Pressure MPa	Energy kW	Unit Energy MJ/t	Remarks
Total							

conversion link back pressure turbines, it is counted as the recovery output item;

(3) The recovered output energy should be calculated as net heat, deducting heat loss from the system pipeline in the plant.

8.3.2.2 Calculation of the Energy Balance of the Whole Plant

The energy balance results of the plant and the system can be summarized according to Table 8.19. The content and calculation are explained as follows.

1. Total energy supply E_p

As a part of energy conversion, the utility unit in a petrochemical plant supplies the energy out often exceeds the energy balance system, such as electricity, water, and steam supplied to neighboring enterprises. Sometimes the fuel is supplied directly to the outside. In this case, the utility system in the plant is taken as a complete conversion link, and the energy supplied out is regarded as the energy supplied output by the conversion link. The resulting work is not only to verify the amount of energy supplied out, but also to determine the conversion energy index of the energy supplied out, share part of the efficiency loss of the conversion link, so that both the supply and the user can get fair benefits in terms of economy and energy consumption.

The total energy supply E_p consists of the following six items:

(1) Boiler fuel E_{PB} (Table 8.17), boiler steam, which is reflected in the boiler fuel consumption in the whole plant (no longer summarized by the steam consumption of each plant);
(2) The heating furnace fuel E_{PF} is obtained by adding the data of each production plant and system. The data should be consistent with the fuel balance of the whole plant;
(3) Catalytic cracking coke burning E_{PG} is summarized by the plant data;
(4) Purchased steam. The steam purchased from the power plant should be calculated into the energy balance according to the steam parameters. For example, when calculating the plant area of the supplier that supplies steam, the loss of this section of the pipe network before entering the plant must be added when calculating the loss of the entire plant's pipe network;

Table 8.19 Summary of whole plant energy balance

Description			Symbol	Unit	Quantity	Energy kW	Unit energy MJ/t	Remarks
Energy supply	Boiler fuel		E_{PB}					
	Furnace fuel		E_{PF}					
	Coke		E_{PG}					
	Purchase steam		E_{PS}					
	Purchase electricity		E_{PE}					
	Self-provided power station fuel		E_{PBF}					
	Internal recycling energy		E_{PH}					
	Reaction exotherm		E_{RE}					
	Total		E_P					
Conversion output	Steam output		E_{BS}					
	Heating output		E_{BH}					
	Electricity output		E_{BE}					
	Total		E_B					
Net energy supply $E_P - E_B$								
Energy loss	Conversion rejection		E_{WX}					
	Conversion heat loss		E_{WD}					
	Ineffective power loss		E_{WP}					
	Total		E_W					
		Effective power of non-process fluids	E_{UO}					
	Effective use of energy		E_U					
	Effective use of energy use link heat dissipation		E_{UD}					
	Heat exchange recovery recycle in		E_{RH}					
	Streams recovery recycling in		E_{RM}					
	Reaction exotherm		E_{RE}					
	Total process energy		E_N					
Thermodynamic energy consumption	physical energy difference of products and feeds		E_{TT}					

(continued)

Table 8.19 (continued)

Description		Symbol	Unit	Quantity	Energy kW	Unit energy MJ/t	Remarks
	Reaction heat	E_{TR}					
	Total	E_T					
	Energy to be recovered	E_O					
	Recover output energy	E_E					
	Recovery recycling energy	E_R					
Energy rejection	Cooling rejection Energy	E_{JC}					
	Heat loss rejection Energy	E_{JD}					
	Stream rejection Energy	E_{JM}					
	Other rejection Energy	E_{JO}					
	Total	E_J					
Unbalanced item		E_{IB}					
Evaluation index	Enterprise test primary energy consumption	E_C					
	Energy conversion efficiency %	η_U					
	Energy recovery efficiency %	η_R					
	Enterprise energy utilization efficiency, %	η					

(5) The purchased electricity is summarized in Table 8.8. For the purchase of water, the actual converted electricity shall also be used for calculation;

(6) The internal recycling supplies energyE_{PR}. The heat of the process streams of each plant (preheating furnace air, preheating boiler feed water, the sensible heat of the hot residual oil greater than the specified temperature), the steam generated from the process streams enters the turbine as power steam, etc., are all recovery recycled energy, it does not directly enter the process utilization link (it is not included in E_{PR}, but is recycled into the energy conversion link). In order to clearly express without making the model too complicated, it is considered that this part of the energy is not only part of E_{PH}, but also recovered and recycled output E_E, it is already handled in each process plant. When performing the whole plant balance, it is only necessary to multiply the data of each plant by the corresponding conversion coefficient to add up.

2. Conversion output energy

Most of the energy output from petrochemical plants outside of the energy balance system is supplied from conversion links and utility units, which is called conversion output energy. Generally, there are four items: water, electricity, steam, and heat. And water can be converted into electricity by the power consumption index of water, so it can be classified into three categories: steam, electricity and heat.

The output steam E_{BS} is two types of boiler steam and back pressure steam.

The output heat E_{BH} is to use the remaining load of the industrial furnace of the entire plant to provide heat energy to plant and scientific research units outside the system.

In the energy balance, it is carried out with the equivalent heat value, without converting to the primary energy, that is, it is calculated with the enthalpy value or energy under the actual parameters. It can be summarized from the data in the electricity, water and steam balance table in this chapter.

3. Net supply energy

The net supply of energy is $E_P - E_B$, which obviously includes the efficiency loss of energy conversion.

4. Direct losses energy E_W, include the following items.

   (1) Conversion link streams rejection energy E_{WX}. It is the sum of exhaust flue gas, blowdown and incomplete combustion losses of each heating furnace (regenerator) and boiler (waste heat boiler) Summarized by Table 8.16.
   (2) Heat dissipation rejection energy E_{WD}: The heat dissipation of each plant conversion link is added, and the heat dissipation loss of the steam pipe network of the whole plant is added, which is also summarized in Table 8.16.
   (3) Power conversion loss E_{WP}: Calculated by the ineffective power of each plant and unit plus the conversion loss of the whole plant, summarized in Table 8.16. However, most of the ineffective power in the water supply (fresh water and cooling water) and air supply units have been counted in the plants and units, and only a few have not been counted. Be careful not to recount. Only the loss of self-use, such as the ineffective power of the part of the water and air volume of the plant that is not counted, are calculated.

5. Effective power of non-process fluids

Refers to effective power such as water and air that are effective in the conversion process, but do not enter the process energy use link and directly enter the recycling link in the balance. The effective power of water supply and air supply systems belong to this category.

6. Effective use of energy

Refers to the energy provided by the conversion link to enter the energy process use link, which can be found by the balance of the conversion link:

$$E_U = E_P - E_B - E_W - E_{UO} \tag{8.8}$$

7. Total process energy E_N

Refers to the total amount of energy used by equipment in the process energy use link in the production process of an enterprise, and its magnitude is directly related to the energy consumption of the enterprise. When the recycling energy is constant, if E_N is large, the energy consumption of the enterprise is high, and vice versa. The total process energy used E_N generally consists of three parts: effective energy supply E_U, recovery and recycling energy E_R (see below), in the energy analysis system excluding the chemical energy of the raw materials, it also includes the reaction heat of the chemical process:

$$E_N = E_U + E_R + E_{RE} \tag{8.9}$$

And check as follows:

$$E_N = \sum (E_{Ni})\delta_i \tag{8.10}$$

Since the effective power provided by the air compressor station and the water supply system does not enter the process energy use link, it does not belong to the total process energy use, so there is no such correction. The two results are consistent, otherwise they should be checked and corrected.

8. Thermodynamic energy consumption E_T

Point out the difference between the total energy of factory products (intermediate products) and incoming raw materials (intermediate products), including chemical energy and physical energy. For the known composition and the reaction heat data at reference temperature can be found from the manual, the reference reaction heat and the difference in physical energy can be directly calculated to obtain the thermodynamic energy consumption. Under normal circumstances, in our energy analysis system that does not consider the chemical energy of raw materials, process streams and products, the chemical energy difference of the product is reflected by the size of the reaction heat. The calculation method has been described in Chap. 5.

$$E_T = E_{TR} + E_{TT} \tag{8.11}$$

E_{TR}—Is the algebraic sum of the reaction heat of the reference temperature of each plant (endothermic is positive, exothermic is negative):

$$E_{TR} = \sum (E_{TR})_i \times \delta_i \tag{8.12}$$

E_{TT}—It is the difference in physical energy of all the products leaving the factory (including the processing loss of self-use fuel) and all the raw materials entering the factory. The calculation method is the same as that of the production plant, and can also be summarized in Table 8.14. For the oil refining industry, when the hot heavy oil is the fuel, the temperature greater than the specified temperature is counted as the output heat. The product energy is only the energy between the reference temperature and the specified temperature.

Thermodynamic energy consumption is the theoretical minimum energy consumption without considering the specific process.

9. Process energy use link heat dissipation E_{UD}

It is the energy lost to the environment in the process equipment, and the heat dissipation of the process equipment will reduce the energy to be recovered, and requires to provide additional energy from Conversion link to compensate heat dissipation. The heat dissipation of process energy use link has been summarized in Table 8.16.

10. Energy to be recovered

Refers to the total energy used in the process, except the thermodynamic energy consumption converted into the product and the heat dissipation of the process equipment, the part that enters the energy recovery and recycling link to be recycled, which can be found by the balance of the links.

$$E_O = E_N - E_T - E_{UD} - E_{RE} \tag{8.13}$$

11. Recovery output energy

Refers to the recovery of energy from the energy to be recovered for effective use outside of the energy use link of the system process. It includes two parts: one is the effective energy output outside the energy balance system; the other is the energy recovered from the energy to be recovered and recycled for the equipment in the system conversion link. Including it to the recovery output energy, correspondingly, it is also included in the energy supply item E_{PR}. This is for the sake of simplifying the energy consumption analysis model, with a certain degree of arbitrariness. The recovered output energy can be summarized in Table 8.17.

12. Recovery and recycling energy E_R

For the whole plant, the energy recovery and recycling link includes two parts, namely, the energy recovered from the energy to be recovered from each production plant and auxiliary unit has been supplied to the plant process energy utilization link, and the heat output between the plants, which is used by the system and other plants in the system. All belong to the Recovery and recycling energy in the system. However, the recovery output heat in each plant includes the energy output to the equipment outside the system and the conversion link should be deducted. The recovery and recycling energy can be calculated by the following formula:

$$E_R = \sum \delta_i (E_{Ri} + E_{Ei}) - E_E \tag{8.14}$$

where: E_E—System recovery output energy;
E_{Ri}—The recycling energy of the plant or unit;
E_{Ei}—The recovery output energy of the device or unit.
The utility system is a conversion link, so there is no recycling energy.

13. Rejection energy E_J

It also refers to the energy that is not recovered in the system and is rejected into the environment. That is, all the energy to be recovered except E_R and E_E are rejected energy. In addition, the effective power of non-process fluids is not used in the process energy use link, but directly enters the recovery and recycling link, and is ultimately lost to the environment in the form of other rejection energy. It can be balanced by the following formula:

$$E_J = E_O - E_R - E_E + E_{UO} \tag{8.15}$$

Rejection energy E_J can also be summed up by each plant and unit, it is better to cross check with the calculation results using Eq. (8.15). The utility system that belongs to the conversion link has no rejection energy items. Rejection energy can be lost to the environment in the following four forms (see Table 8.16):

(1) Cooling energy E_{JC}, has been summarized in Table 8.16;
(2) The heat dissipation rejection energy E_{JD} is summarized by the process equipment and auxiliary unit data;
(3) The stream rejection energy E_{JM} is summarized by the process equipment and auxiliary unit data;
(4) The other rejection energy E_{JO} is summarized by the process equipment and auxiliary unit data.

$$E_J = E_{JC} + E_{JD} + E_{JM} + E_{JO} \tag{8.16}$$

The calculation result and the balance result of Eq. (8.15) are cross checked. When there is an allowable unbalanced item, put it in E_{JO}, and no more items will appear.

8.3.2.3 Evaluation Index of Enterprise Energy Balance

After completing the energy balance of the enterprise and summarizing it in the form of Table 8.19, we can put forward the evaluation index of enterprise energy consumption according to the three-link model of energy utilization:

1. Energy conversion and transmission efficiency

$$\eta = \frac{E_U + E_{UO} + E_B}{E_P} \times 100\% \tag{8.17}$$

Including the externally supplied energy E_B and E_{UO} are both used as effective energy and placed on the Numerator, so as to objectively reflect the actual level of the conversion link.

2. Energy recovery efficiency

It is the ratio of recovered/recycled energy and energy output outside the system to the energy to be recovered.

$$\eta = \frac{E_R + E_E}{E_O + E_{UO}} \times 100\% \tag{8.18}$$

3. The total process energy consumption of enterprise

It is an indicator that reflects the process route and operation level of an enterprise, and is the total amount (net value) of energy that the company's process production process really needs. The calculation has been completed in energy balance.

4. Enterprise effective supplied energy E_{ef}

It refers to the energy that is converted and entered into the production process of the enterprise and supplied out of the system from the various energy actually consumed by the enterprise. For petrochemical enterprises, that is, the effective energy supply of the enterprise energy balance system.

$$E_{ef} = E_U + E_{UO} + E_B \tag{8.19}$$

5. Primary energy consumption during the enterprise test period.

Calculations are the same as in Chap. 5

$$E_C = E_A + \Delta E_m + E_{IC} \tag{8.20}$$

$$\Delta E_m = \sum\nolimits_{i=1}^{m} (\mp E_i \xi_i) \tag{8.21}$$

$$\xi_i = \frac{\textit{Primary energy consumption for producing i energy}}{\textit{Actual enthalpy}} - 1 \tag{8.22}$$

For enterprises, it is generally electricity and purchased steam: $\xi_{elect} = 2.49$; ξ_{steam} can be determined by actual energy consumption and steam parameters.

E_A is the net energy consumption of the enterprise, which refers to the energy actually used and energy lost by the enterprise. The net energy consumption does not include the correction value of the primary energy, which is the actual consumption calculated on the basis of the equivalent heat value. There are two ways to calculate.

(1) From the perspective of process consumption, the net energy consumption is the conversion loss energy, the rejection energy, the heat dissipation of the process equipment, and the thermodynamic energy consumption entering the product.

$$E_A = E_W + E_J + E_{UD} + E_T \tag{8.23}$$

(2) From the perspective of supplying energy, deducting the converted supplying out and recovering output energy from the supplying energy:

$$E_A = E_P - E_B - E_E \tag{8.24}$$

where: E_{IC}—the chemical energy of the raw material at the reference temperature, if the chemical energy is not included, it will be regarded as $E_{IC} \approx 0$;

E_i—i kind of energy or energy-carrying medium for supply energy, see Table 8.19;

ξ_i—Conversion factor for i kinds of energy or energy-carrying working fluid.

6. Enterprise energy utilization efficiency

The energy utilization efficiency of an Enterprise refers to the percentage of the company's effective use of energy to the company's comprehensive energy consumption, namely:

$$\eta = \frac{E_U + E_B + E_{UO}}{E_{PC}} \times 100\% \tag{8.25}$$

The comprehensive energy consumption of the enterprise is the total energy supply plus the primary energy correction item:

$$E_{PC} = E_A + \Delta E_m \tag{8.26}$$

The energy balance result can also be drawn as an energy flow diagram, see Fig. 5.6, and the drawing method see Chap. 5.

References

1. The calculation method of energy consumption in refineries of the Ministry of Petroleum Industry, PRC Ministry of Petroleum Industry of, 1982·11
2. China Petrochemical Corporation Standard, SH2600-92 "Energy Balance Method for Petrochemical Enterprises", 1992

Part III
Energy Analyses and Carbon Reduction Approaches

Chapter 9
Energy Consumption Analysis and Energy-Saving Improvement Methods

Abstract Discussing and analyzing the impact of objective conditions on energy consumption will help to turn unfavorable conditions into favorable conditions in energy-saving work, and identify the root causes for the higher energy consumption. This chapter discusses the objective condition's impact on energy consumption, it includes the Impact of plant size, Influence of ambient temperature; discussed the analysis and estimation of the impact of loading on plant energy consumption in detail (1) Load ratio impact on plant energy consumption and its estimation; (2) Estimate fix energy ratio from energy balance data; (3) The influence of load ratio on heat dissipation fixed energy consumption; (4) The influence of load rate on electricity fixed energy consumption; (5) The influence of load ratio on steam fixed energy consumption; (6) Other fixed energy consumption. Evaluated Inter-equipment and system energy utilization, including process unit benchmark energy consumption method, and energy grade match method between energy supply and energy demand. Then introduced approaches for energy-saving improvement of the production plant, and a large-scale system optimization method and improvement approach from an energy overall point of view.

Keywords Ambient impact · Loading impact · Fix energy-consumption · Benchmark energy-consumption · Energy-saving approach · System optimization

The energy utilization level of a petrochemical plant is restricted by two factors: one is the equipment and system design, operation and management level, which are the main factors affecting the energy consumption of the plant and the whole plant; the impact of objective conditions on energy consumption is also a factor that cannot be ignored. Discussing and analyzing the impact of objective conditions on energy consumption will help to turn unfavorable conditions into favorable conditions in energy-saving work, identify the root of causes for higher energy consumption.

T. A. Chen, *Energy Saving and Carbon Reduction*,
https://doi.org/10.1007/978-981-19-5295-1_9

9.1 The Influence of Plant Size and Ambient Temperature on Energy Consumption

9.1.1 Impact of Plant Size

The scale of the plant mainly affects the unit heat dissipation. Secondly, the pumps and compressors of small-scale plant have low efficiency, which also affects the energy consumption of electricity and steam.

For similar plants, the smaller the scale, the greater the heat dissipation unit consumption, because the larger the heat dissipation surface area corresponding to the unit feed of small plants. Take the oil product pipeline and fractionation tower as an example, assuming the velocity is the same (in fact, the design velocity for small plants is often smaller), then the flow area f per unit feed is the same. For large and small plants, there are:

$$f_1 = \frac{\frac{\pi}{4} D_1^2}{V_1}, \quad and \quad f_2 = \frac{\frac{\pi}{4} D_2^2}{V_2}$$

Because $f_1 = f_2$, then:

$$\sqrt{V_2/V_1} = \frac{D_2}{D_1} \tag{9.1}$$

where: D_1, D_2—internal diameter of large or small plant pipeline and equipment, m;

V_1, V_2—processing capacity of large or small plant, m^3/h;

That is, the diameters of the corresponding pipelines and fractionation towers are proportional to the square root of the processing capacity. Assuming that the fluid temperature t_B of the corresponding parts of the large and small plants is the same, the heat dissipation coefficient α_T is the same, the insulation thickness δ is the same, and the heat dissipation per unit feed (processing capacity) is E_0, then the ratio of the heat dissipation per unit feed for different scales is:

$$\frac{(E_0)_2}{(E_0)_1} = \frac{\pi(D_2 + 2\delta)/V_2}{\pi(D_1 + 2\delta)/V_1} = \frac{V_1(D_2 + 2\delta)}{V_2(D_1 + 2\delta)} \tag{9.2}$$

For equipment (such as fractionation tower) $D \gg 2\delta$, it can be approximately taken as

$$\frac{(D_2 + 2\delta)}{(D_1 + 2\delta)} \cong \frac{D_2}{D_1}$$

For pipelines, when the pipe diameter is small,

$$1 > \frac{(D_2 + 2\delta)}{(D_1 + 2\delta)} > \frac{D_2}{D_1}$$

Substituting the previous expression, eliminate D_2, D_1 to get:

$$\frac{V_1}{V_2} > \frac{(E_0)_2}{(E_0)_1} > \sqrt{V_1/V_2} \tag{9.3}$$

Equation (9.3) can be used as an approximate discriminant for comparing the unit heat dissipation of plants of different scales. For example, under the same insulation level, the heat dissipation unit consumption of the 0.6 million tons/year catalytic cracking unit is at least 1.4 times that of the 1.2 million tons/year unit, and the 0.50.6 million tons/year atmospheric and vacuum unit heat dissipation unit consumption is at least 2.23 times that of the 2.5 million tons/year

9.1.2 *Influence of Ambient Temperature [1, 2]*

The ambient temperature varies with regions (north and south) and seasons (winter, summer), and has a certain impact on the energy consumption of the plant and whole plant. Ambient temperature directly affects the heat dissipation of equipment and pipelines, thereby causing changes in heat preservation and heat tracing conditions, leading to changes in steam and fuel consumption. The ambient temperature indirectly affects the activation time of air-cooling fans and cooling tower fans and the cooling effect of the process cooling system, thereby changing the amount of cooling medium.

The influence of ambient temperature on energy consumption is not only subject to the objective change law, but also closely related to subjective management level, technical measures, equipment conditions, etc.; for example, the oil tracing pipeline of a certain plant is based on its winter temperature and targeted tracing temperature, need heat tracing steam of 0.5 t/h, but the actual steam consumption increased by 2.5 t/h, largely due to poor performance of steam trap and poor steam system management, resulting in a large amount of exhaust "steam" exhausted .

The influence of ambient temperature on heat dissipation energy consumption is discussed below.

When conducting heat dissipation tests and calculations, most of them are determined based on the four parameters of heat dissipation surface area, heat dissipation coefficient, equipment surface temperature and ambient temperature (see Chap. 2 Sect. 2.5).

$$Q_s = \alpha_T F(t_W - t_a)$$

$$\alpha_T = 4.33\frac{\left(\frac{T_W}{100}\right)^4 - \left(\frac{T_a}{100}\right)^4}{T_W - T_a} + 1.47(T_W - T_a)^{1/3} + 4.36\frac{W^{0.6}}{D^{0.4}}$$

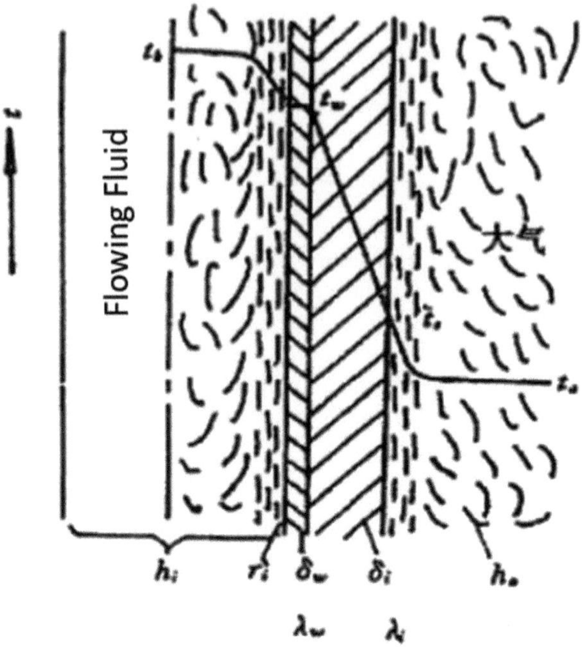

Fig. 9.1 Temperature distribution during surface heat dissipation

From the above two formulas, it is difficult to see the changing law of heat dissipation with the ambient temperature, and it is often overestimated that the heat dissipation changes with the surface temperature.

The total heat transfer condition is determined by five steps (Fig. 9.1). The surface temperature t_W decreases with the decrease of ambient temperature t_a and the increase of heat dissipation coefficient α_T. To make the problem clearer, the change of Q_s with t_a should be analyzed from the total heat transfer equation:

$$Q_s = KF(t_b - t_a) \tag{9.4}$$

In the formula, t_b is the fluid temperature. For equipment with a good thermal insulation layer, Among the five local thermal resistances determining the total thermal resistance K^{-1}, the thermal resistance of the thermal insulation layer δ_i/λ_i is the controlling factor:

$$K^{-1} = h_i^{-1} + \gamma_i + \frac{\delta_w}{\lambda_w} + \frac{\delta_i}{\lambda_i} + h_o^{-1} \tag{9.5}$$

In the formula, K is the total heat transfer coefficient; h_i is the heat transfer coefficient of the fluid inner film; γ_i is the thermal resistance of the tube wall fouling;

Table 9.1 The influence of ambient temperature on total heat transfer coefficient

	Surface of equipment (pipeline) with good insulation			Bare metal surface
	Catalytic reaction oil and gas pipeline	Steam line	Crude oil tank	Steam line
t_b, °C	470	250	50	250
h_i	76.4	232.6	123.8	232.6
h_i^{-1}	0.0131	0.0043	0.0081	0.0043
$\sum \frac{\delta}{\lambda}$	1.43	1.143	1	0.0015
t'_w/t_w	60/87	15/40	−13/23	248.5/249
h'_o/h_o	23.4/22.35	21.37/19.75	18.1/16.7	27.2 /26.0
$(h'_o)^{-1}/(h_o)^{-1}$	0.0427/0.0447	0.0468/0.0506	0.0552/0.0599	0.0368/0.0385
K'/K	0.673 /0.672	0.838/0.835	0.940/0.936	24.27/23.31
K'^{-1}/K^{-1}	1.486 /1.488	1.194/1.198	1.064/1.068	0.0412/0.0429
$\sum(h'_o - h_o)/h_o$, %	4.7	8.2	8.4	4.6
$\sum(K' - K)/K$, %	0.14	0.36	0.43	4.12

Condition: Warm season t_a= 21 °C, Cold season t'_a= −17 °C, average wind speed: W = 3.3 m/s (Summer), 3.7 m/s(winter)

δ_w, δ_i are the tube wall thickness and the insulation thickness; λ_w, λ_i are the thermal conductivity of the tube wall and the insulation layer respectively, h_o is the heat transfer coefficient of the outer film of the insulation layer. Typical data are listed in Table 9.1.

Due to $\sum \frac{\delta}{\lambda} = \gamma_i + \frac{\delta_w}{\lambda_w} + \frac{\delta_i}{\lambda_i}$ does not change, and accounts for most of the total thermal resistance K^{-1}, so as t_a decreases, t_w decreases also, h_o does not change much. The change of K is smaller (< 0.5%) and can be almost regarded as a constant. Therefore, it is explained by formula (9.4): the change of heat dissipation Q_s is only determined by the change of total heat transfer driving force ($t_b - t_a$).

Obviously, the main parameter that affects the change of Q_s with t_a change is the fluid temperature t_b. The higher the t_b is, the smaller the change of the total heat transfer driving force is when the t_a decreases, and the smaller the heat dissipation ratio "ψ" is when the Q_s increases. If the superscript "," is used to indicate cold season conditions, then:

$$\psi = \frac{Q'_s}{Q_s} = \frac{t_b - t'_a}{t_b - t_a} = 1 + \frac{t_a - t'_a}{t_b - t_a} \tag{9.6}$$

where: Q'_s and Q_s are the heat dissipation under corresponding ambient temperature t'_a and t_a respectively; ψ is the ratio of heat dissipation at different ambient temperatures. It can be seen from the above that t_b has a great influence on ψ. The higher the t_b, the ψ is closer to 1, which means that different ambient temperatures will have less impact on heat dissipation. This situation is also reflected in Fig. 9.2, and is confirmed by field data (see Table 9.2). The value of ψ under different conditions

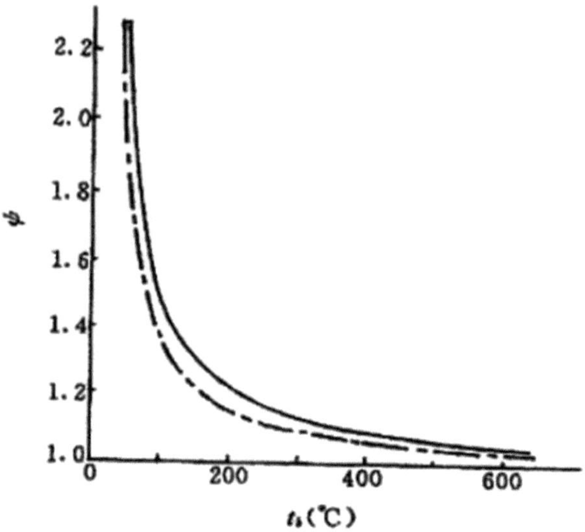

Fig. 9.2 Fluid temperature of insulated equipment on the heat dissipation ratio $-line\ t_a = 21\,°\text{C}, t'_a = -17\,°\text{C}; -\bullet-line, t_a = 21\,°\text{C}, t'_a = -9\,°\text{C}$

Table 9.2 The insulated equipment fluid temperature impacts on the heat dissipation ratio

Equipment name	Fired heater	Steam lines	Oil pipeline	Crude oil tank
t_b (°C)	460	250	130	50
ψ	1.07	1.17	1.35	2·31

The conditions are the same as Table 9.1

of t_b, t_a and t'_a can be calculated by Eq. (9.6) or checked by the calculation diagram 9.3.

As for the exposed surface of the equipment without heat preservation, because the control thermal resistance for heat dissipation is h_o^{-1}, the change of K with h_o is much larger than that of the surface with heat preservation (see Table 8.1), and the value of ψ is also higher. It can be seen that the energy-saving effect of bare surface insulation is more significant in cold regions than in warm regions, especially for equipment with low t_b. For example, the technical and economic evaluation of tank roof insulation shows that for oil tanks with an average annual storage temperature difference $(t_b - t_a) \geq 30\,°\text{C}$, with an insulation layer on the tank roof, the investment payback period is within three years. Take a 10,000 m^3 vaulted oil tank as an example. Although the tank roof without insulation layer only accounts for 30% of the heat dissipation area, it accounts for 57% of the heat dissipation (the surface area of the tank wall accounts for 40%, and the heat dissipation only accounts for 34%. 9% Heat dissipation is at the bottom of the tank) (Fig. 9.3).

Example Known $t_b = 400\,°\text{C}$, $t_a = 15\,°\text{C}$, $T'_a = -20\,°\text{C}$, find ψ?

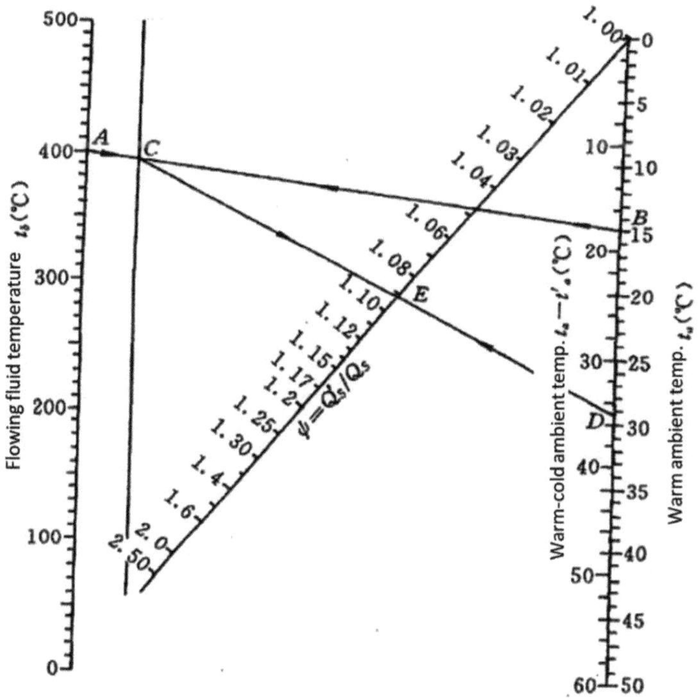

Fig. 9.3 Calculating diagram of heat dissipation with ambient temperature

Solution Line AB, intersect at point C of the minor axis, connect line CD, and intersect at point E of the ψ axis, read $\psi = 1.091$.

9.2 Analysis of the Impact of Loading on Plant Energy Consumption [3]

Due to various factors, the processing rate of the process plant is often inconsistent with the processing capacity of the plant. This situation has a considerable impact on the energy utilization and the actual energy consumption of plant. Analyzing and estimating the law of this influence has very practical significance for the rational use of energy in the design and operation of process plant.

9.2.1 Load Ratio Impact on Plant Energy Consumption and Its Estimation

The load ratio is the ratio of the feed rate or actual load F' (t/h) of the process plant to the processing capacity (load F of the original design or modified design), $L = F'/F$. Its changes affect the energy consumption of the plant through many factors [4]. Amoco first proposed to use the "fixed energy consumption" method to summarize [18]. The fixed energy consumption refers to the energy consumption per unit time (MJ/h), which does not change with the change of the feed volume. For example, the heat loss of a pipeline with a certain operating temperature and a certain surface area will not decrease with the decrease of the fluid flow rate in the pipe. In contrast, another type of proportional reduction as the load rate decreases is called "variable energy consumption". For example, the steam consumption of a steam heater is also proportionally reduced with process stream rate decreases when the process conditions of the inlet and outlet temperature of the heated medium are determined.

However, if expressed in terms of energy consumption per unit of feed, the "fixed energy consumption" E_{CF} (MJ/t feed) obviously increases inverse proportionally with the decrease of the feed (load rate reduction), but "variable energy" E_{CV} (MJ/t) remains unchanged when the feed volume changes. Therefore, if it is assumed that the total energy consumption E_C represented by the unit feed volume under the design load is composed of two parts, the fixed energy consumption E_{CF} and the variable energy consumption E_{CV}, namely:

$$E_C = E_{CF} + E_{CV} \tag{9.7}$$

Let the fixed energy consumption ratio be:

$$\vartheta = E_{CF}/E_C \tag{9.8}$$

When the plant load ratio is L, the total energy consumption of the device E_C' is:

$$E_C' = E_{CF}' + E_{CV} = \frac{E_{CF}}{L} + E_{CV} \tag{9.9}$$

Substituting Eqs. (9.7) and (9.8) into Eqs. (9.9), simplifying, obtain the relationship between total energy consumption and load ratio:

$$E_C' = E_C\left[1 + \vartheta(\frac{1}{L} - 1)\right] \tag{9.10}$$

The only condition for calculating E_C' with this formula is to know the fixed energy consumption fraction ϑ. Amoco has provided ϑ data for several oil refining units [18]: 15% for crude oil distillation unit, 35% for catalytic cracking unit, and

63% for alkylation unit, Coking unit 63% and so on. But in fact, any similar plant (such as atmospheric and vacuum or reforming) is different under different scales, different processes, and different design levels. Even if the same plant has been built, there will be changes before and after the energy-saving modification [5].

With the deepening of energy conservation, a new situation has emerged, that is, appropriate arrangements are made in the design or energy conservation modification so that one or several of the parallel devices can be stopped during a certain low load operation, thereby a certain fixed energy consumption is correspondingly reduced, and then the formula (9.8) will take a new form. After some equipment is shut down, the corresponding reduced energy consumption is E_{CS}. At this time, the fixed energy consumption fraction ϑ^* :

$$\vartheta^* = \frac{E_{CF}^*}{E_C^*} = \frac{E_{CF} - E_{CS}}{E_C - E_{CS}} \tag{9.11}$$

Obviously,

$$E_C^* = E_C - E_{CS} = E_{CF} - E_{CS} + E_{CV} = {E'_{CF}}^* + E_{CV} \tag{9.12}$$

$${E'_C}^* = {E'_{CF}}^* / L + E_{CV} \tag{9.13}$$

Substituting formula (9.11), (9.12) into formula (9.13), there are:

$${E'_C}^* = E_C^* \left[1 + \vartheta^* \left(\frac{1}{L} - 1 \right) \right] \tag{9.14}$$

Obviously, formula (9.14) has the same form as formula (9.10), but the two parameters E_C^* and ϑ^* are calculated according to formula (9.12) and formula (9.11) respectively. If no equipment is out of service, formula (9.14) will be transformed into formula (9.10). For ease of use, the superscript * is omitted and the same form is used. In practice, use formulas (9.12) and (9.11) when there is out-of-service equipment, and use formulas (9.7) and (9.8) if not.

In the above formulas:

E'_C—Actual load operating plant energy consumption, MJ/t;

E_C—Plant energy consumption under design load, MJ/t;

E_{CF}—Fixed energy consumption of the plant under design load, MJ/t;

L—Plant operating load ratio;

ϑ—Fixed energy consumption fraction under design load;

ϑ^*—Fixed energy consumption fraction when some equipment is out of service;

E_{CS}—The corresponding reduction in fixed energy consumption after some equipment is shut down under actual load, MJ/t;

E_{CF}^*—Fixed energy consumption with outage equipment under design load, MJ/t;

E_C^*—plant energy consumption when some equipment is shut down under design load, MJ/t;

E'^{*}_{C}—plant energy consumption of some equipment out of service under actual load, MJ/t.

9.2.2 Estimate ϑ from Energy Balance Data

After the energy balance analysis of the process plant, ϑ can be obtained from the energy balance data.

The key to finding ϑ is to distinguish the "fixed" part E_{CF}in the energy consumption E_C. But the problem is that several components in the energy consumption E_C are not simply either divided into E_{CV} or E_{CF}. Therefore, it is necessary to via analysis, divide the fixed energy consumption into four categories according to heat dissipation, steam, electricity and others, and then analyze one by one, affirm that it belongs to the fixed energy consumption part, and then transform the part that is not completely fixed energy consumption into "fixed energy consumption nature*[1] item". The calculation formula of the fixed energy consumption fraction obtained from the collation results:

$$\vartheta = (E_{FD} + \xi_s E_{FS} + \xi_E E_{FE} + E_{FO} - E_{CS})/(E_C - E_{CS}) \tag{9.15}$$

where: E_{FD}—Fixed energy consumption for heat dissipation, MJ/t;

E_{FS}—Steam fixed energy consumption, MJ/t;

E_{FE}—Electricity fixed energy consumption, MJ/t;

E_{FO}—Other fixed energy consumption, MJ/t;

ξ_s—The primary energy conversion coefficient of steam, for steam from different sources, the weighted average value can be taken, namely:

$$\xi_s = \frac{\text{Weighted average primary energy consumption}}{\text{Steam weighted average enthalpy}}$$

ξ_E—Primary energy conversion coefficient of electricity. For external power supply, take $\xi_E = 3.49$, and take the weighted average value for both self-generation and external power supply:

$$\xi_E = \frac{\text{Weighted average primary energy consumption } \times 4.1868 \text{ kJ }/(\text{kW.h})}{860}$$

[1] According to the definition of fixed energy consumption E_{CF}, the fixed energy consumption $E'_{CF} = E_{CF}/L$ when operating under the load ratio L, which can be obtained:

$$\Delta E_{CF} = E'_{CF} - E_{CF} = E_{CF}(1/L - 1)$$

Therefore, all energy consumption sub-items that conform to this relationship can be considered to have fixed energy consumption properties.

E_{CS} is the corresponding energy consumption reduction after some equipment is out of service, MJ/t (if there is no equipment out of operation at low load, then $E_{CS} = 0$)

$$E_{CS} = E_{CSD} + E_{CSE} + E_{CSS} \tag{9.16}$$

where: E_{CSD}—heat loss energy consumption of out-of-service equipment, MJ/t;

E_{CSE}—power consumption of out-of-service equipment, MJ/t;

E_{CSS}—Steam energy consumption of out-of-service equipment, MJ/t.

Based on the analysis of the influence of the load ratio on the four fixed energy consumption in Eq. (9.15), the estimation formulas of the four fixed energy consumption are derived.

9.2.3 The Influence of Load Ratio on Heat Dissipation Fixed Energy Consumption

In the links of process energy utilization and energy recovery, the heat dissipation of various pipelines, towers, reactors, heat exchangers, and catalytic cracking regenerators are fixed energy consumption, but the heat dissipation of some equipment such as heating furnaces is not completely fixed energy consumption, because when the load ratio decreases, the amount of fuel decreases, and the flue gas temperature in the furnace will change. The process calculation shows that when the heat load of the heating furnace decreases, the flue gas temperature t_p at the outlet of the radiant section usually decreases; under the same conditions, the calculation data shows that the flue gas temperature at the outlet of the radiant section t_p is proportional to the 0.22 power of heating furnace load ratio L. If roughly using t_p to represent the average temperature of the flue gas inside the entire heating furnace, it can be approximated as: 75% of the heat dissipation of the heating furnace is fixed energy consumption, and 25% is variable energy consumption.

If the plant can stop one of the parallel equipment when the load ratio is low to a certain extent, the original unit heat dissipation E_{CSD} of the stopped equipment is included in Eq. (9.16), which is not reflected here. For example, when the coking unit of the double furnace and double coke tower stops one furnace and one tower, it belongs to this.

In summary, the fixed energy consumption calculation formula for heat dissipation of the heating furnace is:

$$E_{FD} = E_{WD} + E_{JD}/\eta\prime + 0.25E_{WDF} \tag{9.17}$$

where: E_{WD}—Conversion link heat dissipation, MJ/t (Available from plant energy consumption balance data);

E_{JD}—Heat dissipation in process energy utilization and energy recovery links, MJ/t;

E_{WDF}—Heating furnace heat dissipation, MJ/t.

The increased heat dissipation in the process energy utilization and energy recovery links will basically be supplemented by increasing the furnace heat supply. The efficiency of the furnace during low-load operation is η', so the increase in fuel is correspondingly greater, but the increased heat dissipation of the furnace itself can be roughly increase the same amount of fuel to compensate.

For plants without heating furnaces and mainly heating by steam, such as gas fractionation, desulfurization, etc., there are:

$$E_{FD} = \xi_s(E_{WD} + E_{JD}) \tag{9.18}$$

For plants with both furnace and steam heating, specific analysis and treatment can be performed according to the above methods.

9.2.4 The Influence of Load Rate on Electricity Fixed Energy Consumption

The power consumption of the plant generally includes the following three parts.

9.2.4.1 Electric Desalination, Lighting, Air Cooling Fan Power Consumption E_{EL}

These are fixed energy consumption. Electric desalination power consumption is determined by the conductivity of the oil, operating voltage and current, and has nothing to do with the load ratio of the desalination tank. Generally, air-cooled fans consume power, when operating at reduced load, in addition to reducing the number of parallel operations, it can be regarded as fixed energy consumption. For air-cooled fans with automatic angle adjustment, the relationship with the load rate is more complicated, and there are many influencing factors, which require specific analysis and processing.

9.2.4.2 Cooling Water Converted Electricity E_{EW}

During low-load operation, the heat exchange depth and final heat exchange temperature may change slightly, but the general situation does not change much. Therefore, the unit energy consumption of product cooling load (MJ/t) should not change with the decrease of the feed volume. However, the reflux, especially the condensing cooling load of the reflux at the top of the tower, is often adjusted increasing due to the need to ensure a good hydrodynamic condition of the tower.

On the other hand, changes in cooling load cannot be equated with changes in cooling water volume, because the actual amount of cooling water depends on the opening of the outlet and inlet valve of the cooler, the two may be consistent only when reasonable operation adjustments are made in time with the decrease in the load ratio.

Therefore, it is necessary to analyze the cooling water converted electricity: generally, the reflux condensing cooling water is a fixed energy consumption; the condensing cooling water of the steam (such as the vacuum top evacuator, the exhaust steam of the condensing turbine, etc.), and as well as the pump cooling water converted electricity can be considered as a fixed energy consumption. However, for a well-designed tower, and the reflux cooling within a certain low load range with practical operation adjustments, the converted electricity E_{CEW} from cooling water of output plant products cooler and other coolers belongs to variable energy consumption, and should be deducted from E_{EW}.

9.2.4.3 Power Consumption of Pump and Compressors

(1) Centrifugal pump driven by non-speed-adjusting motor

According to the characteristic curve of the centrifugal pump, when the flow rate decreases, the resistance of the pipeline decreases, and the pressure head of the pump increases. The increased pressure head is consumed on the valve throttling, and the shaft power of the pump decreases slightly, but it is not proportional decrease to the flow rate; the efficiency will generally decrease. The pump unit flow shaft power $n_S = N_S/G$ will increase, namely:

$$\Delta n_S = n'_S - n_S = \frac{N'_S}{G'} - N_S/G \tag{9.19}$$

where: n_S, n'_S—Pump unit flow shaft power before and after load ratio changes, kW/(t/h);

N_S, N'_S—The shaft power of the pump before and after the load ratio changes, kW;

G, G'—Pump flow before and after load ratio changes, t/h;

Δn_S—Shaft power increment per unit flow before and after load ratio change, kW/(t/h).

It can be proved that when the speed does not change $\Delta n_S > 0$.

Let$\varphi = N'_S/N_S$, and substituting $G'/G = L$, formula (9.19) becomes:

$$\Delta n_S = n_S(\frac{\varphi}{L} - 1) \tag{9.20}$$

φ is determined by the performance of the centrifugal pump and the load ratio L. By summarizing the relationship between the shaft power and the flow rate in the existing centrifugal pump samples, it can be approximately obtained: $\varphi \approx L^{0.47}$, so:

$$\Delta n_S = n_S\left(\frac{1}{L^{0.53}} - 1\right) \tag{9.21}$$

It can be further approximated and summarized as:

$$\Delta n_S = 0.48 n_S\left(\frac{1}{L} - 1\right) \tag{9.22}$$

That is, generally speaking, about 48% of the electrical energy consumption of the centrifugal pump belongs to the fixed energy consumption. Therefore, for the whole plant, the fixed energy consumption of the centrifugal pump driven by the non-speed-adjustable motor of unit feed is:

$$E_{FEP} = 0.48 E_{EP} \tag{9.23}$$

where : E_{EP}—The power consumption of the unit feed for non-speed-regulating centrifugal pump under the design load, it can be obtained from the energy balance data, MJ/t.

(2) Centrifugal compressor driven by non-speed-regulating motor

Centrifugal compressors can adjust the output flow by two methods: discharge and suction pipe valve throttling and anti-surge circulation. When the load ratio decreases and the operation is still far from the surge point, throttling adjustment is adopted, and the power change E_{EP} is proportional to the 0.52 power of the load ratio L.

In the same way, it can be sorted into formula (9.24) (where Δn_S, and n_S have the same meaning as formula (9.22), but here it refers to the centrifugal compressor):

$$\Delta n_S = 0.45 n_S\left(\frac{1}{L} - 1\right) \tag{9.24}$$

Therefore, for the entire plant, the fixed energy consumption of the non-speed-regulated centrifugal compressor per unit of feed is:

$$E_{FEC} = 0.45 E_{EC} \tag{9.25}$$

where: E_{EC}—Electricity consumption of non-adjustable speed centrifugal compressor per unit feed under design load, MJ/t.

If the normal operating conditions are not far from the surge point, and the load ratio reduction is adjusted by increasing the anti-surge circulation, the change in unit shaft power can be completely regarded as fixed energy consumption.

(3) Centrifugal pump driven by speed-regulating motor

The relationship between the change ratio α of the centrifugal pump speed r and the flow rate, pressure head and power are:

$$\frac{G'}{G} = \frac{r'}{r} = \alpha, \frac{H'}{H} = \alpha^2, \frac{N'}{N} = \alpha^3 \tag{9.26}$$

where : G, G'—Pump flow before and after load ratio changes, t/h;

H, H'—Pump head before and after load ratio change, m (liquid column);

N, N'—The shaft power of the pump before and after the load ratio changes, kW.

The application condition of the speed-regulating motor is that not only the flow ratio changes greatly and frequently, but also the system flow resistance accounts for a large proportion of the total pressure, that is, in the equation $H = \Delta Z + \frac{\Delta p}{\gamma} + \sum \Delta h$(where: γ is the fluid density, kg/m^3; $\sum \Delta h$ is the total frictional resistance loss, m liquid column), the head difference ΔZ and static pressure difference Δp are relatively small, while $\sum \Delta h / H$ is bigger. Because only in this circumstance the pressure head change characteristics of the flow system can match the characteristics of the centrifugal pump under variable speed, namely:

$$\frac{G'}{G} = L, \frac{H'}{H} = \frac{(\sum \Delta h)'}{\sum \Delta h} = L^2, \frac{N'}{N} = L^3 \tag{9.27}$$

Under the premise of meeting this condition, the efficiency of the pump does not generally decrease with the decrease of α (see Fig. 9.4).

Substituting formula (9.27) into formula (9.19), the shaft power change per unit flow of the pump can be obtained Δn_S:

$$\Delta n_S = n_S(L^2 - 1) \tag{9.28}$$

According to $\Delta E_{CF} = E'_{CF} - E_{CF} = E_{CF}\left(\frac{1}{L} - 1\right)$ sorted into items with fixed energy consumption future, we can get:

Fig. 9.4 Speed-regulating centrifugal pump system characteristic curve

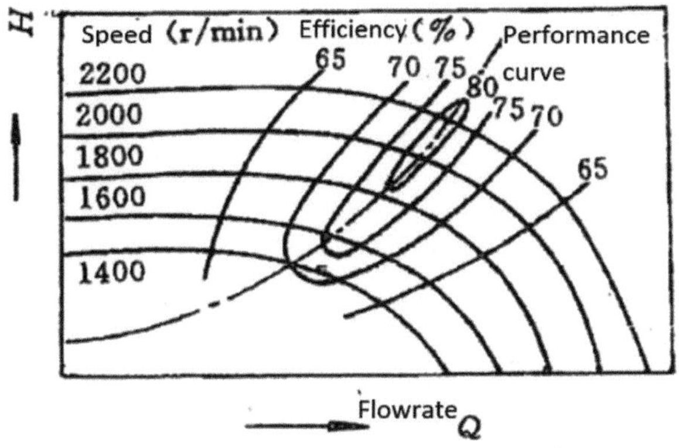

$$E_{FEA} = -E_{EA}L(L+1),\ L < 1 \tag{9.29}$$

where: E_{EA}—Power consumption of a centrifugal pump driven by a speed-regulating motor for unit feed of the plant under design load, MJ/t;

E_{FEA}—Fixed energy consumption of centrifugal pump driven by speed-regulating motor per unit feed., MJ/t.

The meaning of "negative fixed energy consumption" can be seen from Eq. (9.9). When operating at load rate $L < 1$, the unit energy consumption of this item is lower than that at design load. This is an important advantage of speed-regulating motors used in centrifugal pumps.

(4) Reciprocating pumps and reciprocating compressors

The displacement and head of the reciprocating pump piston for one reciprocation is fixed values. When the load ratio is reduced, the flow adjustment must be realized by bypass recirculation or piston movement frequency. For non-speed-regulated motor driving, only the former can be used, so its power consumption belongs to fixed energy consumption.

Although there are many ways to adjust the flow of motor-driven reciprocating compressors, the most commonly used method is the bypass recirculation method, so its power consumption E_{ER} is also a fixed energy consumption.

A reciprocating pump driven by a speed-regulating motor can be considered as variable energy consumption.

In addition, there are several compressors (pumps) running in parallel under normal load. One or more of them can be stopped under low load. At this time, if the original power consumption of the equipment out of service is E_{CSE} (MJ/t), it will also be calculated as the E_{CS} item uniformly, which is not reflected here.

Based on the above factors, the fixed energy consumption E_{FE} of the electricity consumption of the plant can be obtained:

$$E_{FE} = E_{EL} + (E_{EW} - E_{CEW}) + 0.48E_{EP} + 0.45E_{EC} - E_{EA}L(L+1) + E_{ER} \tag{9.30}$$

Each item refers to the power consumption under the design load.

Where: E_{EL}—Lighting, electric desalination, electric refining, air-cooled fans and other electric fixed energy consumption, MJ/t;

E_{EW}—Cooling water power consumption, MJ/t;

E_{CEW}—variable energy consumption part of the energy consumption of cooling water, MJ/t;

E_{EP}—Power consumption of centrifugal pump driven by non-speed-regulating motor, MJ/t;

E_{EC}—Power consumption of centrifugal compressor driven by non-speed regulated motor, MJ/t;

E_{ER}—Power consumption of general motor-driven reciprocating pump, MJ/t;

E_{EA}—Power consumption of centrifugal pump driven by speed-regulating motor, MJ/t.

9.2.5 *The Influence of Load Ratio on Steam Fixed Energy Consumption*

The steam used in the process plant can be summarized into the following eight purposes, with different relationships with load ratios.

1. Stripping steam and reboiler heating steam E_{ST}

(1) There may be two situations for the steam stripping of the fractionation tower

① In the original design, the kinetic energy factor of the valve hole in the stripping section or the velocity of the empty tower in the packing section is relatively high, and it has been considered that when operating under a certain range of low loads, the steam stripping volume is reduced proportionally to still ensure the stripping effect. Then, it can be regarded as variable energy consumption under this certain range of low load.

② The original design kinetic energy factor or velocity is not high anymore. During low-load operation, if the stripping steam is reduced, the tray will leak, and the stripping effect and product quality cannot be guaranteed. Therefore, stripping steam is a fixed energy consumption. The conditions of the settling section and stripping steam of the catalytic cracking reactor are generally similar.

There may also be a situation between the two, and the proportion of the fixed energy consumption in the stripping steam energy consumption should be fixed according to the actual situation.

(2) Reboiler heating steam

Under certain feeding conditions, the reboiler heating steam mainly depends on the reflux ratio in the upper part of the tower. For towers with good design and greater operating flexibility, when the load ratio decreases within a certain range of changes, the reflux ratio can remain unchanged, so the reflux rate decreases. At this time, the heating steam of the reboiler is also reduced proportionally, which is a variable energy consumption. If the load ratio is too low. In order to ensure the yield and quality requirements of the separated product, the reflux ratio must be increased to improve the hydraulic condition of the tray due to low load. Then, the heating steam of the reboiler is basically a fixed energy consumption. Here, the decision should also be made based on actual conditions.

The fixed energy consumption for heat dissipation of the plant reboilers with steam heating has been included in the E_{FD} item, see formula (9.18), and will not be considered again here.

2. Steam for process fluid heating

The amount of heating steam for the raw materials or intermediate products entering the plant is determined by the heating load, that is, the amount of feed, so it belongs to variable energy consumption.

3. Atomization steam E_{SM}

For furnaces equipped with multiple new type of burners, when the load ratio decreases, the amount of atomized steam reduced by some of the burners out of service will be included in item E_{CS} of Eq. (9.16). For furnaces equipped with large burners, the load rate is reduced. After the fuel oil of the furnace is reduced, the oil pressure of the burner inlet is reduced, and the atomization quality becomes worse. If the atomization steam is reduced in proportion, the atomization effect will be much worse (even if the furnace is equipped with a proportional adjustment facility, it is necessary to weigh the pros and cons according to the specific situation and decide if to increase atomized steam). Therefore, it is generally considered that steam for atomization is a fixed energy consumption.

4. Steam used as a reactant to participate in the completion of the reaction process

For example, in the hydrocarbon steam reforming hydrogen production process, steam participates in the reaction as a reactant, and the steam/oil ratio remains unchanged. When the load ratio decreases, the general hydrocarbon conversion volume and steam consumption also decrease, which is a variable energy consumption.

5. Steam for purging, blowing, accidents, firefighting

The characteristics of these steam consumption are intermittent and not directly related to the load rate, which is a fixed energy consumption.

6. Steam for heat tracing and office heating

Heat tracing and heating steam are only related to the local climate and environment at the time, and have nothing to do with the load ratio, and are fixed energy consumption.

7. Inert medium and protective steam

Many chemical reaction processes in the refinery need to be suppressed. Steam acts as an inert medium to inhibit the reaction and protect the equipment. For example, the anti-coking steam for safety valve of the catalytic cracking reactor, and the dilute-phase steam injection of the regenerator belong to this category. It has nothing to do with load ratio and belongs to fixed energy consumption.

Steam injection in the furnace tube of the vacuum furnace, and catalytic cracking of the pre-lifting steam, the purpose of which is to reduce the partial pressure of oil and vapor, increase the flow velocity, and shorten the residence time. During low-load operation, due to the fixed size of the equipment, the steam consumption cannot be reduced, so it is also fixed Energy consumption.

8. Steam used as driven power

(1) Steam reciprocating pump

Steam reciprocating pumps mostly use steam inlet valve throttling to adjust the number of reciprocations of the pump to achieve the purpose of adjusting the flow. Steam consumption decreases as the displacement decreases. At this time, the thermodynamic efficiency of the pump with steam decreases, but it can be roughly regarded as variable energy consumption.

(2) Steam for Condensing Turbine

The condensing steam turbine speed can be adjusted in a wide range. When the load is reduced and the speed is adjusted, the surge point of the centrifugal compressor will also change, and the operating range can be increased, but the pressure ratio will be reduced. If you continue to adjust the speed, the pressure provided will not meet the needs of the process, so after the speed adjustment, the flow rate must be adjusted by throttling. When the throttle is adjusted to be close to the surging point, the anti-surge cycle adjustment is necessary. The energy consumption varies with the load ratio within the scope of these three adjustment methods.

Assuming that the lower limit of the variable speed load ratio of the turbine-driven compressor is L_1, and the lower limit of the load ratio near the surging point is L_0, the three operating conditions of the turbine-driven centrifugal compressor are as follows:

(1) Within the range of available speed adjustment ($L > L_1$), the same formula (9.29), fixed energy consumption:

$$E_{FSA} = -E_{SA}L(L+1), L > L_1 \tag{9.31}$$

(2) Beyond the adjustable speed range, throttle adjustment is required, $L_1 > L > L_0$, fixed energy consumption:

$$E_{FSA} = \frac{E_{SA}}{1-L}\left[0.45L_1^3 + 0.55LL_1^2 - L\right], L_1 > L > L_0 \tag{9.32}$$

(For centrifugal pumps, the coefficients before the brackets are 0.48 and 0.52 respectively)

(3) When the operating conditions are close to the surging point or below, that is, $L \leq L_0$. and the anti-surging cycle adjustment is to be adopted, the fixed energy consumption:

$$E_{FSA} = \frac{E_{SA}}{1-L}\left[0.45L_1^3 + 0.55L_0L_1^2 - L\right], L \leq L_0 \tag{9.33}$$

The only difference between formula (9.32) and formula (9.33) is that the L in the second term in the brackets becomes L_0. When calculating E_{FSA} in practice, according to the relationship between L and L_1, L_0, one of the three equations can be selected respectively.

Among the above formulas:

E_{FSA}—Fixed energy consumption of the centrifugal compressor (pump) driven by the condensing turbine of the plant under design load, MJ/t;

E_{SA}—Energy consumption of the centrifugal compressor (pump) driven by the condensing turbine of the plant under design load, MJ/t.

For the actual load ratio is greater than the lower limit L_0 of the load ratio near the surging point, and there is no speed adjustment can be used, and only the throttling method can be used. The power consumption of the centrifugal compressor is the same as that of the general motor-driven centrifugal compressor, see Eq. (9.24).

(3) Steam jet vacuum pump at the top of the vacuum tower E_{SJ}

The amount of steam used by the Steam jet vacuum pump (constant) is basically a fixed energy consumption. When the load ratio is reduced, the non-condensable gas volume of the vacuum tower will decrease (but not proportionally, because the temperature and pressure conditions remain unchanged. The residence time of the tube and tower bottom is extended). When considering this point, when several sets of parallel Steam jet vacuum pumps are used, when the load ratio is lower than a certain value, one set can be out of service, and the out-of-service steam consumption is included in the E_{CSS} item in formula (9.16), which is not reflected here.

Therefore, from the above analysis, the fixed energy consumption of steam E_{FS} can be obtained by the sum of relevant items:

$$E_{FS} = E_{FST} + E_{SM} + E_{SO} + E_{FSA} + E_{SJ} \tag{9.34}$$

where: E_{FST}—Fixed energy consumption of steam for stripper and reboiler, MJ/t; (estimate the percentage of E_{FST} in E_{ST} according to the specific conditions described in this article);

E_{SM}—Steam energy consumption for atomization, MJ/t;

E_{SO}—Items 5, 6, and 7 of steam energy consumption for purging, heating, heating, protection, etc., MJ/t;

E_{FSA}—Fixed energy consumption of steam for condensing turbine, MJ/t (estimated based on the different relationship between L and L_1, L_0);

E_{SJ}—Unit steam consumption of Steam jet vacuum pump, MJ/t.

9.2.6 *Other Fixed Energy Consumption*

In the production plant of petrochemical plants, there are often other specific energy-consuming items that have the nature of fixed energy consumption, which can also be calculated according to the actual situation and included in the "other fixed energy consumption" item.

1. Fixed energy consumption for the reboiler heat carrier heat supply

The heat load of the reboiler of various towers using heat carrier or other process streams as the heat source is the same as the above-mentioned steam heating. When operating at low load and requiring an increase in reflux ratio to improve the hydraulic of the tray, it can be regarded as a fixed energy consumption.

$$E_{FOH} = E_{UO}/\eta_H \tag{9.35}$$

where: E_{FOH}—Fixed energy consumption for reboiler heating, MJ/t;

E_{UO}—Reboiler heating fixed energy consumption heat load, MJ/t;

η_H—The heat source conversion efficiency of the reboiler, when it comes from external sources $\eta_H = 1$.

2. Fixed energy consumption of back pressure turbine

Since there is no change in the steam mass amount of input and output of the back pressure turbine, the back pressure steam continues to be used in the process, so the fixed energy consumption of steam is not reflected in the steam fixed energy consumption, but the enthalpy drop produced by the steam in the turbine work is also the actual energy consumption. The load ratio change also has an impact on this part of the energy consumption. It's changing law is similar to that of a condensing turbine, and it is included in the "other fixed energy consumption" item.

$$E_{FOB} = -E_{OB}L(L+1), L > L_1 \tag{9.36}$$

$$E_{FOB} = \frac{E_{OB}}{1-L}\left[0.45L_1^3 + 0.55LL_1^2 - L\right], L_1 > L > L_0 \tag{9.37}$$

$$E_{FOB} = \frac{E_{OB}}{1-L}\left[0.45L_1^3 + 0.55L_0L_1^2 - L\right], L \le L_0 \tag{9.38}$$

In the above formulas:

E_{FOB}—Fixed energy consumption of centrifugal compressor (pump) driven by steam back pressure turbine, MJ/t;

E_{OB}—Energy consumption due to enthalpy drop of plant back pressure turbine under design load, MJ/t.

When the actual load ratio L is greater than the lower limit of the near-surging point load ratio L_0, the power consumption when the speed adjustment is not used is the same as that of a centrifugal compressor driven by a general motor, see also Eq. (9.24).

Therefore, other fixed energy consumption can be expressed as:

$$E_{FO} = E_{FOH} + E_{FOB} + \sum (E_{FO})_i \tag{9.39}$$

Table 9.3 Basic data

From energy balance data		From electricity balance and verified pump/compressors data		Balance data from steam production and use	
Description	MJ/t	Description	MJ/t	Description	MJ/t
Recovery Link heat dissipation	63.01	Centrifugal pump power consumption	19.05	Steam for atomization	10.89
Conversion link heat dissipation	22.48	Centrifugal compressor power consumption		Steam for purging, tracing and heating etc.	15.49
Average furnace efficiency	84%	Reciprocating pump power consumption		Stripping steam	58.2
Total power supply	31.19	fans of air cooler and furnace, electric desalination	8.19	steam for condensing turbine	
Plant energy consumption	608.3	Cooling water converted power	4.14	Steam for Jet pump	30.56
Heating furnace heat dissipation	22.48				

where: $(E_{FO})_i$—Other items with fixed energy consumption nature, MJ/t。

Example 9.1 The atmospheric and vacuum unit of a refinery has been calibrated for energy consumption with different processing capacities after the energy-saving modification is completed. Estimate the fixed energy consumption fraction of the plant and the energy consumption of the plant under different load ratios and compare with the measured values.

1. Basic data (under design load) see Table 9.3.
2. Calculating plant fixed energy consumption fraction:

(1) Heat dissipation fixed energy consumption, for plant with the heating furnace-based, it can be calculated from Eq. (9.17)

$$E_{FD} = E_{WD} + \frac{E_{JD}}{\eta'} + 0.25E_{WDF}$$
$$= 22.48 + \left(\frac{63.01}{0.84}\right) - 0.25 \times 22.48 = 91.87\ \text{MJ/t}$$

(2) Fixed energy consumption of electricity, by formula (9.30)

$$E_{FE} = E_{EL} + (E_{EW} - E_{CEW}) + 0.48E_{EP} + 0.45E_{EC} - E_{EA}L(L+1) + E_{ER}$$

The cooling water converted electricity of the plant can generally be regarded as a fixed energy consumption, $E_{CEW} = 0$; there is no centrifuge compressor, reciprocating pump and speed-regulating motor, so $E_{EC} = 0$, $E_{ER} = 0$, $E_{EA} = 0$

$$E_{FE} = 8.19 + 4.14 + 0.48 \times 19.05 = 21.474MJ/t$$

(3) Fixed energy consumption with steam, by formula (9.34)

$$E_{FS} = E_{FST} + E_{SM} + E_{SO} + E_{FSA} + E_{SJ}$$

The design of the plant leaves a lot of margins and can adapt to a 20% increase in processing capacity, so the steam used for stripping can generally be regarded as a fixed energy consumption. which is $E_{ST} = E_{FST}$, no condensing turbine $E_{FSA} = 0$, so

$$E_{FS} = 58.20 + 10.89 + 15.49 + 30.56 = 115.14MJ/t$$

(4) The fixed energy consumption fraction of the plant, by formula (9.15)

$$\vartheta = (E_{FD} + \xi_s E_{FS} + \xi_E E_{FE} + E_{FO} - E_{CS})/(E_C - E_{CS})$$

The steam of this plant is all self-generated, $\xi_s = 1$, no equipment out of service $E_{CS} = 0$, no other fixed energy consumption $E_{FO} = 0$, so

$$\vartheta = \frac{91.87 + 115.14 + 3.49 + 21.474}{608.3} = 0.464$$

3. Comparison of energy consumption prediction and calibration data for different load rates

According to the calibration data of the plant, the energy consumption data under different load rates are estimated according to formula (9.10), see Table 9.4.

It can be seen from Table 9.4 that after determining the fixed energy consumption fraction of the plant, the energy consumption of the plant under different load ratios is predicted according to formula (9.10), and this predicted value is generally consistent with the calibration result. Table 9.4 also lists the reduction value of unit energy consumption predicted by formula (8.10) when the load is higher than the design load, which is also quite consistent with the actual measurement. However, it needs to be explained that this is mainly due to: the requirements for expanding the processing capacity, the accuracy of the actual design calculation method and the selection of equipment series models are restricted, and the role of the design margin is reserved, so there are certain conditions and range, it cannot be universally extended.

Table 9.4 Comparison of energy consumption prediction and calibration value

Description		Case 1	Case 2	Case 3	Case 4
Processing capacity (t/d)		9737.28	8592	10,412.88	11,590.8
Load ratio (%)		100	88.2	107	119
Plant energy consumption (MJ/t)	Test trail	608.3	620.48	589.12	568.15
	prediction	608.3	646.06	589.83	563.23
Relative error (%)		0	4.12	0.14	0.88
Measured fuel consumption kg/t		12.01	12.04[a]	11.57	11.28

[a] Analyzing the calibration results, the low measured value of the fuel oil (the load rate is reduced by 12%, and the fuel oil consumption is roughly unchanged) is the main reason for the large deviation between the predicted energy consumption and the calibration energy consumption

9.3 Assessment of Energy Use Level and Energy Saving Potential

9.3.1 Apply "Benchmark Energy Consumption" to Evaluate the Energy Consumption [6]

In recent years, energy-saving and CO_2 mitigation work has continued to deepen, and the energy balance of plants has been widely deployed in refineries and petrochemical plants; through the energy balance of plants, not only can the energy consumption of the plant be verified, but also the ins and outs of energy consumption and energy loss in the process can be understood. The location and quantity of energy loss can be identified, and put forward the evaluation index of different energy use links (energy conversion, process energy use, recovery and recycling) of the plant, which played an important role in the energy saving work.

However, in the energy utilization analysis and improvement of the plant, it is often necessary to draw on the experience of the analyst and the accumulated data in order to discover the potential in the advanced level of the history of the plant and the specific parameters of similar plants, it brings difficulties to analysis and improvement. At the same time, some production plants in the refinery have already formulated Benchmark Energy Consumption (BEC), and proposed the energy consumption that can be achieved through improvements and efforts under the current technical and economic conditions. Therefore, it can be used as a standard and benchmark for comparison. Discover the potential for improvement in the comparison. Atmospheric and vacuum distillation units, delayed coking units and catalytic cracking units have been developed and put into use. Work on other plants is still in progress. However, due to the different perspectives and ideas of the makers, it is difficult to directly correspond to the energy balance result data, and it needs to be converted before application. This section first discusses the application of benchmark energy consumption for atmospheric and vacuum distillation units.

9.3.1.1 Energy Balance Results Under Benchmark Energy Consumption Conditions

1. Benchmark energy consumption conditions

The atmospheric and vacuum distillation unit uses Daqing crude oil as the benchmark crude oil (specific gravity: d_4^{20}=0.8595, total extraction rate 62.5%), and 15 basic conditions are proposed (see Table 9.5).

2. Energy balance results under benchmark energy consumption

According to the basic conditions of the benchmark energy consumption, the energy balance results of the benchmark energy consumption are determined in Table 9.6.

3. The impact of objective conditions on benchmark energy consumption

According to the stipulated benchmark crude oil and basic conditions, the corresponding benchmark energy consumption is determined, but when the crude oil variety changes and the extraction yield is different, the benchmark energy consumption value changes differently. The load ratio also affects the benchmark energy consumption value of the plant.

(1) Crude oil varieties and extraction yield

The change of crude oil varieties is mainly due to the different relative density of crude oil. When the relative density increases, the yield of light oil decreases, and the yield of heavy oil increases, the tower top products that are difficult to recover and the reflux take less heat, while the tower lower side draw out and bottom oil yield of the tower increase, and the heat removal temperature is high. It is beneficial to energy recovery and can reduce the energy consumption of the plant. When the extraction yield is high, the temperature of the flash section of the tower is required to be high, and the energy consumption is correspondingly increased. The fuel-based wet "benchmark energy consumption" has been regressed with the benchmark energy consumption value that varies with the proportion of China domestic crude oil and the total extraction yield:

$$E = 0.948d^{-0.487}C^{0.522} + 1.75 \tag{9.40}$$

where: d—Relative density of oil at 20 °C, d_4^{20};
C—Total extraction yield, %;
E—Benchmark energy consumption value, MJ/t.

Therefore, the benchmark energy consumption of crude oils with different relative densities and extraction yields can be obtained by adding the correction term ΔE_1 to the results of Daqing Oil's benchmark energy consumption.

$$\Delta E_1 = 0.948d^{-0.487}C^{0.522} - 8.836 \tag{9.41}$$

Table 9.5 Basic conditions for benchmark energy consumption

Description		Fuel type		Lubricating oil type (wet)	Remarks
		Dry	Wet		
Crude oil entering the plant temperature (°C)		40	40	40	
Water content in crude oil after desalination (%)		0.1	0.1	0.1	
Crude oil heat exchange temperature (°C)	Before electric desalination	110	110	110	
	After electric desalination	107~	107~	107~	
	Initial tower bottom oil	300	300	300	
Flash section temperature (°C)	Initial distillation tower	225	225	225	
	Atmospheric tower	355	355	355	
	Vacuum tower	365	385	385	
Temperature difference between flash section and tower bottom	Initial distillation tower	2	2	2	
	Atmospheric tower	5	5	5	
	Vacuum tower	2	10	10	
Over vaporization ratio (%)	Initial distillation tower	0(2)	0(2)	0(2)	The over-vaporized oil from the initial tower is sent to the downstream tower as a product
	Atmospheric tower	0(2)	0(2)	0(2)	
	Vacuum tower	2	2	4	
Reflux heat recovery ratio (%)	Initial distillation tower	0	0	0	
	Atmospheric tower	90	90	90	
	Vacuum tower	90	90	90	
Stripping steam accounts for bottom oil (%)	Initial distillation tower	0	0	0	Including side draw stripping
	Atmospheric tower	1.5	1.5	1.5	
	Vacuum tower		4	4	

(continued)

Table 9.5 (continued)

Description		Fuel type		Lubricating oil type (wet)	Remarks
		Dry	Wet		
Average final temperature after heat changing (°C)	Side draws of atm. tower	70	70		
	Side draws of vacuum tower	90	90		
	Vacuum residue	110	110		
The cold feed accounts for the effective heating load (%)		13	13	13	
Heating furnace thermal efficiency (%)		90	90	90	
Furnace atomization steam to account for fuel (%)		20	20		
Evacuate steam (kg/t)		11	8		
Electricity consumption (MJ/t)		62.8	62.8	62.8	Including electric desalination
Water consumption (MJ/t)		9.21	10.47	13.4	

(2) The influence of load ratio

The load ratio has a certain impact on the energy consumption of the plant, mainly because some of the energy consumption of the plant is fixed energy consumption. According to the concept and calculation method of fixed energy consumption fraction (ratio to plant energy consumption) proposed in Section 2, fixed energy consumption is divided into four types: heat dissipation, electricity, steam, and other four types of calculation to determine energy consumption fraction. The benchmark energy consumption at low load ratios can be modified.

$$\Delta E_L = \vartheta\left(\frac{1}{L} - 1\right) \tag{9.42}$$

ϑ is the fixed energy consumption fraction, which is calculated based on the energy balance data. If there is no data, it can be 0.4.

Therefore, the reference energy consumption after correction is:

$$E' = E + \Delta E_L + \Delta E_1 \tag{9.43}$$

Table 9.6 Energy balance results of BEC of crude distillation unit

Link	Description		Energy MJ /t		
			Fuel type (dry)	Fuel type (wet)	Lubricating oil type
Energy conversion	Total supply	Furnace fuel	349.8	389.8	392.7
		Steam supply	0	0	0
		Electricity supply	20.64	21	21.84
		Supply heat	76	73	72.97
		Sub total	447	483.8	487.5
	Direct loss	Flue gas	30.8	34.3	34.5
		Heat loss	7.7	8.6	8.6
		Ineffective power	8.1	8.1	8.1
		Conversion rejection	46.6	51	51.2
	Efficiently used		397.76	429.8	432.5
Process utilization	Recovery and recycling		653.3	670.9	694.2
	Total process used		1051.1	1100.7	1126.7
	Thermodynamic energy consumption		104.2	104.7	105.1
	Energy to be recovered		829.9	977.2	1001.6
Energy recovery	Recovery and recycling		653.3	670.9	694.2
	Recovery output		75	92.48	71.9
	Rejection	Cooling	162.48	173.97	194.7
		Heat loss	36.2	37.8	37.9
		Stream	1.22	2.05	3.84
		Other	2.64	3	3.84
		Subtotal	202.54	216.82	239.34
Evaluation Indicators	Net energy consumption		371.3	391.36	415.6
	Equivalent to primary energy consumption		424	443.5	472.04
	Energy conversion efficiency %		89.6	89.5	89.5
	Thermodynamic efficiency %		9.91	9.51	9.33
	Energy recovery efficiency %		78.5	78.1	76.5

9.3.1.2 Comparative Analysis of Energy Balance Results of Test Conditions and BEC Conditions

With the benchmark energy consumption data and its balance results, we can compare the energy balance data of the plant with it to discover potential, make improvements, and reduce the energy consumption of the plant. The benchmark energy consumption comparison of the three links in the energy utilization process can be carried out in accordance with the items listed in Table 9.7. When using Table 9.7 for comparative analysis, we can first list the benchmark energy consumption parameters in Table 9.7 according to whether the plant is of fuel type or lubricating oil type, and list the actual parameters of energy balance in the right column, compare link by link and parameter by parameter, and indicate the difference value in the remark's column. Table 8.7 lists the comparison results between the calibration data of a certain plant and the benchmark energy consumption. The ratio of the plant's energy consumption to the benchmark energy consumption is 1.30, which is 30% higher than the benchmark energy consumption. It is the potential for improvement (analysis omitted). Generally, the comparative analysis is carried out in four aspects:

1. Analysis of the overall situation of the plant

Analyze the level of energy consumption from factors such as plant scale, load ratio, crude oil properties and extraction yield, and put forward the overall evaluation results. Calculate the plant's energy consumption factor EF (the ratio of actual energy consumption to benchmark energy consumption).

2. Energy conversion link

(1) Analyze the difference of each supply energy from energy supply perspective of view.
(2) Conversion efficiency analysis, pointing out the reasons for the low conversion efficiency and the potential for improvement.
(3) In the process energy use link, analyzes the total process energy used, the operating conditions and processes of each tower under the calibration conditions, and points out the reasons for the large total process energy used and improvement measures.
(4) Energy recovery and utilization link. The difference between the actual operating energy recovery rate and the specified value of the "benchmark energy consumption" is proposed. Analyze the reasons for the low energy recovery rate and how to improve measures one by one.

For different process plants, different benchmark energy consumption has been developed as a criterion for judging and identifying for the energy saving potential of the plant, which has been adopted by people. However, there is a big difference in the formulation of benchmark energy consumption. At present, the benchmark energy consumption of atmospheric and vacuum plant that has been established in China is based on experience and field data, selecting an operating condition that is considered more reasonable, so as to determine a corresponding energy consumption. Its data

Table 9.7 Comparison results between "BEC" and calibration energy consumption

Description		"Benchmark" conditions	Calibration conditions	Difference
Processing rate (t/h)		312.5	256.625	55.88
Relative density of crude oil		0.8595	0.8861	0.0266
Total extraction yield (%)		62.5	61	−1.5
Conversion link	Furnace fuel (kg/t)	8.3555	11.044	2.6 9
	Electricity supply (including water converted) (kWh/t)	5.73	7.58	1.85
	Supply steam (kg/t)	0	2.22	2.2
	Supply heat (MJ/t)	76.6	67.36	−9.24
	Total energy supply (MJ/t)	447	563.03	116.03
	Heat load of atm. furnace (MJ/t)	249.6	270.96	21.36
	Heat load of vacuum furnace (MJ/t)	96.8	152.19	55.39
	Heating furnace thermal efficiency %	90	83	−7
	Conversion loss energy (MJ/t)	46.6	128.116	81.516
	Effective energy supply (MJ/t)	397.76	434.916	37.156
	Conversion link efficiency %	89.6	77.25	−12.35
Process utilization link	Recovery and recycling energy (MJ/t)	653.3	633.099	−20.201
	Total process energy used (MJ/t)	1051.1	1068.014	16.914
	Initial tower inlet heat (MJ/t)	501.3	474.415	−26.885
	Atm. tower inlet heat (MJ/t)	892.4	891.25	−1.15
	Vacuum tower inlet heat (MJ/t)	620.5	654.46	33.96
	Flash section temperature of initial Tower (°C)	225	220	−5
	Flash section temperature of Atm. Tower (°C)	355	360	5

(continued)

Table 9.7 (continued)

Description		"Benchmark" conditions	Calibration conditions	Difference
	Flash section temperature of Vacuum Tower (°C)	365	385	20
	Steam for stripping (kg/t)	9.71	14.8	5.09
	Equipment heat loss (MJ/t)	17.96	15.512	−2.448
	Thermodynamic energy consumption (MJ/t)	104.2	71.35	−32.85
Energy recovery link	Final temperature after Crude heat change (°C)	300	284	−16
	Total heat recovery (MJ/t)	724.3	701.34	−22.96
	Preheat the air heat (MJ/t)	39	34.9	−4.1
	Crude oil exchange heat (MJ/t)	625.3	564.98	−60.32
	steam generation (MJ/t)	64.7	90.38	25.68
	Energy recovery efficiency, %	78.15	70.28	−7.87
Net energy consumption (MJ/t)		371.3	495.675	124.4
Corrected benchmark energy consumption (MJ/t)		424	433.2	9.2
Plant energy consumption (MJ/t)			564.38	131.2

is relatively reliable and easy to reach. Some energy companies have infiltrated the concept of economy and optimization into the formulation of benchmark energy consumption, UOP uses Pinch technology in continuous reforming and aromatics production, and uses the energy consumption obtained from the pinch point analysis and the comprehensive curve as the benchmark energy consumption [7], Make the benchmark energy consumption an objective energy consumption evaluation scale, which ideas can be borrowed when we develop the benchmark energy consumption in the future. However, the benchmark energy consumption is an approximate optimal value, which is the targeted limit for energy saving. When the energy consumption is lower than the benchmark energy consumption and the energy-saving measures are not worth the loss, it is not economical, but it is not easy to achieve. Appropriate approach should be divided into two levels. One level is the economical optimized energy consumption, that is, the benchmark energy consumption, and the other level is the relative benchmark energy consumption that can be achieved by appropriate energy-saving improvements to improve the operation, so that the energy saving of the plant is not only the ultimate goals, and there are short-term improvement goals.

9.3.2 *Evaluation on Energy Utilization Level and Potential of Equipment and System*

For those that have established the benchmark energy consumption of the plant, we can analyze and evaluate the overall plant and links in accordance with the benchmark energy consumption. However, the energy consumption analysis and evaluation indicators of links and plants are connected with equipment and systems. If we combine the analysis of equipment and systems, the root of causes for the higher energy consumption of links and plants can be identified, the potential solutions will be proposed.

9.3.2.1 Energy Utilization Level and Potential of Equipment

1. The role and function of energy conversion equipment is to convert energy materials and certain forms of energy supplied from the outside into the energy forms required by the process. These energy forms are mainly work and heat. Therefore, it includes thermal energy conversion equipment and work conversion equipment. Generally, refers to fuel boilers, heating furnaces, electric or steam driven pumps/compressors, steam turbine units, gas turbine devices, and the energy obtained is directly carried by the process fluid.

The purpose of energy conversion is to maximize the conversion into the required form of energy. Therefore, its evaluation index is mainly the efficiency of the first law (energy) and the second law (exergy).

Energy efficiency is the ratio of the effective energy supplied by the equipment to the sum of the various energies supplied to the equipment:

$$\eta = \frac{E_{effective}}{E_{in}} \times 100\% = \left(1 - \frac{E_J}{E_{in}}\right) \times 100\% \tag{9.44}$$

Exergy efficiency is also the ratio of effective exergy (used by the process) to total supplied exergy:

$$\eta_x = \frac{EX_{effective}}{EX_{in}} \times 100\% = \left(1 - \frac{D_K + D_J}{EX_{in}}\right) \times 100\% \tag{9.45}$$

The energy efficiency of energy conversion equipment characterizes the extent to which the supplied energy is effectively used. It is currently the main evaluation index in the energy consumption analysis based on the thermodynamic first law. It characterizes the meaning of two aspects, the effective utilization ratio and the ratio of energy loss. Therefore, people are urged to reduce losses and increase efficiency, thereby mitigating the energy efficiency.

Exergy efficiency is an indicator that reflects the change of energy quality. It points out the ratio of conversion into effective exergy in the supply of exergy. The exergy efficiency not only reflects the external loss exergy D_J (that is, the exergy carried in the external energy loss), but also points out the exergy loss D_K caused by the irreversibility of the process. Therefore, the exergy efficiency index allows people to partially turning to reducing the irreversibility of the process from improvement the amount of energy loss, making both of them pay equal attention to improve the process exergy efficiency. The development of steam-powered combined cycle and heating furnace-gas turbine combined system is an effort to improve the efficiency of exergy.

An energy conversion device that converts one form of work to another form of work: the energy efficiency is the same as the exergy efficiency. Such as pumps and compressors that convert electrical energy into fluid energy.

2. Process energy utilization equipment

Like conversion equipment, energy utilization equipment also has two evaluation indicators: energy efficiency and exergy efficiency. Energy utilization equipment includes towers, reactors and other core equipment in the petrochemical production process.

Generally, in the past energy consumption statistics and calculations, little attention was paid to the efficiency of energy-using equipment. In fact, the energy efficiency after deducting the heat dissipation and leakage losses of the equipment exceeds 95%, so it does not seem to be meaningful to investigate the efficiency. In fact, the concept of energy efficiency is put forward to make people pay more attention to the heat preservation of process equipment and reduce heat loss, especially the heat preservation of high-temperature parts, such as the heat preservation at the reactor and bottom of the fractionation tower.

On the other hand, the efficiency of exergy is a very important evaluation index for energy-using equipment. It not only points out the difference between exergy leaving and exergy supplied to the equipment, but also points out the evolution of energy use. Process operating conditions require a certain grade of energy (exergy). In the equipment, due to the irreversible loss, the exergy loss is generated, so that the exergy leaving the equipment is much less than the exergy value supplied to the equipment. This is manifested as the energy grade (T, p) leaving the process equipment is significantly reduced, which brings difficulties to recover the energy from energy-using equipment. In addition, the energy recovery process requires driving force, so it cannot be completely recovered, and there is an energy recovery efficiency.

The means to improve the exergy efficiency of the process equipment is to increase the grade of energy leaving the process equipment to reduce the exergy loss of the process. This will help improving the energy efficiency of the recovery system.

The calculation method of energy and exergy efficiency of process energy equipment is the same as that of energy conversion equipment, that is, the ratio of the energy (exergy) leaving the equipment to the energy (exergy) supplied to the equipment. The function of process equipment processes raw materials into products; therefore, the effective energy is the energy carried by the leaving process stream. In many cases, all the energy carried out cannot be used fully, which is a topic with the energy recovery system.

3. Energy recovery equipment

The energy efficiency and exergy efficiency are calculated with the same formulas (9.44) and (9.45).

Energy recovery equipment is divided into heat recovery equipment and work recovery equipment (hydraulic turbine etc.). In addition to heat dissipation factors, heat recovery equipment has an energy efficiency of nearly 100%. However, the energy recovery ratio of the recovery system is very different, the reason is that the heat transfer temperature difference of each system reflects the exergy loss of the process is very large. When the heat transfer temperature difference is large, as a result, the partial high-temperature energy will inevitably be degraded to the cooling load and lost to the environment. The appropriate heat transfer temperature difference can reduce the cooling load and increase the energy recovery rate. Therefore, it is more appropriate and reasonable to use the exergy efficiency method to evaluate heat exchangers [8]. Reducing the excessive heat transfer temperature difference can improve the efficiency of exergy, improve the grade of discharged energy, increase the recovery efficiency of the energy recovery system, and reduce the cooling rejection energy.

9.3.2.2 Evaluation on Inter-Equipment and System

In addition to energy-using equipment, another aspect of energy consumption in a plant or process is the energy transfer between equipment and between equipment and system. How to evaluate its energy consumption level is also a major content of energy-saving of plant or process.

The energy efficiency of the energy transfer process between equipment and equipment and system is mainly caused by heat dissipation and leakage. In terms of the quantity of energy, heat preservation and leak stoppage methods can be adopted to appropriately improve energy efficiency, but this is very limited. When optimizing and integrating a large system, attention should be paid to whether the equipment and the system and the energy grade between the equipment are matching. The energy grade analysis method [9] can solve this problem. The energy grade analysis method is an energy analysis method proposed by Mr. Yang Donghua. It is also a content of the second law analysis. It is more convenient and intuitive to use between equipment and systems.

The so-called energy grade is the content of exergy in energy:

$$\varepsilon = EX/E \tag{9.46}$$

The energy grade difference $\Delta\Omega$ between energy supply equipment and energy utilization equipment, system and equipment are the potential for energy saving improvement, noted that the energy input and energy required of equipment have the same energy quantity, so:

$$\Delta\Omega = \varepsilon_{in} - \varepsilon_{use} = \frac{EX_{in}}{E_{in}} - \frac{EX_{use}}{E_{use}} = \frac{EX_{in} - EX_{use}}{E_{in}} \tag{9.47}$$

When $\Delta\Omega$ is positive, it means that the supplied energy grade is greater than the required energy grade, and there is potential for energy saving. For example, the steam header supplies 1.0 MPa steam, and when the user needs 0.3 MPa steam, there is an energy grade difference, which can be recovered by setting up a back pressure steam turbine to recover the steam pressure difference to match the energy supply and consumption. Of course, recovering the steam pressure difference must combine engineering and economic conditions. When $\Delta\Omega$ is negative, it indicates that the quality of the supplied energy cannot meet the requirements, and supplied energy grade ε should be increased. The process of transforming high energy grade into low energy grades is called dilution. For example, a heat pump dilutes the energy grade of electricity into a thermal energy grade, and the process of raising the energy grade is called concentration, such as steam turbine power generation.

(1) There is often an energy grade difference between upstream and downstream equipment. This energy grade difference is often an energy-saving potential. For example, there is a large energy grade difference between the catalytic cracking reaction oil and gas and the energy (T, p, phase state) supplied to the fractionation tower. The difference between the energy grade at the outlet of the heating furnace and the energy grade at the entrance of the flash section is often the energy-saving potential.
(2) The energy grade difference between the system and the equipment. The most common is the energy grade difference between the steam header and the steam using equipment. There is a 1.0 MPa steam header, demanded steam pressure only needs 0.3 MPa, obviously there is an energy grade difference, which can be improved through the steam system integration.

The energy grade method to find the energy saving potential of equipment and systems lies in the rational selection of the energy supply grade required by the energy consuming equipment. By comparing the energy grade supplied to the equipment, the potential can be discovered. However, when using the energy grade difference to analyze the potential between the system and the equipment, it is also necessary to combine the energy transfer efficiency to be more comprehensive, and avoid ignoring the potential for heat dissipation and leakage.

9.4 Approaches for Energy-Saving Improvement of Production Plant [21]

After completing the energy balance and exergy balance of the production plant or the whole plant, how to identify improvement directions and measures to achieve the energy saving and CO_2 mitigation based on the existing process conditions is an important and realistic topic for energy-saving workers. This section and section 5 discuss the energy-saving improvement methods and approaches from two aspects: production plant and entire plant/system energy optimization.

The energy utilization three-link model points out the balance and restriction of energy use from a macro perspective. The energy used by the plant runs through the entire process, and the energy grade is reduced. Finally, when the energy can no longer be recovered and recycled practically, it is lost to the environment, resulting in a consumption of energy. It can be seen from the balance model that energy degradation and loss occur gradually in all three different links.

According to the energy utilization analysis model, the energy-saving improvement of petrochemical plant has rules to follow. Therefore, energy-saving improvement should not simply be carried out to improve the efficiency of conversion equipment by energy-saving technological modification (such as heating furnace); and when the process energy use link and energy recovery link are improved, the effective energy that needs to be supplied from the conversion link is reduced, the conversion equipment is still in low-load operation and will not be suitable for the energy-saving modification of the plant, making the improvement of the conversion link futile. Therefore, when embarking on the energy-saving improvement of the plant, the improvement of the process energy use and recovery link should be started first, and the improvement of the conversion link should be considered after the improvement of the two is determined. Otherwise, the investment in energy-saving improvement will not be effective.

9.4.1 Improve Process Conditions to Reduce the Total Energy Consumption of the Process

The total process energy consumption is an important indicator to measure the energy utilization level of the plant. It is the amount of energy required to change the raw materials to the conditions (temperature, pressure) required by the process. Generally speaking, the total process energy used can be divided into three forms: heat, steam and flow work, namely, the heat total process energy used, the steam total process energy used and the power total process energy used.

9.4.1.1 Reduce the Heat Total Process Energy Used

(1) Improving the process

Adopting a new energy-saving process is an important aspect of reducing the heat total process energy used. For example, the atmospheric and vacuum distillation plant in the oil refining industry directly extracts the over-vaporized oil from the initial distillation tower and atmospheric tower, bypassing the heating equipment, avoiding repeated heating, vaporization and condensation of the over-vaporized oil, and reducing heating load of furnaces and the heat exchange load of the bottom oil of the initial distillation tower, thereby reducing the energy brought into the energy-consuming equipment (fractionation tower), so that the total process energy used is reduced. The side draw of the initial distillation tower also plays a similar role in energy saving. A crude oil distillation unit in an oil refinery sends the over-vaporized oil (2.85%) from the initial distillation tower directly to the atmospheric tower in the form of the side draw of the initial distillation tower, reducing the heating furnace and heat exchange load by about 2000 kW, and the total process energy consumption is reduced by 21.8 MJ/t.

Figure 9.5 shows the energy-saving modification of an oil refinery [10] with an annual processing capacity of 3 million tons. The over-vaporized oil from the initial distillation tower is directly taken out into the atmospheric tower, and the over-vaporized oil from the atmospheric tower is taken out directly into the vacuum tower, reducing the process total process energy consumption is 25.1 MJ/t (809 kW), which can save fuel 31.4 MJ/t. Figure 9.6 shows the pre-flash process used in a refinery [11].

(2) Improve the catalyst to reduce the reaction temperature and pressure. For example, the platinum reformer is changed from a single metal catalyst to

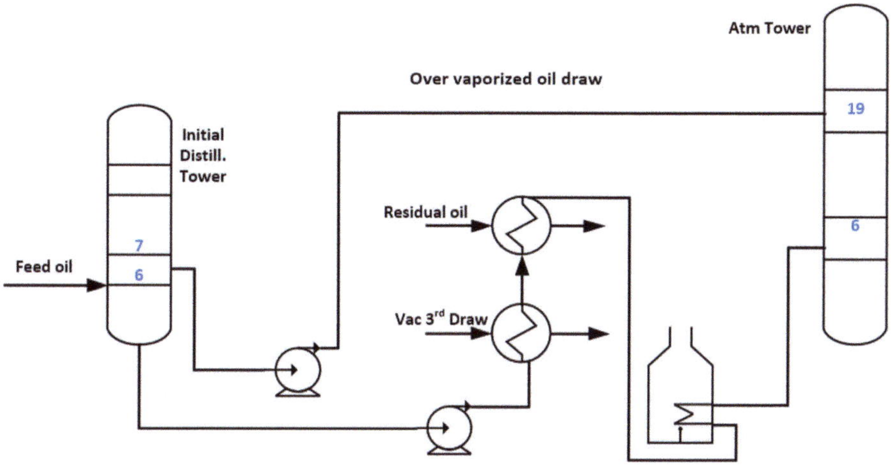

Fig. 9.5 Direct extraction process of over-vaporized oil from initial distillation tower

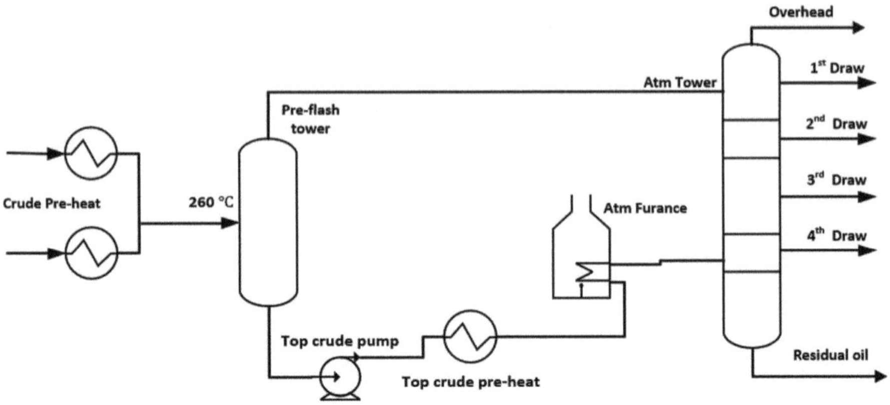

Fig. 9.6 Pre-flash process used in a refinery

a bimetal catalyst and the operating pressure is reduced after the continuous reforming catalyst is used, which saves the total process energy used.

(3) The use of new trays or the use of new packings to reduce the pressure drop of the tower means that the flashing temperature is reduced, and the required furnace outlet temperature is also low, thereby reducing the total process energy used; for absorption mass transfer equipment, reducing the pressure drop also means reducing power consumption and improving separation accuracy. Under the premise of ensuring product quality, the reflux ratio should be reduced, thereby reducing the total process energy used.

(4) Reducing the refining ratio and reflux ratio.

The reflux ratio of the tower is directly related to the total energy consumption of the process. In order to ensure the separation accuracy of the top product, reflux operation is required. However, the condensing and cooling heat discharged from the top of the tower is mostly low-temperature heat, and the tower is required to supply heat at a higher temperature to increase the total energy consumption of the process.

For the reaction process, the total energy consumption of the process increases with the increase in the amount of reaction raw materials and the circulation of unconverted raw materials. The processed fresh raw materials depend on the processing rate. The greater the circulation of unconverted raw materials, the total process energy consumption is also larger, conventionally called the refining ratio, hydrogen-oil ratio, hydrogen-carbon ratio etc.

Reducing the circulation of unconverted raw materials is limited by engineering and technical conditions, such as improving catalyst performance, optimizing operating conditions to increase the conversion ratio of a single pass, etc., there are technical and economic optimization trade off.

9.4.1.2 Reduce the Steam Total Energy Consumption

Steam is a widely used as an energy-carrying fluid in petrochemical production. The use of steam is usually divided into 8 categories [3]: steam for stripping and reboiler heating; steam for process fluid heating; steam for burner atomization; Steam used as a reactant to complete the reaction process, such as hydrocarbon conversion hydrogen production process, CO shift process, etc.; steam used as power, such as steam reciprocating pumps and steam turbines, etc.; inert medium and protection steam; heat tracing and heating steam; steam for purging, blowing, firefighting, etc. Except for atomization and power steam, other forms of steam are used entering the process energy use link, which belongs to the total energy use of the steam process.

Ways to reduce the total energy consumption of steam:

(1) Improving the operation and strengthening the management

Stripping steam is the main content of process steam. Under the premise of ensuring product quality, reducing the amount of stripping steam is the main means to reduce the total energy consumption of steam. This needs to be considered in conjunction with the process operating conditions and the hydraulic of the equipment.

The steam used as a reactant can also be reduced, such as the water-to-carbon ratio of hydrocarbon conversion to hydrogen production. Reducing the water-to-carbon ratio is affected by operating conditions and conversion ratio. When the catalyst remains unchanged, the water-to-carbon ratio drops to 3.0, and energy saving is 1.25 GJ per ton of ammonia synthesis [12].

There is no fixed standard for the amount of steam used for purging, accidents, and firefighting. In practice, steam blowing time and steam consumption are often greater than actual needs, and they can only be reduced by strengthening management. Heat tracing and heating steam are also part of the management content. The phase and temperature of exhaust steam, indoor heating temperature, and the necessity of heat tracing can all be improved by strengthening management.

(2) Replacing steam usage with heating

Steam for stripping can be replaced by a reboiler. For example, the atmospheric tower 1st side draw stripping of atmospheric and vacuum distillation unit can be replaced with a 267 kW of reboiler instead of 0.5 t/h steam, reducing the cooling load on the top of the tower. It also increases the temperature of the side draw stream of 15 °C. Steam for heating and tracing can be replaced by suitable low-temperature heat. Some people propose to use inert gas instead of steam stripping at the bottom of the tower, and use loosing air to replace the losing steam of the catalytic cracking U-tube.

(3) The heat tracing is not required during the transportation of light oil pipelines, such as diesel, and the heat tracing can be completely disabled in some region. Increasing the temperature of the product output plant can cancel the heat tracing of some products.

9.4.1.3 Reduce Power Total Process Energy Use

The effective work of the pump outlet is transformed into heat by overcoming friction during the process energy use and energy recovery process. Except for some rare cases, most of the flow energy is difficult to recover. The ways to reduce the flow energy are:

(1) When selecting a pump, be careful not to leave too much margin, otherwise the head of the pump outlet that is larger than the required part is mostly in the outlet regulating valve throttling loss. For pumps with frequent changes in flow, the speed regulating device can be used to save the head and avoid a large amount of throttling loss;
(2) Throttle valves throughout the system pipeline minimize the pressure drop of the regulating valve while ensuring the regulating quality;
(3) Optimize the design of the pipeline system, select economical pipe diameters, and reduce flow resistance. The economical pipe diameter is obtained by the minimum of the sum of pipeline investment and pressure drop operating costs. For high-temperature fluid transportation pipelines, it must be considered in conjunction with heat preservation; for economical pipe diameters [13] for incompressible fluid transportation, please refer to Sect. 12.3 of Chap. 12 method for calculation. Attention should be paid to the optimization of pressure drop in the optimal design of heat exchangers;
(4) Improve the process and catalyst, reduce the reactor pressure, thereby reducing the system pressure, which can reduce the power total energy consumption. For example, after the platinum reformer is switched to a bimetal or multi-metal catalyst, the system pressure can be reduced from 2.5–3.0 MPa to 1.2–1.5 MPa; while using ultra-low pressure continuous reforming technology, the operating pressure can be reduced to 0.3–0.5 MPa;
(5) Improving the process flow, avoiding repeated pressurization of the streams, throttling; and shortening the process route can all reduce the power total energy consumption;
(6) Reducing the reprocessing rate of unconverted raw materials in the reaction system can reduce the power total energy consumption.

9.4.2 *Reduce Process Exergy Loss in Process Energy-Using Link*

In the case of determining the total energy consumption of the process, reducing the process exergy loss in the process energy using link means that the exergy value of the energy to be recovered can be increased. It is manifested as the increase in temperature and pressure of the discharged stream energy parameters in the process energy using link, which establishes favorable conditions for further recovery in the energy recovery link. First of all, it should be realized that reducing the total process energy consumption reduces the chance of energy usage degradation, which is also beneficial to reducing process exergy loss.

9.4.2.1 Reduce Process Exergy Loss of Fractionation Tower

The exergy loss in the fractionation process is mainly the exergy loss in heat transfer. That is to say, the energy supplied to and from the tower is the same, but the exergy difference is caused by temperature difference of to and from the tower.

The process of improving the fractionation tower is as follows.

(1) Adjust the reflux heat extraction ratio of the tower

Due to the mass transfer characteristics of the tower, the temperature gradually decreases from bottom to top. In the case of a certain amount of heat extraction, which part of the tower to extract heat is an energy-saving issue that should be considered. The more the heat is taken out to the lower part of the tower, the higher the temperature, which can be recovered and recycled, such as preheating the raw materials, reducing the fuel heating load.

The optimized design of the heat extraction ratio of each part of the tower can ensure the product quality and yield, reasonably increase and appropriately distribute the number of plates in each section, so that the reflux heat can be taken out at high temperature as much as possible, creating a way for the recovering & recycling of reflux heat condition. In recent years, the reflux heat extraction of the fractionation towers of atmospheric and vacuum, catalytic cracking, hydrocracking, coking and other plants has been improved, and the cold reflux at the top of the tower has gradually been replaced by the top reflux. But if it is further optimized and improved, the optimization of the tower and the heat exchange network should be considered together.

(2) Reduce the temperature difference between leaving and returning tower

The reflux heat taken from all parts of the tower (except the cold reflux at the top of the tower) is in the liquid phase, and the heat taken out is mostly the sensible heat between the temperature difference of leaving and returning tower. Increase the temperature of the returning tower, and the average temperature increase. Increasing the return flow rate and reducing the temperature difference between the leaving and returning tower are also constrained by technical and economic conditions, and there are optimization and trade off [14]. For atmospheric and vacuum distillation units, crude oil is fractionated into narrow fractions, usually the flow rate of the hot stream is small, and the water equivalent difference between the cold and hot flow is larger. Increasing the reflux in the middle section of the tower reduces the water equivalent difference of the energy recovery system, and can increase the average heat transfer temperature difference. However, after the flow rate increases, the power consumption of the reflux pump increases slightly, which is more mainly restricted by the hydraulic operation of the tower (that is, the tower must be adjusted within the elastic operating range, and it must not cause flooding, and the residence time of the liquid in the downcomer should be less than the specified time related to physical properties, and the tower must be operated outside of mist entrainment). After changing the reflux flow rate, the operation of the tower is still in the operable area. Generally speaking, the recirculation flow rate has little relationship with the change in the gas phase load of the tower, and the key is to prevent liquid flooding.

Therefore, changing the reflux flow to increase the average temperature of heat extraction has an optimized relationship between more heat energy recovery and more power consumption. The production plant should determine the optimal returning flow according to the margin of the pump head.

9.4.2.2 Reduce Process Exergy Loss of Reaction Equipment

The exergy loss of chemical reaction due to entropy increase is not large, especially for complex hydrocarbon petrochemical reactions, the entropy increase is difficult to determine and can be ignored, but the actual chemical reaction has a large exergy loss, which is often due to the irreversible heat transfer in the reaction process, the way to reduce the exergy loss in the reaction process is also to strive to reduce the exergy loss in the heat transfer process of the reaction equipment.

For the endothermic reaction, external heat is required to ensure the temperature conditions of the reaction, and the temperature of the reaction product is often lower than the temperature of the raw materials and the catalyst. When the pressure drop of the reaction equipment and the difference in the number of product molecules are ignored, the exergy loss of the reaction equipment is that the difference between supply in and supply out heat exergy (including heat dissipation), it is a total process exergy loss.

(1) The fluidized catalytic cracking reaction oil and gas are in a high temperature and superheated state, and there is a large temperature difference with the catalytic cracking fractionation tower. The current process is that the reaction oil and gas directly enter the fractionation tower, and the high-temperature oil and gas suddenly drop in temperature, causing heat transfer exergy loss. In addition, the pressure energy of the reaction oil and gas also has a throttling loss, and a large amount of excess heat is taken out by circulating oil slurry, so that the gas phase oil and gas energy at about 460 °C is reduced to 360–280 °C liquid phase oil heat energy, exergy loss is great; how to use reaction oil and gas thermal energy is an energy-saving issue worthy of attention. An effective measure is to feed the raw materials into the fractionation tower to directly heat the raw materials, which not only saves the heat transfer temperature difference, but also greatly increases the preheating temperature of the raw materials, saves the fuel of the heating furnace or takes the heating furnace out of service. At this time, Light oil may be distilled into heavy diesel oil, which affects the yield of light oil, which can be compensated by refining part of heavy diesel oil.

(2) Regarding the pressure energy of reaction oil and gas, due to the high temperature, its expansion work is much greater than the compression work at low temperature of the gas compressor. The recovery and utilization of the pressure expansion work of reaction oil and gas could be developed, but engineering and economic factors should be considered.

9.4.2.3 Use of Streams Depressurization Exergy Loss

Many processes are carried out under a certain operating pressure, and the (reaction) products discharged from the process energy use link often contain pressure energy that can be recovered. Because the gas and liquid phases have different functional relationships with pressure, that is, the pressure energy of the gas phase is proportional to the logarithmic value of the compression ratio, while the liquid phase is proportional to the pressure difference, and because the specific volume of the liquid phase is much smaller than that of the gas; therefore, the liquid phase pressure will not be recovered usually, and only when the operating pressure is high and the recoverable pressure difference is greater than 1 MPa or more, the recovery is considered. For example, the pressure difference between the extraction tower and the stripping tower of a platinum reformer in a refinery can use a hydraulic turbine [15]. A feasible way to select a hydraulic turbine is to use a centrifugal pump, and making it reverse operation, reverse connection of inlet and outlet pipelines is sufficient [19]. Hydrocracking, hydrotreating plants have equipped with hydraulic turbine to utilize the pressure difference between the HP and LP separators.

Since the pressure energy of the gas phase stream is a function of the compression ratio, it is not closely related to the absolute operating pressure of the stream itself, and the expansion work has a very close relationship with the temperature of the stream. In recent years, people have paid attention to the low-pressure and high-temperature catalytic regeneration of flue gas pressure energy use. Flue gas turbines have been used in most of the catalytic cracking units of oil refineries.

In recent years, people have been committed to recovering hydrogen from synthetic ammonia purge gas to avoid waste of hydrogen resources. At the same time, they have also begun to consider the recovery and utilization of purge gas pressure energy [16].

Another aspect of the utilization of pressure difference in petrochemical plants is the utilization of steam pressure. There are many operating pressure levels in petrochemical plants, and it is impossible for utility system to be equipped with corresponding multiple steam pipe network systems. In addition, due to the balance of different levels of steam, there are often energy problems with the use of steam at reduced pressure. 1 MPa of steam is reduced to 0.3 MPa for use, or required steam is 0.3 MPa for process, and 1 MPa steam is used, resulting in a loss of streams pressure exergy. The way to solve this problem is to install back pressure steam turbine recovery work in places where the pressure drop is relatively concentrated and the consumption is large, so that the steam system can be integrated; the second is to use the steam recompression method, that is, use the main steam to compress exhaust steam of $\le$0.3 MPa and discharged for process use. In this way, the amount of steam can be reduced and the steam can be recycled, which is also a practical and feasible energy-saving measure.

9.4.2.4 Avoid Mixed Exergy Loss

Avoid repeated mixing and separation of streams in production. For example, in order to ensure product quality, atmospheric and vacuum distillation requires that the quality index of the distillation product is far lower than the required. When the production fluctuates, sometimes off-spec products are produced and arranged in an oil tank to blend, which will cause qualitative mixed exergy loss.

The mixed exergy loss in the heat transfer process should also be avoided. For example, in the heat exchange system of a multiple heat exchange subsystem of atmospheric and vacuum distillation unit, the final heat exchange temperature of each subsystem is different, and when they merge together, the heat transfer mixed exergy loss will occur.

Some refineries are designed with two sets of atmospheric and vacuum units, the vacuum tower draws the wax oil deeply, and the heavy oil catalytic cracking feed mixes the vacuum residue with the wax oil, resulting in exergy losses. This can be avoided by optimizing production plan.

9.4.3 Improve Energy Recovery Efficiency and Reduce Rejection Energy and Exergy Loss

After the process energy utilization link is improved, the total process energy used is determined, and after the exergy loss of the process is reduced, the energy to be recovered and the exergy to be recovered are determined. The task of energy-saving improvement in the energy recovery link is to recover as much energy as possible from the energy to be recovered and reduce rejection energy.

The energy to be recovered can also be divided into three forms: heat, steam, and power. Generally speaking, it is difficult to use power and steam to be recovered. The flow work of the power energy to be recovered loses its functional capability in overcoming the friction in the energy use and recovery links, and most of it is converted into heat energy at the temperature of the streams, which enters the streams or is lost to the environment. The frictional energy is very small and usually negligible.

For most of the plant with better energy utilization, steam energy to be recovered is discharged in process condensate. It is difficult to recover at low temperature, and it is also discharged in the form of exhausted steam. At this time, the focus of improvement should be on process energy utilization link. Scientific management, less use of steam or replacement (such as low temperature heat instead of steam tracing, etc.), can also enter the exhaust steam network for further use.

Increasing the energy recovery efficiency focuses on improving the heat recovery rate, that is, fully recovering the heat to be recovered, reducing the cooling load, and achieving the purpose of energy saving of the plant.

1. Reduce heat dissipation

At present, the heat preservation of pipeline and equipment is mostly formulated on the basis of heat dissipation, and the factor of fluid temperature is not considered enough. The way to reduce heat dissipation is to improve heat preservation and optimize the heat preservation of equipment, pipelines and valves in accordance with the method of determining the thickness of the economical heat preservation thickness [13]. And pay attention to distinguish the heat dissipation price of different streams temperatures, it is best to use the method of economical insulation thickness. The heat dissipation energy consumption of the plant accounts for about 10–20% of the total energy consumption. Reducing heat dissipation is one of the important energy-saving measures. The investment is not much, but the effect is very significant.

2. Optimize the heat exchange system to reduce heat transfer exergy loss

The heat exchange system is a process that realizes product cooling and cold raw material heating. Recovering energy from the energy to be recovered can reduce the energy provided by the conversion link of the plant and achieve the purpose of reducing the energy consumption of the plant.

The optimization of the heat exchange system, one is the optimization of the equipment structure, so that the equipment transfers heat under the best working conditions, and the logarithmic average temperature difference correction coefficient of the heat exchanger approaches 1. The second is to rationally arrange the heat exchange process to make the cold and hot streams match reasonable and avoid excessive irreversible heat transfer, which is greater than the part of the economic heat transfer temperature difference. The determination of the economic temperature difference depends on the methods and principles of thermo-economics. At present, the average temperature difference of the existing plants is relatively high. As a result, it is difficult for the cold flow to reach a high final heat exchange temperature, and the hot flow leads to a larger amount of low-temperature heat rejection, which reduces the energy recovery rate. Therefore, a reasonable heat exchange temperature difference and network should be determined on the basis of technical and economic conditions.

The optimization of the heat exchange system should have the concept of focusing on the large-scale system. Some plant can use the heat produced by the processed products, such as atmospheric and vacuum distillation plant. There are also many plants that may have excess heat (such as catalytic cracking units). At this time, first consider taking the heating furnace out of service to reduce the fuel consumption of the plant, and then consider outputting heat energy to the outside of the plant to reduce the energy consumption of the external plant or system. In more cases, many low-temperature processing plants lack heat sources, and low-efficiency heating furnaces are used to provide heat sources, which are less efficient. If you stand on a large system, you will consider heat integration, and rationally use the low-temperature heat discharged from other plants that can meet the heat requirements.

Reasonably arrange the water equivalent ratio of the heat exchange system. For plants where cold raw materials are more concentrated and hot products are more dispersed, multiple heat exchanging systems should be used. That is, the cold raw material is divided into multiple strands to exchange heat with the hot product, reducing the temperature difference in heat transfer.

3. Reduce cooling rejection energy

After the heat exchange network is optimized, many low-temperature heat products usually enter the cooling system to remove heat. Generally speaking, the cooling load after the optimization of the heat exchange network is significantly lower than before the optimization, but there is still potential for use.

First of all, for the existing sensible heat and latent heat of the stream above 80 °C, a suitable heat sink should be found to be used as a heat source for preheating boiler feed water, and it can also be used for other purposes. Some refinery and chemical plant in China set up a low-temperature power station to collect the remaining low-temperature heat from the production plants and arrange them for providing a chemical water heat source for the catalyst manufactory. It uses water as the working medium to expand for power generation and realize the cogeneration of heat and power. Of course, it can also be developed towards a combined heating-power generation-refrigeration system, a low-temperature waste heat utilization system for organic Rankine cycle power generation with low-temperature working fluids.

Another measure to reduce the cooling rejection energy is to combine with the system to increase the temperature of the product output plant, which not only reduces the cooling load of the plant and reduces the use of cooling media, but also increases the output heat energy of the plant. For the storage and transportation system, energy consumption for heating and tracing is correspondingly reduced.

4. Pinch technology of heat exchange system

B. Linnhoff's "Pinch Point Technology", which is currently the most valued in practice, is based on the first and second laws of thermodynamics, but does not perform quantitative analysis and calculations, and uses the concept of optimization, but does not require advanced mathematics for optimization technology. The developed super-target optimization software calculates the value of the pinch temperature difference of the heat exchange system based on the plant's cold and hot streams data and the economic temperature difference (ΔT_{mim}) determined by the technical and economic conditions, and points out the smallest cold utility (cooling Load) and hot utilities (heating load outside the system), and extend from the pinch temperature difference to both sides, specifically arrange the optimized synthesis result of the heat exchange network, and improve the energy recovery efficiency. At present, there have been many successful examples [20] of the application of pinch point technology, a special program- Aspen Energy Analyzer can be used for heat integration of process plants.

9.4.4 *Improve the Energy Conversion Link Efficiency to Reduce the Energy Consumption*

After the improvement of the process energy use and energy recovery links, the total process energy use and energy recovery efficiency of the process were determined. After the recovery and recycling energy from the energy to-be-recovered is determined, the energy supplied to the conversion link is determined. The main points of energy consumption analysis are to reduce the total process energy used, reduce the process exergy loss of the process energy utilization link, and increase the recovery efficiency; on the premise of determining the amount of energy that still needs to be supplied to the plant after the improvement of the two links, the improvement measures for the conversion link are proposed.

1. Share heating furnace or take it out of service

When the heat load of the furnace is reduced after the energy saving of the plant is completed, the heating furnace can be taken out of service or shared. For example, when catalytic cracking increases the preheating temperature of the raw materials, improves the atomization quality of the riser reactor feed, and blends the residual oil to meet the heat balance of the reactor, the heating furnace could be disabled. This can not only reduce the flue gas and heat loss of the furnace, but also reduce the total process energy-used and reduce the energy loss of the downstream.

After the improvement of the atmospheric and vacuum distillation plant, the final heat exchange temperature of the raw materials has been increased, and the furnace load has been greatly reduced. It is possible to consider using two atmospheric furnaces together, or even the combined use of atmospheric and vacuum furnaces, which is very energy-saving. However, specific verification should be made. The furnace load has increased, and the efficiency has also increased accordingly.

For heating furnaces that heat low-temperature streams, which have low load, low efficiency, and lower exergy efficiency, you can consider replacing them with steam or low-temperature heat source.

2. Improve furnace efficiency

Improving the efficiency of the heating furnace is the main aspect of the energy-saving modification of the process plant. The thermal efficiency of the large and medium-sized heating furnaces is 80–90%. The continuous improvement of the thermal efficiency is limited by the flue gas acid dew point corrosion and economic exhaust temperature. The temperature of the heated fluid in the furnace is mostly below 400 °C, and the exergy efficiency is as low as 40–50%, indicating that about half of the fuel chemical exergy is lost in the process of converting into heat. Therefore, the focus of improving efficiency is to improve the exergy efficiency of the heating furnace, and the potential is great.

(1) Pay attention to the energy-saving effect of the air preheater. During the fuel combustion process, the air and inert components absorb a large amount of heat

energy from the normal temperature to the combustion temperature, and use flue gas or other heat sources to preheat the air, resulting in saving about the same fuel energy as the preheating load, improves thermal efficiency. From the perspective of exergy analysis, preheating avoids the exergy loss of the mixture of cold air directly with the combustion flue gas, increases the combustion temperature, and helps reduce the exergy loss of the furnace. Therefore, when arranging the heat recovery of the flue gas of the furnace, the air preheating should be prioritized before other measures (such as steam generation, cold feed, pre-heating water, etc.). Secondly, attention should be paid to reducing the excess air coefficient, especially to avoid cold air leakage in, and to strengthen the plugging and management of the fire hole and the convection tube box.

(2) Combination of gas turbine and heating furnace for power supply and heating [17]

In order to improve the exergy efficiency of the heating furnace, the use of gas turbine-heating furnace combined power supply and heating is a major energy-saving measure. Many refineries have realized the combination of gas turbine and heating furnace on the heating furnace of the crude oil distillation unit; in China, the combination of gas turbine and heating furnace on the heat carrier furnace of the platinum reforming unit is under development. The thermal efficiency of the heating furnace is already high, close to the economic limit. The gas turbine is used to first expand the gas at high temperature to do work, and the exhaust flue gas heat plus duct firing fuel if needed are used as the process heat source, which can increase the overall exergy efficiency by about 10%, which can satisfy both heating and power generation. This is the direction for improvement of the heating furnace of petrochemical plants.

3. Use automatic speed control facility

The role of the pump and compressors in the energy conversion link is to convert electrical energy into fluid flow kinetic energy. At present, the actual total efficiency of pump and compressors is less 70%, and the selection of new high-efficiency and energy-saving pumps is an important measure for energy saving.

When the flow rate changes greatly and the frequently, and the system flow resistance (regulating valve, outlet valve, etc.) account for a large proportion of the total head, use the automatic speed-regulating pump to overcome the unfavorable factors of the load ratio change. When operate at low load, the pump head is adapted to the requirements of the piping system, the efficiency is basically unchanged, and the unit energy consumption remains unchanged and slightly reduced. After installing a gas compressor pressure automatic speed controller in a catalytic cracking unit of an oil refinery, it realized automatic speed adjustment, eliminated anti-surge circulation, and saved steam usage greatly.

4. Reasonable choice of steam-powered driving mode

In addition to electric drives, many of the pumps and compressors in the energy conversion link are driven by steam power, such as steam reciprocating pumps, condensing or back pressure steam turbines, and steam jet pumps.

Steam reciprocating pumps have low energy utilization and exhaust steam is difficult to recover, so they are generally not suitable for use.

Condensing steam turbines are commonly used in the drive of catalytic cracking gas compressors. Low-pressure steam condensing steam turbine units have low efficiency, energy utilization efficiency is below 10%, and exergy efficiency is only about 20%.

The back pressure steam turbine uses steam to do work first, and the back pressure steam can still be used as process steam. The imported large-scale ammonia plant has a high level of energy consumption. To a large extent, it is produced by steam, and the utilization is relatively reasonable, and the energy is used in stages, therefore, when choosing a steam-powered drive mode, attention should be paid to the selection of back-pressure steam turbines to optimize the steam-power system.

The energy recovery rate of the steam jet pump is low, and the exergy efficiency is only 15%. Improvements should be made in the following areas:

(1) Set up multiple groups of steam jet pumps to increase the flexibility of starting up and operation. When the load is low or when the non-condensing gas decreases due to other reasons, one or two of the groups can be taken out of service to achieve the purpose of steam saving;
(2) Use mechanical vacuum maker to improve exergy efficiency and avoid latent heat and loss of steam.

5. Recovery and Utilization of flue gas Energy of Catalytic Cracking Regenerator

The catalytic cracking regenerator is responsible for regeneration (process task) and coke burning and transfer of energy to the circulating catalyst (energy conversion). In terms of energy analysis, the regenerator is an energy conversion equipment and has the same function as a heating furnace.

(1) Recovery of pressure energy in the regenerated flue gas. The flue gas pressure of the regenerator of the catalytic cracking unit is 0.2–0.45 kPa, the temperature is 600–700 °C, and the gas pressure exergy is a function of the logarithm of the compression ratio before and after expansion, so despite the operation pressure is not high but it has high energy. The widespread application of flue gas energy recovery units is an important measure for energy saving in catalytic cracking units.

    The installation of a flue gas turbine to recover the flue gas pressure energy can greatly reduce the electricity of the plant, but the unit investment is relatively high.

(2) Recovery of the flue gas carbon monoxide chemical energy in the flue gas

    There are two ways for FCCU catalyst regeneration. One is complete regeneration, that is, all CO in the combustion flue gas is burned to produce CO_2 during regeneration process, and the Arthur's ratio (CO/CO_2) is 0, and all the heat of the coke is released. In the 1980s, CO combustion promoters were gradually promoted and applied, which not only made full use of the CO chemical energy, but also improved the product yield. However, as the secondary processing depth of crude oil increases, the raw material of the plant is changed from wax oil to residual oil or blended with part of the residual oil, as a result, the coke yield increases. In this way, the internal and/or external heat extraction facilities are installed to remove the excess heat, which make the equipment complicated with internal components, coupled with the high cost of platinum combustion promoters and limited regeneration capacity, many plants are still using incomplete regeneration methods.

The traditional method of CO chemical energy recovery in flue gas is to use CO boiler, but a large amount of fuel needs to be supplemented. The more important concern is that the combustion and heat transfer in the combustion chamber occurs simultaneously, it naturally reduces the flue gas temperature near the water wall tube, even if a large amount of fuel is added, it is still difficult to burn out CO at a higher chamber temperature. Using combustion promoter in the regenerator make the higher operation cost, so the CO boilers have been in transition and stagnation, and some have been converted to waste heat boilers.

There are two ways to recover CO chemical energy outside the regenerator:

(1) Install a CO incinerator upstream of the original CO boiler; duct firing a small amount of fuel, or even no fuel, and use its own reaction heat to burn out CO, and the CO burnt gas enter the CO boiler to recover the energy of high-temperature gas. The developed CO incinerator has achieved successful operating experience;
(2) Without a CO boiler, a pipeline CO incinerator can be installed to burn out the CO in the flue gas, and then the sensible heat energy can be recovered by the installed waste heat boiler.
(3) Recovery of sensible heat of flue gas, make full use of flue gas sensible heat and CO combustion heat to generate steam with higher parameters, or as a heating source for other streams.

To sum up, the energy-saving improvement of petrochemical plant has a law to follow. The energy-saving improvement method based on the three-link model of energy analysis reflects this objective law and the mutual restraint relationship between energy-saving improvement and various energy-using links, and inspires to start from the overall situation of the plant and even the entire plant, and identify the solutions and measures for systematic energy-saving improvement.

9.5 Large-Scale System Optimization Method and Improvement Approach

The energy of higher parameters is discharged from the core process equipment (reaction and fractionation, etc.). Except to the recovery of heat and work in the production plant and the reduction of the energy supplied, due to engineering and economic factors, the energy can't be recovered fully. Part of the energy must be rejected; but in another production facility, it may be a suitable heat source. The medium and low temperature heat energy discharged by many plants should be used in the whole plant system, which raises the issue of global system energy utilization optimization in a larger system.

Generally, in petrochemical plants, especially oil refineries, in the process of providing the work and heat required by the process, each production unit forms its own system, and there are fewer optimizations between units and systems from the overall perspective. This may cause unreasonable matching of energy supply and demand in a certain link, leading to an increase in energy consumption of the entire plant.

9.5.1 Improve Production Process

Under the same processing volume, the components and scale of the plant can affect the energy consumption of the whole plant from different aspects. The optimization of the overall production process is based on the maximum profit and tax as the objective function, and the energy consumption of the plant and whole plant affect the objective function in the form of processing costs. For existing enterprises, the components and number of plants have been relatively determined, but the reasonable arrangement of production processes will also play a key role in energy saving and consumption reduction.

1. Plant production load ratio

The energy consumption of the plant is divided into fixed energy consumption and variable energy consumption according to the nature of its change with the production load ratio. In low load ratio production, the fixed energy consumption per unit of raw material (product) increases, which increases the total energy consumption. Therefore, increasing the plant load rate is a subject that should be solved in the production scheduling of the whole plant. For two sets of parallel production units (such as two sets of atmospheric and vacuum units, two sets of catalytic cracking, etc.), one of the low-energy-consumption production unit can be operated in full capacity, and extend the shutdown and overhaul time of the other production unit, so that the two parallel units can be operated at a high load ratio. Although there is energy consumption during start-up and shutdown, avoiding low-load operation in longer period, therefore it is also suitable.

2. Choose a reasonable processing route

With a given plant components and a certain number of raw materials and product varieties, choosing different processing routes will have a certain impact on processing costs and energy consumption, which is a content considered in the overall process optimization. In the case of the same product profit, priority is given to energy-saving production processes with short processing routes and low energy consumption. If there is a hydro-refining and hydrocracking refinery, there are multiple sources of hydrogen. One is to set up a hydrogen production plant, which uses the hydrocarbon steam reforming method to produce hydrogen, which has high energy consumption and cost; the other is reforming by-products. Hydrogen may be enriched by using Prisson membrane separator or pressure swing adsorption for catalytic cracking hydrogen-rich dry gas (or other hydrogen-rich gas), and the energy consumption cost is slightly lower. Which route to choose should be carefully studied and determined according to the specific conditions of the factory and the scale of the supporting facilities. With the rise of heavy oil catalysis, it is economically more attractive to develop hydrogen production from coke in a catalytic process.

3. Reduce unnecessary repetitive processing energy consumption

With the increase in the proportion of residual oil secondary processing, heavy oil catalytic cracking has been constructed and put into use in many refineries. The feed of the heavy oil catalytic unit is not only vacuum residue, but also due to factors such as coke generation rate, it can only be refined in a certain proportion, and a large amount of wax oil needs to be added to improve the feed properties. Such a process brings a waste of energy. Because in the vacuum system, the vacuum degree is desperately increased to increase the extraction oil rate, and the atmospheric residue is separated into vacuum residue and wax oil, which consumes energy. In the heavy oil catalytic plant, two kinds of feeds are processed, many of which are mixed feeds of wax oil and vacuum residue, so that the vacuum part will waste energy. There are also cases in which individual companies have configured new atmospheric and vacuum units for heavy oil catalysis. Therefore, it is recommended that in the case of two sets of atmospheric and vacuum, the vacuum part of the atmospheric and vacuum unit that feeds the heavy oil catalysis can be stopped to avoid the energy consumption of fractionation and then mixing. For the case of only one set of atmospheric and vacuum equipment, the atmospheric tower can also be used to extract part of the regular atm residual. The feed for heavy oil Catalytic Cracking is changed from vacuum residual + wax oil to atm residual + wax oil, but the ratio of wax oil is already reduced greatly, this can reduce the energy consumption of fractionation in the vacuum part.

9.5.2 Heat Integration Among the Plants and System

The heat integration among the plants breaks the situation that energy use is self-contained, so that they can coordinate with each other. This effect is achieved by the combination of atmospheric and vacuum units and catalytic cracking units. In the heat balance of the catalytic cracking unit, the raw materials react at a higher temperature and are in a state of superheated oil and gas entering the fractionation system, resulting in excess heat. The circulating oil slurry is usually used to take out the heat of the high temperature oil and gas. This part of heat has not yet found a suitable use in the plant except for steam generation. The heating of crude oil in the atmospheric and vacuum unit requires a lot of heat. In order to increase the final temperature of the crude oil after heat exchanged, the heat in the high temperature section is lacking. The combination of atmospheric and vacuum unit and catalytic cracking units can take advantage of the high temperature of circulating oil slurry to increase the final heat exchange temperature of crude oil and save furnace fuel. At the same time, atmospheric and vacuum supply of catalytic hot wax oil feed also avoids the cooling and heating of catalytic raw materials, saving heat exchange equipment and energy consumption. The atmospheric and vacuum unit provides hot feed to the coking unit, and the vacuum residual is fed into Coking unit at 200–300 °C, and the same effect can be obtained, avoiding the situation of heat exchange and cooling in atmospheric and vacuum, and heat exchange and fired heating in coking.

In addition to the heat integration between the plants, the heat integration between the storage system and plants also has significant energy-saving effects. The plant product does not need to be overcooled. Within the temperature range allowed by the process, the products in hot state enters the storage system, which saves the cooling energy consumption of the plant and the heating energy consumption of the storage system. For example, heavy diesel oil and residual oil can be stored and transported in hot state.

The low-temperature heat discharged from the plant supplies small plants that require low-grade energy, such as gas separation, alkylation, etc. This kind of heat integration is not only directly heated by a certain stream or heated by heat carrier circulation, but also unified collect the low-temperature heat for overall arrangement of low-temperature heat system.

The realization of heat integration is subject to the constraints of engineering and production conditions: (1) the problem of unsynchronized startup and shutdown should be taken into consideration to make up for it; (2) each plant is generally an independent plant and should be coordinated with each other.

The plant heat integration should focus on the large system, and its rule of thumb are as follows:

(1) Firstly, preheat the raw materials of the plant to reduce the heat load of the furnace and save fuel. When using excess heat to heat the raw materials of this plant, the reasonableness of the heat transfer temperature difference should be maintained. For example, diesel oil (middle reflux oil) to exchange heat with the raw materials is proper in the catalytic cracking unit, if the circulating oil slurry is used to heat the raw materials, the heat transfer temperature difference is too large, which will eventually cause other heat source to be rejected into the cooling load;
(2) Excess heat can be used as the heating source for other plant, saving fuel consumption;
(3) Use the medium and low temperature excess heat of this plant as a heating source (including reboilers) for other small and medium-sized plants to save fuel or steam.

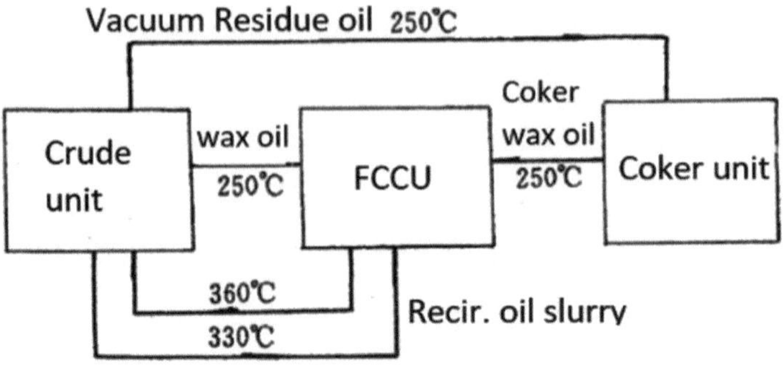

Fig. 9.7 Atmospheric and vacuum-catalysis-coking units heat integration

Figure 9.7 shows the heat integration calibration of atmospheric and vacuum-catalysis-coking units in an oil refinery.

9.5.3 Low Temperature Heat Recovery and Utilization

Due to the characteristics of the energy used in the process, the energy used in the process is composed of two parts: the supply from conversion equipment and the energy recovered and recycled. Generally, recovered and recycling energy can be continuously recycled in the process: entering the process energy utilization link → the energy to be recovered system → energy recovery and recycling → entering the process energy utilization link.

For a stable continuous production process, the recycling energy is almost unchanged. Of course, this is because the conversion equipment continuously supplies energy to keep the energy parameters to be recovered as a constant contribution. In terms of outgoing and incoming, part of the effective energy either enters the product or as energy recovery output for use in other systems or plants, and the rest is completely rejected into the environment.

Generally, there are four types of energy rejection forms: cooling, heat dissipation, streams carry-out, and other rejection. The cooling rejection energy accounts for the largest proportion (about 70%), and it varies with the different energy consumption levels of different plants. After careful analysis, we will find that the temperature of these process streams that need to be cooled is between 80 and 150 °C, or even higher. Under the process conditions of this plant, the significance of recycling is lost, but the "waste heat" can be used for recovery and recycling when incorporated into the whole plant system. Especially in today's energy-saving work and CO_2 mitigation continues to deepen, to reduce the energy consumption of the plant or whole plant, the recovery and utilization of low temperature heat is an essential aspect. However, due to the low temperature of heat source, there are objectively the problems of difficult recovery technology and low economic benefits. Therefore, when formulating a low-temperature heat recovery program, it should be fully analyzed and confirmed.

For low-temperature heat recovery and utilization, it is necessary to establish a supporting recovery system throughout the whole plant, and take out the heat from the plant through an energy-carrying working medium (usually demineralized water). In order to ensure the safe and reliable operation of the plant, the heat can be collected constantly, and the heat load fluctuation can be adjusted by means such as setting a cooler in the recovery system.

Low temperature heat utilization can be divided into two categories of direct use and upgrading use, reader can refer to Sect. 14.7 of Chap. 14 for detail.

9.5.3.1 Same Level Utilization of Low Temperature Heat (Direct Utilization)

According to the temperature of low-temperature heat recovery, selecting the appropriate user not only changes the excessive energy transfer loss caused by using of high and medium-temperature heat sources, but also replaces the high and medium-temperature heat sources, which is a low-temperature heat utilization most attractive solution, energy-saving benefits are particularly significant.

1. Low temperature heat for production

The use of low-temperature heat to replace high- and medium-temperature heat sources in production can not only directly reduce production energy consumption, but also the production heat is mostly continuous and stable in load with larger energy saving and high benefits, priority should be given to the arrangement of low-temperature heating programs. Such heat use includes:

(1) heating of raw materials and tower bottom reboiler heating for low-temperature processing equipment such as gas separation and alkylation units;
(2) heating of washing water in the catalyst plant;
(3) supplementary demineralized water heating of the boiler system of the plant;
(4) oil storage and transportation system, oil tank heating, etc.

2. Heat for living quarter, Labs, offices

At present, with the continuous improvement of welfare facilities offices, living energy consumption has also increased correspondingly. In addition, with the development of enterprises, the energy used by office buildings, education and training systems, and design and research institutes are also increasing. This part of the energy consumption affects the overall energy consumption of the entire plant. If low-temperature heat replaces its energy consumption, not only can the overall energy consumption of the whole plant be reduced, but also the direct production energy consumption can be reduced. This type of heat consumption is generally divided into two categories:

(1) Heating in the office and living quarter (camp) of the factory. The use of low-temperature hot water instead of steam is seasonal, but because steam heating often exceeds national heating standards and there are many wastes in use, the benefits are still considerable;
(2) Domestic hot water. On the one hand, the use of heat improves the welfare of employees, on the other hand, it saves the existing liquefied petroleum gas water heaters for bathing, and reduces the use of liquefied gas. The characteristic is all year round, but the load changes with day and night, and its change pattern is similar to the domestic water system. When formulating the plan, it should be considered how to maintain the balance of the system and extract the heat when the heat consumption is reduced.

9.5.3.2 Low Temperature Heat Upgrade Utilization

After giving priority to users of continuous and stable heat load for low-temperature heat, the upgrading and utilization of excess low-temperature heat should be considered. There are three ways of such use.

1. Heat pump

The heat pump is to increase the low-temperature heat to the parameters that can be used in the process by applying external high-quality energy. Using heat pump technology to increase the temperature of the streams and reuse it in production process is a technical means to effectively utilize low-temperature heat energy.

The working principle of a heat pump is the same as that of refrigeration. It compresses heat from low temperature to high temperature, but the purpose of use is to heat rather than cool. There are two types of heat pumps: compression and absorption heat pumps. In the petrochemical industry, many refinery gas fractionation units and petrochemical units have successfully used compression pumps and achieved good energy-saving effects. Heat pumps are generally used in occasions where there is little temperature difference between the required heat and the low-temperature heat. Excessive temperature differences will be uneconomical, so be careful when making plans.

2. Refrigeration

Refrigeration is a very important aspect of low-temperature heat utilization, and low-temperature heat refrigeration is mainly absorption refrigeration. Steam lithium bromide absorption refrigeration has been widely used. Absorption refrigeration, which uses low-temperature heat instead of steam heat source, has also been put into industrial use. Many petrochemical plants have considered the use of low-temperature heat lithium bromide absorption refrigeration schemes in energy-saving projections. There are two uses for low-temperature heat refrigeration: one is production refrigeration. In the hot southern summer, the ambient temperature and cooling water temperature are high, and the cooling temperature of the product is difficult to meet the requirements, resulting in off-spec products and decline in product yield and increased loss. To solve the problem of stable dry gas absorption in catalytic cracking, in addition to technological improvements, the use of low-temperature heat to produce 8-10 °C chilled water for further cooling can also solve the problem [22]. The second is office and living refrigeration. Using low-temperature heat instead of window air conditioners can reduce electricity consumption.

The principle of ammonia absorption refrigeration is the same as that of lithium bromide absorption refrigeration.

3. Heat transformer

The absorption heat transformer is a low-temperature heat recovery technology developed in recent years, it also called type II Absorption Heat Pump. The low temperature heat (such as 90 °C hot water) is converted into two parts by the heat transformer, one part is converted into higher temperature heat, which can be used as a heating heat source; the other part is degraded into waste low temperature heat, which will be rejected through cooling. Current, producing 150 °C heat using 2-stage lithium bromide heat transformer can be achieved. The target of this upgrading use is to produce heat of around 200 °C. At present, it is generally believed that a mixture containing TFE (2,2,2-trifluoroethanol) is more suitable as a working fluid for achieving this target temperature.

4. Power generation

Low-temperature thermal power generation is an important form of upgrading and utilization. When a large amount of excess low-temperature heat is difficult to find a suitable utilization scheme at the same level, power generation is a choice. The principle of low temperature heat power generation is utilizing the flash stem from hot water to generate power via steam turbine generators [23].

9.5.4 Make Steam System Cascade Utilization

The operating conditions of petrochemical plants are different, and the steam pressure levels of each unit and system in the whole plant are also different. When steam production is arranged, it is impossible to generate steam of various pressure levels to adapt to different units and systems. Based on the principle of using energy according to quality, boilers that generating steam from fuel should be with higher parameters as much as possible, generally medium and high-pressure steam. The pressure difference between steam production and steam demand is an important content of the cascade utilization of steam. In the past, the steam supply parameters were greater than the user's required parameters, install a steam letdown station via throttling to reduce the temperature and pressure, resulting in ineffective exergy loss and a waste of steam. The steps to develop cascade steam utilization are:

(1) The boiler generates high-pressure or medium-pressure steam according to the steam system conditions in the entire plant area to avoid the generation of low-pressure steam;

(2) The excess heat of the plant (unit) should be used to generate the steam of the corresponding parameters on the basis of the optimization of the heat exchange process of the plant, and subject to the steam balance of the whole plant, to avoid the phenomenon of venting and waste caused by the imbalance of production and use; About 3000 kW more power can be recovered from the medium-pressure steam generated by the oil slurry in a typical FCCU than the low-pressure steam generated;
(3) Verify the total steam consumption and parameters of the whole plant, so as to arrange the steam production of the boiler according to the steam consumption and waste heat steam generation;
(4) Investigate the power, parameters, steam consumption, etc. of power machinery that can use steam back pressure turbines;
(5) Develop a cascade utilization plan for steam.
(6) Technical and economic evaluation and optimization of the plan.

Although the required steam parameters of users are different, it is impossible to set up more system headers due to engineering and economic factors. Generally, four system steam headers of 10.0 MPa, 3.5 MPa; 1.0 MPa and 0.3 MPa are set; for enterprises without high-pressure steam, there are only three-level steam headers; some are regional local steam headers. In short, not only cascade utilization, but also practical and engineering factors should be considered in the steam headers setting.

It is very important to verify the steam consumption of various pressure levels, which should often be carried out in conjunction with the entire plant's energy-saving plan. Do not use the current steam consumption arrangement steam cascade utilization, and when the plant and unit energy saving is improved, the reduction of steam consumption will affect the plan, and even the previous efforts will be abandoned.

For occasions where back pressure steam can be used instead of fired heating furnace or to reduce the heating load of the heating furnace, steam should be used to save fuel. The platinum reforming plant of a refinery uses 1.0 MPa steam in the energy saving plan, which reduces the heat load of the original heat carrier furnace, and turns the operation of two furnaces into one furnace operation, which has achieved better energy-saving effects.

The steam cascade utilization is mainly to recover the pressure difference between the steam production and the steam demand to produce work or generate electricity. Directly drive the equipment to do work, with high utilization efficiency and low conversion loss. For example, a catalytic air compressor is directly driven by a back pressure turbine. Back-pressure turbine-driven generators are also commonly used to integrate power generation into the power grid. Most of the self-supplied power stations in many refineries use back-pressure turbine power generation. There are many large-scale water pumps and compressors in fertilizer and chemical systems, therefore, back pressure can be directly used to do work to make cascade utilization. The reason why the energy consumption of the imported ammonia plant to China is low is that the cascade utilization of steam plays an important role. Cogeneration of steam power and heat is used cascade utilization. Condensing turbines should not be used, even if the energy utilization efficiency of medium-pressure condensing turbines is extremely low.

At present, the cascade utilization of steam should pay attention to the utilization of the steam pressure difference between 1.0 and 0.3 MPa. The characteristics of this steam use are relatively dispersed, but for relatively concentrated users of 0.3 MPa, a regional steam network back pressure power generation can be established.

In addition, the development of steam production in petrochemical plants in the direction of high-pressure levels, it is conducive to the cascade use of steam. Figure 9.8 shows the operation diagram of the power station of a high-pressure boiler in an oil refinery.

The above-mentioned four aspects of energy-saving is to propose improvement directions from the perspective of the whole system, combined with the improvement of plant energy-saving, will make the energy consumption of the whole plant tend to be reasonable and achieve energy optimization. However, how to optimize energy use and realize the energy-saving improvement measures in the above-mentioned four aspects, which can be realized through several practices of energy-saving planning for the whole plant, and realize that the optimization of energy use of the whole plant is a system engineering of a large system, combining plants, auxiliary systems,

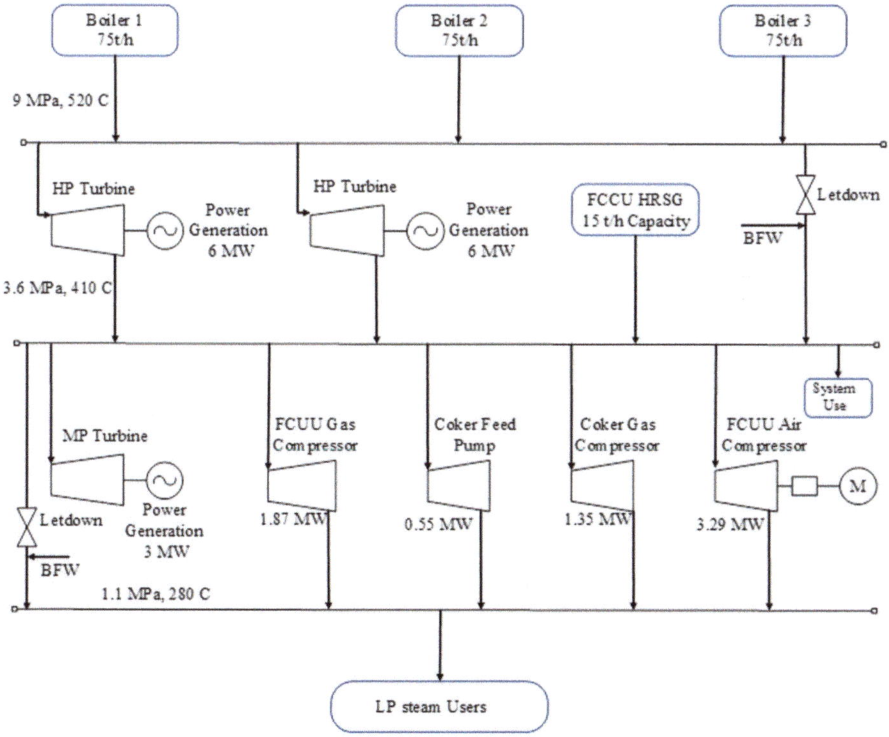

Fig. 9.8 Steam power system of a refinery

utility and living systems are considered in a unified manner. Planning the short-term or medium-term energy saving of the enterprise is a better way to achieve this optimization. The author had led a team implemented many mid-terms and long-term energy-saving projection plans for many oil refineries and petrochemical plants. Based on plant optimization and energy-saving, these four aspects are run through. The energy-saving measures of good results have been achieved, and the energy utilization of enterprises has risen to a new level.

References

1. H. Ben, C. Anmin, Practical calculation and economic analysis of heat dissipation in refineries. Pet. Refin. **2** (1984)
2. H. Ben et al., Several issues of energy conservation in refineries in cold regions. Pet. Refin. **7**, 37–40 (1987)
3. H. Ben, C. Anmin, Analysis and estimation method of the influence of load rate on device energy consumption. Pet. Refin. **1** (1987)
4. "Investigation of Refining Technology Innovation-Crude Oil Distillation Unit", Science and Technology Information Institute of the Ministry of Petroleum (1983)
5. "Benchmark Energy Consumption of Crude Distillation Unit", Oil Refining Technology Center (1984)
6. C. Anmin, T. Hui, Process plant energy consumption level evaluation using benchmark energy consumption. Petrochem. Energy Conserv. Newsletter **3** (1990)
7. Z. Boxi, L. Li, Develop system analysis methods and improve process design level. Pet. Refining. **4**(37) (1990)
8. Z. Yuzhen, J. Baocheng, Analysis of the exergy efficiency of heat exchangers, in *The Fifth Academic Conference on the Analysis of the Second Law of Thermodynamics,* Qinhuangdao, (1990. 7)
9. Y. Donghua, *Exergy Analysis and Energy Grade Analysis* (Science Press, 1986)
10. C. Anmin, Application of exergy analysis in crude plant energy-saving. Atmos. Vac. Crude Distillation **1**, (1989), 33
11. Refinery technology transformation Information Investigation (1st topics of energy saving in refineries)-Crude oil distillation unit, Information Office of SINOPEC, 25 (1983)
12. W. Qiaochan, On energy-saving approaches of fertilizer plants. Petrochem. Energy Conserv. Newsletter **2,** 14 (1990).
13. C. Anmin, Economic pipe diameter and economic insulation thickness for fluid transportation. Pet. Refin. **10**, 12 (1986)
14. S. Baochang et al., Optimization of reflux return to tower in the middle section of crude unit of oil refinery. Pet. Refin. **7**, 12 (1985)
15. J. Zhenduo, Platinum reformation-hydrogenation unit uses hydraulic turbine to recover residual energy. Petrochem. Energy Conserv. Newsletter **2**, 9–10 (1990)
16. Z. Jinhua, Energy utilization device for power generation using purge gas pressure. Petrochem. Energy Conserv. Newsletter **1**, 14–15 (1989)
17. C. Anmin, L. Kun, Thermodynamic analysis of gas turbine-heating furnace combined system. Oil Refinery Des. **4**, 55, 88
18. "External Technical Exchange Materials (Oil and Gas Processing)", No. 11, compiled and printed by the Science and Technology Information Institute of the Ministry of Petroleum (1981)
19. Z. Hongjun, Several issues on energy recovery from hydraulic turbines. Refinery Des. **1**, 34(1987)

20. S. Xiaowen, S. Meisheng, Application of pinch point design method in atmospheric and vacuum crude distillation unit. Oil Refin. Des. **5**, 55–65 (1990)
21. C. Anmin, Methods and approaches for energy-saving improvement of petrochemical plant. Pet. Refin. **4** (1991)
22. S. Peiqi, M. Shumei, Lithium bromide water absorption refrigeration. Refinery Des. **2** (1986)
23. Refinery technology Transformation Information Investigation (Theme of Energy Conservation in Refinery No. 8) - Low Temperature Heat Utilization in Refinery, SINOPEC Information Office, p. 4 (1985)

Chapter 10
CO_2 Capture and Utilization

Abstract This chapter focuses on the energy saving and CO_2 mitigation in the industry's activities in the usage of fossil oil, low-carbon and renewable energy. It covers the topics of CO_2 production, capture and storage in the energy industries. It also discusses CO_2 application in the chemical industries. Ten industrial operation case studies are used in the chapter which have covered different aspects in the CO_2 Generation, CO_2 Capture, Delivery, Storage and Application. The operations are supported with detailed process simulations, and the results are presented for reference by readers. These cases are: Nature gas CO_2 capture via acid gas enrichment—Case study 1; Cascade sour nature gas CO_2 capture—Case study 2; Acid Gas Enrichment and CO_2 Capture—Case Study 3; Syngas CO_2 Capture with single absorber using DEPG solvent—Case study 4; Syngas CO_2 Capture with dual absorber using DEPG solvent—Case study 5; Flue gas CO_2 Capture using DEA solvent—Case study 6; Flue gas CO_2 capture using oxygen instead of combustion air—Case Study 7; CO_2 dehydration and compression—Case study 8; Flue gas CO_2 Capture, compression and dehydration—Case study 9; Flue gas Cryogenic CO_2 Capture—Case Study 10. A newly developed CO_2 Capture Process by Cryogenics is discussed in the chapter. The innovative process will not require a contact liquid. There is no solid formation at the upstream of CO_2-de-sublimation, this eliminates the concern in the sold CO_2 transferring.

Keywords CO_2 capture · CO_2 utilization · Cascade CO_2 capture · Acid-gas enrichment · Cryogenic · Flue gas · Natural gas · Syngas

As mentioned in Sect. 1.5, there are 7 GHG gases, a carbon footprint also includes other greenhouse gases, usually carbon-based, measured in terms of their carbon dioxide equivalence.

CO_2 mitigation/capture is one of the most important factors in slowdown global warming and climate-changing.

CO_2 capture from flue gases and raw natural gases is being driven by multiple factors: reducing greenhouse gas emissions; use it as a raw material making a chemical product; the merchant CO_2 for market needs, enhancing oil/gas recovery, etc.

T. A. Chen, *Energy Saving and Carbon Reduction*,
https://doi.org/10.1007/978-981-19-5295-1_10

As mentioned in Sect 1.5, any effort in elimination of fossil fuel will lead to reduction of CO_2 emission, however, currently the fossil oil is still the mainstream for energy supply, we have to conduct human activities and maintain people's living standards, we have to use energy. The following activities will play important roles to mitigate and capture CO_2.

- Smart using fossil fuels, via efficiency improvement of energy conversion and utilization, improve energy recovery and recycling ratio, as described in previous chapters of this book.
- Using lower carbon fuel instead of higher carbon fuel;
- Using clean energy, renewable energy, and nuclear energy instead of fossil fuel;
- Consuming CO_2 as feed stuff in the manufacture of other chemicals:
- Fertilizer, such as ammonium bicarbonate, carbamide, Ammonium carbamate, etc.
- sodium bicarbonate.
- Methanol instead of CO.
- Enhancing oil and gas production by CO_2 capture, compression, and storage or oil and gas enhancement.

The term climate-neutral reflects the broader inclusiveness of other greenhouse gases in climate change, even if CO_2 is dominant among the GHGs. Carbon dioxide "net zero" is commonly used to describe a broader and more comprehensive commitment to decarbonization and climate action.

If the total greenhouse gases emitted are equal to the total amount avoided or removed, then the two effects cancel each other out and the net emissions are 'neutral'.

Carbon emissions reduction can be achieved by moving towards energy sources and industrial processes that produce fewer greenhouse gases, thereby transitioning to a low-carbon economy. Shifting towards the use of renewable energy such as hydro, wind, geothermal, and solar power, as well as nuclear power, reduces greenhouse gas emissions.

10.1 Fuel Efficiency Improvement, Using Lower-Carbon Fuel and Renewable Energy

10.1.1 Smart Using Fossil Fuel with Improved Energy Efficiency

The energy source is the driving force of completing the desired production rate for all the industrial processes such as oil and gas, oil refining, chemicals, and food processing as well. When we use fossil fuel to keep various production processes in order to maintain the living standard of the human society, smart usage of the fossil fuels in the maximum possible efficiency need to be designed. The fossil fuel demand could be reduced with efficiency improvement of energy source mining, production, and utilization, and improving the energy utilization efficiency for energy conversion, energy utilization, and energy recovery and recycling will reduce the CO_2 emission significantly. This book describes all the energy-saving methods and technologies that will contribute to greenhouse gases reduction. This book describes all the energy-saving methods and technologies leading to GHG reduction.

10.1.2 Using Lower Carbon Fuel Instead of Higher Carbon Fuel

Generally, Fossil fuel is a mixture of carbon and hydrogen elements existing in various chemical structures. Different fossil fuel has different Carbon to Hydrogen ratio, coal and heavy oil have a higher Carbon to Hydrogen ratio, therefore it produces more CO_2 for a required production rate. Natural gas and LNG have lower Carbon to Hydrogen ratio, thus produces lower CO_2 emission.

Per EIA US Energy Information [1] Table 10.1, typical Carbon Dioxide Emission Factors are extracted as in Table 10.1.

From the Table 10.1, coal and petroleum coke will generate the highest CO_2 emission, while the natural gas, ethane, propane, and butane lower CO_2 emissions, in general, lower-carbon fossil fuels will generate CO_2 emissions 20–40% lower than that with the higher-carbon fuels, therefore selecting the type of fuel plays a very important role to reduce CO_2 emission for any green and brownfield projects, it is also providing the direction for existing production to switch to lower carbon type of fuel as practical.

10.1.3 Using Renewable Energy and Nuclear Energy Instead of Fossil Fuel

In addition to using lower-carbon fuel, shifting towards the use of renewable energy such as hydro, wind, geothermal, and solar power, as well as nuclear power reduces greenhouse gas emissions. Theoretically, renewable energy doesn’t produce CO_2 emission, however, preparing for renewable energy use, the consumed material and equipment may require fossil fuel energy use, but compared to renewable energy production, this auxiliary energy demand is negligible. Gradually replacing existing fossil fuels with renewable energy will play a significant role to neutralize CO_2 emissions.

Table 10.1 Carbon dioxide emission factors for stationary combustion[a]

Type of fuel	Fuel	Emission factor	Units
Coal	Anthracite	103.69	kg CO_2/MMBtu
	Lignite	97.72	kg CO_2/MMBtu
Pipeline natural gas[b]	Weighted national average (1029ÂBtu/scf)	53.06	kg CO_2/MMBtu
Petroleum fuels[c]	Middle distillate fuels (no. 1, no. 2, no. 4 fuel oil, diesel, home heating oil)	73.15	kg CO_2/MMBtu
	Jet fuel (Jet A, JP-8)	70.88	kg CO_2/MMBtu
	Kerosene	72.31	kg CO_2/MMBtu
	Heavy fuel oil (no. 5, 6 fuel oil), bunker fuel	78.8	kg CO_2/MMBtu
	Ethane	59.59	kg CO_2/MMBtu
	Propane	63.07	kg CO_2/MMBtu
	Isobutane	65.07	kg CO_2/MMBtu
	n-butane	64.95	kg CO_2/MMBtu
	Unspecified LPG	62.28	kg CO_2/MMBtu
	Refinery (still) gas	64.2	kg CO_2/MMBtu
	Crude oil	74.54	kg CO_2/MMBtu
	Petroleum coke	102.12	kg CO_2/MMBtu

[a] All factors assume 100% combustion except those for MSW, which assume 98% combustion
[b] U. S. Energy Information Administration, Documentation for Emissions of Greenhouse Gases in the United States 2008, DOE/EIA-0638 (2006), October 2008, Table 6.2, p. 183
[c] Energy Information Administration, Documentation for Emissions of Greenhouse Gases in the United States 2005, DOE/EIA-0638 (2005), October 2007, Tables 6.1, 6.2, 6.4 and 6.5

10.2 Properties of CO_2 and Its Distribution

10.2.1 CO_2 Distribution Categories

Take the energy industry as an example, CO_2 produced by the energy industry falls into the following categories:

- **CO_2 comes with a feed stream**

CO_2 is produced with fossil fuel production, such as sour natural gas, produced gas of oil, and condensate production, all this CO_2 comes from raw fossil feed, during the oil and gas processing, acid gas (H_2S and CO_2) is removed, and captured, which accounts for the majority of CO_2 amount for oil and gas production.

The acid gas contains both CO_2 and H_2S, it is usually directed to the sulfur recovery unit to recover sulfur, and the tail gas from the sulfur unit will be incinerated to meet the environmental requirement, however, the CO_2 as inert gas will flow through the sulfur recovery and tail gas treatment units, impact plant performance greatly, in some cases, CO_2 in sour natural gas is dominantly higher than H_2S content, the acid gas enrichment and CO_2 capturing partially will benefit to the sulfur recovery unit operation significantly.

Associated gas production with oil production, usually contains higher CO_2 and H_2S, many oil production facilities inject this sour gas back into oil or gas reservoir, which can maximize the oil or gas production, for example, the KPO condensate facilities and TCO oil production have large sour gas injection system with a high operation pressure of 400–600 atm. This sour gas injection to the reservoir is usually via depleted oil or gas wells.

Some oil and field produce sour gas with a higher CO_2 concentration of over 20%.

- **CO_2 as a byproduct of Chemical processing**

Excessive CO_2 production from the Hydrogen plant, gasification process, etc. this type of CO_2 is usually a high concentration CO_2 stream. Hydrogen plant in refinery and bitumen/heavy upgrading facilities, chemical fertilizer plant discharges the CO_2 stream with a concentration of over 95%, this type of CO_2 stream can be reused or stored (via compression, dehydration, and storage) directly, with no need to further separate CO_2 from the remaining gases.

Mid concentration CO_2 streams, such as Hydrogen plant PSA purge gas, CO_2 content is about 40–60%.

- **Flue gas CO_2 from Fossil fuel combustion**

Flue gas from boilers, fired heaters, conversion furnace in the reformer, Cracking furnace in Ethelene plant, gas turbine exhaust gas, etc. this type of flue gas exists everywhere in the chemical and energy industries.

The low CO_2 concentration in flue gas makes it the most difficult to capture the CO_2. But it will contribute to the CO_2 reduction greatly, it is the area you could not pass it out.

Some of the acid gas recovered are directed to the sulfur recovery unit, concentrated CO_2 will be diluted again via main burners and tail gas incinerator. The solution for that is to enrich the acid gas H_2S content, which can capture the CO_2 and improve sulfur recovery unit performance greatly.

Most of the CO_2 gas is from fossil fuel combustion flue gas, the CO_2 concentration is mainly depending on the combustion situation, for using combustion air as an oxidant, the CO_2 content is in the range of 3–15%.

For lower CO_2 stream, CO_2 capture is a CO_2 concentrating process, one big part of CO_2 capture is the acid gas removal process for sour natural gas, the CO_2 and H_2S will be removed simultaneously, some gas plant keeps up to 3% of CO_2 in treated gas to maintain higher sale gas profit under the treated gas specification. This type of acid gas removal unit is aimed at removing the H_2S and controlling the CO_2 in treated gas within specification.

There are many types of CO_2 capture technologies proposed: such as (1) Absorption; (2) Oxyfuel combustion; (3) Membrane; (4) Cryogenic; (5) Adsorption; (6) Multiphase absorption; (7) Chemical looping combustion; (8) Calcium looping.

Absorption of carbon scrubbing with amines or Selexol is the dominant capture technology in the newly built processing plants. CO_2 Removal and Capture by Sulfinol, Benfield, Rectisol, Catacarb processes are also seen in some chemical plants.

10.2.2 CO_2 Physical Properties

Per Carbon Dioxide-Wikipedia information, the phase diagram of pure CO_2 is shown in Fig. 10.1.

From Fig. 10.1, the CO_2 compression and storage will cross the vapor phase, boundary fall in either liquid phase (< critical temperature) or supercritical liquid phase, depending on the end state of the CO_2 stream after the compression and cooling. For example, if cooled down by an air cooler or water cooler, the temperature can reach 30–40 °C and the CO_2 is in the region of supercritical liquid, in case of a chiller, the end temperature will be below 28 °C in the liquid region; therefore, for injection to adjacent wells, a multiple-stage compressor will be needed, but the last stage could be a pump rather than a compressor.

The actual recovered CO_2 stream usually contain small percentage of impurities that will change the phase diagram slightly. The following sketch is a correction for a CO_2 concentration of 95% (m) in the oxyfuel CO_2 capture option, with a two-phase

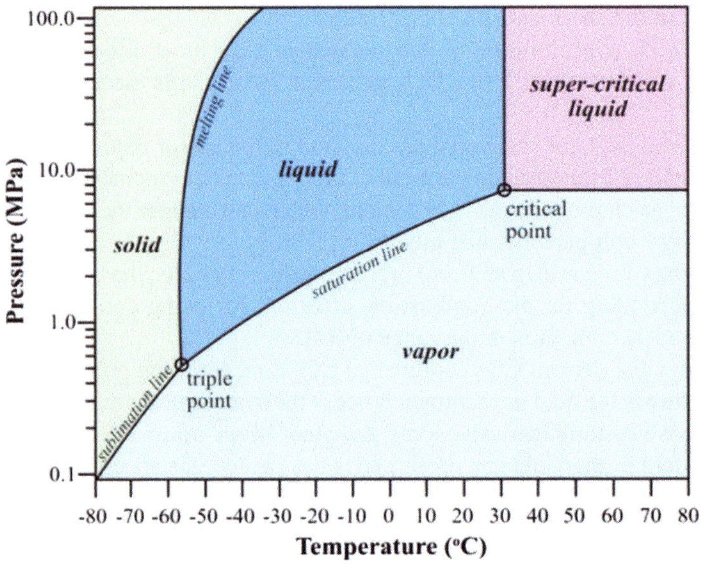

Fig. 10.1 CO_2 Phase diagram

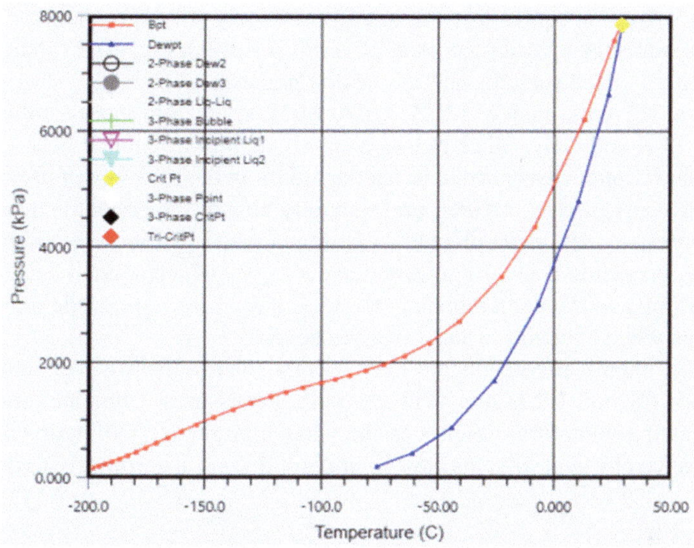

Fig. 10.2 Actual CO_2 gas stream BPT and dew temperature at 95% of CO_2

critical temperature of 28.56 °C, and critical pressure of 78.4 bar, cricondentherm of 28.61 °C, and cricondenbar of 78.6 bar (Fig. 10.2).

A dehydration unit may be added per actual CO_2 stream composition in addition to the compression and cooling.

10.3 CO_2 Capture Approaches

10.3.1 Absorption Solvents

There are many processes for acid gas removal, and they typically fall into five categories: chemical solvents absorption, physical solvents absorption, adsorption, membranes separation, and cryogenic fractionation. For the bulk of CO_2 capture or acid gas removal, the solvent absorption will be more cost-effective compared with adsorption, and membrane separation; and the cryogenic process is an attractive new process that is still under development, it will have the lowest operating and capital costs. The most commonly applicable CO_2 removal/recovery processes in the oil and gas and chemical industry are absorption including chemical solvent absorption and physical solvent absorption.

Generally speaking, the selection of a solvent is based on the balance of the following aspects: technical feasibility; selectivity to gas components; regeneration; demand of final purity of treated gases; capital and operation costs; contamination to environment.

For the solvent absorption process, current widely used are chemical solvents, physical solvents, or hybrid (chemical/physical solvents), Fig. 10.3 summarizes the most commonly used solvents and absorption process.

Chemical solvents, such as MEA, DEA, MDEA, etc., are most commonly used to remove H_2S and CO_2. Since rich solvent regeneration requires heat supply, the operating and capital costs need to be accounted for in the selection of process. Since physical solvents, such as DEPG, preferentially absorb the contaminants at higher operating pressure, the rich solvent can be regenerated by pressure reduction, thus minimizing operating costs. Engineers must consider which solvent has the needed selective affinity to the contaminants and minimum absorption of the main process components when choosing what solvent to be used.

Typically, chemical solvents are most suitable at lower pressures, and physical solvents are favored for higher acid gas partial pressures. Chemical solvents are typically favored in natural gas processing when high recovery of hydrocarbons gas is desired since physical solvents have an unignorable co-absorption of hydrocarbons gas.

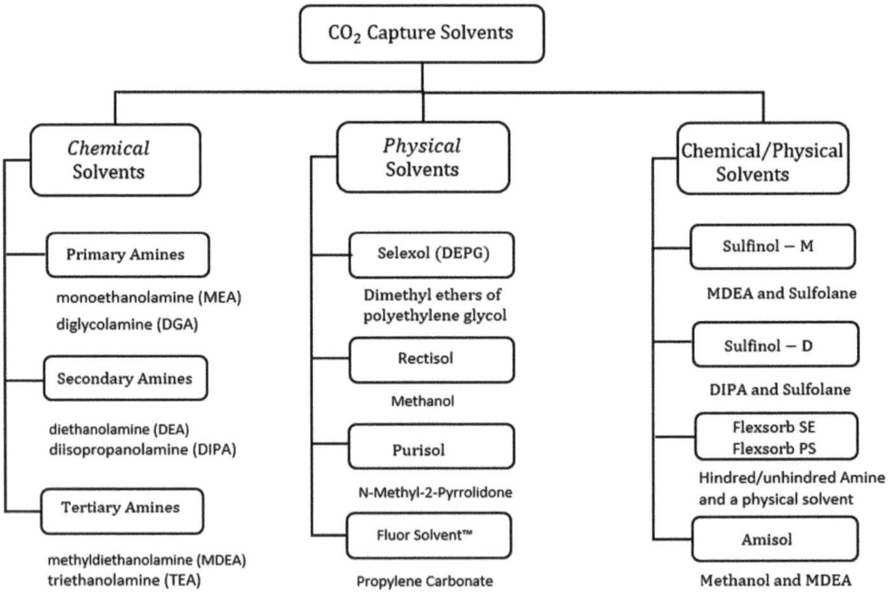

Fig. 10.3 Commonly used chemical and physical solvents

10.3.2 Acid Gas Removal/CO_2 Capture Using Chemical Solvents

10.3.2.1 General Consideration

Amines are widely used to remove CO_2 in various areas ranging from natural gas production to CO_2 capture for flue gas of fired heaters and boilers.

There are multiple classifications of amines, each of which has different characteristics relevant to CO_2 capture. For example, MEA reacts strongly with acid gases like CO_2 and has a fast reaction and strong ability to remove high percentages of CO_2, even at low CO_2 concentrations. Generally, MEA can capture above 90% of the CO_2 from the flue gas of gas-fired heaters, boilers, and coal-fired plants, which is one of the most effective solvents to capture CO_2.

However, primary and secondary amines, for example, MEA and DEA, will react with CO_2 and form degradation products. O_2 from the inlet flue gas will cause degradation as well. The degraded amine is no longer able to capture CO_2, which decreases the overall carbon capture efficiency, therefore installation of a reclaimer will be needed for unit stable operation.

Solvent selection is highly dependent on the raw gas to be treated, the component, and partial pressure of CO_2, challenges of amine CO_2 capture include:

- Difficulties in absorption of CO_2 due to low pressure of gases to be treated;
- Amine degradation and acid formation caused by excess O_2 in flue gas
- CO_2 degradation of primary (and secondary) amines
- High energy consumption for regeneration and gas compression
- CO_2 sink for industrial or commercial utilization
- Disposal of the removed CO_2.

The amine concentration in the absorbent solvent is also an important parameter in the design and operation of an amine CO_2 capture process. Depending on which one of the following four amines the unit selected and what feed gases will be treated to capture CO_2 or remove the acid gas, the followings are some typical amine concentrations in weight percent [2]:

- MEA: For removing H_2S and CO_2, usually 20% MEA solvent, and about 32% for removing only CO_2.
- DEA: About 20–25% for removing H_2S and CO_2
- MDEA: About 30–55% for removing H_2S and CO_2
- DGA: About 50% for removing H_2S and CO_2

Another factor to be considered in choosing an amine concentration is the relative solubility of H_2S and CO_2 in the selected amine. The type of amine will affect the required amine solution circulation rate, the reboiler duty of the regenerator, and the selectivity in removing either H_2S alone or CO_2 alone if desired.

For both chemical solvent and physical solvent, the partial pressure of CO_2 is the driving force to transfer CO_2 into the solvent. Under low partial pressure of

CO_2, achieving the CO_2 capture target usually requires higher reboiler duty for the chemical solvent, which will result in higher operating costs.

10.3.2.2 Amine Absorption Chemistry

Generally, the Amine acid gas (H_2S, CO_2) removal Chemistry are.

Primary Amine-MEA formular:

$$C_2H_7NO, R_1H\text{-}N\text{-}H, R_1 = CH_3CH_2OH$$

Secondary Amine-DEA formular:

$$C_4H_{11}NO_2, R_1R_2N\text{-}H, R_1 = CH_3CH_2OH, R_2 = CH_3CH_2OH$$

Tertiary Amine-MDEA formular:

$$C_5H_{13}NO_2, R_1R_2R_3N, R_1 = CH_3CH_2OH,\ R_2 = CH_3CH_2OH, R_3 = CH_3.$$

Amines remove H_2S and CO_2 in two steps:

- Gas dissolves in solvent (physical absorption)
- Dissolved gas (a weak acid) reacts with weakly basic amines.

H_2S reaction

$$H_2S \leftrightarrow H^+ + HS^- \tag{10.1}$$

$$R_1R_2R_3N + H_2S \leftrightarrow R_1R_2R_3NH^+ + HS^- \tag{10.2}$$

CO_2 reacts two ways with amine (with water):

$$CO_2 + H_2O \leftrightarrow H^+ + HCO_3^- \tag{10.3}$$

$$CO_2 + H_2O + R_1R_2R_3N \leftrightarrow R_1R_2R_3NH^+ + HCO_3^- \tag{10.4}$$

CO_2 reaction rate is much slower than H_2S reaction.

10.3.2.3 Typical Amine Absorption Process

Sour gas is normally the natural gas from the gas field and the associated gas from the oil or condensate fields.

Removal of sulfur and CO_2 via amine absorption process are completed in two independent sections. Sulfur content is "zero" when the feed gas enters the carbon removal section.

The typical chemical solvent absorption process is shown in Fig. 10.4. Individual CO_2 capture process varies slightly with feed gas conditions, treated gas spec, selected solvent, etc.

The sour feed gas is sent to a liquid KO drum to remove any entrained liquid or particles if any, and then fed to the bottom of the amine absorber column. The absorber can be either a tray or a packed tower, packing is usually preferred due to its high capacity and better options for construction materials.

The feed gas then flows upward, counter-current to the lean solvent solution which is introduced to the top of the amine absorber. The treated gas meets the sweet gas specification of CO_2 and H_2S and exits the top of the column to downstream facilities or grid as a sales product. The rich solvent with the absorbed CO_2 or acid gas is directed to a flash drum at reduced pressure, the flash gas can be directed to the fuel gas system for use.

The flashed rich solvent is sent to the Lean/rich amine heat exchanger to raise solvent temperature, then enters the regenerator for regeneration using a reboiler, the regenerated lean amin discharged from regenerator bottom enters the lean/rich amine heat exchanger to recover the stream heat, then lean solvent is cooled down further via a cooler, the cooled lean amine is pumped to the top of the absorber, completed a solvent recycle loop. The lean solvent concentration and recycle rate will be monitored and maintained via fresh amine makeup.

Due to formation of heat-stable salts by the side reaction of CO_2 with Amine, solvent degradation will happen in normal operation; therefore, a reclaimer is usually required for primary amine-based systems, such as MEA and DGA, the secondary amine system may also need a reclaimer to be installed; the reclaimer helps removing degradation products from the solution, i.e., heat-stable salts, suspended solids, acids, and iron compounds.

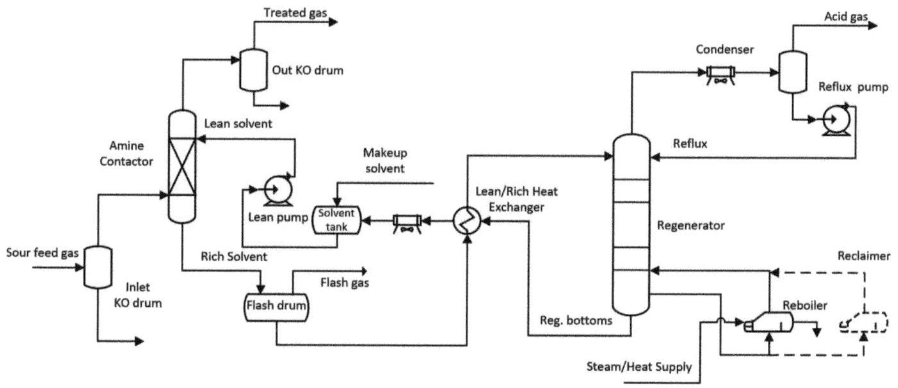

Fig. 10.4 Typical chemical absorption process flow diagram

The removed acid gas (CO_2 and H_2S) usually is treated in two ways, one is being reinjected back to the reservoir for oil and gas production enhancement, which mitigates the CO_2 production; the other is being directed to the sulfur recovery unit to recover H_2S and produce sulfur byproduct, the CO_2 is eventually diluted with combustion air during the operation of main burners and tail gas incinerator in the sulfur recovery process. It will not contribute to CO_2 mitigation at all.

For CO_2 capture and acid gas H_2S enrichment, the post acid gas removal unit, an additional facility may be installed to capture the CO_2 and enriches the H_2S concentration, which not only captures the CO_2 but also benefits the sulfur recovery unit operation significantly.

10.3.3 Acid Gas Removal/CO_2 Capture Using Physical Solvent

10.3.3.1 General Consideration

The physical solvent is also widely used in CO_2 capture and acid gas removal, it is mostly applied for purification of the feed gas with higher acid gas content (higher partial pressure), but it may not be suitable for natural gas stream due to the unignorable solubility of C_2 + hydrocarbon gases.

As shown in Fig. 10.3 commonly used chemical and physical solvents, there are many physical solvents which are very popular for the CO_2 capture or acid gas removal processes, such as DEPG (Selexol™ or Coastal AGR), NMP (Purisol), Methanol (Rectisol), and Propylene Carbonate (Fluor Solvent™).

Physical solvents can usually be stripped of impurities by reducing the pressure without the additional demand for heat supply; furthermore, physical absorption processes are attractive because of the following advantages:

- Regeneration at low temperatures with an inert stripping gas instead of reboilers. it requires little or no energy for stripping;
- Rich solvent can be set to multi-stage flashing to low pressures;
- In general, physical solvents are capable of removing COS, CS_2, and mercaptans organic sulfur compound;
- Solvent degradation caused by oxygen and CO_2 is manageable;
- Some physical solvent (DEPG) acts as a dehydrator, which can eliminate the additional dehydrator; it may require additional equipment and additional energy to dry the solvent.
- The processes can operate at an ambient or lower temperature to enhance the solubility of the acid gases. Chemical losses are low due to low solvent vapor pressure.
- The solvents are relatively non-corrosive so carbon steel can be used.

Physical solvents will absorb heavy hydrocarbons from the gas stream resulting in high hydrocarbon content in the acid gas stream as well as possibly significant hydrocarbon losses. This disadvantage limits its application in natural gas processing.

It is suitable for the following

- Low concentration of heavy hydrocarbon in the feed gas
- Removal of the bulk of acid gas
- Selective removal of H_2S.

Different physical solvent requires a different process, for example, due to the high vapor pressure of Methanol, Rectisol process operates at extremely low temperature (< -50 °C, varying pending on operation pressure), a refrigeration system is required.

The flashed gases from the multi-stage flash drums are in higher CO_2 content, that could be collected for reuse or storage directly.

The solvent's capacity for absorbing acid gases increases as the temperature is decreased.

10.3.3.2 Gases Solubilities in Physical Solvent

Barry Burr and Lili Lyddo provided gases solubilities relative to CO_2 in the physical solvents table [3] and Aspen tech has done the same but using the solubilities relative to Methane in DEPG solvent.

The results of "Solubilities of Gases in DEPG Relative to Methane" are summarized by Aspen Tech in Table 10.2 [4].

From Table 10.2, the ethane has 7.2 times CH_4 relative solubility, the propane has equivalent solubility as CO_2, and even the methane relative solubility is 5 times higher than H_2, the solubility of hydrocarbons in physical solvents increases with the molecular weight of the hydrocarbon. Thus, hydrocarbons above the ethane are also absorbed to a large extent and flashed from the solvent along with the acid gas. At lower pressure, physical solvent processes are generally not economical for the

Table 10.2 Solubilities of gases in physical solvents (DEPG) relative to CH_4

Component	Solubility to CH_4	Component	Solubility to CH_4
H_2	0.2	COS	35
N_2	0.3	NH_3	73
CO	0.43	H_2S	134
CH_4	1.00	CH_3SH	340
C_2H_6	7.2	CS_2	360
CO_2	15.2	SO_2	1400
C_3H_8	15.4	C_2H_2S	8200
$i-C_4H_{10}$	28	H_20	11,000
$n-C_4H_{10}$	36	HCN	19,500

natural gas processing that contain a substantial amount of ethane plus hydrocarbons. Therefore, the physical solvent is not suitable for natural gas processing but the physical solvent is great for syngas CO_2 capture. Subject to a pressure boosted to above 10 bars, CO_2 capture from flue gas is also feasible.

From Table 10.2, you can also see the very higher solubility for Mercaptans, COS, CS_2, and SO_2 in this physical solvent. Therefore, these compounds are removed to a large extent along with the acid gases.

10.3.3.3 Typical Physical Solvent (DEPG) Absorption Process

A. Single absorber process

For the acid gas containing CO_2 with minor H_2S, the single absorber will be good enough, the typical single absorber DEPG physical solvent absorption process is shown in Fig. 10.5.

A single absorber system can be used for the following scenarios:

- The feed gas has the bulk of CO_2 gas content without H_2S;
- Lower H_2S content in feed or H_2S content is already meeting the treated gas spec; after treatment, both the sweet gas and acid gas meet the H_2S spec;
- Both H_2S and CO_2 content are Higher, no sulfur recovery is required, and the acid gas will be used for reinjection back to the reservoir to enhance oil production.

The feed gas is sent to a liquid KO drum to remove any entrained liquid or particles if any, and then fed to the bottom of the absorber column. The absorber can be either a tray or a packed tower, packing is usually preferred due to its high capacity and better options for construction materials.

The feed gas then flows upward, counter-current to the lean solvent solution which is introduced in one or more entrances around the top of the absorber. The treated gas meets the sweet gas specification of CO_2 and H_2S and exits the top of the column to downstream facilities or grid as a sales product. The rich solvent with the absorbed

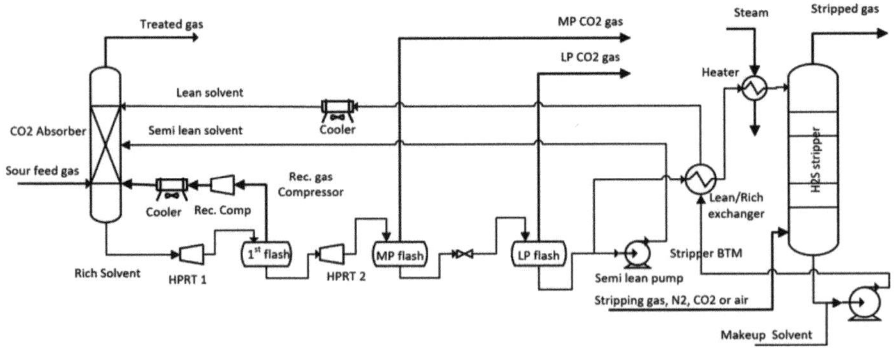

Fig. 10.5 Typical physical solvent absorption (DEPG) PFD-single absorber

CO_2 or acid gas is directed to multiple flash drums with reduced pressure, before entering 1st flash drum, higher pressure rich solvent via an HPRT (hydraulic turbine) to recovery the liquid power, the pressure of 1st flash drum should set for fully flash the main gas stream (CH_4 or H_2), flashed gas from 1st flash drum will be compressed, cooled and sent it back to the absorber, and then via the second HPRT to recover the rich solvent power further to MP flash drum and LP flash drum, MP flash drum operate at about 3–6 bar, the MP CO_2 gas can be directed to the second stage of CO_2 gas compressor to reduce the power demand of CO_2 gas compression in downstream CO_2 compression unit. LP flash drum usually operates at 0.5 barg pressure to fully flash the CO_2 gas.

Depending on the H_2S amount of feed gas, an H_2S stripper may be needed, it is preferred to heat the flashed solvent to reduce the stripping gas demand. In case of a stripper is needed, usually split the flashed solvent, take about 20–30% of solvent to stripper, lean solvent from stripper's bottom is directed to Lean/rich solvent heat exchanger, then via a cooler to cool down to the desired temperature; then enters the top of the absorber. The remaining semi lean solvent (flashed) is pumped and sent to the absorber at the location below the lean solvent entrance.

Physical solvent regeneration is completed by reducing the pressure in a couple of stages, unless deep cleaning of H_2S or CO_2 is required, in which case, a stripper column will be used.

When H_2S is present in significant amounts, thermal regeneration is usually necessary to accomplish the thorough stripping of the solvent needed to reach stringent H_2S purity requirements.

B. Dual absorber process

For the acid gas containing both CO_2 and H_2S, usually, two absorbers will be considered, one for H_2S absorption, one for CO_2 absorption, the typical dual absorber DEPG physical solvent absorption process is shown in Fig. 10.6.

In addition to the single absorber process description, the dual absorber process provides the capability to remove the H_2S and CO_2 separately.

The feed gas is sent to a liquid KO drum to remove any entrained liquid or particles if any, and then fed to the bottom of the H_2S absorber column. The feed gas then flows upward, counter-current to the semi-rich solvent solution which is introduced to the top of the absorber. Gas from the H_2S absorber is directed to the CO_2 absorber to capture the CO_2 fully, and then flows upward, counter-current to the semi lean and lean solvent sequentially, the treated gas from the CO_2 absorber meets the sweet gas specification of CO_2 and H_2S, and exits the top of the column to downstream facilities or grid as a sales product. The semi-rich solvent with the absorbed CO_2 or acid gas will be split (20–30%) and sent to the H_2S absorber, the remaining is directed to multiple flash drums with reduced pressure, before entering 1st flash drum, higher pressure rich solvent via an HPRT (hydraulic turbine) to recovery the liquid power, the pressure of 1st flash drum should set for fully flash the main gas stream (CH_4 or H_2), flashed gas from 1st Flash drum will be compressed, cooled and sent it back to CO_2 absorber, and then the liquid solvent will flow through a second HPRT to recover

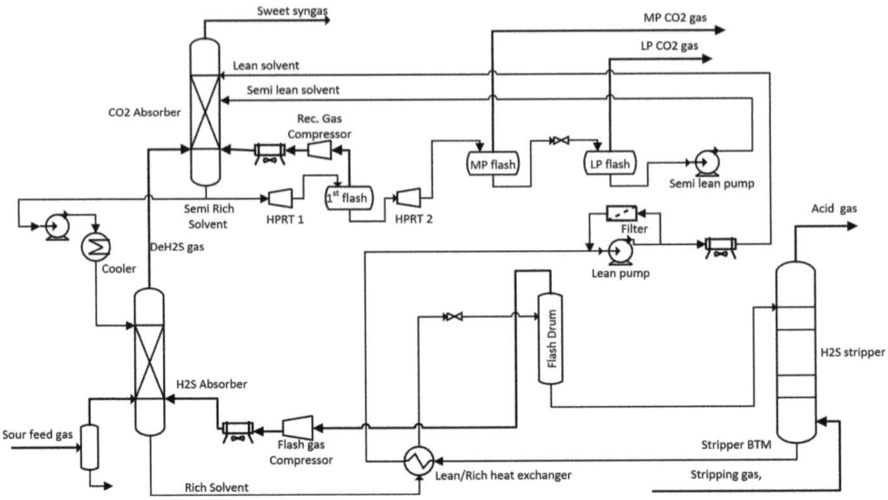

Fig. 10.6 Typical physical solvent absorption (DEPG) PFD-dual absorber

the rich solvent power further, then sent to MP flash drum and LP flash drum, MP flash drum operate at about 3–6 bar, the MP CO_2 gas can be directed to the second stage of CO_2 gas compressor to reduce the power demand of CO gas compression downstream of CO_2 capture unit. LP flash drum usually operates at 0.5 barg pressure to fully flash the CO_2 gas. The flashed semi lean solvent then be pumped back to the CO_2 absorber at the location below the lean solvent entrance.

The rich solvent from the H_2S absorber is directed to the Lean/rich solvent heat exchanger, then to the H_2S stripper, the lean solvent from the stripper bottom exchanges heat in the lean/rich exchanger with rich solvent, via the lean pump and a cooler, with the desired pressure and temperature enters the top of the CO_2 absorber.

The solvent's capacity for absorbing acid gases increases as the temperature is decreased. A decrease in temperature can reduce the circulation rate, thus reducing operating costs. The solubility of CH_4, H_2, and CO shows little change with temperature, so the absorption of acid gas is more selective.

10.4 Sour Natural Gas CO_2 Capture

This section presents the case studies of CO_2 capture from sour natural gas and acid gas with detail simulation results support; Sects. 10.5 and 10.6 present the simulation results of some case studies, including CO_2 capture from Syngas and flue gases, which have covered most of the CO_2 capture scenarios. This is an effort to provide references to readers in their feasibility studies; however, some processes may be a licensed process, for the actual project, please contact related licensor accordingly.

Sour gas includes natural gas, produced gas from oil filed, refining gas, etc. it contains either H_2S or CO_2 or both. the gas containing dominated CO_2 and/or H_2S is called acid gas, current, the acid gas removed from sour gas is directed to the sulfur recovery unit to meet the strict environmental requirements of SO_2 emission; CO_2 usually as inert gas flows through the sulfur recovery unit and tail gas treatment unit, eventually vent to atmosphere after diluted with combustion gases, therefore, the CO_2 in acid gas is not captured.

H_2S and CO_2 content in sour gas vary greatly, for the sour gas with higher H_2S content and lower CO_2 content, the acid gas removed can be directed to the sulfur recovery unit without any concerns, however, for lower H_2S and higher CO_2 sour gas, it is required to enrich the H_2S content in acid gas.

Therefore, the operation of sour gas CO_2 capture requires separation of CO_2 and rich H_2S acid gas independently, it may consist of two solvent loops depending on the solvent and the process selected.

10.4.1 Nature Gas CO_2 Capture via Acid Gas Enrichment—Case Study 1

This sour natural gas contains higher concentration of both H_2S and CO_2, the acid gas removed may not be in an ideal H_2S/CO_2 ratio for the sulfur recovery unit, acid H_2S enrichment with CO_2 capture will improve the sulfur recovery unit performance; therefore, the natural gas CO_2 capture usually will consist of two steps, using DEA to removal the H_2S and CO_2, then using MDEA solvent for H_2S enrichment and CO_2 Capture.

DEA chemical solvent is selected due to its high capability to move CO_2 and H_2S, while degradation caused by CO_2 is much mild compared with primary solvent MEA.

10.4.1.1 Gas Composition and Product Spec.

A natural gas plant processes a sour feed natural gas with a capacity of 2500 kgmole/h, sour natural gas contains both CO_2 and H_2S in 10.9% and 9.7% respectively, the sour feed gas composition is shown in Table 10.3.

The majority of sweet gas will be directed to the sales gas pipeline, small portion of the sweet gas will be used as fuel gas for plant internal use. The specification of treated sweet gas is set at 4 ppm or less for H_2S and 2% (mol) or less for CO_2.

Table 10.3 Sour feed gas composition of an acid gas removal unit

Molar flow	2500.0	kgmole/h
Mass flow	53,444.2	kg/h
Molecular weight	21.38	
Composition		
H_2O (%)	0.00	mole %
CO_2 (%)	10.91	mole %
H_2S (%)	9.71	mole %
Methane (%)	76.02	mole %
Ethane (%)	2.96	mole %
Propane (%)	0.50	mole %

10.4.1.2 Process Description

Sour natural gas CO_2 capture using DEA and MDEA solvents process flow diagram is show as Fig. 10.7.

Acid Gas Removal

The sour natural gas is sent to a liquid KO drum to remove any entrained liquid or particles if any, and then fed to the bottom of the amine absorber column; The gas then flows upward, counter-current to the lean solvent solution which is introduced to the top of the amine absorber. The treated gas meets the sweet gas specification of CO_2 and H_2S and exits the top of the column to downstream facilities or grid as a sales product. The rich solvent with the absorbed acid gas is directed to a flash drum

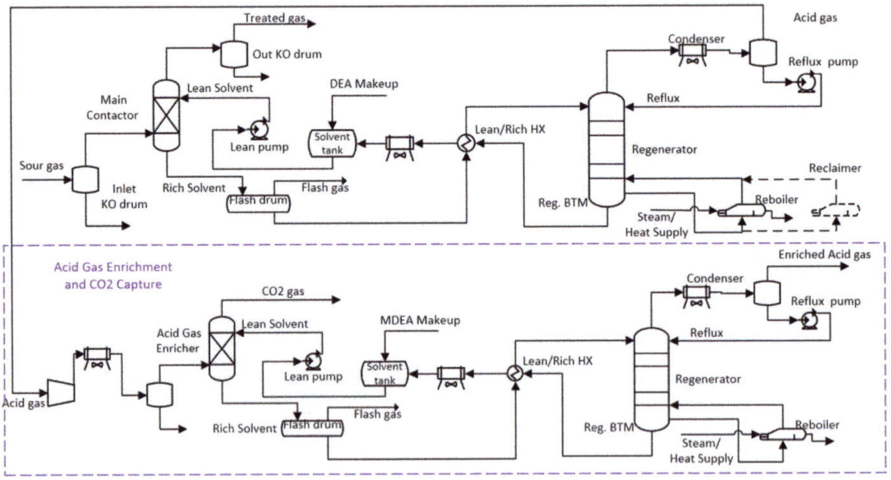

Fig. 10.7 Natural gas CO_2 capture-acid gas enrichment

at reduced pressure, the flash gas can be directed to the fuel gas system for use or merged with the acid gas system.

The flashed rich solvent will be directed to a lean/rich amine heat exchanger to increase the rich solvent temperature, then the heat exchanged rich solvent enters to regenerator for regeneration using a reboiler, the lean amine discharged from regenerator bottom enters the lean/rich amine heat exchanger to recover the stream heat, then to be cooled down further via a cooler, the cooled lean amine enters the top of the absorber, completed a solvent recycle loop; lean solvent concentration and recycle rate will be monitored and maintained via fresh amine makeup.

Due to solvent degradation in presence of CO_2 component in the amine-based system, a reclaimer may be required for DEA solvent, the reclaimer helps to remove degradation products from the solution and also aids in the removal of heat-stable salts, suspended solids, acids, and iron compounds.

Acid Gas Enrichment and CO_2 Capture

The removed acid gas contains about 50% of CO_2 and 47.6% of H_2S, to capture the CO_2 in acid gas, an additional facility for CO_2 capture and H_2S enrichment will be installed. This CO_2 capture of this case study takes the advantage of a MDEA solvent in its fast-absorption of H_2S, the acid gas can be separated into CO_2 gas and enriched acid gas.

The pressure of acid gas from the main regenerator is about 2.5 bar, to improve the absorption efficiency, an acid gas compressor is required to boost the acid gas pressure, the boosted acid gas then enters to Acid Gas Enricher, flows upward, counter-current to the lean MDEA solvent which is introduced to the top of the acid gas enricher. The CO_2 gas from the top of the acid gas enricher contains 98.7% of CO_2, which can be directed to the CO_2 pipeline for use or storage.

The rich solvent enriched with acid gas is directed to a flash drum at reduced pressure, the flash gas is an acid gas, it can be merged with enriched acid gas as the feed for the sulfur recovery unit.

The flashed rich solvent is directed to a Lean/rich amine heat exchanger, then enters the regenerator, enriched acid gas will merge with flashed gas and be directed to the sulfur recovery unit. The lean MDEA solvent discharged from the regenerator bottom enters the lean/rich amine heat exchanger to recover the stream heat, then is cooled down further via a cooler, and pumped to the top of the absorber, completing a solvent recycle loop.

10.4.1.3 H&M Balance

Figure 10.7 shows a typical process flow diagram of CO_2 Capture and Acid Gas Enrichment, a simulation model has been built using a commercial simulator.

Acid Gas Removal

Figure 10.8 is a typical HYSYS simulation sheet for sour natural gas CO_2 capture using DEA + MDEA solvents. The sub flowsheet of Acid gas enrichment is shown in Fig. 10.9.

The H&M balance-key stream information is also shown in Table 10.4.

Acid Gas Enrichment

Figure 10.9 indicates a typical HYSYS simulation sheet for the acid gas enrichment using MDEA solvent.

The key stream information of acid gas enrichment is also shown in Table 10.5.

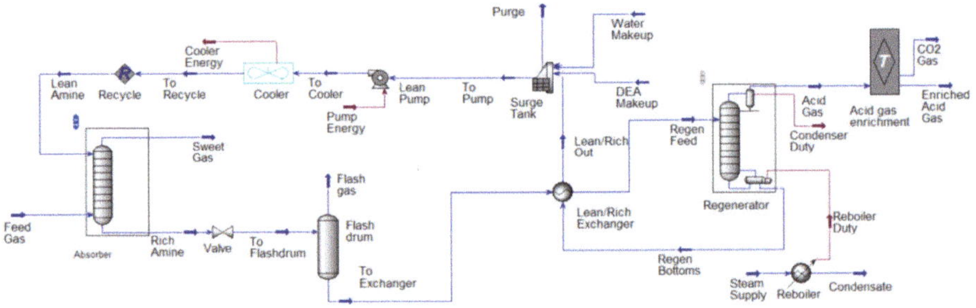

Fig. 10.8 Nature gas acid gas removal using DEA solvent simulation sheet

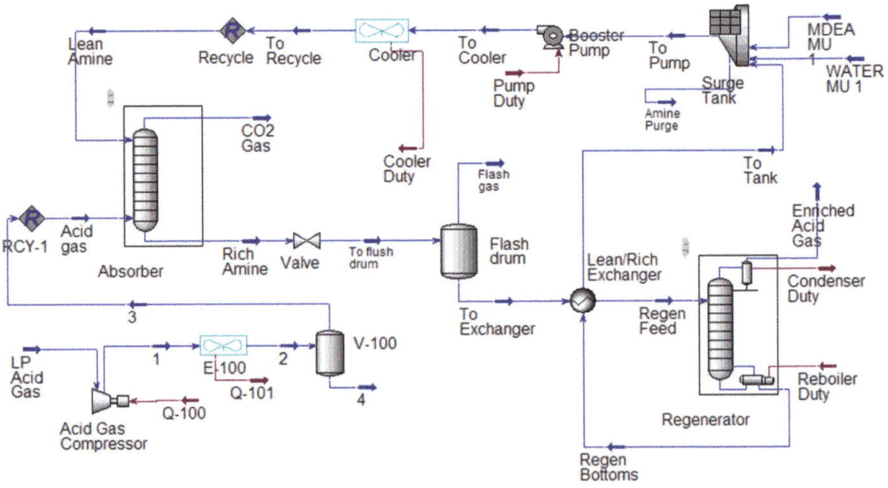

Fig. 10.9 Acid gas enrichment using MDEA solvent simulation sheet

Table 10.4 Case study 1 acid gas removal stream information

Name	Feed gas	Lean amine	Sweet gas	Rich amine	To flash drum	Flash gas	To exchanger	Regen feed	Acid gas	Regen bottoms	Lean/rich HX out	CO_2 gas	Enriched acid gas
Vapour fraction	1.000	0.000	1.000	0.000	0.001	1.000	0.000	0.005	1.000	0.000	0.000	1.000	1.000
Temperature [°C]	37.8	35.0	35.3	60.3	60.6	60.6	60.6	93.3	35.0	130.7	98.5	40.6	35.0
Pressure [kPa]	3618	3618	3535	3549	483	483	483	448	241	262	228	830	185
Molar flow [kgmole/h]	2500	12,900	1999	13,401	13,401	12	13,390	13,390	496	12,893	12,893	175	290
Mass flow [kg/h]	53,160	293,333	33,631	312,862	312,862	287	312,575	312,575	19,171	293,405	293,405	7657	10,393
Liquid volume flow [m^3/h]	136.2	287.2	109.2	314.2	314.2	0.6	313.5	313.5	26.2	287.4	287.4	9.3	15.2
Molecular weight	21.3	22.7	16.8	23.3	23.3	24.4	23.3	23.3	38.6	22.8	22.8	43.7	35.8
Mole frac (DEAmine)	0.0000	0.0541	0.0000	0.0520	0.0520	0.0000	0.0521	0.0521	0.0000	0.0541	0.0541	0.000000	0.0000
Mole frac (H_2O)	0.0000	0.9453	0.0018	0.9097	0.9097	0.0401	0.9104	0.9104	0.0240	0.9446	0.9446	0.009232	0.0311
Mole frac (CO_2)	0.1091	0.0003	0.0093	0.0193	0.0193	0.1660	0.0191	0.0191	**0.4984**	0.0007	0.0007	0.986679	**0.2242**
Mole frac (H_2S)	0.0971	0.0003	0.0000	0.0184	0.0184	0.1830	0.0183	0.0183	**0.4762**	0.0007	0.0007	0.000046	**0.7447**

(continued)

Table 10.4 (continued)

Name	Feed gas	Lean amine	Sweet gas	Rich amine	To flash drum	Flash gas	To exchanger	Regen feed	Acid gas	Regen bottoms	Lean/rich HX out	CO_2 gas	Enriched acid gas
Mole frac (Methane)	0.7640	0.0000	0.9518	0.0006	0.0006	0.5858	0.0001	0.0001	0.0014	0.0000	0.0000	0.003875	0.0000
Mole frac (Ethane)	0.0298	0.0000	0.0371	0.0000	0.0000	0.0251	0.0000	0.0000	0.0001	0.0000	0.0000	0.000168	0.0000
Mole frac (Propane)	0.0000	0.0000	0.0000	0.0000	0.0000	0.0000	0.0000	0.0000	0.0000	0.0000	0.0000	0.000000	0.0000

Indicate the chnges befoe after acid gas enrichemnt

Table 10.5 Case study 1 acid gas enrichment stream information

Name	LP acid gas	Acid gas	Lean amine	CO_2 Gas	Rich Amine	To flush drum	Flash gas	To Exchanger	Regen Feed	Enriched Acid Gas	Regen Bottoms	To Pump
Vapour fraction	1.000	1.000	0.000	1.000	0.000	0.006	1.000	0.000	0.039	1.000	0.000	0.000
Temperature [°C]	35.0	35.0	40.0	40.6	69.2	66.5	66.5	66.5	98.9	35.0	122.0	71.7
Pressure [kPa]	241	840	1618	830	840	267	267	267	267	185	200	200
Molar flow [kgmole/h]	496	488	5037	175	5350	5350	32	5317	5317	290	5027	5037
Mass flow [kg/h]	19,171	19,019	129,180	7657	140,542	140,542	1150	139,392	139,392	10,393	128,999	129,180
Liquid volume flow [m^3/h]	26.2	26.0	127.8	9.3	144.5	144.5	1.6	142.8	142.8	15.2	127.6	127.8
Molecular weight	38.6	39.0	25.6	43.7	26.3	26.3	35.4	26.2	26.2	35.8	25.7	25.6
Mole frac (MDEAmine)	0.0000	0.0000	0.0753	0.0000	0.0709	0.0709	0.0000	0.0714	0.0714	0.0000	0.075475	0.0753
Mole frac (H_2O)	0.0240	0.0074	0.9242	0.0092	0.8706	0.8706	0.0926	0.8753	0.8753	0.0311	0.924043	0.9242
Mole frac (CO_2)	**0.4984**	0.5069	0.0004	0.9867	0.0143	0.0143	0.2848	0.0126	0.0126	**0.2242**	0.000382	0.0004
Mole frac (H_2S)	**0.4762**	0.4843	0.0001	0.0000	0.0443	0.0443	0.6226	0.0407	0.0407	**0.7447**	0.000100	0.0001
Mole frac (methane)	0.0014	0.0014	0.0000	0.0039	0.0000	0.0000	0.0000	0.0000	0.0000	0.0000	0.000000	0.0000
Mole frac (ethane)	0.0001	0.0001	0.0000	0.0002	0.0000	0.0000	0.0000	0.0000	0.0000	0.0000	0.000000	0.0000

(continued)

Table 10.5 (continued)

Name	LP acid gas	Acid gas	Lean amine	CO_2 Gas	Rich Amine	To flush drum	Flash gas	To Exchanger	Regen Feed	Enriched Acid Gas	Regen Bottoms	To Pump
Mole frac (propane)	0.0000	0.0000	0.0000	0.0000	0.0000	0.0000	0.0000	0.0000	0.0000	0.0000	0.000000	0.0000

Indicate the chnges befoe after acid gas enrichemnt

10.4.1.4 Results and Recommendation

Per the simulation result, the DEA chemical solvent is capable for remove both CO_2 and H_2S of this case study, the overall unit performance summary is shown in Table 10.6.

Per the simulation result, the MDEA chemical solvent is capable for remove CO_2 from acid gas, that is, the MDEA can remove all the H_2S and partially CO_2, unabsorbed gas will be the CO_2 gas containing 98.7% of CO_2; the acid gas enrichment performance summary is shown in Table 10.7.

From Table 10.6, using DEA solvent, 2.1 ppm of H_2S content and 0.93% of CO_2 content in the sweet gas can be achieved, it meets sale gas specification; the lean solvent recirculation rate is 290 m^3/h, and DEA strength is 25% (w), rich DEA amine loading is 0.724 mol acid gas/mole amine. H_2S and CO_2 in acid gas are 47.6% and 49.8% respectively.

From Table 10.7, with acid gas enrichment using MDEA solvent, the H_2S content in the enriched acid gas achieves 74.2%, and the CO_2 content in the enriched acid gas falls to 22.4%, about 70% of CO_2 is captured in a purity of 98.7%.

H_2S content of 46 ppm in CO_2 gas can be achieved, the lean MDEA solvent recirculation rate is about 125 m^3/h, and MDEA strength is 35% (w); In case the CO_2 gas is for chemical usage, it is possible to achieve a H_2S content of 4 ppm or below through adjustment of the recirculation rate; the rich amine loading achieved is 0.825 mol acid gas/mole amine.

Table 10.6 Case study 1 overall performance summary

Description	Value	Unit	Description	Value	Unit
Feed gas flowrate	2500	kgmole/h	Rich amine loading	0.7240	
Feed gas CO_2 mole fraction	10.91	%	Amine strength	25	% (w)
Feed gas H_2S mole fraction	9.71	%	Amine recirculation rate	285	m^3/h
Sweet gas CO_2 composition	0.9299	%	CO_2 gas mass flow	7657	kg/h
Sweet gas H_2S composition	2.09	ppm	CO_2 gas comp mass flow (CO_2)	7616	kg/h
Acid gas mass flow	19,171	kg/h	CO_2 gas comp mole frac (CO_2)	98.67	wt%
Acid gas molar flow	496	kgmole/h	Enriched acid gas CO_2	22.42	%(m)
Acid gas comp mass flow (CO_2)	10,888	kg/h	Enriched acid gas H_2S	74.47	%(m)
Acid gas mole frac (CO_2)	49.84%	%	Enriched acid gas molar flow	290	kgmole/h
Acid gas mole frac (H_2S)	47.62%	%	Acid gas CO_2 captured	70	%

Table 10.7 Case study 1 acid gas enrichment performance

Description	Value	Unit	Description	Value	Unit
Acid gas flow rate	488.0	kgmole/h	Enriched acid gas molar flow	290.2	kgmole/h
Acid gas H_2S mole fraction	0.4843		Enriched acid gas mole frac (CO_2)	0.2242	
Acid gas CO_2 mole fraction	0.5069		Enriched acid gas mole frac (H_2S)	0.7447	
Acid gas comp mass flow (CO_2)	10,887.4	kg/h	Lean amine loading	0.0063	
CO_2 gas comp mass flow (CO_2)	7615.7	kg/h	Rich amine loading	0.8252	
CO_2 gas H_2S	45.9	ppm	Amine strength	35	% (w)
CO_2 gas CO_2	98.7	mole %	Lean amine liq. std. vol. flow	125.0	m^3/h
Enriched acid gas mass flow	10,393	kg/h	CO_2 captured efficiency	70	%

Therefore, it is capable to remove the acid gas and capturing about 60–80% of the acid gas CO_2 using both DEA and MDEA solvent,

If a plant has a sour gas or acid gas reinjection facility, there may be no need for CO_2 capture and acid gas enrichment. Acid gas from the main regenerator contains both the CO_2 and H_2S, depending on the plant's overall arrangement, it can be either directed to the sulfur recovery unit or injected it to a reservoir to enhance oil recovery.

10.4.2 Cascade Sour Nature Gas CO_2 Capture—Case Study 2

The natural gas of this study 2 contains H_2S and a higher CO_2 content, the acid gas removed contains very high CO_2 content, which will be difficult for the main burner operation in the sulfur recovery unit. A cascade acid gas removal method will be used for this case study 2; that is, it removes the acid gas (H_2S and CO_2) and CO_2 in two steps using different solvents and systems; therefore, the natural gas CO_2 capture usually will consist of two steps: MDEA chemical solvent is selected as absorption solvent to remove all the H_2S and partial CO_2; to enhance CO_2 capture and protect the solvent from thermal and oxidative degradation, a 2–5% (w) of piperazine is blended with 35% (w) of MDEA solvent to remove the CO_2 in the de-H_2S gas in a separate system.

Amine blends of piperazine are used extensively in commercial CO_2 removal for carbon capture, advantageously allowing for protection from significant thermal and oxidative degradation at typical natural gas or flue gas conditions. The thermal degradation rates for MDEA and piperazine (PZ) are negligible, and PZ protects MDEA from oxidative degradation. This increased stability of the MDEA/PZ solvent

Table 10.8 Sour feed gas composition of an acid gas removal unit

Molar flow	2500.0	Kgmole/h
Mass flow	53,444.2	kg/h
Molecular weight	21.38	
Composition		
H_2O	0.00	mole %
CO_2	13.12	mole %
H_2S	4.04	mole %
Nitrogen	0.50	mole %
Methane	75.70	mole %
Ethane	5.12	mole %
Propane	0.51	mole %

blend over MDEA and other amine solvents provides for greater capacity for capturing a given amount of CO_2 [5, 6].

10.4.2.1 Gas Composition and Product Spec.

A natural gas plant processes a sour feed natural gas with a capacity of 5000 kgmole/h, sour natural gas contains both CO_2 and H_2S, it contains 13.12% of CO_2 content and 4.04% of H_2S, the sour feed gas composition is shown in Table 10.8.

The majority of the sweet gas will be directed to the sales gas pipeline, a small portion of the sweet gas will be used as a fuel gas for plant internal use. The specification of sweet gas is set at 4 ppm or less for H_2S, and not be higher than 3% (mole) for CO_2 content.

10.4.2.2 Process Description

Sour gas CO_2 capture using DEA and MDEA solvents process flow diagram is show as Fig. 10.10.

H_2S Removal

This case study takes advantage of MDEA solvent for its feature of fast-absorbing H_2S, all the H_2S and partial CO_2 will be separated from the sour feed gas. The removed acid gas has an optimal H_2S to CO_2 ratio as a feed to the sulfur recovery unit.

The sour natural gas is sent to a liquid KO drum to remove any entrained liquid or particles if any, and then fed to the bottom of the H_2S absorber column; The gas then flows upward, counter-current to the lean MDEA solvent solution which is introduced to the top of the amine absorber. All the H_2S and partial CO_2 will be removed, the de-H_2S gas exits the top of the column to the CO_2 removal section.

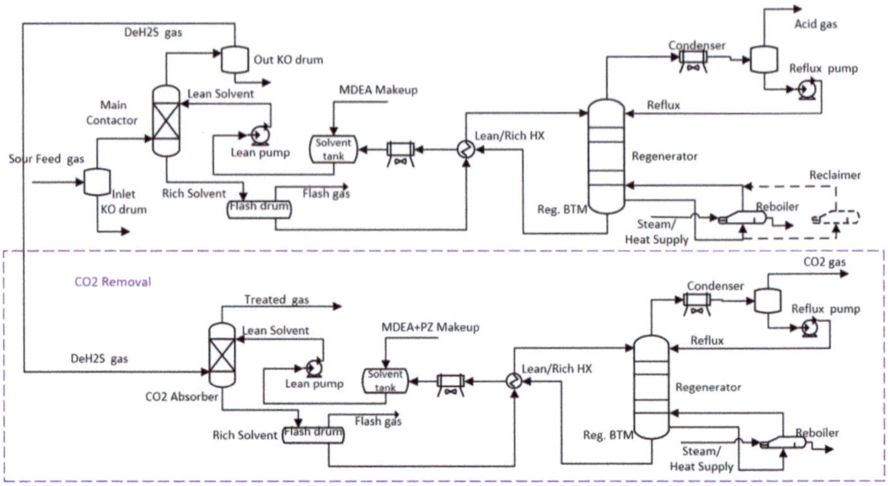

Fig. 10.10 Natural gas cascade CO_2 capture

The rich solvent with the absorbed acid gas is directed to a flash drum at reduced pressure, the flash gas can be directed to the fuel gas system for use or merged into the acid gas system.

The flashed rich solvent is directed to a Lean/rich amine heat exchanger to increase the rich solvent temperature, then the heat exchanged rich solvent enters the regenerator for regeneration using a reboiler, the lean solvent discharged from the regenerator bottom enters the lean/rich amine heat exchanger to recover the stream heat, then the lean solvent is cooled down further via a cooler, the cooled lean amine enters the top of the absorber, and a solvent recycle loop is completed; the lean solvent concentration and recycle rate will be monitored and maintained via fresh amine makeup.

Gas Sweetening and CO_2 Removal

To remove the CO_2 and meet the sales gas specification, MDEA and piperazine solvent will be used.

The de-H_2S gas contains about 11% of CO_2 and a trace amount of H_2S, enters the CO_2 absorber, the de-H_2S gas then flows upward, counter-current to the lean solvent which is introduced to the top of the CO_2 absorber. The treated gas from the top of the CO_2 absorber meets the sweet gas specification of CO_2 and H_2S, and it is directed to the fuel system or grid as a sales product.

The rich solvent is directed to a flash drum at reduced pressure, the flash gas is a fuel gas, and it can be directed to the fuel gas system.

The flashed rich solvent is directed to a Lean/rich amine heat exchanger, then enters the regenerator, the CO_2 gas from the regenerator is directed to the CO_2 gas disposing system, the lean solvent discharged from the regenerator bottom enters the lean/rich amine heat exchanger to recover the stream heat, then the lean solvent

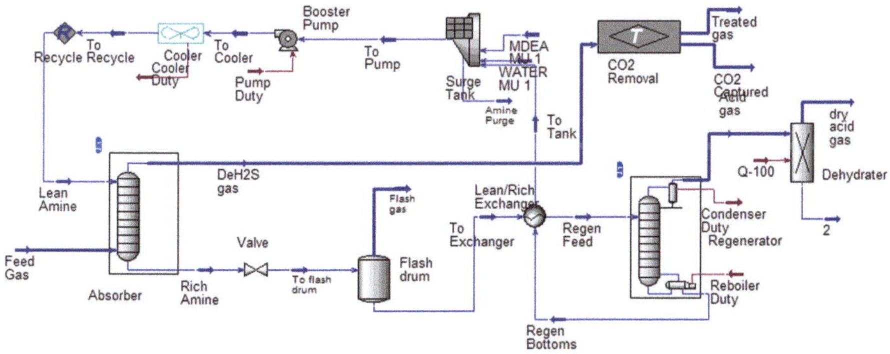

Fig. 10.11 Nature gas acid gas removal—DeH_2S

to be cooled down further via a cooler, the cooled lean amine enters the top of the absorber, completed a solvent recycle loop.

10.4.2.3 H&M Balance

Figure 10.10 shows the configuration of a simulation model which was built using a commercial simulator.

H_2S Removal
Figure 10.11 indicates a typical HYSYS simulation sheet for the sour natural gas cascade CO_2 capture using MDEA + piperazine solvents.

The DeH_2S section H&M balance-key stream information is also shown in Table 10.9.

CO_2 Removal
Figure 10.12 indicates a typical HYSYS simulation sheet for the sour natural gas CO_2 capture using MDEA and Piperazine solvents.

The key stream information of CO_2 removal section is also shown in Table 10.10.

10.4.2.4 Results and Recommendation

Per the simulation results, the MDEA and Piperazine chemical solvents are capable for remove CO_2 and H_2S in this cascade CO_2 capture case study, to remove the acid gas (H_2S and CO_2) and CO_2 separately, two independent solvent systems will be needed. The overall unit performance is summarized in Table 10.11.

From Table 10.11, using MDEA solvent, 86 ppm of H_2S content and 11.45% of CO_2 content in the De-H_2S gas can be achieved, the lean solvent recirculation rate is 360 m^3/h, and the MDEA strength is 35% (w); the De-H_2S gas will be directed to CO_2 removal section to remove the trace amount of H_2S and majority of CO_2.

Table 10.9 Natural gas DeH_2S unit stream information—case study 2

Name	Feed gas	Lean amine	DeH_2S gas	Rich amine	To flash drum	Flash gas	To exchanger	Regen feed	Acid gas	Regen bottoms	To tank	Dry acid gas	CO_2 captured	Treated gas
Vapour fraction	1.000	0.000	1.000	0.000	0.000	1.000	0.000	0.001	1.000	0.000	0.000	1.000	1.000	1.000
Temperature [°C]	20.0	40.0	40.3	47.4	47.6	47.6	47.6	98.9	40.0	126.7	77.7	4.5	4.5	35.8
Pressure [bar]	16.2	16.2	16.2	16.2	2.7	2.7	2.7	2.7	2.3	2.3	2.3	2.3	2.3	16.2
Molar flow [kgmole/h]	5000	14,507	4696	14,810	14,810	5	14,805	14,805	331	14,474	14,474	321	434	4250
Mass flow [kg/h]	108,221	372,080	96,438	383,862	383,862	95	383,767	383,767	12,261	371,506	371,506	12,076	19,058	77,135
Liquid volume flow [m^3/h]	278.3	368.1	261.3	385.2	385.2	0.3	384.9	384.9	17.3	367.5	367.5	17.2	23.2	237.6
Molecular weight	21.6	25.6	20.5	25.9	25.9	19.0	25.9	25.9	37.1	25.7	25.7	37.7	43.9	18.1
Master comp mole frac (MDEAmine)	0.000000	0.075335	0.000000	0.073790	0.073790	0.000001	0.073815	0.073815	0.000000	0.075503	0.075503	0.000000	0.000000	0.000000
Master comp mole frac (H_2O)	0.000000	0.924069	0.004634	0.903657	0.903657	0.037828	0.903950	0.903950	0.032572	0.923870	0.923870	0.001681	0.001677	0.003534
Master comp mole frac (CO_2)	0.131218	0.000422	0.114508	0.008402	0.008402	0.021248	0.008397	0.008397	0.356885	0.000431	0.000431	0.368280	0.995109	0.024696

(continued)

Table 10.9 (continued)

Name	Feed gas	Lean amine	DeH_2S gas	Rich amine	To flash drum	Flash gas	To exchanger	Regen feed	Acid gas	Regen bottoms	To tank	Dry acid gas	CO_2 captured	Treated gas
Master comp mole frac (H_2S)	0.040375	0.000175	0.000086	0.013774	0.013774	0.029965	0.013769	0.013769	0.607493	0.000196	0.000196	0.626891	0.000904	0.000001
Master comp mole frac (nitrogen)	0.005047	0.000000	0.005370	0.000001	0.000001	0.002216	0.000000	0.000000	0.000005	0.000000	0.000000	0.000005	0.000003	0.005931
Master comp mole frac (methane)	0.757025	0.000000	0.804921	0.000331	0.000331	0.806877	0.000057	0.000057	0.002568	0.000000	0.000000	0.002650	0.001881	0.888158
Master comp mole frac (ethane)	0.051195	0.000000	0.054392	0.000035	0.000035	0.079733	0.000008	0.000008	0.000379	0.000000	0.000000	0.000391	0.000334	0.059946
Master comp mole frac (propane)	0.015141	0.000000	0.016089	0.000010	0.000010	0.022130	0.000002	0.000002	0.000098	0.000000	0.000000	0.000101	0.000093	0.017733

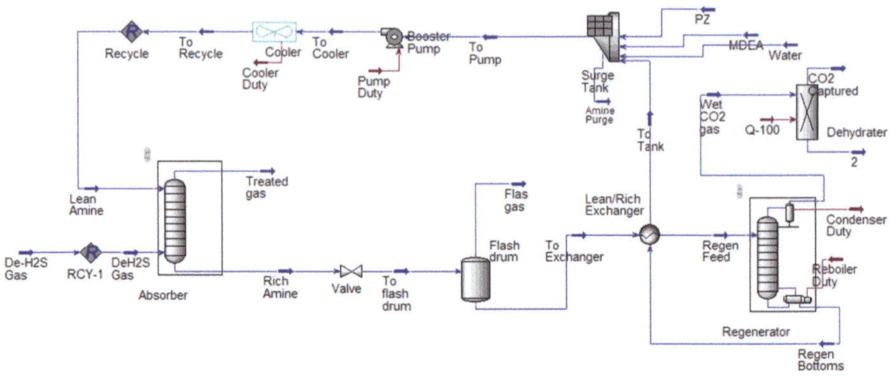

Fig. 10.12 CO_2 Removal unit simulation sheet

The treated gas has 1.24 ppm of H_2S of and 2.5% of CO_2, which meets the sale gas specification; rich MDEA amine loading achieved is 0.3 mol acid gas/mole amine. H_2S and CO_2 in acid gas achieved are 62.7% and 36.7% respectively.

Per the simulation result, the solvent of MDEA and piperazine mixture is capable for CO_2 removal from the acid gas; the CO_2 removal section performance summary is shown in Table 10.12.

From Table 10.12, using a mixture solvent of MDEA and Piperazine, the CO_2 captured is about 80%, the captured CO_2 gas has a CO_2 purity of 99.5% and 900 ppm H_2S, in case of strict H_2S content requirement in the CO_2 gas, 4 ppm or below of H_2S content can be achieved easily by adjusting the operation conditions in the H_2S removal section operation.

The mixed solvent recirculation rate is 125 m^3/h, and the amine strength is 35% (w), rich amine loading achieved is 0.3 mol acid gas/mole amine.

Please note, this process is a combination of CO_2 capture and H_2S removal, without CO_2 capture (ref to Fig. 10.11), the de-H_2S gas is not good as a sweet product.

If the plant has a sour gas or acid gas reinjection facility, there may be no need for CO_2 capture. Acid gas from the main regenerator contains both CO_2 and H_2S, depending on the plant's overall arrangement, it can be either directed to the sulfur recovery unit or injected all the acid gas to the reservoir to enhance the oil recovery.

10.4.3 *Acid Gas Enrichment and CO_2 Capture—Case Study 3*

10.4.3.1 Background

A gas processing facility use MDEA solvent to process the sour gas with 4% of H_2S and CO_2 of 13%, and acid gas from the Acid Gas Removal Unit has 65% of CO_2 and

Table 10.10 CO_2 capture stream information—case study 2

Name	DeH_2S Gas	Lean amine	Treated gas	Rich amine	To flash drum	Flas gas	To exchanger	Regen feed	Wet CO_2 gas	Regen bottoms	To tank	CO_2 captured
Vapour fraction	1.000	0.000	1.000	0.000	0.000	1.000	0.000	0.000	1.000	0.000	0.000	1.000
Temperature [°C]	40.3	35.0	35.8	59.4	59.6	59.6	59.6	98.9	40.0	126.6	89.1	4.5
Pressure [bar]	16.2	16.2	16.2	16.2	2.7	2.7	2.7	2.7	2.3	2.3	2.3	2.3
Molar flow [kgmole/h]	4696	15,087	4250	15,534	15,534	6	15,528	15,528	448	15,080	15,080	434
Mass flow [kg/h]	96,439	415,304	77,135	434,608	434,608	120	434,488	434,488	19,308	415,181	415,181	19,058
Liquid volume flow [m_3/h]	261.3	420.5	237.6	444.2	444.2	0.3	443.9	443.9	23.5	420.4	420.4	23.2
Molecular weight	20.5	27.5	18.1	28.0	28.0	19.3	28.0	28.0	43.1	27.5	27.5	43.9
Mole frac (MDEAmine)	0.000000	0.080851	0.000000	0.078527	0.078527	0.000003	0.078558	0.078558	0.000000	0.080892	0.080892	0.000000
Mole frac (H_2O)	0.004637	0.898917	0.003534	0.873513	0.873513	0.066494	0.873835	0.873835	0.032506	0.898828	0.898828	0.001677
Mole frac (CO_2)	0.114511	0.001056	0.024696	0.028890	0.028890	0.043313	0.028884	0.028884	0.964379	0.001094	0.001094	0.995109
Mole frac (H_2S)	0.000085	0.000002	0.000001	0.000027	0.000027	0.000076	0.000027	0.000027	0.000876	0.000002	0.000002	0.000904
Mole frac (N_2)	0.005370	0.000000	0.005931	0.000001	0.000001	0.001919	0.000000	0.000000	0.000002	0.000000	0.000000	0.000003

(continued)

Table 10.10 (continued)

Name	DeH_2S Gas	Lean amine	Treated gas	Rich amine	To flash drum	Flas gas	To exchanger	Regen feed	Wet CO_2 gas	Regen bottoms	To tank	CO_2 captured
Mole frac (methane)	0.804916	0.000000	0.888158	0.000362	0.000362	0.776082	0.000053	0.000053	0.001823	0.000000	0.000000	0.001881
Mole frac (ethane)	0.054392	0.000000	0.059946	0.000044	0.000044	0.086941	0.000009	0.000009	0.000323	0.000000	0.000000	0.000334
Mole frac (propane)	0.016089	0.000000	0.017733	0.000013	0.000013	0.025171	0.000003	0.000003	0.000090	0.000000	0.000000	0.000093

Table 10.11 Overall performance summary—case study 2

Description	Value	Unit	Description	Value	Unit
Feed gas flow rate	5000	kgmole/h	Treated gas comp mole frac (CO_2)	0.0247	
Feed gas H_2S mole fraction	4.04%	%	Treated gas H_2S composition (ppm)	1.24	ppm
Feed gas CO_2 mole fraction	13.12%	%	CO_2 captured mass flow	19,058	kg/h
Feed gas comp mass flow (CO_2)	28,874	kg/h	CO_2 captured comp mass flow (CO_2)	19,013	kg/h
De-H_2S gas H_2S (ppm)	86.45	ppm	CO_2 captured comp mole frac (CO_2)	99.51	%
De-H_2S gas CO_2 (mole %)	11	%	CO_2 captured comp mole frac (H_2S)	0.09	%
De-H_2S gas comp mass flow (CO_2)	23,667	kg/h	Rich amine loading	0.3005	
Dry acid gas mass flow	12,076	kg/h	Amine strength	35	(w) %
Dry acid gas molar flow	321	kgmole/h	Lean amine std. vol. flow	360	m^3/h
Dry acid gas mole frac (CO_2)	36.83	%	CO_2 captured ratio (to De-H_2S gas)	80.33%	%
Dry acid gas mole frac (H_2S)	62.69	%	CO_2 captured ratio (to feed gas)	65.85%	%

Table 10.12 CO_2 removal section performance—case study 2

Description	Value	Unit	Description	Value	Unit
DeH_2S gas H_2S mole fraction	0.00008		Dry CO_2 gas comp mass flow (CO_2)	19,013	kg/h
DeH_2S gas CO_2 mole fraction	0.11451		Dry CO_2 gas mole frac (CO_2)	0.9951	
DeH_2S gas comp mass flow (CO_2)	23,668	kg/h	Dry CO_2 gas mole frac (H_2S)	0.0009	
DeH_2S gas flow rate	4696	kg/h	Rich amine loading	0.2977	
Treated gas CO_2	2.5	mole %	Amine strength	41	(w)%
Treated gas H_2S (ppm)	1.2	ppm	Lean amine std. vol. flow	400.0	m^3/h
Dry CO_2 gas mass flow	19,058	kg/h	CO_2 captured efficiency	80.3	%
Dry CO_2 gas molar flow	434.1	kgmole/h			

32% of H_2S. The CO_2 content will have significant impact on the H&M balance and main burner performance in the sulfur recovery unit; the CO_2 will be functioning as inert gas flow through the sulfur recovery unit and tail gas treatment unit, eventually vent to the atmosphere, causing the huge CO_2 emission. It is desirable to have H_2S content above 60%.

Acid gas enrichment can improve the operation of the sulfur recovery unit, but also captures about 70% of the CO_2, thus mitigates the CO_2 emission greatly. There are two options for the processing:

- modify the gas processing unit with a new solvent, or add a new solvent system, and keep the existing MDEA loop to remove the H_2S fully; the new solvent loop will capture the CO_2 in natural gas, the gas volumetric flow will be almost the same as the feed gas, full size of equipment will be required.
- Alternatively, remove the CO_2 content in the acid gas. Due to the smaller acid gas flow, the column and associated equipment will be smaller in size.

Case study 2 uses MDEA + piperazine solvent to enrich the acid gas H_2S and capture the CO_2, the desired results can be achieved; this case study uses DEPG physical solvent to enrich the acid gas H_2S and capture the CO_2. The H_2S can be absorbed faster than CO_2, so H_2S is separated at the first, the left-over is the CO_2 to be recovered.

Please note that the DEPG can be acted a dehydrator to remove the water, therefore, the regenerator may be needed to remove the absorbed water.

Reboiler temperature should not be higher than 175 °C to avoid DEPG solvent degradation.

10.4.3.2 Gas Composition and Product Spec.

Acid gas from the acid gas removal unit with a flow rate of 2230 kgmole/h, the acid gas contains 64.57% of CO_2 content and 32.28% of H_2S, the acid gas composition is shown in Table 10.13.

This acid gas contains too high concentration of CO_2, it will impact main burners operation and H&M balance results in the sulfur recover unit, the CO_2 will act as an inert gas flowing through the sulfur recovery unit and tail gas treatment unit, and eventually vent to atmosphere, causes a huge CO_2 emission, therefore, developing an acid gas enrichment process is desirable. The desired acid gas composition is H_2S to CO_2 ratio of around 60:40, the specification of H_2S in captured CO_2 gas depends on the usage of CO_2 gas, for storage or enhancement of oil/gas recovery, a less than 1% of H_2S content in CO_2 gas is acceptable.

Table 10.13 Sour feed gas composition of an acid gas removal unit

Molar flow	2230.0	Kgmole/h
Mass flow	89,179	kg/h
Molecular weight	39.99	
Composition		
H_2O (%)	3.14	mole %
CO_2 (%)	64.57	mole %
H_2S (%)	32.28	mole %

10.4.3.3 Process Description

For this case study, DEPG physical solvent is used for the acid gas enrichment, all the H_2S and partial CO_2 will be absorbed, the remaining gas will be the CO_2 rich gas, it can be collected as a CO_2 stream for CO_2 storage or enhancement of the oil and gas production.

The flow diagram of the acid gas enrichment using DEPG solvent process is show in Fig. 10.13.

The feed acid gas contains about 64.6% of CO_2 and 32.3% of H_2S, to capture the CO_2 in acid gas, an additional facility for CO_2 capture and H_2S enrichment will be installed. There is no C_2 + hydrocarbons in the acid gas, so we can use DEPG physical solvent for this case study without concern of ethane and propane loss. Due to the water content in feed acid gas, the DEPG physical solvent will acts as a dehydrant, therefore, a regenerator is required to remove the water in the solvent.

The higher pressure of absorption will be a benefit to physical solvent absorption, the pressure of acid gas from the main regenerator is about 2.5 bar, to improve the absorption efficiency, a two-stage acid gas compressor is required to boost the acid gas pressure to around 15 bar, the boosted acid gas enters to the absorber, the acid gas then flows upward, counter-current to the lean solvent which is introduced to the top of the acid gas enricher. The CO_2 gas from the top of the acid gas absorber contains 99.2% of CO_2, which can be directed to the CO_2 pipeline for use or storage.

The rich solvent is directed to a flash drum at reduced pressure, the flash gas is an acid gas, it can be directed to the bottom of the regenerator and plays the role of stripping gas.

The flashed rich solvent will be directed to a lean/rich amine heat exchanger, then enters the regenerator, the enriched acid gas will merge with the flashed gas and flow to the sulfur recovery unit. the regenerator bottom enters the lean/Rich amine heat exchanger to recover the stream heat, then the lean solvent is cooled down further via a cooler, and the cooled lean amine enters the top of the absorber, completing a solvent recycle loop.

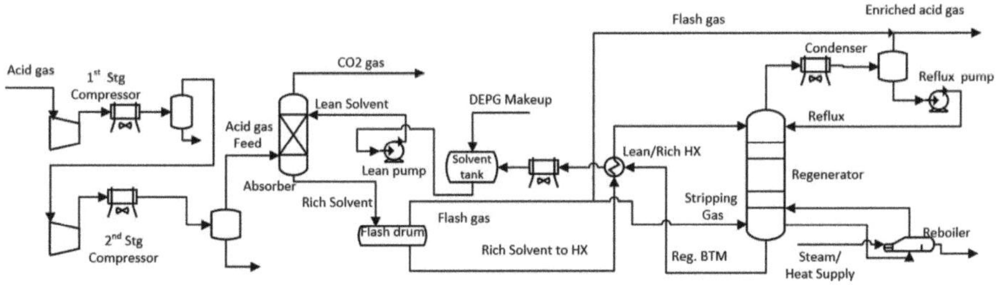

Fig. 10.13 Acid gas enrichment using DEPG solvent PFD

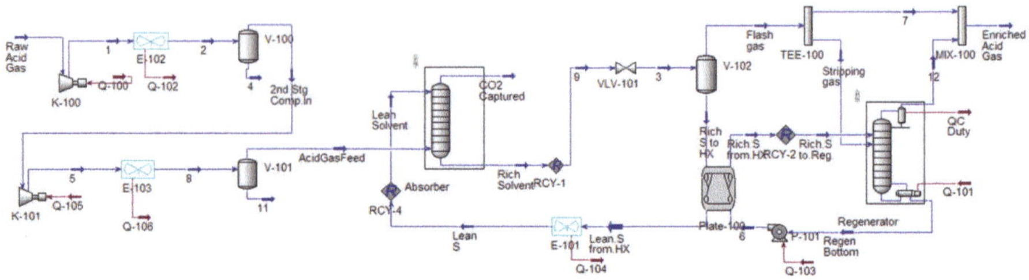

Fig. 10.14 Acid gas enrichment simulation sheet

10.4.3.4 Simulation and H&M Balance

The simulation model has been built to predict the unit performance; the acid gas enrichment simulation sheet using DEPG solvent is shown in Fig. 10.14.

Acid gas enrichment simulated stream data is summarized in Table 10.14.

10.4.3.5 Results and Recommendation

Per the simulation result, the DEPG solvent is capable of removing CO_2 from acid gas and enriching acid gas; the overall performance is summarized in Table 10.15.

From Table 10.15, at 321 m^3/h of lean DEPG recirculation rate, the process enriches the feed acid gas to 65.1% of H_2S and 34.1% of CO_2 from 32.3% of H_2S and 64.6% of CO_2, which is a great ratio to sulfur recovery unit feed gas. A 74.4% CO_2 in feed acid gas has been captured, the purity of the CO_2 captured reaches 98.6%; the total $H_2S + CO_2$ loading in the rich solvent is about 0.933 mol/mol, and the CO_2 loading and H_2S loading are 0.313 and 0.62 mol/mol respectively.

The captured CO_2 contains 1.39% H_2S, which is good for reservoir injection to enhance the oil/gas production; If the CO_2 stream is used for chemical utilization, additional H_2S polishing unit will be needed.

It is noted that a physical solvent to be selected for acid gas enrichment is for its higher selectivity of H_2S over CO_2, but the acid gas needs to be boosted for absorption under high pressure, due to the physical solvent acting as a dehydrant, a regenerator is also needed to remove the water in the lean solvent. This process is technically feasible, but may not be an economic option, using amine to enrich the acid gas, as case study 1 indicated is recommended.

Table 10.14 Acid gas enrichment simulated stream data

Description	Raw acid gas	2nd stg comp. in	Acid gas feed	Lean solvent	CO_2 captured	Rich solvent	Flash gas	Rich S to HX	Rich S to Reg	Stripping gas	Regen bottom	Lean S from HX	Enriched acid gas
Vapour fraction	1.0000	1.0000	1.0000	0.0000	1.0000	0.0000	1.0000	0.0000	0.1595	1.0000	0.0000	0.0000	1.0000
Temperature [°C]	35.0	35.0	35.0	30.0	40.4	52.4	36.0	36.0	129.0	36.0	136.0	37.3	32.1
Pressure [kPa]	180	570	1590	1630	1520	1540	180	180	130	180	130	1650	120
Molar flow [kgmole/h]	2230	2182	2168	1293	1086	2375	836	1539	1539	42	1293	1293	1082
Mass flow [kg/h]	89,179	88,316	88,061	336,625	47,648	377,038	31,879	345,159	345,159	1594	336,625	336,625	40,413
Master comp mole frac (DEPG)	0.0000	0.0000	0.0000	0.9230	0.0000	0.5025	0.0000	0.7752	0.7752	0.0000	0.9230	0.9230	0.0000
Master comp mole frac (CO_2)	0.6457	0.6598	0.6640	0.0032	0.9857	0.1571	0.4124	0.0185	0.0185	0.4124	0.0032	0.0032	0.3410
Master comp mole frac (H_2S)	0.3228	0.3298	0.3319	0.0276	0.0139	0.3117	0.5856	0.1631	0.1631	0.5856	0.0276	0.0276	0.6512
Master comp mole frac (H_2O)	0.0314	0.0103	0.0040	0.0461	0.0003	0.0287	0.0019	0.0432	0.0432	0.0019	0.0462	0.0462	0.0077

Table 10.15 Acid gas enrichment and CO_2 capture performance data

Description	Value	Unit
Acid gas molar flow	2230	kgmol/h
Acid gas mass flow	89,179	kg/h
Acid gas molar Frac H_2S	32.28	%
Acid gas molar Frac CO_2	64.57	%
Acid gas mass flow (CO_2)	63,369	kg/h
Acid gas mass flow (H_2S)	24,537	kg/h
CO_2 captured mass flow	44,244	kg/h
CO_2 captured mole frac (CO_2)	98.57	%
CO_2 captured mole frac (H_2S)	1.39	%
CO_2 captured mass flow (CO_2)	47,122	kg/h
Enrich AG mole frac (CO_2)	34.10	%
Enrich AG mole frac (H_2S)	65.12	%
Lean solvent liq. Std. vol. flow	321.3	m^3/h
Lean solvent water content	0.32	(w) %
Rich solvent CO_2 loading	0.313	kgmol/kgmol
Rich solvent H_2S loading	0.620	kgmol/kgmol
CO_2 captured ratio	74.36	%

10.5 Syngas CO_2 Capture

10.5.1 Syngas CO_2 Capture with Single Absorber Using DEPG Solvent—Case Study 4

10.5.1.1 Feed Gas Composition and Product Specification

A syngas acid gas removal and CO_2 capture unit are proposed to capture CO_2 for IGCC project, the syngas feed rate to CO_2 capture is 5000 kgmole/h at a pressure of 66 brag, and temperature of 35 °C, the syngas composition is shown in Table 10.16.

Due to the syngas containing a bulk of CO_2 with higher pressure, DEPG physical solvent will be the best solvent for this application, the syngas also contains minor H_2S, therefore using a single absorber process will achieve the required treated syngas specification (4 ppm H_2S, and CO_2 mole 2%).

10.5.1.2 Process Description

The sour syngas CO_2 capture using DEPG solvent process flow diagram is show in Fig. 10.15.

Table 10.16 Syngas composition

Composition	% (mole)
Hydrogen	52.975
Nitrogen	0.905
CO	0.801
CO_2	43.996
Methane	0.066
Argon	0.579
H_2S	0.667
COS	0.001
Ammonia	0.000
Oxygen	0.000
SO_2	0.000
H_2O	0.010
HCN	0.000

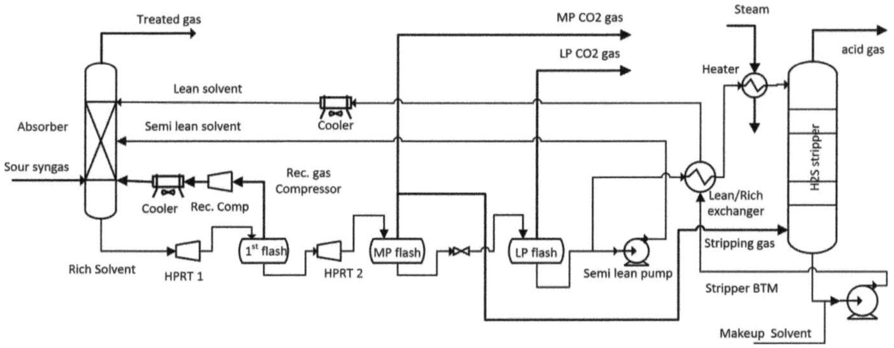

Fig. 10.15 Sour syngas CO_2 capture using DEPG solvent PFD

The sour syngas enters the bottom of the absorber column, then flow upward, counter-current to the lean/semi lean solvent which is introduced in one or more entrances around the top of the absorber. The treated gas meets the sweet syngas specification of CO_2 and H_2S. The rich solvent with the absorbed CO_2 or acid gas is directed to multiple flash drums at reduced pressures, before entering 1st flash drum, higher pressure rich solvent via an HPRT (hydraulic turbine) to recovery the liquid power, the pressure of 1st flash drum should set for fully flash of the main gas stream, flashed gas from 1st Flash drum will be compressed, cooled and sent back to the absorber, while the rich solvent flows via the second HPRT to recover the hydraulic power further then to MP flash drum and LP flash drum sequentially, MP flash drum operate at about 3–6 bar, the MP CO_2 gas can be directed to the second stage of CO_2 gas compressor to reduce the power demand of CO_2 gas compression

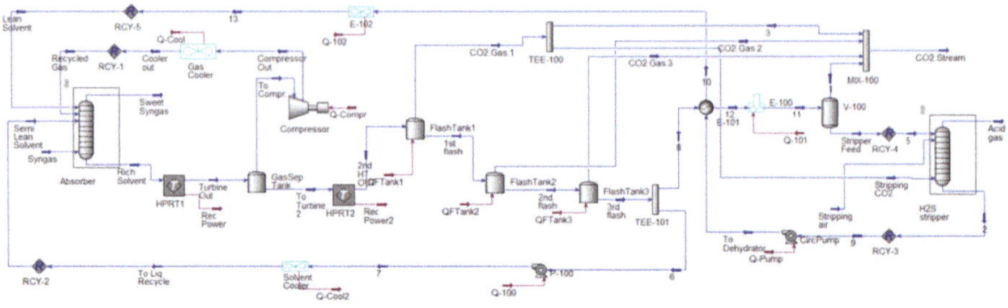

Fig. 10.16 Typical syngas CO_2 capture using DEPG solvent simulation sketch—single absorber

in downstream CO_2 compression unit. LP flash drum usually operates at 0.5 barg pressure to fully flash the CO_2 gas.

To enhance the CO_2 capture effect, about 20% of the flashed semi lean solvent has been split to the stripper, it is preferred to heat the flashed solvent to reduce the stripping gas demand. The semi lean solvent exchanges heat with the lean solvent from the stripper's bottoms, preheated up to about 90 °C, then enters the solvent stripper. The lean solvent leaving the lean/semi lean solvent heat exchanger, cooling down to the desired temperature via a cooler, then enters the top of the absorber. The remaining semi lean solvent (flashed) is pumped and sent to the absorber at the location below lean solvent entrance.

This case study considers an H_2S stripper instead of re-boiling regeneration, which has the advantage of lower energy demand. Regenerator with reboiler is also suitable for DEPG solvent, however, to keep it in mind, DEPG is suitable to operate up to 175 °C, degradation could happen at higher a temperature, sometimes the hybrid option (reboiler and stripping) man be considered.

10.5.1.3 Simulation and H&M Balance

Based on Fig. 10.15 configuration, a single absorber simulation model has been developed, the simulation sheet is shown in Fig. 10.16.

Per process simulation results, the main stream information is summarized in Table 10.17, it indicates the preliminary H&M balance for this CO_2 capture case study.

10.5.1.4 Results and Conclusion

Per the simulation result, the DEPG physical solvent is capable for capturing the CO_2 from Syngas, the results of overall unit performance by simulation are summarized in Table 10.18.

From Table 10.18, the treated syngas CO_2 content is about 1.45%, 95.8 t/h of CO_2 are captured from syngas, Efficiency in CO_2 capture achieves 97.6% with CO_2 purity

Table 10.17 Case study 4 key simulation stream

Name	Syngas	Recycled gas	Lean solvent	Semi lean solvent	Rich solvent	Sweet syngas	Turbine out	To compr.	To Turbine 2	Compressor out	CO_2 stream	Acid gas	Stripper feed	Stripping air
Vapour fraction	1.000	1.000	0.000	0.000	0.000	1.000	0.009	1.000	0.000	1.000	1.000	1.000	0.000	1.000
Temperature [C]	35.0	20.1	30.0	7.2	26.5	30.5	26.4	26.4	26.4	58.7	13.4	89.6	90.0	25.0
Pressure [kPa]	6600.0	6620.9	6948.9	6895.0	6600.0	6560.0	4911.0	4911.0	4911.0	6620.9	140.0	140.0	140.0	400.0
Molar flow [kgmole/h]	5000.0	50.6	673.1	2909.6	5893.4	2739.9	5893.4	50.6	5842.8	50.6	2242.8	127.9	687.6	110.8
Mass flow [kg/h]	106,900	933	187,352	755,767	1,040,592	10,361	1,040,592	933	1,039,659	933	95,821	3928	187,366	3200
Liquid volume flow [m^3/h]			178.4	722.4	1024.5		1024.5		1021.8				178.6	
Composition, % (mole)														
DEPG	0.000%	0.000%	99.365%	91.520%	56.532%	0.000%	56.532%	0.000%	57.022%	0.000%	0.001%	0.047%	96.900%	0.000%
H_2O	0.010%	0.001%	0.496%	0.509%	0.316%	0.000%	0.316%	0.001%	0.319%	0.001%	0.008%	0.217%	0.526%	0.000%
CO	0.801%	1.116%	0.000%	0.000%	0.057%	1.361%	0.057%	1.116%	0.048%	1.116%	0.124%	0.000%	0.000%	0.000%
CO_2	43.996%	36.761%	0.000%	6.904%	40.375%	1.453%	40.375%	36.761%	40.406%	36.761%	95.719%	10.198%	1.896%	0.000%
H_2S	0.667%	0.179%	0.001%	1.067%	1.094%	0.000%	1.094%	0.179%	1.102%	0.179%	1.273%	3.636%	0.677%	0.000%
Hydrogen	52.975%	59.836%	0.000%	0.000%	1.523%	94.504%	1.523%	59.836%	1.018%	59.836%	2.651%	0.000%	0.000%	0.000%
Nitrogen	0.905%	1.179%	0.131%	0.000%	0.046%	1.607%	0.046%	1.179%	0.036%	1.179%	0.095%	67.745%	0.000%	79.000%
Methane	0.066%	0.102%	0.000%	0.000%	0.010%	0.100%	0.010%	0.102%	0.009%	0.102%	0.024%	0.000%	0.000%	0.000%

Table 10.18 Case study 4 simulated overall performance

Description	Value	Unit	Description	Value	Unit
Sour syngas flow	5000	kgmole/h	CO_2 stream (H_2S)	1.27	mole %
Sour syngas H_2S	6668	ppm	CO_2 capture efficiency	97.59	mole %
Sour syngas CO_2	44.00	mole %	H_2S lean loading	0.012	mole/mole
Sweet gas H_2S	3.93	ppm	CO_2 lean loading	0.075	mole/mole
Sweet gas CO_2	1.45	mole %	Rich S H_2S loading	0.019	mole/mole
Sweet syngas pressure	6560	kPa	Rich S CO_2 loading	0.714	mole/mole
CO_2 stream mass flow	95,821	kg/h	Semi S. recirc. std. flow	712.9	m^3/h
CO_2 stream (CO_2)	95.72	mole %	Lean solvent std. flow	176.5	m^3/h

of 95.7% (mole). Rich solvent CO_2 loading achieved is 0.714 mol CO_2 gas/mole DEPG. Small amount of acid gas from the stripper also contains CO_2 and H_2S also, it needs to be disposed properly. It is noted that the H_2S content in CO_2 stream is 1.27%, it is good for CO_2 storage and enhancing oil & gas recovery, in case of lower H_2S (<4ppm), a dual-absorber configuration process will be needed.

10.5.2 *Syngas CO_2 Capture with Dual Absorber Using DEPG Solvent—Case Study 5*

10.5.2.1 Feed Gas Composition and Product Specification

A syngas acid gas removal and CO_2 capture unit are proposed to capture CO_2 for an IGCC project, it is different from Case study 4, the syngas feed gas contains 2.55% of H_2S and 44.47% of CO_2, processing rate is 5000 kgmole/h at a pressure of 66 barg, and temperature of 35 °C, the syngas composition is shown in Table 10.19.

Considering the high pressure and Bulk of CO_2 content in the syngas, DEPG physical solvent will be the best solvent for this application, the syngas also contains H_2S, therefore a dual absorber process is selected to achieve the required treated syngas specification (4 ppm H_2S, and CO_2 mole 2%).

10.5.2.2 Process Description

The sour syngas CO_2 capture with dual absorbers using DEPG solvent process flow diagram is show as Fig. 10.17.

The dual absorber process provides the capability to remove the H_2S and CO_2 sequentially.

The sour syngas is sent to a liquid KO drum to remove entrained liquid or particles if any and then fed to the bottom of the H_2S absorber column. The syngas then flows

Table 10.19 Feed syngas composition

Composition	mole %
Hydrogen	50.50
Nitrogen	0.92
CO	0.81
CO_2	44.47
Methane	0.07
Argon	0.59
H_2S	2.55
COS	0.00
Oxygen	0.00
SO_2	0.00
H^2O	0.10
HCN	0.00

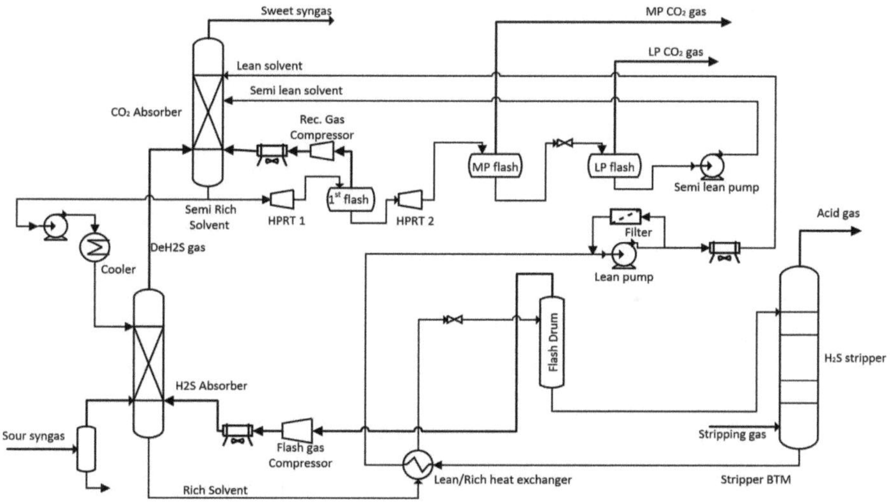

Fig. 10.17 Sour syngas CO_2 capture with dual absorbers

upward, counter-current to the semi-rich solvent solution which is introduced to the top of the absorber. De-H_2S gas from the H_2S absorber is directed to the CO_2 absorber to capture the CO_2 fully, and then flows upward, counter-current to the semi lean and lean solvent sequentially, the treated gas from the CO_2 absorber meets the sweet gas specification of CO_2 and H_2S, and exits the top of the column to downstream facilities or grid as a sales product. The semi-rich solvent with the absorbed CO_2 or acid gas will be split (20–30%) and sent to the H_2S absorber, the remaining semi-rich solvent is directed to multiple flash drums with reduced pressures, before entering 1st flash drum, the higher pressure semi-rich solvent via an HPRT 1 (hydraulic turbine) to recovery the liquid power, the pressure of 1st flash drum should set for fully flash of the main gas stream (CH_4 or H_2), flashed gas from 1st Flash drum will be compressed, cooled and sent back to CO_2 absorber, and then the liquid solvent will flow through a second HPRT 2 to recover the solvent power further, then sent to MP flash drum and LP flash drum sequentially, MP flash drum operate at about 3–6 bar, the MP CO_2 gas can be directed to the second stage of CO_2 gas compressor to reduce the power demand of CO_2 gas compression in downstream CO_2 compression unit. LP flash drum usually operates at 0.5 barg pressure to fully flash the CO_2 gas. The flashed semi lean solvent then is pumped back to the CO_2 absorber at the location below the lean solvent entrance.

The rich solvent from the H_2S absorber is directed to the Lean/rich solvent heat exchanger, then to the H_2S stripper, the lean solvent from the stripper bottom exchanges heat with rich solvent, then via cooler to cool down to the desired temperature by a cooler and enters the top of the absorber.

10.5.2.3 H&M Balance for Typical Application

A simulation model for the sour syngas CO_2 capture with dual absorbers has been developed; the simulation sheet is shown in Fig. 10.18.

This case study selects H_2S stripper instead of re-boiling regeneration, the stripping gas is MP CO_2 gas; which has the advantage of lower energy demand. Regenerator with reboiler is also suitable for DEPG solvent, however, to keep it in mind. DEPG is suitable to operate up to 175 °C, degradation could happen at higher temperature, sometimes the hybrid option (reboiler and stripping) may be an option.

Per process simulation results, the following information of the mainstream is summarized in Table 10.20, it indicates the preliminary H&M balance for this CO_2 capture and H_2S removal case study.

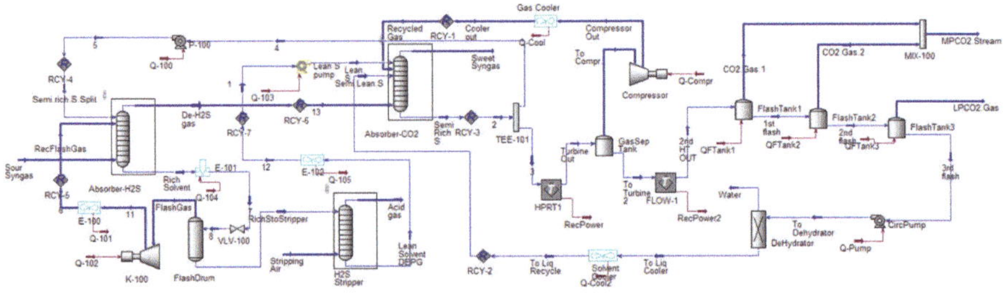

Fig. 10.18 Syngas gas CO_2 capture with dual absorbers simulation sheet

Table 10.20 Case study 5 key stream data

Name	Sour syngas	De-H_2S gas	Sweet syngas	Semi lean S	Lean S	Rich solvent	Rec flash gas	Rich S to stripper	Acid gas	Semi rich S split	Turbine out	Recycled gas	$LPCO_2$ gas	$MPCO_2$ stream	Stripping air
Vapour fraction	1.000	1.000	1.000	0.000	0.000	0.000	1.000	0.000	1.000	0.000	0.010	1.000	1.000	1.000	1.000
Temperature [°C]	35.0	36.1	37.3	7.2	37.1	40.8	40.0	77.9	71.5	29.5	29.2	20.1	13.4	16.4	40.0
Pressure [kPa]	6600	6560	6460	6895	6800	6600	6680	1379	115	6800	4811	6521	150	450	350
Molar flow [kgmole/h]	5000	5358	2626	3052	807	1987	884	1103	427	1461	5181	50	400	1673	131
Mass flow [kg/h]	111,022	127,238	10,031	804,345	224,307	272,827	36,721	236,104	15,571	252,322	894,586	1056	17,616	71,513	3800
Liquid volume flow [m^3/h]				768.1	213.5	276.8		229.4		248.8	881.9				
Molecular weight	22.2	23.7	3.8	263.6	278.0	137.3	41.5	214.1	36.4	172.7	172.7	20.9	44.0	42.7	28.9
Composition, mole frac															
DEPG	0.000000	0.000000	0.000000	0.930331	0.992166	0.402963	0.000025	0.726055	0.000122	0.548058	0.548051	0.000000	0.000000	0.000000	0.000000
H_2O	0.001018	0.000003	0.000009	0.000014	0.006807	0.003271	0.000235	0.005705	0.001872	0.000832	0.000832	0.000002	0.000019	0.000007	0.000000
CO	0.008098	0.007711	0.014495	0.000000	0.000000	0.000398	0.000876	0.000014	0.000037	0.000574	0.000573	0.010987	0.000007	0.001442	0.000000
CO_2	0.444684	0.501606	0.013703	0.068915	0.000000	0.454936	0.830028	0.154186	0.397935	0.434145	0.434139	0.428217	0.998313	0.966621	0.000000
H_2S	0.025506	0.000590	0.000000	0.000737	0.000001	0.126543	0.142412	0.113803	0.293709	0.000805	0.000818	0.000151	0.001604	0.000600	0.000000
Hydrogen	0.505013	0.475181	0.943289	0.000000	0.000000	0.011134	0.024778	0.000194	0.000502	0.014545	0.014544	0.539946	0.000021	0.028739	0.000000
Nitrogen	0.009151	0.008665	0.017022	0.000000	0.000980	0.000336	0.000742	0.000009	0.240297	0.000468	0.000468	0.011488	0.000003	0.001100	0.787000
Methane	0.000665	0.000648	0.001083	0.000000	0.000000	0.000062	0.000135	0.000004	0.000009	0.000102	0.000102	0.001041	0.000007	0.000283	0.000000

Table 10.21 Case study 5 simulated overall performance

Description	Value	Unit	Description	Value	Unit
Sour syngas molar flow	5000	kgmole/h	Acid gas mass flow	15,571	kg/h
Sour syngas (CO_2)	44.5	mole %	Acid gas (CO_2)	39.8	mole %
Sour syngas (H_2S)	2.6	mole %	Acid gas (H_2S)	29.4	mole %
Sweet gas H_2S	0.3071	ppm	Semi lean H_2S loading	0.0008	mole/mole
Sweet gas CO_2	1.37	mole %	Semi lean CO_2 loading	0.0741	mole/mole
LP CO_2 gas mass flow	17,616	kg/h	Rich S CO_2 loading	1.1290	mole/mole
LP CO_2 gas (CO_2%)	99.8	mole %	Rich S H_2S loading	0.00000	mole/mole
MP CO_2 gas mass flow	71,513	kg/h	Semi lean recirc. std. flow	758.6	m^3/h
MP CO_2 gas (CO_2%)	96.7	mole %	Lean recirc. std. flow	211.3	m^3/h
Syngas comp. mass flow (CO_2)	97,852	kg/h	MP CO_2 comp mass flow (CO_2)	71,174	kg/h
LP CO_2 comp. mass flow (CO_2)	17,593	kg/h	CO_2 Capture efficiency	90.7	%

10.5.2.4 Discussion and Recommendations

Per the simulation result, the DEPG physical solvent is capable of capturing the CO_2 from Syngas, the overall unit performance summary is shown in Table 10.21.

From Table 10.24, the CO_2 content of 1.23% and H_2S content of 0.7 ppm in the treated syngas have been achieved, 95.8 t/h of CO_2 gas has been captured from feed syngas, CO_2 capture efficiency achieves 97.6% with CO_2 purity of 96.7% (MP CO_2 gas) and 9.8% (LP CO_2 gas). Rich solvent CO_2 loading is 1.129 mol CO_2 gas/mole DEPG.

A small amount of acid gas from the stripper contains CO_2 and H_2S also, it needs to be disposed of properly.

10.6 Flue Gas CO_2 Capture

10.6.1 Flue Gas CO_2 Capture Using DEA Solvent—Case Study 6

10.6.1.1 Gas Composition and Product Spec.

An fuel fired heater of a refinery discharges 1000 kgmole/h of flue gas, the flue gas currently is discharged to the atmosphere at 160 °C, to avoid paying a high carbon tax, protect the environment, it has been decided to install a CO_2 capture trail unit to capture CO_2 from the. Flue gas feed to the CO_2 capture unit is shown in Table 10.22.

Table 10.22 Flue gas composition

Composition	Value (%)
Nitrogen	78.32
H_2O	8.03
CO	0.00
CO_2	12.55
Oxygen	1.10
H_2S	0.00

The captured CO_2 gas should have then CO_2 purity of 99%, and CO_2 capture efficiency should be 92% or above. The nitrogen recovered can be used in a plant utility system, it requires a nitrogen purity of 96% or above.

10.6.1.2 Process Description

Flue gas has low CO_2 content and low pressure, and therefore, the chemical absorption method is considered suitable for separating and recovering CO_2 from flue gas. This CO_2 Capture unit will use DEA chemical solvent, the proposed process flow sheet is shown in Fig. 10.19.

The flue gas from fired heaters, boilers, or process furnaces usually contains about 11–15% of CO_2, the majority of flue gas is nitrogen, to capture the CO_2 in flue gas and improve the absorption efficiency, a two-stage acid gas compressor is required to boost the flue gas pressure to absorber operating pressure.

The flue gas from a fired heater at 160 °C enters to flue gas cooler to cool down to 35 °C, then is directed to a flue gas K.O. drum to separate the water vapor condensate, and enters a 2-stage flue gas compressor to be boosted the flue gas pressure to 8 barg, and fed to the bottom of the absorber column. The absorber can be a tray or packed tower, the feed gas then flows upward, counter-current to the lean amines which are introduced to the top of the absorber. The treated flue gas venting at the top of the column is a nitrogen stream with a purity of 96%, it can be used as utility inert gas in

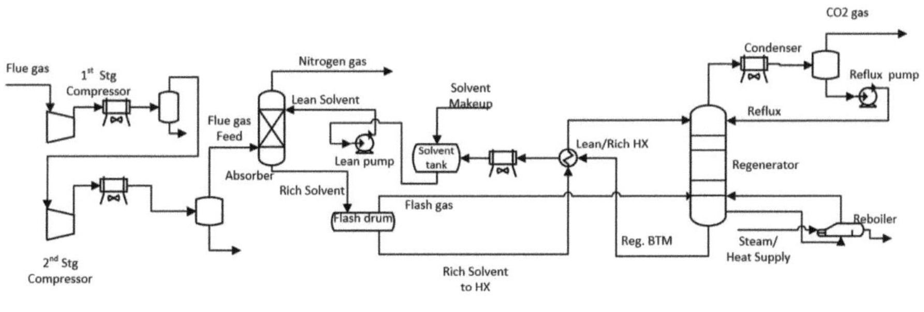

Fig. 10.19 Flue gas CO_2 capture using DEA solvent process flow sheet

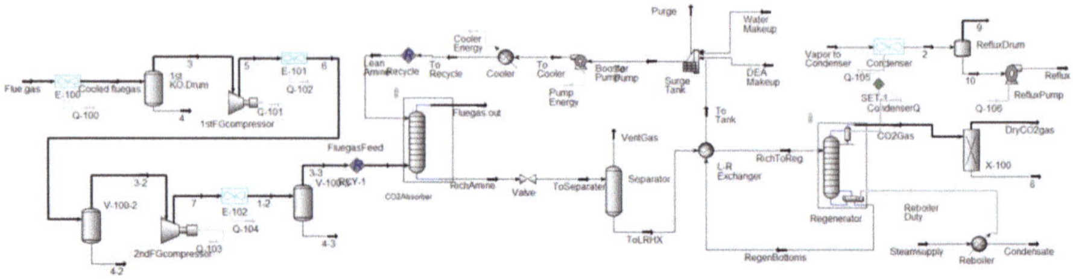

Fig. 10.20 Typical flue gas CO_2 capture using DEA Amine simulation sketch

this plant or vent to the atmosphere. The rich solvent is directed to a flash drum and then to the heat exchanger between lean and rich solvent to recover rich solvent heat, reduce the lean solvent temperature, the heated rich solvent is then sent a regenerator to be regenerated using heating, usually steam. The CO_2 gas from the regenerator is routed to the CO_2 gas compression and dehydration unit.

The regenerator bottom stream enters the lean/Rich amine heat exchanger to recover the stream heat, then the lean solvent is cooled down further via a cooler, cooled lean amine enters the top of the absorber, thus completing a solvent recycle loop.

Flash gas from the flash drum is mainly a nitrogen component, it can be vented to the atmosphere directly.

10.6.1.3 Simulation and H&M Balance

The simulation model has been built to predict the unit performance; The flue gas CO_2 capture simulation sheet using DEPG solvent is shown in Fig. 10.20.

Per process simulation results, the mainstream information is summarized in Table 10.23, it indicates the preliminary H&M balance for this CO_2 capture case study.

10.6.1.4 Results and Recommendation

Per the simulation result, the DEA chemical solvent is capable of capturing the flue gas CO_2, the overall unit performance is summarized in Table 10.24.

From Table 10.18, CO_2 content of the treated flue gas (nitrogen gas) is about 0.9%, the nitrogen purity of treated flue gas is 96.8%, it can be used as inert gas for plant utility system; CO_2 capture efficiency achieves 94% with CO_2 Purity of 99.5%. this CO_2 gas can be either used for chemical byproduct production or injected to the reservoir for enhancing oil recovery or storage. Rich amine loading achieved is 0.49 mol CO_2 gas/mole amine.

Flash gas from flash drum is mainly a nitrogen component, it can be vented to atmosphere directly without impact to environment.

Table 10.23 Simulation summary of key stream data

Name	Flue gas	Cooled flue gas	Flue gas feed	Lean amine	Flue gas out	Rich amine	To separater	Vent gas	To LR HX	Rich to reg.	CO_2 gas	Dry CO_2 gas	Regen Bottoms	Steam supply
Vapour fraction	1.000	0.973	1.000	0.000	1.000	0.000	0.000	1.000	0.000	0.000	1.000	1.000	0.000	1.000
Temperature [°C]	160	35	35	40	41	59	59	59	59	93	43	43	116	180
Pressure [kPa]	105	103	890	3618	850	890	483	483	483	448	150	150	180	1000
Mass flow [kg/h]	29,262	29,262	27,927	124,666	22,739	129,855	129,855	5	129,850	129,850	5337	5209	124,513	11,899
Liquid flow [m_3/h]	50.7	50.7	49.4	122.3	43.0	128.6	128.6	0.0	128.6	128.6	6.5	6.3	122.1	11.9
Molecular weight	29.3	29.3	30.2	22.9	28.1	23.3	23.3	28.6	23.3	23.3	42.4	43.9	22.9	18.0
Composition, mole frac														
H_2O	0.080326	0.080326	0.006788	0.940592	0.009021	0.920562	0.920562	0.037574	0.920588	0.920588	0.059485	0.003152	0.940501	1.000000
CO	0.000030	0.000030	0.000032	0.000000	0.000037	0.000000	0.000000	0.000039	0.000000	0.000000	0.000000	0.000000	0.000000	0.000000
CO_2	0.125509	0.125509	0.135532	0.004988	0.009171	0.026102	0.026102	0.058173	0.026101	0.026101	0.938849	0.995082	0.004994	0.000000
H_2S	0.000000	0.000000	0.000000	0.000000	0.000000	0.000000	0.000000	0.000000	0.000000	0.000000	0.000000	0.000000	0.000000	0.000000
Nitrogen	0.783176	0.783176	0.845812	0.000000	0.968226	0.000063	0.000063	0.888905	0.000037	0.000037	0.001616	0.001713	0.000000	0.000000
Oxygen	0.010960	0.010960	0.011836	0.000000	0.013544	0.000002	0.000002	0.015309	0.000001	0.000001	0.000050	0.000053	0.000000	0.000000
DEAmine	0.000000	0.000000	0.000000	0.054420	0.000000	0.053271	0.053271	0.000000	0.053273	0.053273	0.000000	0.000000	0.054505	0.000000

10.6.2 *Flue Gas CO_2 Capture Using Oxygen Instead of Combustion Air—Case Study 7*

Fossil fuel's final combustion products are CO_2 and H_2O, CO produced is based on the burning conditions, such as excess air coefficient and flame temperature. For the combustion air oxidant, combustion gas contains a majority of nitrogen, the nitrogen, and impurity of air dilutes the flue gas significantly, as a result, the CO_2 concentration is usually below 16% (V), if oxygen is used instead of air as oxidant, about 95% of CO_2 concentration can be achieved depending on the oxygen stream purity.

10.6.2.1 Oxygen Source and Study Basis

Using oxygen instead of air as a combustion oxidant will be a costly option, due to separating air into N_2 and O_2 requires additional capital and operating costs. It will have excess new energy consumption, and also produce new CO_2 stream.

However, many oil and gas facilities, LNG, and oil refineries are equipped with nitrogen unit to produce nitrogen for purging and blanketing use, some nitrogen unit is already using air separation unit, the higher purity of oxygen will be the byproduct of nitrogen unit. Instead of venting the oxygen stream, using oxygen for fossil fuel combustion replacing combustion air will be an attractive option for CO_2 capture and operating cost saving.

This case study uses a 30 t/h MP steam boiler (40 bar MP steam) as an example, boiler feed water (BFW) temperature is 120 °C, BFW is supplied by recovered condensate and the oxygen removed demineralized water makeup. Oxygen is taken from the nitrogen unit (air separation) with 99.5% of oxygen purity and 0.5% of nitrogen. fuel gas composition as shown in the H&M balance table.

10.6.2.2 Process Description

Flue gas CO_2 Capture using oxygen as oxidant process flow sheet is shown in Fig. 10.21.

The oxygen stream from the air separation unit enters the boiler combustion chamber with fuel gas supplied, the excess oxygen is controlled at 0.8% at the outlet of combustion gas; to control the flame temperature below 1300 °C, the cooled flue gas from the economizer is partially recycled back to the combustion chamber. The combustion gas then enters to boiler radiation and convection chamber to generate MP steam at 40 barg pressure, then the flue gas is cooled in a flue gas cooler to 35 °C. To condense the water vapor generated in combustion, the cooled flue gas enters a plate frame heat exchanger. A chilled water at 10 °C as coolant to chill down the flue gas to 12 °C. then enters to a water separator to separate the combustion water, the dry flue gas contains 95% of CO_2 then be directed CO_2 compression and dehydration outside the boiler system.

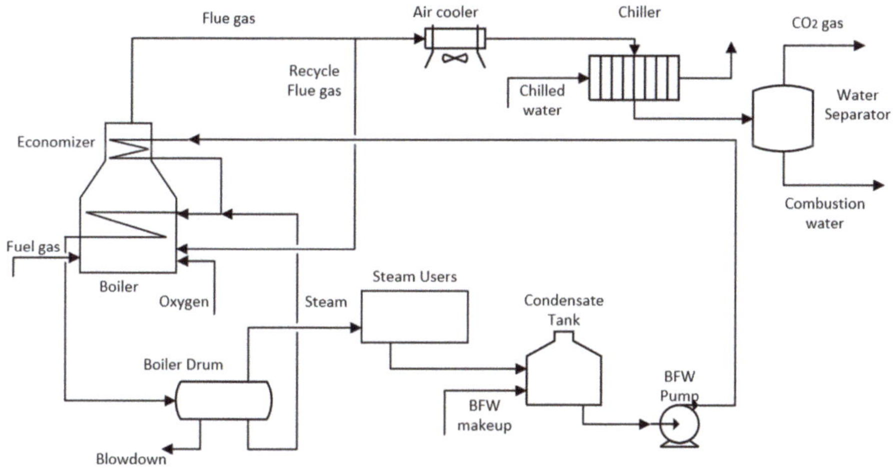

Fig. 10.21 Flue gas CO_2 Capture using oxygen as oxidant process flow sheet

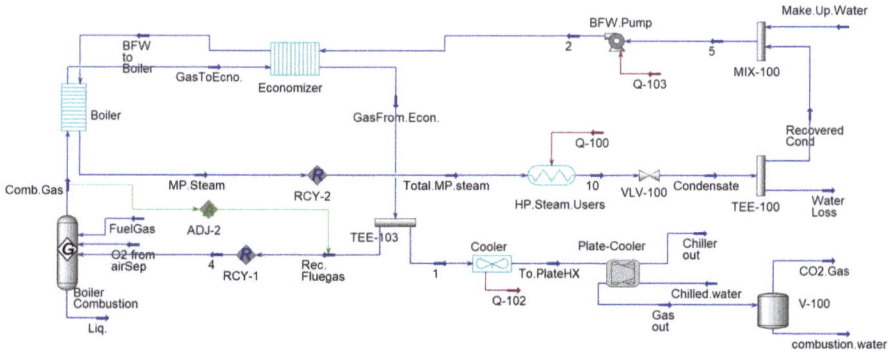

Fig. 10.22 Flue gas CO_2 capture using O_2 as oxidant

10.6.2.3 Simulation and H&M Balance

A typical simulation model has been developed using Aspen HYSYS commercial simulator, the simulation sheet is as shown in Fig. 10.22.

Per process simulation results, the main stream information is summarized in Table 10.25, it indicates the preliminary H&M balance for this CO_2 capture case study.

10.6.2.4 Results and Conclusion

Per simulation results, compared with the air oxidant configuration, oxygen oxidant configuration efficiency increases by 5% at the same flue gas temperature of 169 °C,

Table 10.25 Key stream data of flue gas CO_2 capture using oxygen as oxidant

Name	Fuel gas	O_2 from air sep.	Rec. flue gas	Comb. gas	Gas to econ	Gas from econ	BFW to boiler	MP steam	Condensate	Water loss	combustionwater	To plate HX	Chilled water	CO_2 gas
Vapour fraction	1.000	1.000	1.000	1.000	1.000	1.000	0.000	1.000	0.000	0.000	0.000	0.383	0.000	1.000
Temperature [°C]	20	20	169	1287	310	169	177	253	121	121	12	35	10	12
Pressure [kPa]	106	301	106	106	106	106	4201	4101	900	900	104	105	350	104
Mass flow [kg/h]	1500	5892	31,267	38,659	38,659	38,659	30,148	30,148	30,148	603	3095	7392	25,000	4297
Actual liquid flow [m^3/h]	0.0	0.0	0.0	0.0	0.0	0.0	34.2	0.0	32.4	0.6	3.0	3.0	24.5	0.0
Molecular weight	19.2	32.0	27.3	27.3	27.3	27.3	18.0	18.0	18.0	18.0	18.0	27.3	18.0	43.2
H_2O	0.000004	0.000000	0.637609	0.637609	0.637609	0.637609	1.000000	1.000000	1.000000	1.000000	0.999337	0.637609	1.000000	0.013403
CO	0.000000	0.000000	0.000053	0.000053	0.000053	0.000053	0.000000	0.000000	0.000000	0.000000	0.000000	0.000053	0.000000	0.000144
CO_2	0.000000	0.000000	0.350243	0.350243	0.350243	0.350243	0.000000	0.000000	0.000000	0.000000	0.000662	0.350243	0.000000	0.953488
H_2S	0.000004	0.000000	0.000001	0.000001	0.000001	0.000001	0.000000	0.000000	0.000000	0.000000	0.000000	0.000001	0.000000	0.000003
Nitrogen	0.005063	0.005051	0.004895	0.004895	0.004895	0.004895	0.000000	0.000000	0.000000	0.000000	0.000000	0.004895	0.000000	0.013341
Oxygen	0.000000	0.994949	0.007199	0.007199	0.007199	0.007199	0.000000	0.000000	0.000000	0.000000	0.000000	0.007199	0.000000	0.019621
Methane	0.812468	0.000000	0.000000	0.000000	0.000000	0.000000	0.000000	0.000000	0.000000	0.000000	0.000000	0.000000	0.000000	0.000000
Ethane	0.149351	0.000000	0.000000	0.000000	0.000000	0.000000	0.000000	0.000000	0.000000	0.000000	0.000000	0.000000	0.000000	0.000000
Propane	0.030883	0.000000	0.000000	0.000000	0.000000	0.000000	0.000000	0.000000	0.000000	0.000000	0.000000	0.000000	0.000000	0.000000

this is because that the oxygen continent in air oxidant is only about 21%, the nitrogen as an inert gas flows the boiler system, as a result, flue gas rate of air oxidant configuration will be 3.7 times higher than oxygen oxidant configuration, rejecting more fuel gas combustion heat. The impurities in the CO_2 gas are mainly the excess oxygen and nitrogen impurity in oxygen feed.

Oxygen oxidant configuration can achieve a fully CO_2 capture of 95% of CO_2 concentration without a CO_2 capture unit to be installed, only quite few pieces of equipment are needed, the economic results are highly dependent on the existing utility system configuration, installing a new air separation unit will not be considered due to the economic reason.

With oxygen oxidant, the adiabatic combustion temperature will be extremely high, it could reach at above 2600 °C, at current technical–economic conditions, the equipment and system are not suitable for this condition, two options could be considered: one option is partially recycling back the cooled flue gas to the combustion chamber, which can control the flame temperature as desired; alternatively, combustion chamber must be equipped with heat removal system to control the chamber temperature below 1300 °C, otherwise, significant amount of CO will be generated which lower the efficiency and pollute the environment.

The oxygen oxidant configuration can be used in fired heaters; it is expected to have similar results.

To implement the oxygen oxidant combustion CO_2 capture, further study and detailed engineering will be needed.

10.7 CO_2 Compression and Dehydration

10.7.1 Background Information

Except for a small amount of CO_2 demand for chemical and food processing use, the majority of captured CO_2 will be compressed and dehydrated for oil production enhancement or storage, then injected to underground (reservoir), the pressure required is depending on the geological and geographical conditions and the pressure of the reservoir, most of the cases require a pressure of above 150 barg. The specification of CO_2 gas stream usually requires water content less than 100 ppm to prevent Hydrate formation.

No matter using a chemical solvent or physical solvent, the CO_2 stream captured is usually at a very lower pressure, around 0.5 barg pressure. A multiple-stage compressor is required to increase CO_2 stream pressure.

Because DEPG physical solvent itself is capable of dehydration, the CO_2 captured using DEPG physical solvent usually already met the CO_2 gas specification for water content, in case of more strict water content spec applies, an additional polishing drier may be needed.

Triethylene glycol (TEG) is the most common glycol used in these absorption systems. TEG dehydration is the most common dehydration process that can be used to meet CO_2 gas specifications.

More important, the TEG decomposition (degradation) temperature is 206.7 °C, considering a 2 °C margin, the reboiler temperature should not be higher than 204.7 °C, for most cases, the hybrid option that is the combination of stripping gas and reboilering will be attractive.

10.7.2 Gas Stripping in Dehydration Regenerator

The rich TEG solution is normally regenerated at low pressure and high temperature. In an atmospheric regenerator operating at a decomposition temperature of 206.7 °C, the maximum concentrations achievable is 98.5–99 (w) %.

The dew point of the corresponding dry gas is above -10 °C. Generally, the common TEG dehydration unit can meet the CO_2 gas Specification, To meet the dew point specification If the required lean glycol concentration at the absorber is higher than the above maximum concentrations, further increase of the glycol concentration at the regenerator is under consideration. All of the approaches for lowering the partial pressure of water in the glycol solution either by making a vacuum on the regenerator or by using stripping gas into the regenerator, will enhance the dehydration significantly.

Any inert gas, fuel gas, the dehydrated CO_2 gas are suitable for stripping gas, the economic tradeoff is needed in stripping gas selection; fuel gas is an effective gas; however, the stripper off-gas will contain the flammable fuel gas, which could not be vented directly; nitrogen is an ideal stripping gas but with high product cost, the flash gas from rich solvent flash drum could be used for stripping gas, it is very convenient but increases the CO_2 emission slightly.

The quantity of stripping gas required is small, the stripping gas can be injected directly into the reboiler. Adding gas to a packed unit between the reboiler and surge tank could result in a lower stripping gas rate [7].

10.7.3 CO_2 Dehydration and Compression—Case Study 8

10.7.3.1 Case Study Basis

The CO_2 captured from Syngas has LP CO_2 gas stream and MP CO_2 gas stream, the composition and operating condition of two feed CO_2 gases are shown in Table 10.26.

The CO_2 stream should be delivered in liquid phase, under pressure of 85 bar at plant battery of limit. The water content in CO_2 stream should be less 30 ppm.

Table 10.26 Two feed CO_2 gases composition and operating condition

Description	LP CO_2 gas	MP CO_2 gas
Vapor fraction	1.0000	1.0000
Temperature [°C]	35	35
Pressure [kPa]	150	450
Molar flow [kgmole/h]	411.5	1407.4
Mass flow [kg/h]	18,058.8	60,498.0
Molecular weight	43.9	43.0
Mole frac (CO_2)	0.9875	0.9696
Mole frac (CO)	0.0000	0.0012
Mole frac (H_2S)	0.0124	0.0050
Mole frac (hydrogen)	0.0000	0.0222
Mole frac (H_2O)	0.0000	0.0000
Mole frac (nitrogen)	0.0000	0.0009
Mole frac (argon)	0.0000	0.0010
Mole frac (oxygen)	0.0000	0.0000
Mole frac (methane)	0.0000	0.0002

10.7.3.2 CO_2 Compression and Dehydration Process

A typical CO_2 compression and dehydration process flow sketch is shown in Fig. 10.23.

LP CO_2 gas from CO_2 capture unit is fed to inlet K.O. drum then enters to 1st stage compressor to booster the CO_2 gas pressure from 150 to 450 kPa, after cooling and liquid separation, the CO_2 gas stream enters second stage compressor, in case

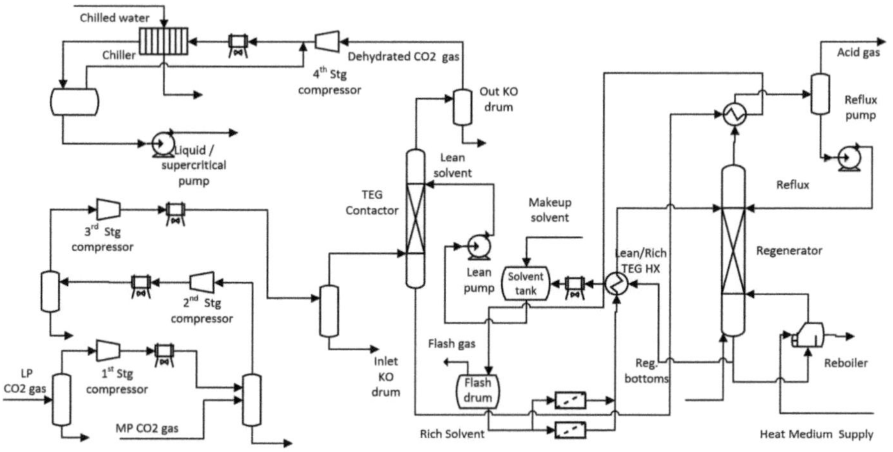

Fig. 10.23 Typical CO_2 compression and dehydration PFD

may have MP CO_2 stream at around pressure of 400–500 kPa, join in this stage to save energy, the combined CO_2 gas stream then enters to 3rd stage compressor, the compressed CO_2 gas from 3rd stage compressor via aftercoolers and KO drum enters to TEG contactor to remove its water content. The regenerated TEG (lean TEG) is pumped to the top tray of the contactor (absorber). The glycol absorbs water as it flows down through the contactor counter-current to the gas flow. Water-rich glycol leaves the bottom of the contactor, passes through the reflux condenser coil, flashes off most of the soluble gas in the flash drum, and flows through the rich-lean heat exchanger to the regenerator. In the regenerator, the absorbed water is distilled from the glycol at lower pressure by application of heat, which could be a fired reboiler or a steam heating reboiler. The regenerated lean glycol flows through the rich-lean exchanger and glycol cooler exchanger for cooling down before being recirculated to the contactor by the glycol pump.

To reduce CO_2 loss, the flash gas from the rich TEG flash drum will be recycled back to the 2nd stage compressor suction drum, the recycled flash gas contains about 90–96% of CO_2.

The dehydrated CO_2 gas enters the 4th stage compressor, please note that the 4th stage compressor will cross the critical condition (31 °C at 73.8 bar for pure CO_2 gas), when the CO_2 gas pressure is higher than 73.9 bar and temperature higher than 31.1 °C, the CO_2 gas falls in the supercritical condition.

10.7.3.3 Simulation

A typical simulation model has been completed using Aspen HYSYS commercial simulator, the case study simulation sheet is shown in Fig. 10.24.

For the dry CO_2 gas composition of this case study, and the critical conditions by simulation are shown in Tables 10.27, and 10.28 (Fig. 10.25).

Per Tables 10.27 and 10.28, after considering the system equipment and pipeline pressure drop, the 4th stage compressor discharge pressure should be slightly above True P_c(83 bar) to lower the transportation power, it is preferred to chill the CO_2 gas to below the Critical temperature, which can make the CO_2 gas a liquid CO_2 stream, and liquid CO_2 pump could be used for pumping the CO_2 stream to required pressure level.

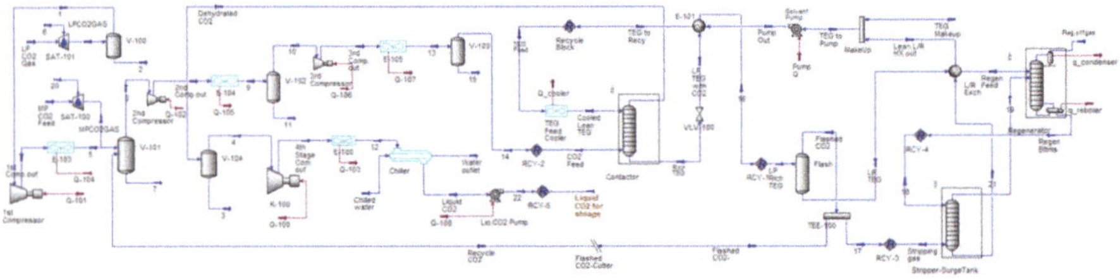

Fig. 10.24 Case study CO_2 compression and dehydration process simulation sheet

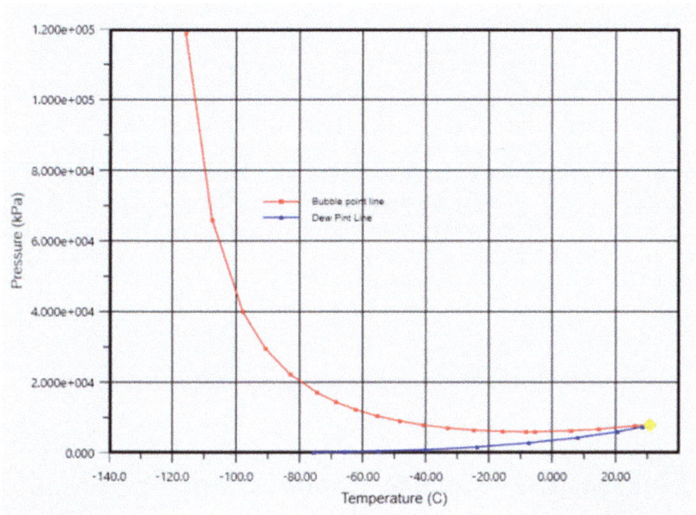

Fig. 10.25 CO_2 PT phase diagram. Note: due to the hydrogen in the CO_2 gas, the simulated pressure is extremely high

Table 10.27 CO_2 gas true critical conditions

Stream: 3rd Comp. out — Select Stream...

Properties

True Tc [C]	34.25
Pseudo Tc [C]	29.09
True Pc [kPa]	8299
Pseudo Pc [kPa]	7376
True Vc [m3/kgmole]	9.813e-002
Pseudo Vc [m3/kgmole]	9.273e-002
True Zc	0.3186
Pseudo Zc	0.2722

10.7.3.4 Study Results and H&M Balance

Per process simulation results, the mainstream information is summarized in Table 10.29, it indicates the preliminary H&M balance for this CO_2 compression and dehydration case study.

From the Table 10.28, without gas stripping, the TEG concentration achieves 98.7% at a reboiler temperature of 204.7 °C. A small portion of flashed gas is directed to the regenerator as stripping gas, It makes the stripping process more effective; a split flash gas flow of 430 Sm^3/h is taken for this study, the TEG concentration of 99.7% (w) can be achieved, the water content in dehydrated CO_2 gas is about 26 ppm.

Table 10.28 Two phase properties

Stream: 3rd Comp. out

Critical Values

Two-Phase Critical Temperature	30.68 C
Two-Phase Critical Pressure	7950 kPa
Three-Phase Critical Temperature	<empty>
Three-Phase Critical Pressure	<empty>

Maxima

Cricondentherm	30.69 C
Cricondenbar	7995 kPa

10.7.4 Flue Gas CO_2 Capture, Compression and Dehydration—Case Study 9

10.7.4.1 Study Basis

The feed flue gas composition, flow rate, and operating conditions are summarized in Table 10.30.

To capture CO_2 from the flue gas, the CO_2 stream will be delivered in liquid CO_2 at a pressure above CO_2 critical pressure and the temperature below the critical temperature, the CO_2 purity should be no less than 98%.

To deliver the liquid CO_2, it requires the combination of the flue gas CO_2 capture, CO_2 compression, and dehydration processes.

10.7.4.2 Process Description

The PFD of the combined processes in traditional CO_2 capture and compression using a chemical solvent is shown in Fig. 10.26.

Flue Gas CO_2 Capture

The flue gas from fired heaters, boilers, or process furnaces usually contains about 11–15% of CO_2, a two-stage acid gas compressor is required to boost the flue gas to absorb operating pressure.

The flue gas from a fired heater at about 160 °C enters to flue gas cooler to cool the flue gas to 35 °C, the cooled flue gas is then directed to a flue gas K.O. drum to separate the water vapor condensate, then enters a 2-stage flue gas compressor to booster the flue gas pressure to 8 barg, and flows in the bottom of the absorber column. the feed gas flows upward, counter-current to the lean amines which are introduced to the top of the absorber. The treated flue gas exits the top of the column with nitrogen purity of 96%, it can be used as utility inert gas in this plant or vent to the atmosphere. The

Table 10.29 Simulation summary of CO_2 compression and dehydration key stream data

Name	LP CO_2 gas	MP CO_2 gas	CO_2 feed	Cooled lean TEG	Dehydrated CO_2	Regen feed	Stripping gas	Reg. offgas	Recycle CO_2	Liquid CO_2	Chilled water	4th stage com. out
Vapour fraction	1.000	1.000	1.000	0.000	1.000	0.019	1.000	1.000	1.000	0.000	0.000	1.000
Temperature [°C]	35	35	35	30	40	130	50	124	50	31	15	107
Pressure [kPa]	150	450	4530	4480	4420	421	490	101	490	8650	600	8700
Molar flow [kgmole/h]	428	1426	1954	460	1777	480	18	38	139	1777	2995	1777
Mass flow [kg/h]	18,349.6	60,835.4	84,292.1	67,532.9	76,740.5	68,273.5	800.0	1520.9	6042.6	76,740.5	53,951.1	76,740.5
Liquid volume flow [m^3/h]	22.2	74.5	103.0	59.9	93.8	60.8	1.0	1.8	7.3	93.8	54.1	93.8
Molecular weight	42.9	42.7	43.1	146.8	43.2	142.3	43.6	40.0	43.6	43.2	18.0	43.2
Mole frac (CO_2)	0.950231	0.956830	0.969812	0.001448	0.974631	0.026175	0.962150	0.776773	0.962016	0.974631	0.000000	0.974631
Mole frac (H_2S)	0.011941	0.004934	0.009188	0.000172	0.005196	0.006257	0.036975	0.094701	0.036962	0.005196	0.000000	0.005196
Mole frac (H_2O)	0.037762	0.013140	0.002634	0.024098	0.000026	0.033117	0.000383	0.126228	0.000532	0.000026	1.000000	0.000026
Mass frac (TEGlycol)				0.996569	0.000001	0.986210	0.000007	0.007498		0.000001	0.000000	0.000001

Table 10.30 Flue gas composition, flow rate, and operating conditions

Name	Flue gas	Unit
Vapor fraction	1.0000	
Temperature	160	°C
Pressure	106	kPa
Mass flow	145,663	kg/h
Molecular weight	29.1	
Comp, H_2O	10.00	mole %
Comp, CO_2	13.00	mole %
Comp, nitrogen	76.00	mole %
Comp, oxygen	1.00	mole %

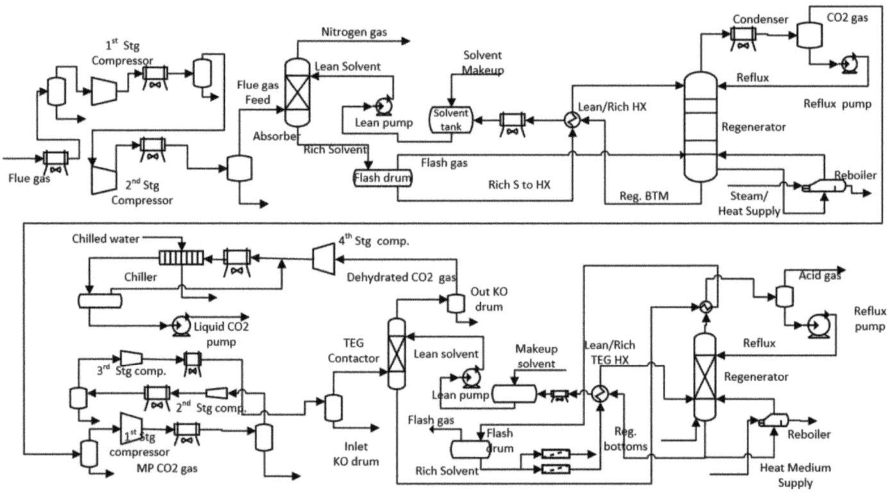

Fig. 10.26 Traditional CO_2 capture and compression combined process PFD using a chemical solvent

rich solvent is directed to a flash drum and then to the heat exchanger between lean and rich solvent to recover rich solvent heat, decrease lean solvent temperature, the heated rich solvent is then sent a regenerator to be regenerated using heating, usually steam. The CO_2 gas from the regenerator is routed to the CO_2 gas compression and dehydration sections.

The regenerator bottom stream enters the lean/Rich amine heat exchanger to recover the stream heat, then is cooled down further via a cooler, the cooled lean amine enters the top of the absorber, completing a solvent recycle loop.

Flash gas from the flash drum is mainly a nitrogen component, it can be vented to the atmosphere directly.

CO_2 Compression and Dehydration

CO_2 captured is fed to inlet K.O. drum then enters to a 3-stage compressor to boost the CO_2 gas pressure to 40–45 bar, the compressed CO_2 gas from 3rd stage compressor via aftercoolers and KO drum enters to TEG contactor to remove water content in CO_2 gas stream. the regenerated TEG (lean TEG) is pumped to the top tray of the contactor. The glycol solvent absorbs water as it flows down through the contactor counter-current to the gas flow. Water-rich glycol leaves the bottom of the contactor, passes through the reflux condenser coil of the regenerator to the rich TEG flash drum, flashes off most of the soluble gas in the flash drum, and flows through the rich-lean heat exchanger to the regenerator. In the regenerator, the absorbed water is distilled from the glycol at lower pressure by application of heat, which could be a fired reboiler or a steam heating reboiler. The regenerated lean glycol flows through the rich-lean heat exchanger and glycol cooler for cooling before being recirculated to the contactor by the TEG pump.

The dehydrated CO_2 gas enters the 4th stage compressor, please note that the 4th stage compressor will cross the critical condition (31 °C critical temperature at 73.8 bar for pure CO_2 gas), when the CO_2 gas pressure is higher than 73.9 bar and temperature higher than 31.1 °C, the CO_2 gas falls in the supercritical condition. CO_2 can be liquified via a chilling water cooler, then pumped to the network.

10.7.4.3 Process Simulation

The overall process simulation sheet of the flue gas CO_2 capture is shown in Fig. 10.27.

The stream data in the overall process simulation for the flue gas CO_2 capture are summarized in Table 10.31.

The process simulation sheet for the CO_2 compression and dehydration section is shown in Fig. 10.28.

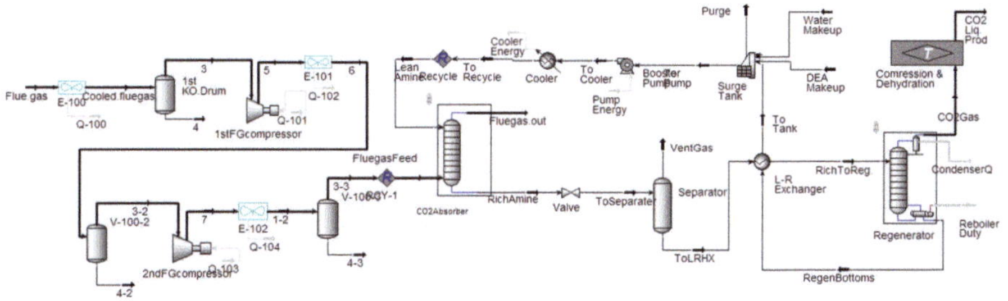

Fig. 10.27 Traditional flue gas CO_2 capture overall process simulation sheet

Table 10.31 Flue gas CO_2 capture overall process simulation stream data

Name	Flue gas	Cooled flue gas	Flue gas feed	Lean amine	Fluegas out	Rich amine	To separater	Vent gas	To LRHX	Rich to reg.	CO_2 gas	Regen bottoms	CO_2 liq. Prod.
Vapour fraction	1.000	0.952	1.000	0.000	1.000	0.000	0.000	1.000	0.000	0.000	1.000	0.000	1.000
Temperature [°C]	160	35	35	40	41	61	61	61	61	93	43	116	31
Pressure [kPa]	105	103	890	3618	850	890	483	483	483	448	150	180	8650
Mass flow [kg/h]	145,665	145,665	137,208	540,233	110,965	566,475	566,475	25	566,450	566,450	26,991	539,459	24,909
Liquid flow [m^3/h]	250.0	250.0	241.6	529.8	209.6	561.8	561.8	0.0	561.7	561.7	32.7	529.0	30.3
Molecular weight	29.1	29.1	30.3	22.9	28.2	23.4	23.4	29.7	23.4	23.4	42.4	22.9	44.0
Composition, mole frac													
H_2O	0.099997	0.099997	0.006790	0.940579	0.009087	0.917289	0.917289	0.042351	0.917320	0.917320	0.059521	0.940472	0.000024
CO	0.000028	0.000028	0.000030	0.000000	0.000035	0.000000	0.000000	0.000034	0.000000	0.000000	0.000000	0.000000	0.000000
CO_2	0.129996	0.129996	0.143444	0.005000	0.013346	0.029563	0.029563	0.130099	0.029559	0.029559	0.939177	0.005008	0.998498
H_2S	0.000000	0.000000	0.000000	0.000000	0.000000	0.000000	0.000000	0.000000	0.000000	0.000000	0.000000	0.000000	0.000008
Nitrogen	0.759979	0.759979	0.838700	0.000000	0.964840	0.000062	0.000062	0.814060	0.000033	0.000033	0.001264	0.000000	0.001419
Oxygen	0.010000	0.010000	0.011035	0.000000	0.012691	0.000001	0.000001	0.013456	0.000001	0.000001	0.000037	0.000000	0.000051
DEAmine	0.000000	0.000000	0.000000	0.250000	0.000000	0.238419	0.238419	0.000000	0.238429	0.238429	0.000000	0.250359	0.000000

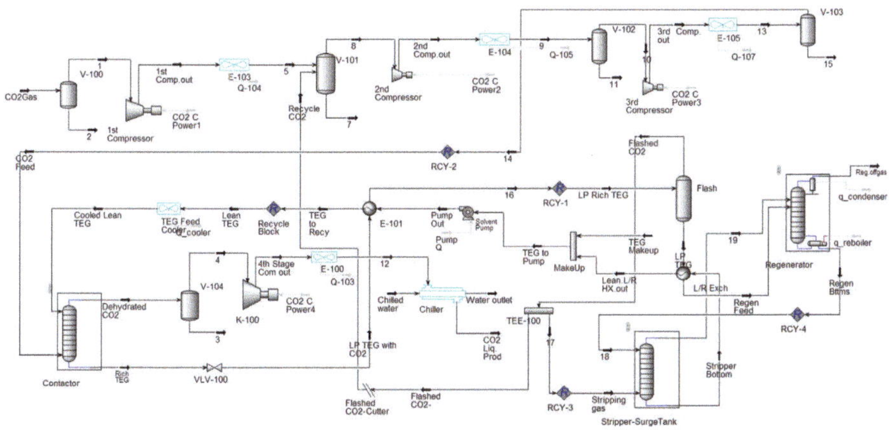

Fig. 10.28 Process simulation sheet for CO_2 compressions and dehydration section

The stream data in CO_2 compression and dehydration section are summarized in Table 10.32.

10.7.4.4 Performance and Study Results

The performance data by overall simulation of flue gas CO_2 capture process are summarized in Table 10.33.

From Table 10.33, the CO_2 content of the treated flue gas (nitrogen gas) is about 1.3%, the nitrogen purity of treated flue gas is 96.5%, it can be used as inert gas for plant utility system; CO_2 capture efficiency achieves 92% with wet CO_2 Purity of 99.5%. This CO_2 gas is compressed and dehydrated; the dried CO_2 gas has a purity of 99.8%; rich amine loading achieved is 0.56 mol CO_2 gas/mole amine.

Flash gas from the flash drum is mainly a nitrogen component, it can be vented to the atmosphere directly without impact on the environment.

CO_2 capture efficiency is 91.9%, after dehydration, the total efficiency is about 87%, CO_2 lost in off-gas of TEG regenerator, including the 800 kg/h of Flash gas as Stripping gas.

Captured 25 t/h liquid CO_2 product with a purity of 99.8%, 55 t/h steam required for Amine reboiler of this case study, 8 t/h MP steam required for TEG regenerator.

A small portion of flashed gas is used as stripping gas, It is an effective way to operate the regenerator; for this study, using 800 kg/h of flashed gas, the TEG concentration of 99.7% (w) can be achieved at a reboiler temperature of 204.7 °C, the water content in dehydrated CO_2 gas is about 26 ppm.

Table 10.32 CO_2 compression and dehydration section stream data

Name	CO_2 gas	1st comp. out	Recycle CO_2	2nd comp. out	3rd comp. out	CO_2 feed	Cooled lean TEG	Dehydrated CO_2	4th stage comp. out
Vapour fraction	1.000	1.000	1.000	1.000	1.000	1.000	0.000	1.000	1.000
Temperature [°C]	43	154	50	147	163	35	30	39	105
Pressure [kPa]	150	460	490	1350	4560	4530	4480	4420	8700
Mass flow [kg/h]	26,991	26,986	5723	32,250	32,115	32,011	67,534	24,909	24,909
Liquid flow [m^3/h]	32.6	32.6	6.9	39.0	38.9	38.8	59.9	30.2	30.2
Molecular weight	42.4	42.5	44.0	43.6	43.8	43.9	146.8	44.0	44.0
H_2O	0.059521	0.059148	0.000435	0.016540	0.006576	0.002668	0.024174	0.000024	0.000024
CO	0.000000	0.000000	0.000000	0.000000	0.000000	0.000000	0.000000	0.000000	0.000000
CO_2	0.939177	0.939550	0.996915	0.981877	0.991824	0.995364	0.001497	0.998497	0.998497
H_2S	0.000000	0.000000	0.002610	0.000459	0.000463	0.000817	0.000012	0.000008	0.000008
Nitrogen	0.001264	0.001265	0.000033	0.001092	0.001103	0.001109	0.000000	0.001419	0.001419
Oxygen	0.000037	0.000037	0.000006	0.000033	0.000034	0.000041	0.000000	0.000051	0.000051
TEG mass frac							0.996582	0.000001	0.000001

(continued)

Table 10.32 (continued)

Name	CO_2 liq. prod.	Chilled water	Rich TEG	LP rich TEG	Flashed CO_2	Regen feed	Reg. offgas	Regen Bttms	Stripping gas	Stripper bottom
Vapour fraction	0.000	0.000	0.000	0.238	1.000	0.017	1.000	0.000	1.000	0.000
Temperature [°C]	31	15	38	50	50	130	124	205	50	196
Pressure [kPa]	8650	600	4460	490	490	421	101	103	490	105
Mass flow [kg/h]	24,909	9652	74,636	74,638	6523	68,115	1391	68,624	800	67,524
Liquid flow [m^3/h]	30.2	9.7	68.5	68.5	7.9	60.6	1.7	60.9	1.0	59.9
Molecular weight	44.0	18.0	119.9	119.9	44.0	143.6	42.6	139.1	44.0	146.8
H_2O	0.000024	1.000000	0.020961	0.020944	0.000435	0.027359	0.059074	0.083482	0.000435	0.024177
CO	0.000000	0.000000	0.000000	0.000000	0.000000	0.000000	0.000000	0.000000	0.000000	0.000000
CO_2	0.998497	0.000000	0.258262	0.258289	0.996913	0.027266	0.930946	0.000237	0.996913	0.001497
H_2S	0.000008	0.000000	0.000959	0.000960	0.002609	0.000445	0.007865	0.000002	0.002609	0.000012
Nitrogen	0.001419	0.000000	0.000008	0.000008	0.000033	0.000000	0.000018	0.000000	0.000033	0.000000
Oxygen	0.000051	0.000000	0.000001	0.000001	0.000006	0.000000	0.000004	0.000000	0.000006	0.000000
TEG mass frac	0.000001	0.000000	0.901755	0.901746	0.000007	0.988106	0.007360	0.989114	0.000007	0.996582

Table 10.33 Flue gas CO_2 capture process overall simulated performance

Description	Value	Unit	Description	Value	Unit
Flue gas molar flow	5000	kgmole/h	Amine recirculation rate	520.0	m^3/h
Flue gas mole frac (CO_2)	0.13		Lean amine loading	0.0919	
Flue gas comp molar flow (CO_2)	28,605.6	kg/h	Rich amine loading	0.5569	
Dry flue gas flowrate	4530.68	kgmole/h	Regenerator feed temperature	93.3334	°C
Dry flue gas CO_2 mole fraction	0.14		Regenerator reflux ratio (mole basis)	0.6	
Treated flue gas CO_2	1.3	mol%	Lean amine temperature	40.00	C
CO_2 gas mole frac (CO_2)	0.9392		Reg steam mass flow	54,998.6	kg/h
CO_2 gas molar flow	636.0	kgmole/h	Product comp mass flow (CO_2)	24,885.5	kg/h
CO_2 gas mass flow	26,991	kg/h	CO_2 capture efficiency	91.9	%
Amine strength	25.0	%	Final CO_2 captured efficiency	87.0	%

10.8 CO_2 Cryogenic Separation

From the CO_2 thermodynamics properties and phase diagram (Fig. 10.1), when lowering the CO_2 stream temperature and increasing pressure, the CO_2 stream phase could be changed from a gas phase to a liquid even into a solid phase.

When the CO_2 stream falls below the critical point, the CO_2 phase changes with reference to the CO_2 saturation line, when pressure falls below the saturation line, the CO_2 stream stays in the gas phase, when the pressure increases to above the saturation line, CO_2 becomes a liquid phase.

When the CO_2 stream temperature and pressure fall below the triple point, the CO_2 phase changes with reference to the CO_2 sublimation line; when its pressure falls below the sublimation line, the CO_2 stream stays in the gas phase, when the pressure increases to above the sublimation line, CO_2 become a solid phase. Therefore, it is a practical approach to change the CO_2 phase by changing temperature and/or pressure.

In the natural gas or flue gas stream, the component of nitrogen, oxygen, and methane has a much different phase diagram, all of them have much lower boiling points compared with the CO_2 component, which provides an opportunity to separate the CO_2 from CO_2 rich streams.

Cryogenic CO_2 separation is not suitable for the gas containing C_3 plus hydrocarbons, in the case of existing ethane components, due to the closer properties between CO_2 and ethane, it may cause the ethane loss in the CO_2 stream.

The phase diagram for the CO_2 rich stream will be much different from the pure CO_2 phase diagram, except experimental methods can be used to generate the CO_2 stream phase diagram, there is a fast approach using a commercial simulator such as Aspen HYSYS to produce the dewpoint line, and freeze outline (sublimation line) of the CO_2 stream. Unfortunately, simulating the CO_2 stream sublimation/de-sublimation phase equilibrium is not available in lower version of Aspen HYSYS.

10.8.1 Actual Rich CO_2 Gas Stream Gas Liquid Dewpoint and Freezeout Curve

The composition of flue gas from fired heaters, boilers, and gas turbines varies depending on the type of fuels, fuel composition, excess oxygen amount, etc., using air as a combustion oxidant, the typical flue gas composition is shown in Table 10.34.

From Table 10.34, the wet flue gas contains about 10% combustion water, this vapor water will be knocked down when the flue gas gets colder and pressure increases.

When the flue gas is chilled to 1 °C of, the water content drops to about 200 ppmw.

For the chilled flue gas with CO_2 composition of 14.4% (mole), the flue gas dewpoint and freezeout points could be simulated using Aspen HYSYS; the dew point data is simulated using the HYSYS property table at a vapor fraction of 0.9999, and the freezeout curve is simulated using CO_2 freezeout utilities at different Stream pressure. The simulated dew points curve and CO_2 freezeout curve are shown in Fig. 10.23.

From Fig. 10.29, at CO_2 partial pressure of 830 kPa or below, the freezeout temperature by simulation is higher than the dew point temperature, which indicates that CO_2 in flue gas will be de-sublimated to solid CO_2 rather than to the liquid phase; the CO_2 rich gas will be frozen out first during the cryogenic process; when

Table 10.34 Typical flue gas composition under operating conditions

Name	Flue gas	Cold FG	Chilled FG
Vapor fraction	1.0000	1.0000	1.0000
Temperature [°C]	160	35	0
Pressure [kPa]	106	104	2590
Mass flow [kg/h]	145,663	141,287	136,650
Molecular weight	29.1	29.7	30.4
Comp. mole % (H_2O)	10.00	5.40	0.03
Comp. mole % (CO_2)	13.00	13.66	14.44
Comp. mole % (nitrogen)	76.00	79.88	84.45
Comp. mole % (oxygen)	1.00	1.05	1.11

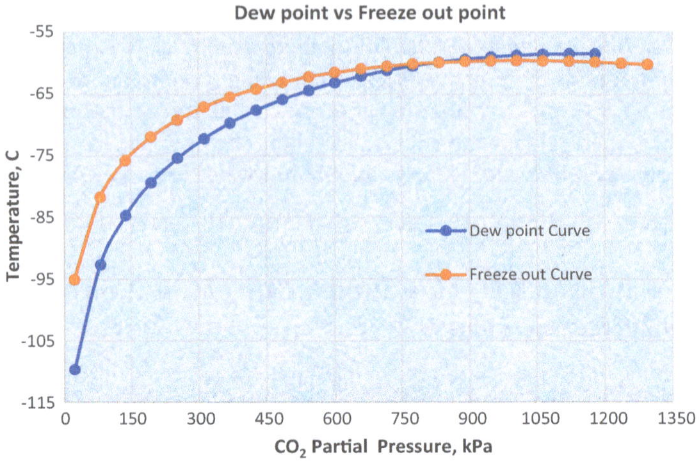

Fig. 10.29 Flue gas dew point versus freeze out point

the CO_2 partial pressure reaches 830 kPa, the two curves are overlapping, the curves become flat.

Figure 10.29 also indicates that the freeze-out temperature increases with CO_2 partial pressure, and reaches the turnaround point at partial pressure of 830 kPa, then varies following the flat cure; for this typical flue gas (14.4% m CO_2), it requires the flue gas overall pressure of 58 bar. Therefore, increasing the CO_2 rich gas stream pressure will increase cryogenic temperature, trading off between increasing CO_2 rich stream pressure and lowing the cryogenic temperature is needed before proceeding with the selected cryogenic CO_2 capture process.

10.8.2 Cryogenic Temperature Selection

The CO_2 freeze-out temperature is only the starting point of the CO_2 de-sublimation process, to complete the de-sublimation of the most CO_2 in the flue gas and achieve the CO_2 recovery over 90%, it requires an additional temperature approach (about 18–20 °C).

Aspen HYSYS simulator can be used to forecast the freeze-out temperature at different treated flue gas CO_2 compositions, and then estimate the CO_2 partial pressure at the pressure of de-sublimation.

Per simulation results of rich CO_2 flue gas, the freezeout temperature and CO_2 partial pressure relationship is shown in Fig. 10.30. At the overall de-sublimation pressure of 180 kPa, to achieve a 92% CO_2 removal efficiency, the residue CO_2 in the treated flue gas is about 1.1% (mole), it is estimated that the required freeze-out temperature is about − 117 °C.

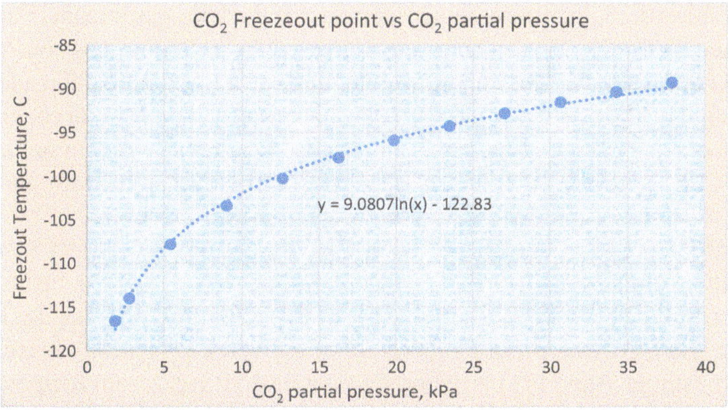

Fig. 10.30 Flue gas freezeout point

The freezeout temperature and CO_2 partial pressure relationship can be regressed as Eq. (10.5).

$$t_{freeze} = 9.0807 \ln\left(p_{CO_2}\right) - 122.83 \tag{10.5}$$

where: t_{freeze}—freeze out temperature of CO_2 rich gas stream, °C

p_{CO_2}—CO_2 partial pressure of treated flue gas, kPa.

To estimate the required freezeout temperature, regress the simulated freezeout temperature information from Aspen HYSYS simulator into the following equation:

$$P_{desubl} = 729956e^{0.1099t} \tag{10.6}$$

where: P_{desubl}—CO_2 de-sublimation operating partial pressure, kPa.

t—CO_2 freezeout temperature, °C

Equation (10.6) is regressed from HYSYS simulation, due to the CO_2 partial pressure that enters regression, thus eliminates the CO_2 concentration and stream pressure, therefore it can be used to estimate the required freezeout temperature for any different scenarios, the estimate results match the Eq. (10.7), proposed by Jensen et al. [8, 9]

$$P_{CO_2}^{subl} = \exp\left(57.52 - \frac{3992.84}{T} - 4.9003 \ln(T) + 2.415 \times 10^{-15} T^6 + \frac{8125.6}{T^2}\right) \tag{10.7}$$

where: $P_{CO_2}^{subl}$—CO_2 sublimation pressure, Pa.

T—CO_2 sublimation, K.

From Table 10.35, the freezeout temperature of the CO_2 gas stream (flue gas) can be estimated using Eq. 10.5, which was regressed from S7 flue gas composition in Table 10.34 at treated flue gas conditions, the impact of the composition and pressure are eliminated by the use of flue gas CO_2 partial pressure, therefore for rough estimate purpose, the Eq. (10.5) can be used universally. Freezeout temperature calculated using Eq. (10.5) matches the Aspen HYSYS simulation results with an error less than 1%.

CO_2 partial pressure calculated using Eq. (10.6) matches HYSYS simulation results with errors less than 11%. The calculation results by the Eq. (10.6) matches Eq. (10.7) results well with an average error of 6%. Therefore, we can use Eq. (10.5) to estimate the CO_2 freeze-out temperature at a given CO_2 partial pressure, or use Eq. 10.6 to estimate the CO_2 partial pressure at desired freeze-out temperature.

The first step is to set the CO_2 removal efficiency, then determine the treated flue gas CO_2 composition, calculate the CO_2 partial pressure of the treated flue gas, calculate the required freeze-out temperature per CO_2 partial pressure using Eq. (10.5), the cryogenic temperature will be the estimated freezeout temperature plus required temperature approach of (18–20 °C) to reach the higher CO_2 removal efficiency due to the lower Residual CO_2 in treated flue gas.

After estimating the CO_2 freeze-out temperature at a given CO_2 partial pressure you can estimate the required cryogenic temperature at a given partial pressure, generally, the following steps should be followed to determine the required cryogenic temperature:

- Collect the flue gas composition
- Remove the combustion water by lowering the flue gas temperature to about 0 °C
- Screening cryogenic processes, determine the treated flue gas pressure
- Estimate the CO_2 partial pressure per required CO_2 removal efficiency;
- Estimate the CO_2 freeze-out temperature per estimated CO_2 partial pressure;
- Impose a temperature approach (18–20 °C) on CO_2 freeze-out temperature to achieve the required CO_2 removal efficiency.

10.8.3 Cryogenic Technology Development

A Cryogenic CO_2 capture development report has been submitted to the US Department of Energy by Sustainable Energy Solutions Company in Sept. 2019 [10]. Stated:

Pilot skid installed at the Hunter Power Plant. 35 h of Multiple tests were completed, with a total of over 600 h of CO_2 capture completed using flue gas at the Hunter Power Plant with an average CO_2 capture rate of over 91% during those hours. The Cryogenic carbon capture technology has evolved with claims [11]: Separate 90% + of the CO_2 in flue gas and prepare it for storage; 40% lower cost of electricity.

Prediction and validation for a full-scale Power plant report completed by Brigham Young University, Provo, Utah, confirmed its feasibility and economic compatibility [12].

Table 10.35 HYSYS simulation and literature calculation comparison

HYSYS simulation			Equations (10.5) and (10.10–10.6)				Equation (10.7)			
Treated FG CO_2 (mole) %	Freeze out T, °C	Partial pressure, kPa	Freeze out temperature		Partial pressure		Temperature K	Partial pressure		
			°C	Error, %	kPa	Error, %		Pa	kPa	Error, %
1.0	− 116.58	1.80	− 117.49	0.78	1.99	10.57	156.57	2041	2.04	13.37
1.5	− 114.01	2.70	− 113.81	− 0.18	2.64	− 2.20	159.14	2826	2.83	4.65
3.0	− 107.82	5.40	− 107.52	− 0.28	5.21	− 3.48	165.33	5914	5.91	9.51
5.0	− 103.37	9.00	− 102.88	− 0.47	8.50	− 5.51	169.78	9716	9.72	7.96
7.0	− 100.29	12.60	− 99.82	− 0.46	11.93	− 5.29	172.86	13,486	13.49	7.03
9.0	− 97.90	16.20	− 97.54	− 0.37	15.51	− 4.27	175.25	17,236	17.24	6.39
11.0	− 95.94	19.80	− 95.72	− 0.24	19.23	− 2.87	177.21	20,973	20.97	5.92
13.0	− 94.28	23.40	− 94.20	− 0.08	23.10	− 1.29	178.87	24,701	24.70	5.56
15.0	− 92.82	27.00	− 92.90	0.09	27.10	0.39	180.33	28,423	28.42	5.27
17.0	− 91.53	30.60	− 91.76	0.26	31.25	2.12	181.62	32,140	32.14	5.03
19.0	− 90.36	34.20	− 90.75	0.44	35.53	3.89	182.79	35,854	35.85	4.84
21.0	− 89.29	37.80	− 89.85	0.62	39.95	5.68	183.86	39,566	39.57	4.67

A patent application [13], CO_2 Separation and Liquefaction System and Method, has been filed in Canadian Intellectual Property Office. Which claimed that no external refrigeration was needed, only containing the 3 stage flue gas compressors and a two-stage cold flue gas compressor with 4 expanders.

10.8.4 Flue Gas Cryogenic CO_2 Capture—Case Study 10

The CO_2 in the gas stream can be captured using the cryogenic method even at very low CO_2 concertation, such as flue gas from fired heaters or boilers, generally speaking, the flue gas mainly contains CO_2 (10–15% mole) and nitrogen, the CO_2 cryogenic capturing is not only capturing the CO_2 but also purifying the nitrogen content, it can generate the nitrogen with a purity over 97%, which can meet the requirement of production for purging, blanketing usage.

10.8.4.1 Problem Definition

The feed flue gas composition, flow rate, and operating conditions are same as Case study 9 Table 10.30.

For a typical fired heater flue gas, take 5000 kmole/h (145,663 kg/h) of raw flue gas at 160 °C, in the case study, the CO_2 Content is about 10–13% (mole) with about 10% (mole) combustion water.

To capture the CO_2, the flue gas has to be cooled down to 35 °C from 160°C using an air cooler, a compressor will be installed to boost the flue gas pressure.

The pressure of flue gas is determined by trading off between the power demand in the flue gas compressors and mixed refrigerant compressors, after running the simulations comparison for power demands, the flue gas pressure of 25 barg is selected for this case study. Literature [11] states that using 10 bar pressure but consuming more power for MR compression at the same required cryogenic temperature for desirable CO_2 removal efficiency.

Boosting the flue gas pressure provides an opportunity to utilize the John Thomson effect/Turbo expander to lower the flue gas temperature, significantly saving the mixed refrigerant compressor's power demand.

The process simulators currently have no such function of predicting the equipment performance on CO_2 de-sublimation and displaying the gas–liquid-solid phase behaviors. Aspen Plus simulators claimed to be able to predict the CO_2 solid phase using the Gibbs reactor with minimizing the free energy. Aspen HYSYS can predict the CO_2 freezing out temperature only. It is expected that the process simulators will be upgraded with a convenient method for CO_2 de-sublimation process in the near future.

To achieve the target of refrigeration temperature, the refrigeration process uses the mixed refrigerant (MR), which is a mixture of propane, ethane, and methane.

A new cryogenic CO_2 capture process BFD is shown in Fig. 10.31.

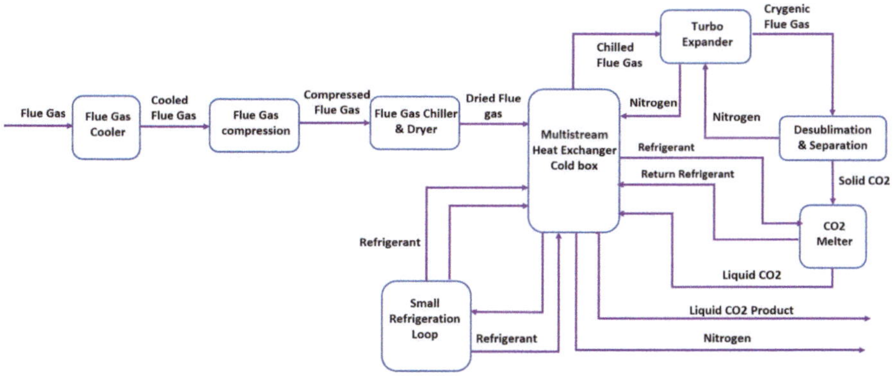

Fig. 10.31 A new cryogenic CO_2 capture process block flow diagram

10.8.4.2 Process Description

Please note that this new process is a pending patent, the flue gas CO_2 cryogenic capture process flow diagrams and the process description are not shown, please contact the author for detailed information.

10.8.4.3 H&M Balance

In order to evaluate the feasibility of the cryogenic flue gas CO_2 capture process and operation, a simulation model was built using commercial simulation software. The technical feasibility and outstanding energy-saving effect of the new process are confirmed; As the new process is patent pending, the process simulation sheet and detail H&M balance data will not be described here, and readers who are interested in detailed information may contact the author.

The stream data by simulation are summarized in Table 10.36. Note that the new process is patent pending, Table 10-36 does not include operating conditions, only lists material balance and composition data.

From Table 10.36, the captured CO_2 stream has a purity of 99.7%, and the CO_2 recovery efficiency is about 92%. It can be used for either chemical utilization or enhancement of oil/gas recovery or storage.

Treated flue gas has a nitrogen purity of 97%, the nitrogen purity is highly dependent on the excess oxygen in the flue gas, and the residue CO_2 in the treated flue gas. The excess oxygen in flue gas used in this case study is 1%, and residue CO_2 in the nitrogen stream is 1.4%.

The stream data of MR refrigeration loop is shown in Table 10.37.

From Table 10.37, the mixed refrigerants have been cold exchanged with cold streams of nitrogen and CO_2 liquid, the MR temperature is chilled down to − 30 °C

Table 10.36 Stream data in CO_2 capture from flue gas

Description	S1	S2	S3	S4	S5	S6	S7	S8	S9	S10	S11	S12	S20	S21	S22
	Raw flue gas	Cooled FG	1st stg. FG	2nd stg. FG	3rd stg. FG	Dried FG	Chilled FG	Expander inlet FG	Expander outlet FG	N_2 expander out	Warmed nitrogen	Nitrogen product	Liquid CO_2	Pumped liq. CO_2	Liq. CO_2 product
Vapour fraction	1.0000	1.0000	1.0000	1.0000	1.0000	1.0000	1.0000	0.8901	0.8662	0.9867	1.0000	1.0000	0.0274	0.0000	0.0000
Mass flow [kg/h]	145,663	141,287	138,193	137,213	136,892	136,650	136,650	136,650	136,650	110,264	110,264	110,264	26,386	26,386	26,386
Molecular weight	29.1	29.7	30.1	30.3	30.3	30.4	30.4	30.4	30.4	28.3	28.3	28.3	43.8	43.8	43.8
Comp mole frac (H_2O)	0.1000	0.0540	0.0186	0.0069	0.0030	0.0000	0.0000	0.0000	0.0000	0.0000	0.0000	0.0000	0.0000	0.0000	0.0000
Comp mole frac (CO_2)	0.1300	0.1366	0.1417	0.1434	0.1440	0.1444	0.1444	0.1444	0.1444	0.0141	0.0141	0.0141	0.9879	0.9879	0.9879
Comp mole frac (nitrogen)	0.7600	0.7988	0.8287	0.8387	0.8420	0.8445	0.8445	0.8445	0.8445	0.9731	0.9731	0.9731	0.0118	0.0118	0.0118
Comp mole frac (oxygen)	0.0100	0.0105	0.0109	0.0110	0.0111	0.0111	0.0111	0.0111	0.0111	0.0128	0.0128	0.0128	0.0003	0.0003	0.0003

Table 10.37 Stream data in CO_2 capture by cryogenics with mixed refrigerant

Description	S31	S32	S33	S34	S35	S36	S37	S38
	MR to MP comp.	MP MR	Chilled MR	MR after valve	MR after LNG HX	MR to LP comp.	LP MR comp. out	MP MR comp. out
Vapor fraction	1.0000	1.0000	0.3800	0.3069	0.8716	1.0000	1.0000	1.0000
Temperature [°C]	2.0	30.0	– 30.0	– 107.5	– 47.6	25.0	93.0	73.9
Pressure [kPa]	790.0	2480.0	2450.0	350.0	310.0	280.0	850.0	2500.0
Mass flow [kg/h]	14,334	14,334	14,334	14,334	14,334	14,334	14,334	14,334
Molecular weight	28.7	28.7	28.7	28.7	28.7	28.7	28.7	28.7
Comp. mole frac (methane)	0.4000	0.4000	0.4000	0.4000	0.4000	0.4000	0.4000	0.4000
Comp. mole frac (ethane)	0.3000	0.3000	0.3000	0.3000	0.3000	0.3000	0.3000	0.3000
Comp. mole frac (propane)	0.3000	0.3000	0.3000	0.3000	0.3000	0.3000	0.3000	0.3000

from 35°C in LNGHX-02, then flows to the solid CO_2 melter and achieves a temperature of – 76 °C, after the control valve throttling and evaporation, MR temperature falls to – 107 °C

The cryogenic temperature of – 120 °C or below benefits from the cold integration on cold recovery, turbo expanding, and MR refrigeration.

The Flue Gas knock out water Data is shown in Table 10.38.

The raw flue gas contains about 10% of combustion water, the water in flue gas can be knocked out through the flue gas chilling, the water content of 200 ppmw can be achieved at 1 °C, to remove all the trace amount of free water, the silicon gel or molecular sieve drier might be needed. The total water knocked out is 9 t/h for this case study.

The power demand for cryogenic CO_2 capture is shown in Table 10.39.

The power demand for the cryogenic CO_2 capture is 16.8 MW, and the turbo expander can recover 2.5 MW, so the overall power demand is 14.4 MW.

10.8.4.4 Conclusion and Recommendation

The predicted performance data of CO_2 capture by cryogenics is shown in Table 10.40.

Table 10.38 Water data in flue gas knock out

Description	S40	S41	S42	S43	S44	S45	S46
	Raw FG KO water	1st KO water	2nd KO water	3rd KO water	Chiller FG KO water	Water from drier	Total water
Mass flow [kg/h]	4376	3094	980	321	210	33	9014
Molecular weight	18.0	18.0	18.0	18.1	18.1	18.0	18.0
Comp mole frac (H_2O)	0.9999	0.9998	0.9994	0.9984	0.9965	1.0000	0.9997
Comp mole frac (CO_2)	0.0001	0.0002	0.0005	0.0013	0.0031	0.0000	0.0003
Comp mole frac (nitrogen)	0.0000	0.0000	0.0001	0.0002	0.0004	0.0000	0.0000
Comp mole frac (oxygen)	0.0000	0.0000	0.0000	0.0000	0.0000	0.0000	0.0000

Table 10.39 Power demand for cryogenic CO_2 capture

Description	1st comp.	2nd comp.	3rd comp.	LP refrig.	MP refrig.	Liq. CO_2 pump	Power demand	FG expander	N_2 expander
Power [kW]	5546	5147	5078	508	456	165	16,901	1326	1148

From Table 10.40, the CO_2 recovery is about 91.5% with CO_2 purity of 99%, the nitrogen can be recovered with a purity of 97.3%. The unit CO_2 power demand is about 0.55 MWh/ton. The nitrogen stream can be directed to the plant utility system for usage in purging, blanketing, or delivered to the users outside the plant.

This cryogenic and turbo expanding process eliminates the de-sublimination column, contact liquid recirculation system, and slurry pumping and transferring system, and makes sure that the heat exchangers are operating at above CO_2 de-sublimination temperature, thus resolving the concerns of CO_2 freezing on the heat transfer area, and plugging in the de-sublimination column, pump, and piping system. This process uses cryogenic heat exchangers- cold boxes for cold exchange efficiently at lower temperature approaches.

The design avoids introduction of the contact (dilution) liquid, simplifies the cryogenic CO_2 capture system, avoids the liquid loss to the CO_2 stream.

It is recommended to install pilot equipment to gain the design and operating experience, then scaled up to the industrial application.

Table 10.40 Performance data of CO_2 capture by cryogenics

Description	Unit	Value
Dry FG comp. mole frac (N_2)	%	84.4%
Dry FG comp. mole frac (CO_2)	%	14.4%
Dry FG comp. mass flow (CO_2)	kg/h	28,601
Dry FG comp. mass flow (N_2)	kg/h	106,449
N_2 stream mole frac (CO_2)	%	1.4%
N_2 stream mole frac (N_2)	%	97.3%
N_2 stream comp mass flow (N_2)	kg/h	106,249
CO_2 stream mole frac (CO_2)	%	98.8%
CO_2 stream mole frac (N_2)	%	1.2%
CO_2 stream: comp. mass flow (CO_2)	kg/h	26,180
CO_2 recovery	%	91.5%
N_2 recovery	%	99.8%
Power demand	kW	16,901
N_2 expander recovery	kW	1148
FG expander recovery	kW	1326
Total power demand	kW	14,427
Power demand per ton CO_2	MWh/tone	0.551

10.8.5 Comparison Between Conventional CO_2 Capture and a Cryogenic CO_2 Capture

Case study 9 and Case Study 10 use the same flue gas composition, flow rate, and feed conditions, and both deliver a liquid CO_2 product under almost the same conditions, therefore, we can compare the two case studies from operating cost and capital cost point of view.

10.8.5.1 Traditional CO_2 Capture Process Operating Cost

The high-level operating costs estimation is mainly based on the utility demand. Per simulation results of case study 9, the power demand for the traditional CO_2 process is summarized in Table 10.41, and the heat demand is summarized in Table 10.42.

Assumed the Fuel gas price is \$4.39/GJ, and the power price is \$0.07/kWh, therefore, from Tables 10.41 and 10.42, the traditional CO_2 capture process operating cost is about \$65/ton CO_2 captured (electricity, \$42.5, heat \$22.4).

This high-level cost estimate doesn't include the chilling water demand, amine and TEG makeup, etc.

Table 10.41 Power demand for the traditional CO_2 process

Description	1st FG Comp. P	2nd FG Comp. P	CO_2 C Power1	CO_2 C Power2	CO_2 C Power3	CO_2 C Power4	Amine pump	TEG pump	Total power demand
Power [kW]	6545	4832	777	860	922	318	691	157	15,104
Unit power demand @ 24.885 t/h CO_2	263	194	31	35	37	13	28	6	607
Power cost @ $0.07/kwh	18.41	13.59	2.19	2.42	2.59	0.89	1.95	0.44	42.49

Table 10.42 Heat demand for the traditional CO_2 process

Description	Amine reboiler	TEG reboiler	Total heat demand
Heat flow [GJ/h]	112.26	14.71	126.96
Unit heat demand @ 24.885 t/h CO_2	4.51	0.59	5.10
Fuel gas cost @ $4.39/GJ	19.80	2.59	22.40

10.8.5.2 Operating Cost of New Cryogenic CO_2 Capture Process

The cryogenic CO_2 process doesn't require steam or heat supply but only power demand, per simulation results of case study 10, the power demand for the new cryogenic CO_2 process is summarized in Table 10.43.

From Table 10.43, the operating cost of new cryogenic CO_2 capture process is $38.6/ton CO_2 captured, which has been reduced by 41%.

Table 10.43 Power demand for the new cryogenic CO_2 capture process

Description	1st com. power	2nd com. power	3rd com. power	LP refrig. power	MP refrig power	Liq. CO_2 P power	Power demand	FG exp. power	N_2 exp. power	Total power demand
Power [kW]	5546	5147	5078	508	456	165	16,901	– 1326	– 1148	14,427
Unit power demand @ 26.18 t/h CO_2	212	197	194	19	17	6	646	– 51	– 44	551
Power cost @ $0.07/kwh	14.83	13.76	13.58	1.36	1.22	0.44	45.19	– 3.55	– 3.07	38.58

10.8.5.3 Capital Cost Comparison-Number of Equipment

For this high-level estimate, per the simulation results of both studies, the equipment required in these two processesare are summarized in Table 10.44.

From Table 10.44, the traditional process has a higher number of equipment (47), two absorbers, and two regenerators; the cryogenic flue gas CO_2 capture process requires only 23 pieces of equipment, reduced by 51%.

It is recommended to have a detailed estimate per vendor's input for your project, it is expected that the capital cost will reduce by 30–50% compared with that in the traditional CO_2 capture process.

10.8.5.4 Conclusion and Recommendations

Using the new cryogenic CO_2 capture process, the utility demand is electricity only, no steam or heat supply is needed. It is expected that the operating cost will reduce by around 40%, requiring only electricity, 0.55 MW/t CO_2 capture.

The traditional CO_2 capture process power demand is slightly higher than the new cryogenic CO_2 capture process. The unit power demand is 0.607 MW/ton CO_2 captured, but requires heat supply of 5.1 GJ/ton CO_2 captured for amine and TEG regenerator reboilers.

Per simulation results, the traditional process has a long list of equipment demand in total of 49, including two absorbers and two regenerators; the cryogenic flue gas CO_2 capture process requires only 23 pieces of equipment, reduced by 53%. It is expected that the capital cost will reduce by 30–50%.

The new CO_2 capture process has the following characteristics.

Captured CO_2 will be delivered out as liquid product directly; no Compression and dehydration systems are needed.
The feasibility is confirmed by a detailed simulation result;
No contact liquid loop is needed; There are no cold flue gas compressors needed;
Instead, a much smaller external refrigeration loop is required for operation and startup;
Uniquely design special equipment-melting gas-solid separator are applied;
The de-sublimation conditions are optimized to cross the gas-solid sublimation line.

The new Cryogenic CO_2 capture process is recommended for the following reasons:

CO_2 will be separated as in liquid phase and delivered directly;
A simple CO_2 capture process can replace the traditional CO_2 capture, CO_2 compression, and dehydration process units, eliminates CO_2 gas compression and dehydration process;
The operation cost and capital cost will be reduced 40%, 30–50% respectively;

Table 10.44 The required number of the equipment for the both processes

Traditional process equipment list				Cryogenic process equipment list	
1	CO_2Absorber	25	Flash Drum @TPL1	1	Flue gas K.O. drum
2	Flash drum	26	1st K.O. drum@TPL1	2	1st K.O. drum
3	Flue gas K.O. drum	27	2nd K.O. drum@TPL1	3	2nd K.O. drum
4	1st K.O. drum	28	3rd K.O. drum@TPL1	4	3rd K.O. drum
5	2nd K.O drum	29	Dehydrate feed K.O. drum@TPL1	5	Chilled Flue gas K.O. drum
6	Reflux drum	30	4th K.O. drum@TPL1	6	Feed flue gas cooler
7	L-R exchanger	31	Regenerator @TPL1	7	1st Flue gas compressor after cooler
8	Regenerator	32	TEG feed cooler @TPL1	8	2nd Flue gas compressor after cooler
9	Surge tank	33	1st CO_2 gas compressor after cooler @TPL1	9	3rd Flue gas compressor after cooler
10	Booster pump	34	2nd CO_2 gas compressor after cooler@TPL1	10	Refrigerant compressor after cooler 1
11	Reflux pump	35	3rd CO_2 gas compressor after cooler@TPL1	11	Refrigerant compressor after cooler 2
12	Lean Amine Cooler	36	4th CO_2 gas compressor after cooler@TPL1	12	1st FG compressor
13	Feed flue gas cooler	37	CO_2 gas cooler @TPL1	13	2nd FG compressor
14	1st Flue gas compressor after cooler	38	Contactor @TPL1	14	3rd FG compressor
15	2nd Flue gas compressor after cooler	39	Stripper-surge tank @TPL1	15	refrig. compressor 1
16	Condenser	40	1st CO_2 gas compressor @TPL1	16	refrig. compressor 2
17	Reboiler	41	2nd CO_2 gas compressor @TPL1	17	Expander 1
18	1s tFG compressor	42	3rd CO_2 gas compressor @TPL1	18	Expander 2
19	2nd FG compressor	43	4th CO_2 gas compresso@TPL1	19	Multi-stream heat exchanger
20	L/R exch. @TPL1	44	Condenser @COL2	22	Dehydrator
21	Rich TEG Pre-heater @TPL1	45	Reboiler @COL2	23	Melting separator
22	Chiller @TPL1	46	TEG reflux drum		
23	TEG Tank @TPL1	47	TEG reflux pump		
24	Solvent pump @TPL1				

The Utility demand is only electricity with 0.55 MW/ton CO_2 captured, which is slightly lower than the power demand of the traditional CO_2 capture process, there is no fuel gas and steam demand.
No contact liquid loop is needed, the contact liquid separation process is eliminated;
The optimized configuration avoids solid formation upstream of CO_2 De-sublimation and separation, thus eliminating the equipment CO_2 freezing concerns;
No cold flue gas compressors is needed;
No CO_2 solid transferring equipment is needed.

Suggestion.

Formally complete the case studies with economic evaluation on a new process and a traditional process for your proposed project;
Install a pilot plant based on one of the existing flue gas systems, give the pilot a trial, and scale up to full size for all flue gas systems.

10.9 CO_2 Chemical Utilization and Storage

CO_2 captured from fossil fuel combustion or oil and gas production could be utilized as a raw material in the chemical industry for producing commercial products, for example, CO_2 and ammonia as stuffs in the manufacture of fertilizers, such as ammonium bicarbonate, Ammonium carbamate, carbamide, etc.; furthermore, using sodium compound and CO_2 to produce the sodium carbonate or sodium bicarbonate; using rich CO_2 syngas to produce methanol; other examples are the production of vulcanized rubber, polyurethane foam, and polycarbonates that are inert and long-lived. methanol production.

Please note that chemical utilization can consume only a limited amount of CO_2. The majority of CO_2 captured should be used to enhance oil production or storage.

10.9.1 Urea Production Using CO_2

Urea is a very important nitrogen fertilizer widely used in the agricultural field because it contains a high percentage of nitrogen. Urea dissolves very well in water. Urea is called also ammonium carbamide, which is an organic compound with a chemical formula of $CO(NH_2)_2$.

Urea manufacturing is a conventional technology and widely used in the chemical fertilizer industries, the CO_2 raw material is taken usually from the CO_2 removal unit in Ammonia plant, the key to this section is using CO_2 captured from oil and gas as well as chemical industries, and ammonia as raw materials to an integrated facility for urea production.

As the results of commercial use of a hydrogen production technology, such as electrolysis of water, it is possible to get rid of the hydrogen produce process from the fossil fuel, such as steam reforming, which may lower even eliminate CO_2 produced in an ammonia plant, providing an opportunity for reusing CO_2 captured such as from oil and gas industries.

Raw materials of urea manufacturing are:

- Liquid ammonia—NH_3
- Carbon dioxide—CO_2.

Ammonia is manufactured by an ammonia plant, it can be purchased from the market.

Carbon dioxide (CO_2) has been captured from the CO_2 capture unit in the oil and gas and chemical industries, with compression, liquid CO_2 or supercritical state CO_2 will be supplied as a raw material to produce Urea.

Ammonia reacts with carbon dioxide in a urea reactor at high temperature and high pressure. Urea is formed in two-step reactions (ref to Fig. 10.32).

First step, ammonia and carbon dioxide reaction.

Ammonia and carbon dioxide react together and give ammonium carbamate (NH_2COONH_4) at high temperature and pressure. It is a fast, exothermic reaction.

$$2NH_3(g) + CO_2(g) \Leftrightarrow NH_2COONH_4(l) \quad (10.8)$$

Second Reaction, ammonium carbamate dehydration to form urea and water, which is a slow endothermic reaction.

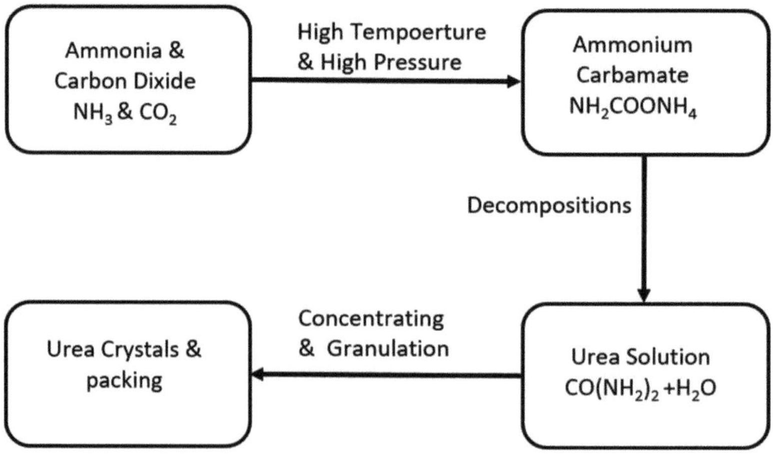

Fig. 10.32 Urea formation steps

$$NH_2COONH_4(l) \Leftrightarrow CO(NH_2)_2(aq) + H_2O(l) \tag{10.9}$$

The reaction (10.8) can be carried out at room temperature and is fast. The reaction (10.9), that is, the dehydration reaction of ammonium carbamate proceeds very slowly in the solid phase, and must be in the liquid state to have a significant speed. Equations 10.8 and 10.9 are completed at the Urea Synthesis Tower, the reaction heat released from 10.8 support the demand for heat of Eq. 10.9.

Manufactured urea contains unreacted ammonia, carbon dioxide, and ammonium carbamate. Ammonium carbamate is decomposed to ammonia and CO_2 by reducing pressure and heating, and recovered as NH_3 and CO_2. When heating the product stream, ammonia and carbon dioxide is separated from the product mixture. The advantage of this process is ammonia and carbon dioxide can be recycled back into the process.

Urea is obtained as a solution, but that solution is concentrated to give 99.6% of molten urea, and granulated for use as fertilizer.

From the above chemical reaction, theoretically, every kmol urea produced will consume one kmol CO_2 and 2 kmol ammonia, which indicates that **1** tone of Urea produced will consume **0.73** tons of CO_2.

A typical Urea plant process flow sheet [14] is shown in Fig. 10.33.

Liquid ammonia is pumped via the multistage pump which maintains the reaction pressure in the vertical stainless steel vessel; Carbon dioxide from the CO_2 capture unit is fed to the urea plant, the CO_2 should be compressed before reaching the urea plant, in case that CO_2 is supplied by ammonia plant, it is taken directly from the CO_2 removal unit of the ammonia plant, and raised to the required pressure in urea synthesis by a CO_2 compressor; Urea synthesis tower is lined with a film of

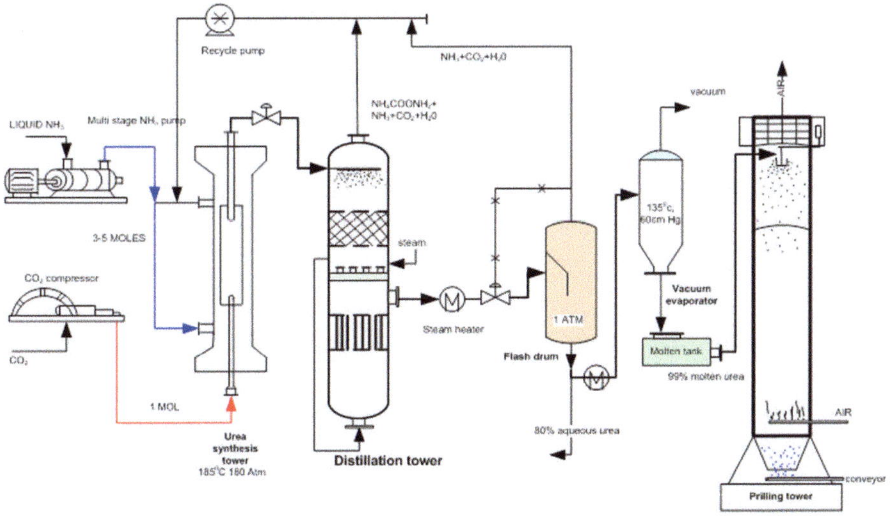

Fig. 10.33 Typical urea production process

oxides to protect from corrosion. The tower is maintained at pressure of 180–200 bar and temperature of 180–200 °C. Plug flow operation takes place and the urea and Carbamate solution is discharged from the top of the tower; high-pressure urea and carbamate solution is let down to distillation tower and Flash drums to decompose carbamate and recover ammonia and CO_2. After flash and separation under MP, LP processing, the solution is fed to vacuum evaporator for concentrating the slurry to < 0.3% of water; urea melt product is passed from the top of the tower into a granulation section. The granules from the granulation section are sent by the conveyor to the bagging section.

10.9.2 Ammonia Bicarbonate Production Using CO_2

Like urea production, ammonium bicarbonate can also be produced using carbon dioxide and ammonia:

$$CO_2 + NH_3 + H_2O \rightarrow (NH_4)HCO_3 \quad (10.10)$$

Ammonia bicarbonate production is a matured conventional technology, from Eq. (10.10), producing **1** kmol ammonium bicarbonate requires **1** kmol CO_2, that is, producing **1** ton of Ammonia bicarbonate consumes **0.57** tone of CO_2.

There were many small scales of ammonium bicarbonate/carbonate manufacturers in China in the 1970s–1980s as low grade of nitrogen fertilizer supply to agricultural use, these ammonium bicarbonate plants are usually integrated with a small scale of ammonia plant, use both CO_2 and ammonia from the ammonia plant.

Because of the high energy demand and operating cost in maintaining a small size of ammonia plant, and the low grade of fertilizers, this type of fertilizer plant has been phased out in favor of urea for quality and stability.

Ammonium bicarbonate can be used as a component in the production of fire-extinguishing compounds, pharmaceuticals, dyes, and pigments, and it is also a basic fertilizer, being a source of ammonia. Ammonium bicarbonate is still widely used in the plastics and rubber industry, in the manufacture of ceramics, in chrome leather tanning, and for the synthesis of catalysts.

10.9.3 Sodium Bicarbonate Production Using CO_2

Sodium bicarbonate is known as a natural ingredient in hot springs and is a safe substance. As the environmental impact becomes a more immediate problem, its gentle effects and natural qualities have been reassessed and it has come to be used in a variety of fields.

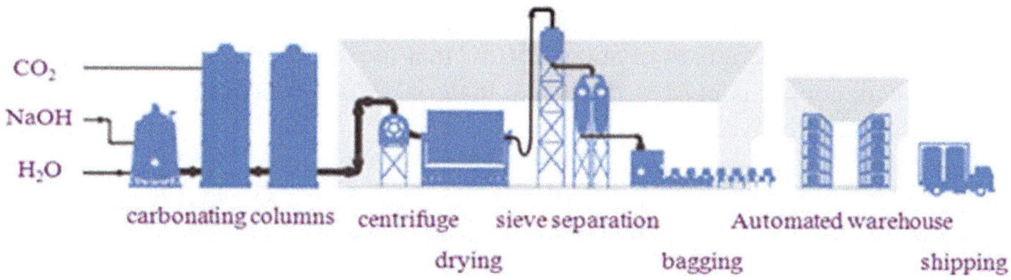

Fig. 10.34 AGC sodium bicarbonate production using caustic soda and CO_2

Sodium bicarbonate has a multitude of uses from everyday goods to pharmaceuticals and beyond. It can be made either from CO_2 and sodium carbonate or CO_2 and sodium hydroxide.

AGC chemical, the largest manufacturer of sodium bicarbonate in Japan has a commercial sodium bicarbonate production facility using sodium hydroxide and CO_2 [15]. Its chemical reaction is as Eq. 10.11.

$$CO_2 + N_aOH \rightarrow NaHCO_3 \tag{10.11}$$

From Stoichiometric formula 10.11, producing 1 kmol N_aHCO_3 requires 1 kmol CO_2, or producing **1** tone N_aHCO_3 will consume **0.52** tone CO_2.

There are 2 types of manufacturing processes in processing for sodium bicarbonate: Wet and dry. AGC uses a wet method to manufacture its sodium bicarbonate, which sodium bicarbonate is precipitated as crystals by infusing carbon dioxide gas into caustic soda derived by electrolysis to cause a reaction.

AGC sodium bicarbonate process [15] is shown in Fig. 10.34.

10.9.4 Using CO_2 as Feed to Produce Methanol

Currently, the feed gas for methanol production is mainly syngas, which are produced from natural gas using a steam reformer, generally, the syngas contains H_2, CO, and CO_2. Both CO and CO_2 can be hydrogenated to make methanol, but carbon monoxide is preferable because the current methanol synthesis catalysts are designed for CO hydrogeneration, using the same catalyst, the CO_2 hydro generation rate will be much slower.

The Methanol Synthesis Chemistry is shown in the following:

$$CO + 2H_2 \rightarrow CH_3OH \tag{10.12}$$

The formation of methanol is a reversible exothermic reaction, the reaction (10.12) is carried out at temperatures of about 250 °C that uses catalysts, such as Cu–ZnO–Al_2O_3, where copper is used as the active material which acts as catalysts to speed up the reaction effectively.

The reaction (10.12) is widely recognized for Methanol synthesis, but many studies and production information have indicated that CO_2 hydrogenation for methanol synthesis is feasible with Cu based catalyst.

$$CO_2 + 3H_2 \rightarrow CH_3OH + H_2O \tag{10.13}$$

Per chemistry formulas (10.12) and (10.13), CO_2 hydrogeneration making methanol requires 50% more hydrogen than CO hydrogeneration, therefore for single syngas to methanol synthesis, there seems to be no advantage for CO_2 hydrogenation, because, for natural gas steam reformer, both CO_2 or CO comes from C atom in natural gas, 1 Carbon atom, produce 1 CO or CO_2, and CO_2 and CO can be shifted each other reversely:

$$\mathrm{CO} + \mathrm{H_2O} \leftrightarrow \mathrm{CO_2} + \mathrm{H_2} \tag{10.14}$$

10.9.4.1 Supply External CO_2 Captured to Fully Utilize the Hydrogen Produced

However, depending on the feed natural gas composition, for lower-carbon natural gas, such as methane feed, the hydrogen in syngas will be in higher concentration, the CO_2 plus CO kg-moles produced will be much lower than Hydrogen kg-moles produced in the steam reformer. As a result, it requires un-balanced or unreacted hydrogen to be recircled, and eventually purged (Ref to Fig. 10.35), it causes lower efficiency and high energy demand.

Therefore, it also provides an opportunity to supply the external or internal supplemental CO_2 to fully utilize the Hydrogen, for example, set a CO_2 capture unit to capture reformer flue gas CO_2, to increase existing plant methanol production capacity. Optimize the plant H_2, and CO/CO_2 balance to minimize the purge of hydrogen.

Methanol Institute (QAFAC's Methanol Plant) [16] also proposed a CO_2 recovery and utilization scheme, the CO_2 in flue gas from the reformer burner is captured, and supplied to syngas compressor suction, mix with syngas for methanol conversion, it is expected to achieve the followings.

- 500 MTPD of CO_2 is recovered from the flue gas using MHI's proprietary KS–1™ solvent and injected into the synthesis loop for boosting Methanol production.
- The capacity of the Methanol Plant has increased by 300 MTPD with the addition of CO_2 in the synthesis gas mixture as excess H_2 is available for the methanol reaction.

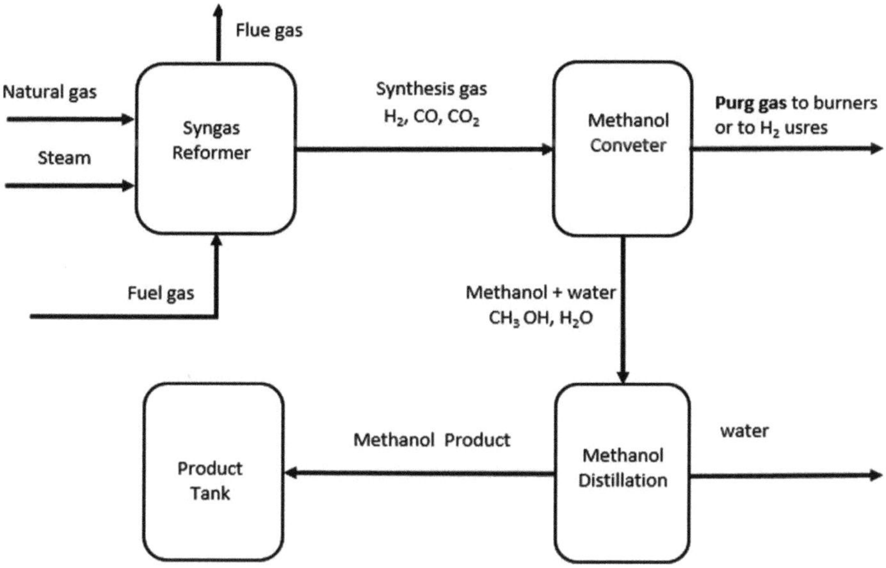

Fig. 10.35 Simple BFD of current methanol production process

- Thus, QAFAC's Methanol Plant became Self–sufficient for raw material (CO_2).

10.9.4.2 Develop a CO_2 Hydrogenation Making Methanol Process

From a thermodynamic point of view, manufacture of methanol by CO_2 hydrogenation is feasible, however, the current methanol conversion catalyst is not fully suitable for CO_2 hydrogenation operation, and it could be the barrier to the CO_2 hydrogenation process development. The existing commercial methanol synthesis catalyst offers a lower efficiency for CO_2 hydrogenation, due to the low conversion of CO_2 and its deactivation resulting from high water content during the conversion process.

Produce of methanol by CO_2 hydrogenation will be one of the effective CO_2 chemical utilizations, it is the direction for the development of CO_2 utilization, the study on commercial application is moving ahead. Two concerns have to be figured out to move forward with the actual application.

1. Hydrogen source

There are three scenarios to acquire hydrogen:

(1) Current, hydrogen is produced using the natural gas steam reforming approach, for this process, CO and CO_2 are also produced with hydrogen in syngas, the composition in the syngas is a large extent depending on the operating conditions and the H/C ratio of natural gas feed, there may be a hydrogen surplus resulting in hydrogen purge loss, optimization of the plant $H_2/CO/CO_2$ balance, and supply of external CO_2 to expend the methanol production will be the best choice.

(2) have additional hydrogen byproduct from some special facilities, for example, the reformer unit of a refinery, some petrochemical unit with dehydrogenation rection, etc. for this scenario, an integrated plant with methanol production to recover both hydrogen and CO_2 will be a great choice.
(3) New technology for sustainable Hydrogen production, such as electrolysis, biomass conversion or photolysis, etc. With this sustainable hydrogen-making commercial development, CO_2 captured can be utilized to a large extent.

2. Competitive catalyst of CO_2 hydrogenation

A. **Three stage heat exchanger reactors** [17]

To overcome the shortage of the current commercial catalyst ($CuO/ZnO/Al_2O_3$), an efficient process consisting of three-stage heat exchanger reactors was proposed for CO_2 hydrogenation [17], these three reactors are connected in series for CO_2 hydrogenation.

The catalyst volume in the conventional methanol reactor (CR) is divided into three sections to load reactors. The product stream of each reactor is conveyed to a flash drum to remove methanol and water from the unreacted gases (H_2, CO, and CO_2). Then, the gaseous stream enters the top of the next reactor as the inlet feed. This novel configuration increases CO_2 conversion almost twice compared to a one-stage reactor. This paper indicates that conversion of CO_2 and methanol production rate increase about 50% and 103% respectively compared to one-stage conversion reactor.

B. **New Catalyst of CO_2 hydrogenation** [18]

CO_2 hydrogenation to make methanol using CO_2 captured from oil and gas, as well as the chemical industry and a renewable energy-based "green hydrogen" source, is one of the sustainable approaches to achieving the goal of carbon neutrality.

Carbon capture Journal [18] reported a "New catalyst boosts hydrogenation of CO_2 to methanol". This study led by Prof. Deng Dehui from the Dalian Institute of Chemical Physics (DICP) of the Chinese Academy of Sciences (CAS), achieved CO_2 hydrogenation to methanol at low temperature and with high efficiency.

This study indicates that traditional metal oxide catalysts for CO_2 to methanol reaction typically require a high temperature, which causes undesired reverse water–gas shift (RWGS) side reactions, thus produces a large amount of CO as the by-product.

The introduction of transition metal components onto metal oxides can promote the activation of H_2, thereby reducing the reaction temperature, but this also facilitates excessive hydrogenation of CO_2 to CH_4, leading to lower methanol selectivity. In this study, the sulfur vacancy-rich few-layered MoS2, as the catalyst, could simultaneously activate and dissociate CO_2 and H_2 at low temperatures and even at room temperature, thereby facilitating the hydrogenation of CO_2 to methanol with high activity and selectivity.

Therefore, CO_2 hydrogenation making Methanol will be a mature technology for commercial application in the near future, which provides an opportunity to utilize the captured CO_2 gas more efficiently and contributes to achieving the goal of carbon neutrality.

10.9.5 CO_2 Storage

AS mentioned earlier, CO_2 chemical utilization can take only a limited amount of CO_2. The majority of CO_2 captured will be used to enhance oil/gas production or storage.

Following the CO_2 capture and compression/dehydration process, the liquid or supercritical CO_2 stream is delivered via pipeline to store in an underground geological formation. The pressure of the CO_2 stream required should be the pressure of the reservoir layer to be stored plus the pressure drop to overcome the friction loss of pipeline and injection tubes. It can be also combined with the process in the enhancement of the oil/gas recovery to eliminate CO_2 emissions and improve revenue.

CO_2 Storage and enhancement of oil/gas recovery is widely used for decades, this book will not discuss in detail this topic, but a brief introduction.

There are many oil and gas field that are reinjected sour produced gas from oil production and condensate production to some depleted oil and gas wells, including, CO_2, H_2S, and hydrocarbon gas, the purpose is to enhance oil or condensate production, this injection eliminates sulfur recovery facility, reduces the CO_2 and SO_2 emission, injection pressure usually is very high, some injection facilities require 450–600 barg pressure (such as KPO and TCO facilities in Kazakhstan) with multiple stage compressors.

Storage media include geologic sinks and the deep ocean [19]. Geologic storage options include deep saline formations (subterranean and subsea bed), depleted oil and gas reservoirs, formations for enhanced oil/gas recovery operations, and unminable coal seams. Deep ocean storage approaches include direct injection of liquid carbon dioxide into the water column at intermediate depths (1000–3000 m), or depths greater than 3000 m.

In most CO_2 EOR projects, much of the CO_2 injected into the oil reservoir may be temporarily stored. This is because the decommissioning of an EOR project usually involves the "blowing down" of the reservoir pressure to maximize oil recovery. This blowing down results in CO_2 being released, with a small but significant amount of the injected CO_2 remains dissolved in the immobile oil. If there will be no blowdown phase to be planned, the CO_2 storage will be the permanent storage.

Typical CO_2 storage and enhance gas recovery [20, 21] are consist of CO_2 capture, CO_2 compression, CO_2 transportation and CO_2 injection.

References

1. Carbon dioxide emissions coefficients detail factors, Table 1. https://www.eia.gov/environment/emissionsa/archive/coefficients.php#tbl2
2. A. Kohl, R. Nielson, *Gas Purification*, 5th edn. (Gulf Publishing, 1997). ISBN 0-88415-220-0
3. B. Burr, L. Lyddo, *A Comparison of Physical Solvents for Acid Gas Removal* (Bryan Research & Engineering, Inc.)
4. J. Dyment, S. Watanasiri, *Acid Gas Cleaning Using DEPG Physical Solvents: Validation with Experimental and Plant*, white paper (Aspen technology Inc., 2015)
5. L. Li, A.K. Voice, H. Li, O. Namjoshi, T. Nguyen, Y. Du, G.T. Rochelle, Amine blends using concentrated piperazine. Energy Proc. **37**, 353–369 (2013)
6. F. Closmann, T. Nguyen, G.T. Rochelle, MDEA/piperazine as a solvent for CO_2 capture. Energy Proc. **1**(1), 1351–1357 (2009)
7. J.M. Campbell, Gas conditioning and processing, 9th edn, vol. 2, in *The Equipment Module*, 2nd printing (J. M. Campbell and Company, Norman, Oklahoma, USA, 2014), p. 358
8. M.J. Jensen, C.S. Russell, D. Bergeson, C.D. Hoeger, D.J. Frankman, C.S. Bence, L.L. Baxter, Prediction and validation of external cooling loop cryogenic carbon capture (CCC-ECL) for full-scale coal-fired power plant retrofit. Int. J. Green-house Gas Control **42**, 200–212 (2015)
9. L.A. Pellegrini, G. De Guido, S. Ingrosso, Thermodynamic framework for cryogenic carbon capture, in *Proceedings of the 30 European Symposium on Computer Aided Process Engineering (ESCAPE30)*, May 24–27, 2020, Milano, Italy
10. Cryogenic carbon capture development final/technical report. Sustainable Energy Solution, 2019-09-28. https://www.osti.gov/biblio/1572908
11. PDF slides cryogenic carbon capture. Sustainable Energy Solution. https://www.arpa-e.energy.gov/sites/default/files/documents/files/CO2_Workshop_SES.pdf
12. C. Hoeger et al., Cryogenic carbon capture™ technoeconomic analysis, in *15th International Conference on Greenhouse Gas Control Technologies*, GHGT-15, March 2021
13. Patent, CO_2 separation & liquefaction system and method, Canadian Intellectual Property Office
14. Engineering guide, flow diagram of urea production process from ammonia and carbon dioxide. https://enggyd.blogspot.com/2010/09/flow-diagram-of-urea-production-process.html
15. AGC chemical website, products and service, sodium bicarbonate, manufacturing process. https://www.agc-chemicals.com/jp/en/products/detail/index.html?pCode=JP-EN-C010
16. 03 methanol production. Methanol institute. https://www.coursehero.com/file/24991286/MI-Combined-Slide-Deck-MDC-slides-Revisedpdf/
17. F. Samimi, M.R. Rahimpour, A. Shariati, *Development of an Efficient Methanol Production Process for Direct CO_2 Hydrogenation Over a Cu/ZnO/Al_2O_3 Catalyst*, Nov 2017. https://www.researchgate.net/publication/320941810
18. New catalyst boosts hydrogenation of CO_2 to methanol. Carbon Capture J. (2021). https://www.carboncapturejournal.com/news/new-catalyst-boosts-hydrogenation-of-co2-to-methanol/4559.aspx
19. H. Herzog, D. Golomb, Carbon capture and storage from fossil fuel use. Encyclopedia Energy (2004). https://www.academia.edu/1885399/Carbon_capture_and_storage_from_fossil_fuel_use
20. M.O. Saar et al., Assessment and optimization of carbon storage and combined EGR-CPG development from high-temperature natural gas reservoirs, 2016. https://geg.ethz.ch/project-cpg-egr/
21. B. Smit, Carbon capture and storage: introductory lecture, June, 2016, Faraday discussion. https://pubs.rsc.org/en/content/articlehtml/2016/fd/c6fd00148c

Chapter 11
Technical and Economic Evaluation of Energy-Saving Measures

Abstract Once you have completed energy and exergy balance, identified the energy-saving and carbon mitigation potential, proposed the solutions of energy-saving and carbon mitigation, it is needed to have a technical and economic evaluation of energy-saving projects, this chapter introduces the time value of capital, as well as the static evaluation method and dynamic evaluation method for the energy-saving project; proposed estimation methods of economic benefits of energy-saving measures based on currency balance, it includes (1) determination of fuel price, (2) determination of steam and thermal energy price, (3) determination of the steam and power price of backpressure power generation; (4) determination of the price of electricity, and water, etc.

Keywords Capital time-value · Payback period · Steam price · Power-steam cogeneration · Energy-saving benefits

When the energy consumption analysis and calculation of the plant are completed, and the directions and measures for energy-saving improvement are proposed, it is needed to think about the economics of the energy-saving measures, which raises the question of the technical and economic evaluation of energy-saving measures. We know that the purpose of energy saving is not only to reduce energy consumption, and CO_2 mitigation,

When the energy utilization analysis and calculation of the plant are completed, and the directions and measures for energy-saving improvement are proposed, it is needed to think about the economics of the energy-saving measures, which raises the question of the technical and economic evaluation of energy-saving measures. We know that the purpose of energy saving is not only to reduce energy consumption, and CO_2 mitigation, but ultimately still comes down to obtaining the greatest economic benefits with as less energy, resources and capital investment as possible. Therefore, the technical and economic evaluation of energy-saving measures is an important part of energy-saving work. Technical economy is a specialized subject. Only the concepts and evaluation indicators [5, 6] involved in the evaluation of energy-saving measures and the methods for determining energy-saving benefits are discussed here.

T. A. Chen, *Energy Saving and Carbon Reduction*,
https://doi.org/10.1007/978-981-19-5295-1_11

11.1 Time Value of Capital

In technical and economic analysis and evaluation, the time value of funds must first be considered. The so-called time value of funds refers to the objective law that funds can continue to increase in value in the process of capital circulation. But if a sum of money is stored, no matter how long it passes, the amount of money will not change. An enterprise can build plants and purchase raw materials to create profits if it has funds. To a certain extent, bank interest reflects the time value of funds.

11.1.1 Interest

We deposit a sum of money in the bank, and after a period of time, we can get a certain amount of new currency. Generally, the original deposit currency is called principal, and the excess currency is called interest. The ratio of interest to principal in a unit of time (that is, an interest calculation cycle) is called the interest rate. Similarly, if you borrow money from a bank, you will have to pay more than the original loan amount when you repay it. The overpayment is also called interest. There are several types of interest, such as simple interest, compound interest and continuous compound interest. The most common methods in China are simple interest and compound interest:

1. Simple interest

Only the principal generates interest, and the interest is no longer calculated, that is, the calculation method that does not accumulate the previous interest into the principal is called simple interest. The magnitude of interest has a linear relationship with time, and the calculation formula for the sum of principal and interest is:

$$F_n = P(1 + n.i) \tag{11.1}$$

where: F_n—Sum of principal and interest at the end of the n cycle;

P—principal;

i—interest rate;

n—The number of interest calculation cycles (month, quarter, half year, year, etc.).

2. Compound interest

In addition to generating interest on the principal, interest is also calculated, that is, the calculation method of accumulating the previous interest to the principal is called compound interest. The calculation formula for the sum of costs and interests is:

$$F_n = P(1 + i)^n \tag{11.2}$$

The symbol in the formula is the same as before.

Unless otherwise stated, the interest calculation cycle is generally one year. In this case, compound interest is called annual compound interest and interest rate also refers to annual interest rate.

3. Short-term compound interest and real interest rate

In many cases, the interest calculation cycle is not one year, but every six months, quarterly, monthly or even daily. At this time, the number of interest calculations per year is correspondingly changed to 2, 4, 12, or 365. For convenience, the interest rate is generally expressed as an annual interest rate, which is called the nominal interest rate. The interest rate for each interest accrual period is the annual interest rate (nominal interest rate) divided by the number of interest accruals per year.

Obviously, the monthly interest rate is higher than the annual interest rate. This kind of actual interest rate period is shorter than one year and the resulting interest rate is called the actual interest rate. The actual interest rate is mainly caused by compound interest, which is generally higher than the nominal interest rate. The shorter the interest rate period, the higher the interest rate.

If r is the nominal interest rate, i is the actual interest rate (converted to the annual interest rate), and m is the number of interest accruals, then at the end of the year:

$$S = P(1 + r/m)^m \tag{11.3}$$

$$\text{Actual interest rate} \quad i = \frac{P\left(1+\frac{r}{m}\right)^m - P}{P} = \left(1+\frac{r}{m}\right)^m - 1$$

$$\text{Or nominal interest rate} \quad r = m\left[(1+i)^{1/m} - 1\right]$$

The above formula shows the relationship between the nominal interest rate and the actual interest rate.

Example 11.1 Deposit 1000 yuan in the bank, when the interest calculation period is 1 year, and the annual interest rate is 12%, find the sum of principal and interest after 1 year under the condition that the interest calculation period is 1 month.

$$\text{Solution:} \quad S = 1000 \times \left(1 + \frac{0.12}{12}\right)^{12} = 1126.8\ \text{Yuan}$$

$$\text{Real interest rate} \quad i = \left(1 + \frac{0.12}{12}\right)^{12} - 1 = 0.1268$$

4. Continuous compound interest

In short-term compound interest, the interest accrual period m keeps increasing until it reaches infinity, which is called continuous compound interest. At this time, the actual interest rate:

$$i = \lim_{m\to\infty}\left(1+\frac{r}{m}\right)^m - 1 = \lim_{m\to\infty}\left[\left(1+\frac{r}{m}\right)^{\frac{m}{r}}\right]^r - 1$$

Because it has been proved in calculus that the limit of the function $(1+\alpha)^{1/\alpha}$ is e, where $\alpha = \frac{r}{m}$, the actual interest rate becomes:

$$i = \lim_{m\to\infty}\left[\left(1+\frac{r}{m}\right)^{\frac{m}{r}}\right]^r - 1 = e^r - 1 \tag{11.4}$$

When the principal is P and the annual interest rate is r, the sum of principal and interest after n years of continuous compounding is:

$$S_n = P(1+i)^n = P\left(1+e^r-1\right)^n = Pe^{rn} \tag{11.5}$$

The significance of continuous compound interest is that in actual economic activities, cash income and expenditure do not always occur at the beginning and end of the interest calculation cycle, but occur at any time every day. Currency flows in and out almost continuously, and profits continue to benefit Investing in corporate activities. The use of continuous compound interest is more accurate and realistic.

11.1.2 Currency Equivalent, Present Value and Future Value

1. Equivalent

Equivalence when considering the time value of money, that is, currencies with different absolute values but equal value at different times. For example, with an annual interest rate of 10%, 100 yuan this year is equivalent to 110-yuan next year and 121 yuan the year after next year. From the point of view of the absolute value of the currency, the two are different, but at 10% interest rate calculated by compound interest, that is, 100 yuan this year and 121 yuan the year after next year are equivalent. Currency equivalence is a very important concept in technical economy, and this concept is often used in program comparison. Currency equivalence is determined by three factors: (1) the size of the amount: (2) the time when the amount occurs; (3) the size of the interest rate.

2. Present value

It refers to the monetary value at the present moment, or to convert the monetary value at a certain moment into the current value. You can convert 121 yuan two years later into 100 yuan now. This conversion process is called discounting, and the value converted to the present moment is the present value, also known as the "initial value".

3. Future value

Under the condition of compound interest, the "present value" can be converted to the value at a certain moment in the future at a certain interest rate, which is also called the “end value” or “final value”. The relationship between future value and present value is:

$$\text{Future value} = \text{present value} + \text{compound interest}$$

11.1.3 Equivalent Value Calculation of Funds

From the concept of time value of funds, it can be seen that during the movement of funds, as time goes by, the absolute amount of funds will change, that is, increase in value at different moments. Different absolute amounts of money at different time points may have equal actual economic utility under certain time value or interest rate conditions. This is the concept of equivalent value of money.

The calculation of the equivalent value of funds is based on the compound interest calculation formula. In the calculation of the equivalent value, the reference point must first be selected. Generally, the starting point of the calculation period (usually the moment of initial investment or deposit or borrowing) is used as the reference point. A sum of money at the reference point is called the present value. Calculated at a certain interest rate, the value at a certain time in the future after a certain time interval is called the final value or future value.

1. Compound interest final value formula

$$F = P(1+i)^n \tag{11.6}$$

where: P—Present value of funds;

F—The final value of funds;

i—Interest;

n—Number of interest accrual periods.

In the formula $(1+i)^n$ is called the final value coefficient of compound interest. Since P and F are paid in one time, they are also called one-time payment (or single payment) compound interest terminal value coefficient.

2. Compound interest present value formula

When F is known and P is desired, the formula (11.6) can be shifted to:

$$P = F(1+i)^{-n} \tag{11.7}$$

The above formula can be used to convert funds at a certain moment in the future into present value. $(1+i)^{-n}$ is called the present value coefficient.

3. Equal series final value formula

For the same number of deposits at the end of each year, find the sum of the principal and interest to be withdrawn at the end of the period. At this time, the following formula should be used:

$$F = A\left[\frac{(1+i)^n - 1}{i}\right] \tag{11.8}$$

The above formula is called the final value formula of the equal amount series, $\left[\frac{(1+i)^n-1}{i}\right]$ is called the final value coefficient of the equal amount series, A is the payment amount of the equal amount series, and is generally called an annuity when the interval is 1 year. Therefore, the above formula is also called the final value formula of annuity.

4. Equivalent series savings formula

If F is known and A is desired, then

$$A = F\left[\frac{i}{(1+i)^n - 1}\right] \tag{11.9}$$

This formula is often used to accumulate a certain amount of funds, and the amount should be stored in equal amounts in each period, which is called the equal amount series deposit formula. $\left[\frac{i}{(1+i)^n-1}\right]$ is called the equal series deposit coefficient.

5. Equal series fund recovery formula

If P is known and A is desired, substituting $F = P(1+i)^n$ into formula (11.9), then:

$$A = P\left[\frac{i(1+i)^n}{(1+i)^n - 1}\right] \tag{11.10}$$

This formula is often used for a current investment P, which will be recovered in equal amounts at the end of several years in the future, to find the amount of funds A. recovered each year, so $\left[\frac{i(1+i)^n}{(1+i)^n-1}\right]$ is called the equal series fund recovery coefficient. In the technical and economic evaluation of energy-saving measures, when the economic life n of the equipment is determined, at a realistic interest rate, for an investment P, how much capital A should be recovered every year in order to fully recover the investment, refer to the above formula.

6. Equal series present value formula

$$P = A\left[\frac{(1+i)^n - 1}{i(1+i)^n}\right] \tag{11.11}$$

This formula is used to find the present value when the installment is paid (received), $\left[\frac{(1+i)^n-1}{i(1+i)^n}\right]$ is called the current value coefficient of the equal series. Formula (11.11) is also known as the formula of annuity annual value.

In the evaluation of energy-saving measures, when the annual net benefit A of energy-saving measures is known, this formula can be used to estimate the present value of the total benefit at the end of the period.

Table 11.1 Commonly used fund equivalent value calculation formula table

Order	Formula name	known	Find	Equation	Formula code
1	Compound interest final value formula	i, n, P	F	$F = P(1+i)^n$	$F = P(F/P, i, n)$
2	Compound interest present value formula	i, n, F	P	$P = F(1+i)^{-n}$	$P = F(P/F, i, n)$
3	Annuity Final Value Formula	i, n, A	F	$F = A\left[\frac{(1+i)^n-1}{i}\right]$	$F = A(F/A, i, n)$
4	Equal savings formula	i, n, F	A	$A = F\left[\frac{i}{(1+i)^n-1}\right]$	$A = F(A/F, i, n)$
5	Equivalent fund recovery formula	i, n, P	A	$A = P\left[\frac{i(1+i)^n}{(1+i)^n-1}\right]$	$A = P(A/P, i, n)$
6	Annuity present value formula	i, n, A	P	$P = A\left[\frac{(1+i)^n-1}{i(1+i)^n}\right]$	$P = A(P/A, i, n)$

The calculation coefficients for the equivalent value of funds for general compound interest are summarized in Table 11.1.

For the calculation of the equivalent value of continuous compound interest, based on the continuous compound interest formula $F = Pe^r$, results similar to those in Table 11.1 can be derived. However, it is not commonly used in the evaluation of energy-saving measures, especially in petrochemical industries.

11.2 Static Evaluation Method

Funds are limited, in order to save limited investment, economic efficiency must be emphasized. Before making an investment decision, a feasibility study must be carefully conducted, and the economic benefits of the investment project must be calculated and analyzed. When there are more than one options available, the economic benefits of each option must be compared and selected. This process of analysis and demonstration is called project economic evaluation. The purpose of economic evaluation is to provide a scientific basis for investment decision-making, reduce and avoid investment decision-making mistakes, and improve economic benefits.

Project economic evaluation is divided into financial evaluation and national economic evaluation. Financial evaluation is to examine the project's cost, profit, benefit and loan repayment ability from the perspective of the enterprise under the current national fiscal and tax system and price conditions, that is, financial feasibility. National economic evaluation is to calculate and analyze the net benefits brought by the project to the national economy from the perspective of the country as a whole. Energy-saving measures generally only conduct corporate financial evaluations.

Project financial evaluation is divided into two categories: static evaluation and dynamic evaluation according to whether it considers the relationship between funds and time. The following introduces several static economic evaluation indicators and their calculation methods commonly used in the evaluation of energy-saving measures.

In the financial evaluation of investment projects, if the time value of funds is not considered, it is called a static evaluation method. The commonly used static price indicators mainly include the investment profit rate, the payback period and the annual cost comparison method.

11.2.1 Investment Profit Rate

1. Investment profit rate

For an investment project, the ratio of its annual profit or average annual profit to the project investment is called the investment profit rate. It reflects the ability of a unit investment to make a profit each year. The calculation formula is

$$E_A = \frac{R}{I} \times 100\% \tag{11.12}$$

where: E_A—Investment profit rate;

R—Annual profit rate;

I—Investment amount.

The static investment profit rate refers to the ratio (percentage or fraction) of the total annual sales profit of the project in a normal year after the project reaches the designed production capacity and the total project investment (including construction period interest). For projects with large changes in the total annual profit during the production period, the annual average total profit during the production period shall be calculated.

2. Investment profit and tax rate

From a corporate perspective, tax is a kind of expenditure, so companies and investors care about the profitability of their investment after paying taxes. But from a national perspective, taxes are also part of the project's profitability. Therefore, the investment profit and tax rate are another indicator to measure the profitability of the project.

The so-called investment profit and tax rate refers to the ratio of the annual profit, total tax, or the average annual total profit and tax during the project's production period to the total investment in a normal production year after the project reaches the designed production capacity. The calculation formula is:

$$Investment\ profit\ and\ tax\ rate = \frac{Total\ annual\ profit\ and\ tax}{Total\ investment} \times 100\% \tag{11.13}$$

3. Benchmark investment profit rate

Total annual profit and tax = annual sales revenue—total annual cost—annual technology transfer fee—annual non-operating net expenditure

If the total annual profit and total investment of the project are known, the profitability of the project can be calculated by formula (11.12). For example, is 10% or 0.1 of this investment profit rate high or low? Is this item desirable? This is to be compared with a certain standard, which is the benchmark investment profit rate. It is a quantitative standard or boundary for measuring the desirability of an investment project. Only when the investment profit rate of the project to be selected is higher than the benchmark investment profit rate, the project is feasible.

The benchmark investment rate of return is set by each company, and is called the minimum allowable rate of return. Generally speaking, it is a function of industry type and risk and is related to the company's current economic conditions (how much capital, bank interest rates) and investment policies.

11.2.2 Payback Period

1. Payback period

The payback period is the time required for the project's net profit to offset the investment amount, expressed in years. When the profit in each year is the same or close to or can be averaged:

$$P_t = \frac{I}{R} \tag{11.14}$$

where: P_t—payback period;

I—investment amount;

R—Annual net profit.

Like the investment profit rate, there can be different understandings of the annual net income (or profit) and the amount of investment. In addition, there is also the question of when the investment payback period is calculated, that is, from the date when the project investment starts, or from the time when the project is completed and put into production. If calculated from the date of commissioning, it can be seen from Eq. (11.14) that the payback period can be regarded as the reciprocal of the investment profit rate.

It is stipulated that the investment payback period generally starts from the beginning of the project investment, that is, includes the construction period. However, at the same time, the investment payback period from the start of production should be stated.

The premise of formula (11.14) is that the net income of each year is the same. When the net income of each year is different, the calculation formula is (11.16).

When using the investment payback period indicator to evaluate the project, it is hoped that the investment is recovered as soon as possible, but like the investment profit rate, it should be compared with a standard payback period. The standard payback period is generally the reciprocal of the standard investment profit rate. The standard payback period for energy-saving technical measures is generally 3–5 years.

The payback period is an important indicator that reflects the financial solvency of the project. When investors are very concerned about the speed of investment recovery, the payback period is easy to give people a deep impression. Its advantages are simple, intuitive and easy to understand, so it is widely used. But its main disadvantage is that it does not consider the length of the project's life span and the benefits after the current period, which may lead to erroneous conclusions.

2. Annual conversion cost method

In many cases, energy-saving measures are only partial, and it is difficult to accurately estimate the overall economic benefits. There are also some environmental protection projects that have no economic benefits but only social benefits. For this type of project, if each option has the same function and meets the same process purpose, then only the cost can be compared, and the annual conversion cost can be used to select the options.

The essence of this method is to proceed from an economic point of view. It is believed that the economics of an option is determined by the size of the one-time expense (investment) and the amount of operating expenses (annual operating expenses) that frequently occur after the plan is realized. When choosing a plan, always strive to choose the smallest sum of the above two costs. However, the investment (one-time expenditure) and the cost cannot be simply added. The total investment should be allocated to the annual investment recovery cost. The other part is the project's annual operating or maintenance costs.

$$Z = CR + OM \tag{11.15}$$

where: Z—Annual conversion cost;

CR—Annual investment recovery cost, $CR = E_A \times I$;

OM—Annual operating and maintenance costs.

E_A is the conversion coefficient for converting investment into annual investment expenses. Generally, the benchmark investment profit rate (the reciprocal of the standard investment payback period) can be used as the value of the conversion coefficient.

11.2.3 Cash Flow and Cash Flow Curve

1. The concept of cash flow

The construction of a project must raise funds, invest in labor, pay for construction, purchase machinery and equipment and other infrastructure costs. For engineering projects, these costs are cash outflows. When the project is completed, it is necessary to invest working capital to purchase raw materials, auxiliary materials and pay workers' wages. These are all expenditures from the project and are cash outflows.

When the project is completed and put into production, product sales can get sales income. These are the income items of the project, which are called "cash inflows".

In technical and economic evaluation, "cash inflow" is expressed as a positive value, and "cash outflow" is expressed as a negative value. "Cash inflow" and "cash outflow" are commonly referred to as cash flow. The difference between "cash inflow" and "cash outflow" is called net cash flow.

Usually, the inspection of an engineering project takes the construction period and the entire life span after it is put into operation as the inspection object.

In the economic evaluation of construction projects, cash inflow mainly refers to the project's product sales income in the enterprise, recover residual value of fixed assets and recover working capital. However, when calculating the residual value of fixed assets, various removal, decontamination pollutant treatment and environmental restoration expenses should be deducted.

"Cash outflow" refers to the investment, operating costs, working capital, and various taxes paid to the state during the construction and production of the project.

2. Cash flow curve

After calculating the net cash flow, with time as the horizontal axis and cumulative net cash flow as the vertical axis, the cumulative cash flow curve or discounted cash flow curve of an engineering project can be drawn. The former is static and does not take into account the time factor, while the latter is dynamic and takes into account the time value of funds. For conventional investment projects, the cash flow curve is shown in Fig. 11.1.

Before the start of the project, its cash position is zero, (point A). Development, design and other preparations in the initial stage of the project will spend funds, and the cash flow will be negative AB. After the construction investment period, plants, production facilities and other equipment will be built, and investment will still be required until the trial production is completed (The curve drops to point D), the income derived from product sales exceeds production costs and other business expenses, and the cash flow curve rises until the break-even point, when all income and expenditures for this project are exactly balanced. From now on, the cash flow curve continues to rise, until the end of the economic life of the project, the cash flow curve gradually declines.

As shown in Fig. 11.1, curve (1) is a cumulative cash flow curve drawn without considering the time factor. Curve (2) is the discounted cash flow curve.

In this figure, the characteristic points of curve (1) are as follows:

AB is the preliminary research fee (feasibility study, consulting fee);

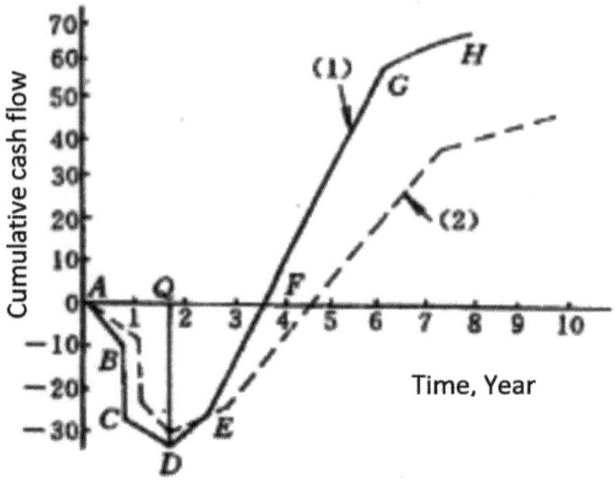

Fig. 11.1 Cash flow curve

BC	is the project construction infrastructure investment and construction cost;
CD	is preparing expenses for trial production;
DE	is the trial production stage (part of the sales revenue has already been made);
EFGH	is profitable production;
AF	is the static payback period;
QD	is the cumulative maximum investment (or cumulative liabilities);
Q	is the peak time of debt;
F	is the break-even point.

Correspondingly, various characteristic points can be found on the curve (2) in Fig. 11.1. The intersection of curve (2) and the horizontal axis is the dynamic payback period.

3. The cash flow expression of static investment payback period

The above curve intuitively and comprehensively indicates the desirability of an engineering project. When comparing plans, it is easy to judge the pros and cons of each plan from the cash flow curve.

With the concept of cash flow, the calculation expression for the static payback period can be written as

$$\sum_{t=0}^{P_t}(CI - CO)_t = 0 \quad (11.16)$$

where: P_t—Payback period; CI—Cash inflow; CO—Cash outflow; $(CI - CO)_t$—Net cash flow in year t.

From the cash flow chart, the payback period is AF from the beginning of the investment, and QF from the beginning of production.

11.3 Dynamic Evaluation Method

The biggest disadvantage of the static evaluation method of investment benefits is that it does not consider the relationship between capital and time, that is, it does not consider the time value of capital and the mode of cash flow occurrence (time, amount distribution). In order to carry out comprehensive investment analysis and decision-making, it is necessary to convert cash flows occurring at different times into equivalent value conversions, and then compare and evaluate them on the same benchmark. This evaluation method that considers the time value of funds is called a dynamic evaluation method.

11.3.1 Dynamic Payback Period

Investors are generally very concerned about the rate of return on investment. It is hoped that the earlier the repayment, so the payback period is a commonly used economic evaluation indicator. The dynamic payback period makes up for the shortcomings of the static payback period that does not consider the time value of funds.

The present value P of a future fund F is:

$$P = F\left[\frac{1}{(1+i)^n}\right]$$

where i—Discount Rate; n—Years.

Therefore, the net cash flow in year t, $R_t = (CI - CO)_t$, which can be converted to its present value $(R_{DCF})_t$ is called discounted cash flow:

$$(R_{DCF})_t = R_t\left[\frac{1}{(1+i)^n}\right] \tag{11.17}$$

According to the aforementioned definition of the static payback period, the dynamic payback period is the time required for the project from the beginning of the investment to the cumulative discounted cash flow equal to zero. The calculation formula is:

$$\sum_{t=0}^{P_t} R_t\left[\frac{1}{(1+i)^t}\right] = 0 \tag{11.18}$$

where: $R_t = (CI - CO)_t$—Net cash flow in year t; i—Benchmark discount rate; P_t—Dynamic payback period.

Example 11.2 The net cash flow of a chemical plant in each year is as follows (ref to Table 11.2). Try to draw a cumulative cash flow diagram and a cumulative discounted cash flow diagram. Calculate the static and dynamic payback period. Let the discount rate be 0.1.

Solution From the above table, the cash flow of each year can be calculated. The cumulative cash flow and the cumulative discounted cash flow are shown in Table 11.3 and Fig. 11.2.

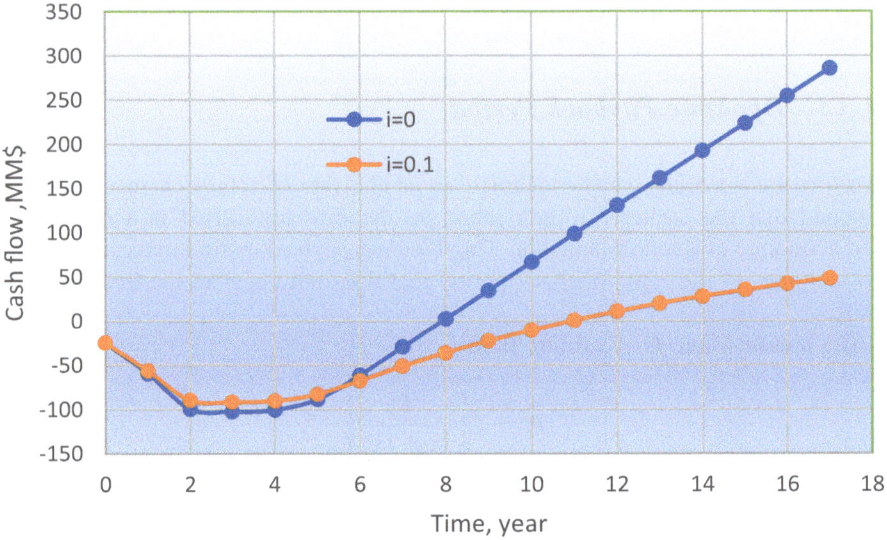

Fig. 11.2 Example 11.2 cumulative cash flow diagram

Table 11.2 Example data sheet (MM$)

Year	Fixed investment	Working capital	Net income	Year	Fed investment	Working capital	Net income
0	25			9			32
1	35			10			32
2	40			11			32
3		5	2	12			32
4			12	13			31
5			17	14			31
6			27	15			31
7			32	16			31
8			32	17			31

Table 11.3 Example 11.2 calculation results (MM$)

Year	Case flow	Discount factor	Discounted cash flow	Cumulative cash flow	Cumulative discounted cash flow
0	– 25	1.000	– 25.00	– 25.00	– 25.00
1	– 35	0.909	– 31.82	– 60.00	– 56.82
2	– 40	0.826	– 33.06	– 100.00	– 89.88
3	– 3	0.751	– 2.25	– 103.00	– 92.13
4	2	0.683	1.37	– 101.00	– 90.76
5	12	0.621	7.45	– 89.00	– 83.31
6	27	0.564	15.24	– 62.00	– 68.07
7	32	0.513	16.42	– 30.00	– 51.65
8	32	0.467	14.93	2.00	– 36.72
9	32	0.424	13.57	34.00	– 23.15
10	32	0.386	12.34	66.00	– 10.81
11	32	0.350	11.22	98.00	0.40
12	32	0.319	10.20	130.00	10.60
13	31	0.290	8.98	161.00	19.58
14	31	0.263	8.16	192.00	27.74
15	31	0.239	7.42	223.00	35.16
16	31		6.75	254.00	41.91
17	31	0.198	6.13	285.00	48.04
Total	285		48.04		

It can be seen from Table 11.3 and cumulative cash flow chart 11.2 that the static payback period is about 8 years, and the dynamic payback period is about 11 years. All of the above includes a 3-year construction period.

11.3.2 Simplified Calculation Equation for Dynamic Payback Period

The above list algorithm is relatively cumbersome and is suitable for the situation where the annual cash flow is uneven equal.

When the initial investment can be simplified as a one-time investment I (or discounting each year's investment into I), and the annual income after production is equal to the annuity A, most of the energy-saving projects and measures fall in this case, as shown by the formula (11.10). It can be seen that the relationship between A and I is

$$A = I\left[\frac{i(1+i)^n}{(1+i)^n - 1}\right]$$

When the investment payback period is represented by t:

$$A = I\left[\frac{i(1+i)^t}{(1+i)^t - 1}\right] = \frac{I.i(1+i)^t}{(1+i)^t - 1}$$

or

$$A\left[(1+i)^t - 1\right] = I.i(1+i)^t$$

$$(1+i)^t = \frac{A}{A - I.i} = \frac{1}{1 - \left(\frac{Ii}{A}\right)}$$

Logarithm of both sides

$$t.\lg(1+i) = lg1 - lg\left(1 - \frac{Ii}{A}\right)$$

Therefore

$$t = \frac{-\lg(1 - Ii/A)}{\lg(1+i)} \tag{11.19}$$

The above formula is a commonly used formula for calculating the dynamic investment payback period. For the known annual energy-saving benefits, one-time investment I (if the instalment investment can be discounted) and the annual interest rate i, we can directly calculate the investment payback period.

Example 11.3 The energy-saving renovation of the heating furnace of a petrochemical plant requires a one-time investment of 1.76 MM$. The construction period is 2 years, and after the third year of production, the net energy-saving benefit can be 0.71 $MM per year. Try to estimate the payback period. The benchmark discount rate is taken as 0.1.

Solution Static investment payback period: 1.76/0.71 = 2.5 years.

In order to use the simplified formula, the investment must be converted from year 0 to year 1, $I = 1.76 \times 1.1 = 1.936$ MM$. Therefore, the dynamic payback period is:

$$t = \frac{-\lg\left(1 - \frac{193.6 \times 0.1}{71.2}\right)}{\lg(1 + 0.1)} = 3.3 \text{ years}$$

From the calculation results, the dynamic payback period is 0.8 years longer than static payback period.

11.3.3 Net Present Value Method

1. The meaning of net present value and its calculation

Convert the net cash flow $R_t = (CI - CO)_t$. of an investment project in each year during its life cycle into the base moment (usually the project start time, that is, 0 years), the sum of which is called the net present value (NPV) of the project. The calculation formula is:

$$NPV = \sum_{t=0}^{n} (CI - CO)_t \left[\frac{1}{(1+i)^t} \right] \quad (11.20)$$

where: n—Project construction and production service years;

i—Set discount rate or benchmark investment profit rate;

t—years;

$(CI - CO)_t$—Net annual cash flow in t year.

Calculation method: (1) List the year-by-year cash flow and cash outflow. Both take the positive value and the difference between the two is the annual net cash flow. (2) Multiply the net cash flow of each year by the corresponding year's discount rate. The discount factor is used to obtain the discounted cash flow year by year; (3) the net present value of the project is obtained by adding the discounted cash flow of each year.

As a result of the calculation, if the net present value is 0, it means that in the sense of time value equivalent, the cash inflow of the project is just enough to offset t cash outflow of the project, that is, the final cumulative discounted cash flow of the project is equal to 0, that is, no profit or no deficit. If the net present value is greater than 0, it means the profitability of the project at a given discount rate. The greater the net present value, the greater the profitability. Conversely, if the net present value is less than 0, it means that the investor will not only make no profit, but it is equivalent to an immediate loss of that amount of funds, and the project is obviously undesirable. Therefore, the net present value is a measure of the profitability of an investment project. When using the net present value method to evaluate a single plan, if the discount rate is based on the benchmark investment return rate, then when:

When $NPV > 0$, The solution is feasible,
When $NPV \leq 0$, The solution is infeasible.

2. Determination of discount rate

The value of the discount rate has a great influence on the project's net present value. As can be seen from the following example, for the same project, the lower the discount rate, the greater the net present value. Conversely, the higher the discount rate, the smaller the net present value.

Example 11.4 A project started with an investment of $1000, and it began to make profits after 3 years, and the annual income was $400 for 4 consecutive years. Ask what is the net present value of the project when the discount rate is $i = 0, i = 0.05, i = 0.1, i = 0.15$?

Solution The cash flow diagram of the project is

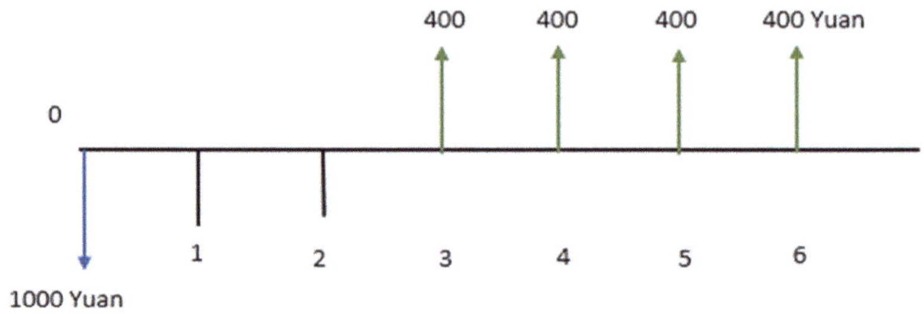

$$i = 0, \quad NPV = 1000 + 400 + 400 + 400 + 400 = 600(\$)$$

$$i = 0.05, \quad NPV = 1000 + \frac{400}{(1.05)^3} + \frac{400}{(1.05)^4} + \frac{400}{(1.05)^5} + \frac{400}{(1.05)^6} = 286(\$)$$

$$i = 0.10, \quad NPV = 1000 + \frac{400}{(1.1)^3} + \frac{400}{(1.1)^4} + \frac{400}{(1.1)^5} + \frac{400}{(1.1)^6} = 47.9(\$)$$

$$i = 0.15, \quad NPV = 1000 + \frac{400}{(1.15)^3} + \frac{400}{(1.15)^4} + \frac{400}{(1.15)^5} + \frac{400}{(1.15)^6} = -136.5(\$)$$

It can be seen from this example that the higher the discount rate, the smaller the net present value, or even a negative value. When the investment is all loans, if the discount rate is based on the loan interest rate, the net present value is actually the net profit after repaying all the principal and interest of the loan. The higher the discount rate, the higher the loan interest or investment cost, and the lower the net profit of course. Therefore, using the net present value method to evaluate investment plans must correctly specify the discount rate, otherwise it will lead to wrong conclusions. The discount rate determined according to the investment effect pursued by the enterprise is called the benchmark discount rate.

The basic principles for determining the benchmark discount rate are:

(1) If the source of investment is a loan, the benchmark discount rate should be higher than the interest rate of the loan. The margin depends on the size of the project risk and the requirements of investors, and it is generally about 5%. If the loan interest rate is 6%, the benchmark discount rate should be around 11%.
(2) If the source of investment is self-owned funds, the benchmark investment rate of return is generally used as the benchmark discount rate. Because if you don't invest in this project, the funds can be invested in other projects and get normal returns.
(3) If the investment sources are both self-owned funds and loans, the weighted average of the proportions of the two and their interest rates should be calculated as the benchmark discount rate.

The main disadvantage of the net present value is that it compares several plans with different investment amounts, which cannot reflect how many investment assets the net present value is derived from. For this reason, people propose to use the net present value ratio to determine the order of the pros and cons of the plans.

3. Net present value ratio

Net present value ratio (NPVR) is also called net present value rate, present value index or investment present value rate. It is the ratio of net present value divided by investment present value. In fact, it represents the net present value of unit investment, and the calculation formula is

$$NPVR = NPV/I_P \tag{11.21}$$

where: I_P—Present value of investment I.

In the case where the life periods of the various plans are not equal, the same calculation period should be used, or the least common multiple of each life period should be used as the calculation period for comparison.

11.3.4 Internal Rate of Return

1. The meaning of internal rate of return

The net present value of the project varies greatly with the value of the discount rate. The relationship between the project's net present value and the discount rate is shown in Fig. 11.3.

It can be seen from Fig. 11.3 that under a given service life and cash flow model, the project's net present value changes drastically with the value of the discount rate. The smaller the discount rate, the larger the net present value, and the larger the discount rate, the lower the net present value. When the discount rate reaches a certain value (for example, between 0.10 and 0.15), that is, $i = i^*$. The net present

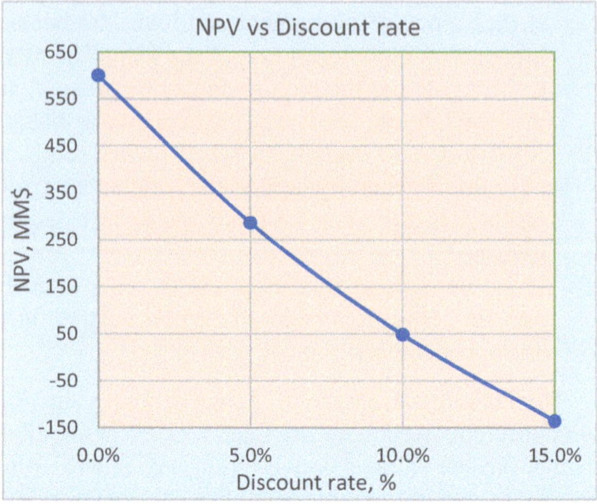

Fig. 11.3 Relationship between net present value and discount rate

value $NPV = 0$, the discount rate continues to increase, and the net present value becomes negative.

The discount rate ($i*$) when the cumulative present value of the net cash flow of each year of the investment project during its life (including the construction period and the production service period) is equal to zero, or to put it simply, discount rate ($i*$) of making net present value of the investment project equal to zero is the project's discounted cash flow rate of return, or internal rate of return (IRR).

The meaning of the discounted cash flow rate of return is that if the investment of the project is all financed by borrowing, when the loan interest rate i is equal to IRR or i^*, the entire income during the entire life span is just enough to repay the principal and interest of the loan. Therefore, IRR represents the highest interest rate or capital cost that the project can afford. If all the funds required for the project are self-owned funds, IRR represents a measure of the project's ability to generate income. If the IRR is greater than the benchmark investment return rate, it indicates that the project's return exceeds the average investment return rate of the industry.

Therefore, when evaluating a single plan, the project's internal rate of return i^* should be greater than the loan interest rate, or should be equal to or greater than the benchmark investment rate of return, which is economically desirable. The larger IRR, the better the plan. When comparing multiple plans, under other conditions that are similar, you can choose the plan with the highest discounted cash flow rate of return.

2. Calculation of internal rate of return

By definition, the internal rate of return is i^* that satisfies the following formula:

$$\sum_{t=0}^{n} \frac{(CI - CO)_t}{(1 + i^*)^t} = 0 \tag{11.22}$$

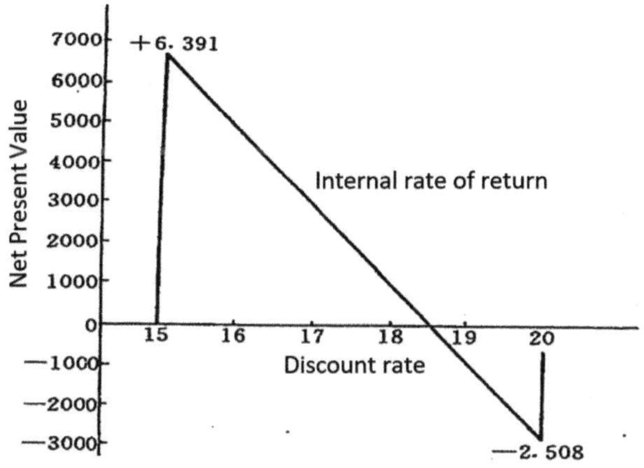

Fig. 11.4 Interpolation method for internal rate of return

The meaning of each symbol in the formula is the same as formula (11.20).

The above formula is inconvenient to solve due to higher-power equations. Generally, can be calculated by the following method.

(1) The graphical method. Assumes several different discount rates in Eq. (11.20), and calculates the corresponding net present values respectively. Then, using NPV as the ordinate and i as the abscissa, draw the relationship curve shown in Fig. 11.4. The discount rate when the curve passes through the NPV $= 0$ horizontal line is the desired i^*.
(2) Trial and error method. First, suppose an initial value i_0, Substitution to formula (11.22), if the net present value is positive, then increase the value of i, if the net present value is negative. Then decrease the value of i until the net present value is 0 or close to 0, and the value of i at this time is the desired i^* value.

For one-time investment (or can be converted into one-time investment) and the same annual energy saving benefit, the following trial difference can be used. Given the annual income A, investment I and project service life t, the discount rate that satisfies the following formula is the internal rate of return, as shown in formula (11.19).

$$t = \frac{-lg(1 - Ii/A)}{\lg(1 + i)}$$

When using trial and error method, first take the value of i, calculate whether the value of t is close to the value given in the question, and then change i until the two matches.

(3) Interpolation method

The trial-and-error method mentioned above is tedious. Usually when the trial calculation finds two i values i_1 and i_2 that are relatively close, and the corresponding two net present values NPV_1 and NPV_2 are around 0, which are positive and negative values separately, interpolation can be used to quickly solve the problem. Because at this time, the relationship between i_1, i_2, NPV_1, NPV_2 and i^* will be shown in Fig. 11.4. When the values of i_1 and i_2 are very close (generally, the difference is no more than 1–2%, and no more than 5% at most), the AB connection can be regarded as a straight line. So, there is

$$\frac{\|NPV_1\|}{\|NPV_2\|} = \frac{i^* - i_1}{i_2 - i^*}$$

or

$$\frac{\|NPV_1\|}{\|NPV_2\| + \|NPV_1\|} = \frac{i^* - i_1}{i_2 - i_1}$$
$$i^* = i_1 + \frac{\|NPV_1\|}{\|NPV_2\| + \|NPV_1\|}(i_2 - i_1) \tag{11.23}$$

The above formula can be expressed as

$$Internal\ Rate\ of\ Return = Low\ discount\ rate + \frac{\text{Net present value with low discount rate} \times \text{The difference between the two discount rates}}{\text{The sum of the absolute value of the net present value of the two discount rates}}.$$

The economic evaluation of energy-saving measures generally uses evaluation indicators such as the investment payback period (static or dynamic) investment profit and tax rate, internal rate of return, etc., and the corresponding evaluation method should be determined according to the specific situation.

11.4 Estimation of Economic Benefits of Energy-Saving Measures

After completing the energy consumption analysis of a production plant and proposing relevant energy-saving measures, how economical are these energy-saving measures, the advantages and disadvantages of energy-saving measures should be determined through technical and economic analysis and evaluation. Technical economic evaluation is actually the calculation and comparison of the economic benefits of various technical programs and measures.

Although the energy-saving measures of some units are determined through certain optimization methods, a large number of them are proposed based on the potential of energy consumption analysis and people's experience. Even if the determined energy-saving measures are optimized, corresponding evaluation indicators need to be proposed formally for approval by higher authorities. Therefore, the technical and economic evaluation of energy-saving measures is very important. The technical and economic evaluation of energy-saving measures is divided into three parts: benefit estimation, investment estimation and evaluation index.

The effect of energy-saving measures, are generally attributed to physical energy consumption savings. After the technical plan was determined, the savings in physical quantity were proposed. These physical materials are mainly in the form of fuel, electricity, steam, water, etc. The key to energy benefit estimation is to determine the price of these energy sources.

11.4.1 Determination of Fuel Price [7]

The fuel used in petrochemical plants is mainly fuel oil and fuel gas. Generally, the fuel balance of petrochemical plants is to balance the fuel gas first, and the insufficient part uses fuel oil. This is because the use and storage of fuel gas is not as convenient as fuel oil. Therefore, the physical savings of fuel can be considered as fuel oil. In fact, in the fuel balance of the whole plant, the fuel gas should be burned first, and the insufficient part should be supplemented with fuel oil on the premise of not venting and not burning in the flare. The saved fuel oil can be sold as commercial heavy oil or used as raw material for further secondary processing.

In fact, not all of the fuel oil saved by the factories is delivered as commercial heavy oil. What is really provided to the society is the products transformed from it. Therefore, it is reasonable to determine the fuel oil price according to the processing efficiency. On the other hand, the evaluation of energy-saving measures requires a comprehensive calculation of the processing benefits of raw materials, and sometimes it may not have the conditions, especially for chemical plant and other plants fuel-saving, it is inconvenient to trace the processing benefits of the refinery. Therefore, under the current price system, a unified fuel oil price or the estimated energy-saving economic benefits based on the local heavy oil price at that time can basically reflect the actual situation.

Current, many countries have a regulations and policies of greenhouse gas reduction, carbon tax may applied to CO_2 emission from chemical, petrochemical plants and energy industries, burning more fossil fuel will result in paying the more carbon tax; generally, burning one tone fuel (gas or oil) will produce 2.5–3.0 tones CO_2 varies with H/C ration of fuel; the carbon tax rate varies with countries, in the range of \$50–\$80 per ton CO_2 emission, any fuel gas or fuel oil saving will eliminate the carbon tax paid the fuel saving will have also the Carbon tax saving benefits.

11.4.2 Determination of the Price of Steam and Power of Back Pressure Power Generation [8]

Steam is the most commonly used energy working fluid in petrochemical plants. The price of steam mainly depends on the price of fuel, plus the cost of water, electricity and equipment depreciation required to generate steam; in addition, the carbon tax may be applied depend on the Carbon reduction policy of project located country.

There are many types of steam in petrochemical plants, especially in the deepening energy saving of enterprises, the level of steam header tends to increase, generally four level of steam header (10, 3.5, 1.0, 0.3 MPa), determine the steam of different level of steam header price is an important part of energy benefit estimation.

1. Model of steam power system in petrochemical industry

For large and medium-sized petrochemical companies, seldom use fuel to directly generate 1.0 MPa or lower pressure steam. Some of thermal power generation plants have high-pressure steam systems, and a large number of them generate 3.5 MPa steam from the primary boilers. Generally, the back pressure turbines are used to generate electricity (work) first, and then the back pressure steam is used as process steam. Steam is sometimes back pressured multiple times. In addition, chemical reaction heat and waste heat are used to generate steam in many production processes, such as shift gas boilers in chemical fertilizers, catalytic cracking flue gas waste heat boilers, and recirculation slurry oil steam generators. For a large number of process production systems that use 0.3 MPa steam, low-pressure backpressure turbines are often set to recover 1.0 MPa expansion work. Therefore, the production model of the steam power system is shown in Fig. 11.5.

2. Determination of the steam price of the fuel steam production system

For the fuel steam system to do currency balance, as shown in Fig. 11.6, the price of steam in the steam production system can be determined. The fuel steam production is generally medium or high-pressure steam. The target product of the steam generation system is steam, and other energy costs such as flue gas and heat dissipation loss are paid, and there is no cost, but the CO_2 emission from flue gas may trigger carbon tax, the carbon tax paying increase operating cost, usually you could add carbon tax up into fuel cost. Generally, companies use the boiler plant as their accounting unit, which includes the water treatment system. The energy supplied to the steam production system is fuel, water and electricity consumption and all have the nature of cash flow. In addition, there are depreciation costs for the equipment in the steam production system, ignoring the labor management costs. Therefore, there are:

$$FC_F + WC_W + EC_E + P_B N_B = SC_S \tag{11.24}$$

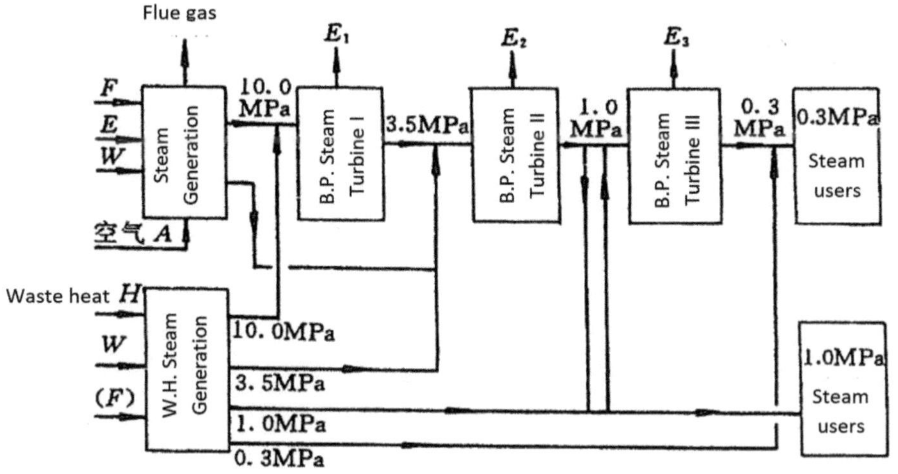

Fig. 11.5 Steam system model of petrochemical enterprise

Fig. 11.6 Cost balance of steam production system

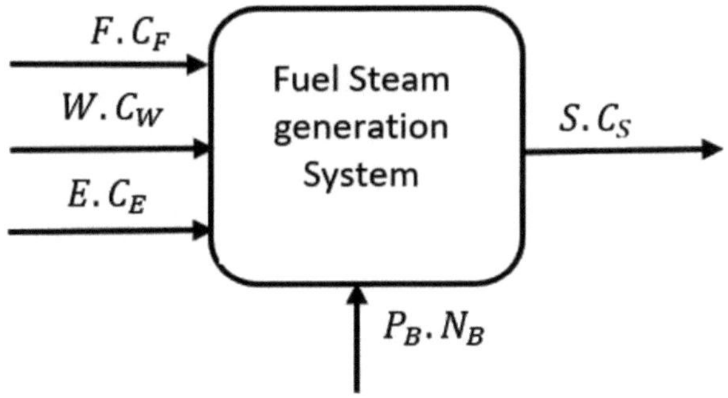

The unit price of steam for any parameter of the steam generation system is:

$$C_S = (FC_F + WC_W + EC_E + P_B N_B)/S.$$

Suppose the unit fuel consumption of steam generation is f, the unit consumption of water is w, and the unit consumption of electricity is e:

$$C_S = fC_F + wC_W + eC_E + (P_B N_B)/S \tag{11.25}$$

Take $\rho = (P_B N_B)/S$, which is the depreciation cost of each ton of steam investment in the steam generation system, yuan/t, then the above formula can be changed

$$C_S = fC_F + wC_W + eC_E + \rho \tag{11.26}$$

where: F—Fuel consumption, t/a,

E—Electricity consumption of steam production system, kWh/a;

C_F—Unit price of fuel, \$/t; If Carbon tax is applicable, the fuel price should include the Carbon tax.

C_E—Electricity unit price, \$/kWh;

W—Total water consumption for steam production (including water treatment water), t/a;

C_W—Unit price of fresh water, \$/t;

S—System steam production, t/a;

C_S—Steam price, \$/t;

P_B—One time investment in steam production system, \$;

N_B—The compensation coefficient of equal investment in the steam production system, when compound interest is calculated, is:

$$N_B = i(1+i)^n/\left[(1+i)^n - 1\right]$$

n—Service life of steam production system, Year;

i—Bank loan interest rate.

Generally speaking, there are two types of content in determining the unit price of steam: one is operating costs, that is, the cost of fuel, electricity, water consumption, plus carbon tax if applicable. The fuel is generally oil, which can be determined according to the previous content. When the boiler system burns coal, it can also be changed to coal price. The second is the depreciation cost of equipment investment, the former accounts for the largest proportion, of which fuel constitutes the main part of the steam price. Fuel consumption varies slightly depending on the pressure of steam generated. The life of the steam generation system is more than 15 years, and the depreciation cost of equipment investment generally accounts for less than 2% of the steam price, which can be ignored in rough estimation.

Example 11.5 For a 3.5 MPa steam generation system in a refinery, the unit fuel consumption is 79.16 kg/t, the unit water consumption is 1.57 t/t, and the unit power consumption is 8.1 kWh/t, of which the unit price of fuel, water and electricity are \$66/t, \$0.05 yuan/t and \$0.12/kWh respectively, and the one-time investment in the steam production system is \$MM 1.5. The steam output is 52t/h, and find the price of this steam, no carbon tax on this project.

Solution From Eq. (11.26), ignore the depreciation cost of steam equipment per ton,

$$C_S = fC_F + wC_W + eC_E + \rho$$

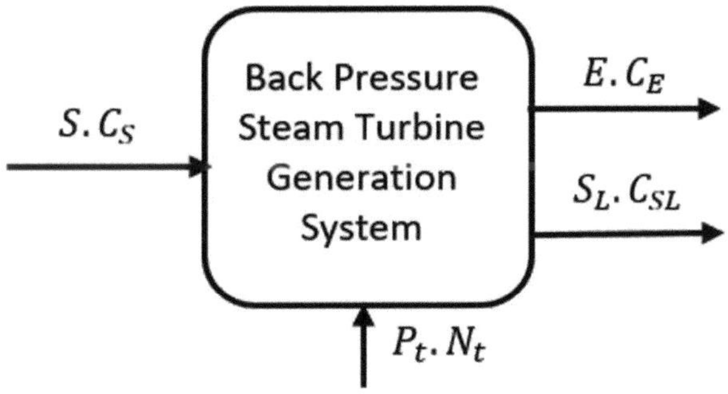

Fig. 11.7 Cost balance of back pressure power generation system

$$= 0.07916 \times 66 + 1.571 \times 0.05 + 8.1 \times 0.12 + 0 = 6.28(\$/\text{t}).$$

3. Determination of the price of back pressure steam

The products of the steam back pressure system are back pressure steam and electricity, which are two different types of energy, and the determination of steam prices is often more complicated. The cost balance of the steam back pressure power generation system is shown in Fig. 11.7. The price of steam supplied to the back pressure system should be determined first. Obviously, the 1st level back pressure inlet steam has been determined by the fuel steam boiler system. For 2nd level and 3rd level back pressure, the back pressure steam price determined by the back pressure steam of the previous level can also be used as the price of inlet steam for the back pressure system at this level. Many researchers have proposed to use exergy parameters as a bridge to determine the price of back pressure steam and electricity, and proposed the rule of setting the electricity price based on steam price [1]. The determination of the price of steam and electricity for the back pressure system should generally be carried out in accordance with the specific conditions of the enterprise as follows.

(1) Set the price of back pressure steam based on the external electricity price

This method uses electricity as a by-product of the back pressure system, and the external electricity price has been determined. The back pressure system generates or performs work, which can reduce the amount of outsourced electricity. Taking the external electricity price as the price of back pressure power generation, the price of back pressure steam is thus determined. Therefore, the investment in the back pressure system is borne by the product back pressure steam.

$$SC_S + P_t N_t = EC_E + S_L C_{SL} \tag{11.27}$$

$$C_{SL} = (SC_S + P_t N_t - EC_E)/S_L \tag{11.28}$$

where: P_t—One time investment in back pressure turbine, \$;

N_t—Equal investment repayment coefficient;

S_L—Back pressure turbine pressure steam discharge, t/a;

E—The amount of power or work done by the back pressure system, kWh;

C_{SL}—Back pressure steam price, \$/t.

When ignoring the leakage loss between the main steam and the back pressure steam, and the amount of water injection that reduces the temperature and pressure before the back pressure, the main steam volume is the same as the back pressure steam volume.

$$C_{SL} = C_S - \left(\frac{E}{S_L}\right) C_E + P_t N_t / S_L \tag{11.29}$$

Taking $\beta = S_L/E$ as the steam consumption per unit of back pressure power generation (kg/kWh), $q = P_t N_t / S_L$ is the annual depreciation cost of the back pressure turbine per ton of steam input, then:

$$C_{SL} = C_S - C_E/\beta + q \tag{11.30}$$

(2) Steam exergy price equalization method to determine the price of steam and electricity

Steam and electricity are different energy sources. Although exergy is thermodynamically equivalent, economically, electricity exergy and thermal exergy are not equivalent. Back pressure steam and main steam belong to the same type of energy. The main steam does work (generation) in the back pressure system. After that, the back pressure steam has a reduced performing work capacity. Although the steam rate remains unchanged, the exergy value drops. It is believed that the exergy price of steam is equal, which reflects the price of different quality steam, which is basically reasonable. Obviously, the back pressure steam parameter drops, the exergy value is low, and the unit steam price is lower than the main steam.

The exergy value of steam can be determined by thermodynamic parameters such as enthalpy and entropy. For the main steam exergy

$$EX_I = \Delta H_I - T_0 \Delta S_I \tag{11.31}$$

It has been determined that the main steam price of the fuel steam production system is C_S (\$/t), from which the exergy price of the main steam can be determined, C_{SX} (\$/GJ),

$$C_{SX} = C_S / \left[(\Delta H_I - T_0 \Delta S_I) \times 10^3 \right] \tag{11.32}$$

Assuming that no matter how high or low the steam parameter is, the price of the steam unit exergy is equal (in fact, the high-parameter steam exergy value is higher, and the price is higher when converted to a ton of steam, and vice versa):

$$C_{SX} = C_{SXL} = C_{SXH}$$

Then determine ΔH_{II}, $T_0 \Delta S_{II}$ according to the back pressure steam parameters to determine the price of back pressure steam

$$C_{SL} = C_{SX} \times 10^3 (\Delta H_{II} - T_0 \Delta S_{II}) \tag{11.33}$$

where: EX_I—Back pressure turbine inlet steam exergy, kJ/kg;

ΔH_I—Back pressure turbine inlet steam enthalpy difference, kJ/kg;

ΔS_I—Entropy difference of back pressure turbine inlet steam, kJ/(kgK);

C_{SX}, C_{SXL}, C_{SXH}—Exergy prices for main stream, low pressure steam and high pressure steam respectively, \$/GJ;

ΔH_{II}—Back pressure steam enthalpy difference, kJ/kg;

ΔS_{II}—Back pressure steam entropy difference, kJ/(kgK);

T_0—Reference temperature, K.

Substituting the formula (11.32) into the above formula, we get:

$$C_{SL} = C_S \times (\Delta H_{II} - T_0 \Delta S_{II})/(\Delta H_I - T_0 \Delta S_I) \tag{11.34}$$

This formula is the equation for determining the price of back pressure steam. As long as the difference in enthalpy and entropy of the steam before and after the back pressure turbine is found, the price of back pressure steam can be determined from the inlet steam price.

The price of backpressure turbine power generation can be determined by Eq. (11.27) according to the currency balance relationship of backpressure turbine. Back pressure turbine power generation is the main purpose product, which bears the depreciation cost of the investment, and the determined electricity price is:

$$\begin{aligned} C_E &= (SC_S - S_L C_{SL} + P_t N_t)/E (S \approx S_L) \\ &= [(S(C_S - C_{SL}) + P_t N_t)]/E \\ &= \beta (C_S - C_{SL}) + P_t N_t / E \end{aligned}$$

Taking $\frac{P_t}{E} = \gamma$ as the investment cost per kilowatt and N_t as the repayment coefficient (see above), then:

$$C_E = \beta (C_S - C_{SL}) + \gamma N_t \tag{11.35}$$

For other sub-level back pressure systems, the price of back pressure steam and electricity can also be determined according to the above method. It is just that the inlet steam is the back pressure steam (the price is known) from upstream back pressure turbine system. The method of determining the price of steam and electricity for the back pressure system can generally be selected according to the specific conditions of the factory. The first method is to determine the electricity price first, and the steam price is determined by the electricity price. The result is that the electricity price of the whole plant is unified, and the steam price varies with different parameters; the second method is that the electricity price of the whole plant is not uniform, and the back pressure power generation price is related with the steam price, in addition, it is also related to the scale and efficiency of equipment. Generally speaking, it is reasonable to use exergy parameters to determine different steam prices.

4. Determination of waste heat steam price

Regarding the use of waste heat and waste heat to generate steam in the conversion link and energy recovery and recycling link of the plant, especially when 3.5 or 10.0 MPa steam is generated, the fuel consumption of the power station is reduced, and the same effect as the steam generation of the boilers is achieved. The generated low-pressure steam can also have the effect of back pressure steam with the same parameters. Therefore, in the case of steam generation from waste heat, the price is compared with the price of steam or back pressure steam with the same parameters. This is a basic principle for determining the price of self-produced steam.

11.4.3 Determination of the Price of Electricity and Water

Due to the status of power industry development, both the normal electricity price and peak electricity price systems coexist in the power consumption of enterprises. In the peak price system, there are also cases of excessive use of fines. Generally, parity electricity consumption can only meet part of the electricity consumption of enterprises, and the shortfall is made up by peak priced electricity. With the development of enterprises, the number of production plants increases. The electricity gap is even greater, so the electricity saved in energy-saving improvements is the peak price part, so electricity saving or power generation should be determined according to the peak price system. The electricity price of different enterprises could be different. For enterprises with self-supplied power stations and thermal power plants, the cost accounting of power generation systems and electricity industry policy determines the price of electricity.

The determination of the water price mainly depends on the electricity price. The water used by petrochemical companies is generally fresh water and cooling water. Fresh water refers to the primary water used from the water source to the enterprise. The determination of the price of fresh water depends on three factors: one is the cost of power consumption for the pump to boost the water pressure. The power consumption per ton of water for each plant varies with specific conditions, ranging from 0.4 to 0.8 kWh/t. This part of the cost can be determined according to the price of electricity. Others, such as the use of heat and steam during the heating period of the water source station, are used intermittently, and the amount is small, so it is not considered; the second is the depreciation cost of the investment in pump equipment and pipelines, etc. It is generally not large and can be ignored in rough estimation; the third is the water resource fee levied by the urban construction department. Water resources fees vary with different regions (water shortage or not) and different water sources (reservoirs, rivers, groundwater, etc.). All enterprises should determine the price of fresh water in accordance with the water resource fee collection standards in the region. Cooling water is recycled and used as a self-contained system. Generally, there is no water resource fee issue. The price is determined by the cost of power consumption of the pump, the cost of making up fresh water and the cost of chemicals added to improve the quality of circulating water, and the depreciation cost of ton of water invested in the cooling water system.

After the price of energy source is determined, the annual energy saving benefit can be calculated based on the actual savings.

11.4.4 Other Benefits of Energy-Saving Measures

In the energy-saving modification of equipment, plants and systems in petrochemical plants, in addition to energy-saving benefits, there are also costs for saving raw materials and reagents. At this time, the savings benefits should be determined at the corresponding prices. In addition, many energy-saving modification are often combined with the revamp of plant, which will bring product upgrades after commissioning and increase the benefits of product sales revenue. This situation is more complicated. It should be determined by the method of determining the annual profit (or annual profit and tax) of the industrial project, deducting various costs and taxes, etc. [2, 3].

11.5 Cost Estimate and Technical Economic Evaluation

As energy-saving work continues to deepen, the difficulty of energy-saving has become more and more difficult, and the investment required to save 1t of standard oil has increased year by year [4], some payback period of many major energy-saving measures calculated using fuel oil are more than 3 years. There are two factors contributing to the increase in the investment payback period: one is the price increasing in equipment and material, labor cost etc., which objectively caused the increase in investment; the other is the continuous reduction in the driving force of energy-saving measures, further increasing the final heat exchange temperature will require greater equipment investment than before. The cost of the efficiency improvement of the heating furnace and other measures are similar. Large-scale system optimization measures require mutual coordination and increase a lot of production control and adjustment costs.

11.5.1 Classification of Energy-Saving Measures

Energy-saving measures are divided into three categories according to their functions: improving the process flow and operating conditions to reduce the total process energy-using; improving the efficiency of energy conversion and transmission; improving the energy recovery rate and reducing cooling and heat dissipation.

According to the advantages and disadvantages of its energy-saving effects and the amount of investment, it is divided into four types: improving operation management and energy saving without investment; applying mature conventional energy-saving technology requires certain investment; adopting advanced energy-saving new technology, but mature and feasible, costing investment more; combined energy-saving modification and production expansion. Investment and economic evaluation should adopt different levels of complexity in accordance with the different situations of these four types of measures.

References

1. D. Yang, Thermal Economics, East China Institute of Chemical Technology Press, 1990.5.
2. State Planning Commission (ed.), *Economic Evaluation Methods and Parameters of Construction Projects* (China Planning Press, Beijing, 1987)
3. W. Resnick, translated by Su Jianmin, *Chemical Process Analysis and Design* (Chemical Industry Press, 1985)
4. B. Wang, Deep energy saving potentials and increasing social supply (summary speech at the energy saving experience exchange conference), in *Compilation of Papers of the Second Energy Saving Exchange Conference of China Petrochemical Corporation* (Editorial Department of Petroleum Refining, 1989)
5. J. Su. (ed.), *Chemical Technology and Economics* (Chemical Industry Press, 1990)
6. H. Li (ed.), *Technical Economics of Petroleum Industry* (Petroleum University Press, 1991)
7. J. Wang, Z. Cai, Discussion on several issues of economic evaluation of refinery energy conservation. Refin. Des. **3** (1985)
8. A. Chen, Calculation Method of Steam and Electricity Price in Cogeneration System, Petroleum Engineering, 5, 1992

Part IV
Energy Utilization Optimization and Key Energy Saving Technologies

Chapter 12
Optimization on Pipeline and Equipment

Abstract The purpose of energy-saving improvement is not only to save energy but also to bring economic benefits, any energy-saving will contribute also to CO_2 reduction. Generally, saving energy can reduce costs including saving Carbon Tax. Investment costs and energy-saving benefits are a pair of complementary contradictions; under specific conditions, to organically unify investment and benefits to produce maximum benefits is what we call optimization methods. This Chapter discussed the relationship between process rate and exergy loss, including heat transfer process, flow flowing process, mass transfer, and chemical reaction processes, especially discussed identifying the effectiveness of the driving force, proposed the driving force efficiency estimate method; following the optimization method, proposed the following process optimization approach and results: Economical insulation thickness of hot fluid pipeline; Economical pipe diameter and insulation thickness for fluid transportation, Optimization of heat exchange equipment; Economical thermal efficiency of the heating furnace.

Keywords Process rate · Driving force · Economic diameters · Economic insulation-thickness · Equipment optimization · Economic efficiency

Energy saving is a multi-disciplinary and multi-professional system. The purpose of energy-saving improvement is not only to save energy, but also to bring economic benefits, any energy saving will contribute also to CO_2 reduction. Generally, saving energy consumption can reduce costs including saving Carbon Tax, and at the same time pay the investment cost for energy saving. Investment costs and energy-saving benefits are a pair of complementary contradictions; under specific conditions, to organically unify investment and benefits to produce maximum benefits is what we call optimization methods. The optimization of petrochemical plants is not only a matter of energy conservation and investment, but its benefits often include a wider range of content, that is, product sales revenue, and energy consumption is only an important factor that affects production costs. The optimization of the overall production process of the refinery developed and applied in recent years is proposed to solve this problem. However, in order to accomplish the same production tasks and functions, a large number of unit processes and equipment have contradictions in

T. A. Chen, *Energy Saving and Carbon Reduction*,
https://doi.org/10.1007/978-981-19-5295-1_12

energy conservation and investment. Its optimization is an important part of energy-saving improvement. This chapter discuss the individual equipment and pipeline optimization, heat exchange network optimization will be discussed in Chap. 13.

12.1 Process Rate and Exergy Loss

Through energy balance and exergy balance, the amount and location of process energy and exergy loss can be found. There is a limit to reducing the process exergy loss, because the process exergy loss is related to the process driving force. Reducing the process exergy loss will affect the process driving force and reduce the process rate. In order to complete certain production tasks, it is necessary to spend greater investment on increasing the size of the equipment.

The exergy analysis calculation is based on the second law of thermodynamics, taking the reversible process as the limit, and pointing out the maximum working capability of the system, thereby determining the energy grade of the system. In fact, petrochemical processes are all irreversible processes, which will inevitably produce process exergy losses. In order not to be contrary to classical thermodynamics, irreversible process thermodynamics have detoured into heat transfer, mass transfer, fluid flow and reaction kinetics, that is so-called "three transfer and one reaction" unit operation process [16]. The petrochemical process is nothing more than the combination, crossover, and penetration of these four types of unit operation processes. Therefore, energy-using analysis is not only a matter of energy balance and exergy balance, but also process energy use “kinetics”. The operation process of any unit has the following relationship:

$$\text{Process Rate} = \text{Driving Force}/\text{Resistance} \tag{12.1}$$

In the above formula, the process rate is the purpose of the process requirements. In order to complete a given production task, a certain process rate must be maintained. And to maintain a certain process rate, it is necessary to pay the price of driving force, that is, energy use and consumption. On the other hand, the resistance increases to a certain extent as the driving force of the process increases. For example, in the process of fluid flow, fluid flow resistance will inevitably increase when the pressure difference increases. As the driving force increases, the process rate increases, but not proportionally. The reason is that the resistance is also changing with the driving force. The purpose of energy saving is to appropriately reduce the driving force of the process, strengthen the process, reduce the resistance, increase investment when necessary, and keep the yield per unit time unchanged. Therefore, there is an optimization between investment and energy-saving benefits.

12.1.1 Heat Transfer Process

The basic equation of heat transfer in the heat transfer process:

$$Q = KA\Delta t_m = \Delta t_m / \left(\frac{1}{KA}\right) \tag{12.2}$$

where: Q—Heat transfer rate, kW;

K—Total heat transfer coefficient, W/(m^2 °C);

A—Heat transfer area, m.

For the heat transfer process per unit area, it is expressed as the thermal intensity relationship:

$$q = \Delta t_m K^{-1} \tag{12.3}$$

q is an index indicating the rate of heat transfer process, Δt_m is the heat transfer driving force, and K^{-1} is the heat transfer resistance. For a certain heat transfer load Q, the size of the thermal intensity q means the size of the required heat transfer area, so it also involves the amount of investment and required space.

To a large extent, energy saving lies in reducing the driving force Δt_m of the heat transfer equipment, thereby reducing the heat transfer exergy loss and increasing the energy recovery rate. Generally, when Δt_m decreases, q is slightly lower, but it can also be compensated by reducing the heat transfer resistance (increasing the heat transfer coefficient K value). From Eq. (12.2), increasing the heat transfer area A can also keep the total heat transfer Q unchanged. The driving force Δt_m is:

$$\Delta t_m = F_t.\Delta t_{\ln} \tag{12.4}$$

Directly related to energy saving and system energy recovery rate is the logarithmic mean temperature difference $\Delta t_{\ln}$ of the cold and hot streams. In order to use the temperature difference more effectively, the temperature difference correction factor F_t of the heat exchange equipment should be as close to 1 as possible. This requires reasonable selection of the equipment structure to avoid the inaction loss of the temperature difference. Increase F_t so that exergy loss is really used to promote the heat transfer process.

Exergy loss of heat transfer, when ignoring heat loss of equipment,

$$D_{KH} = QT_0(T_h - T_c)/(T_h T_c) \tag{12.5}$$

From Eq. (12.5), it can be seen that the heat transfer exergy loss is proportional to the difference between the average temperature of cold and hot flow T_h, T_c, and inversely proportional to the product of the average temperature of cold and heat flow. When the temperature difference is not large, it can be proved that $(T_h - T_c)$ is close to the logarithmic average temperature difference, that is, $\Delta t_{\ln} \approx T_h - T_c$.

Because when the ratio of the temperature difference at the two ends $\Delta t_1/\Delta t_2 < 2$, the arithmetic mean temperature difference can be used to replace the logarithmic mean temperature difference [1], and the arithmetic mean temperature difference:

$$\begin{aligned}\Delta t_m &= (\Delta t_1 + \Delta t_2)/2 \\ &= [(T_{h1} - T_{c2}) + (T_{h2} - T_{c1})]/2 \\ &= [(T_{h1} + T_{h2}) - (T_{c1} + T_{c2})]/2\end{aligned}$$

And because, when the heat capacity of the heat exchange stream is constant

$$T_h = (T_{h1} + T_{h2})/2, T_c = (T_{c1} + T_{c2})/2$$

$$\Delta t_m = T_h - T_c \tag{12.6}$$

Therefore, from the formula (12.5), and the formula (12.6):

$$D_{KH} = QT_0 \Delta t_m/(T_h T_c)$$

It shows that process exergy loss is proportional to the heat transfer temperature difference of the equipment. Therefore, when other conditions remain unchanged, reducing Δt_m can also reduce D_{KH}.

For the basic heat transfer equation $Q = KA\Delta t_m$, when we reduce Δt_m in order to reduce the process exergy loss, the process rate Q may be reduced and the production task cannot be completed. At this time, it can be compensated by increasing the heat transfer coefficient K or increasing the heat transfer area A. Increasing the K value and enhancing heat transfer are more economical, and can generally be solved by improving the flow state and other factors. Increasing the heat transfer area means increasing investment and needs to be optimized.

12.1.2 Fluid Flow Process [1]

According to the Bernoulli equation of the fluid flow process, when ignoring the position head and velocity head, for a given pipeline operating system for an incompressible fluid, and the power consumption provided by the pump is not included, then

$$-(\Delta p/\rho) = \sum h_f = \lambda[(L + L_e)/d]u^2/2 \tag{12.7}$$

Therefore, the flow rate (flow velocity) of the flow process can be expressed as:

$$u^2 = (p_1 - p_2)/\left[\lambda\left(\frac{\rho}{2d}\right)(L + L_e)\right] \tag{12.8}$$

For example, the square of the flow velocity u^2 represents the process rate, and the driving force is the pressure difference $(p_1 - p_2)$, and the process resistance is $\lambda(\frac{\rho}{2d})(L + L_e)$ (λ is the pipeline friction coefficient; ρ is the fluid density, kg/m^3; d is the inner diameter of the pipe, m; L is the length of the pipeline, m; L_e is the equivalent pipe length, m).

It can be seen from the above formula that to increase the process rate (flow velocity), it is necessary to increase the driving force (pressure difference) and reduce the flow resistance. Both can receive the effect of increasing the process rate. However, the pressure difference is provided by the conversion of electric energy, and increasing the pressure difference means that more electric energy is consumed and a price is paid. However, we can reduce the friction coefficient ($\lambda = 8\tau/\rho u^2$) and reduce the friction per unit area (i.e., shear stress), which can be solved by reasonable selection of pipe materials. On the other hand, the pipe diameter can be increased, the equivalent pipe length of the fittings can be reduced, and the resistance can be reduced. Increasing the pipe diameter can reduce resistance, but it means increased investment. We should balance investment and benefit. Reducing the equivalent pipe length of the piping system is an important part of reducing resistance. Specially to reduce the throttling loss of the valve. Reduce the loss of throttling at satisfying the process control and regulation, to a certain extent, it does not affect the process rate; while keeping the flow rate constant, it can reduce the power consumption of the pump, which is one of the important measures for energy saving.

For the incompressible fluid flow process, the exergy loss expression is:

$$D_{KL} = \left(\frac{T_0}{T}\right).V(p_1 - p_2) \tag{12.9}$$

If V_m is used to represent the average specific volume of the compressible fluid before and after throttling, then V in Eq. (12.9) is replaced by V_m, which can also be applied to compressible fluid. It can be seen that the flow exergy loss is a function of the pressure difference. When the driving force (pressure difference) increases, the process exergy loss increases. Theoretically, it seems that the process rate should also increase, in fact, the flow velocity depends on the conditions required by the process, the increase in pressure difference is mostly throttled through the regulating valve of the pipeline system and the control valve, the equivalent pipe length of the pipeline increases, and the final process rate does not increase. Therefore, when the pump head is too high, the increased pressure drop will account for the throttling loss. Therefore, under a given flow and pipeline conditions, increasing the pressure difference can only increase the exergy loss, and the process rate remains unchanged.

12.1.3 Mass Transfer and Chemical Reaction Process [2]

The mass transfer process includes distillation, gas absorption, liquid–liquid extraction and so on. Distillation is often a process that combines heat and mass transfer. Taking gas absorption as an example, the mass transfer rate equation can be expressed as Fick's law:

$$J_A = \frac{-D_{AB}(dC_A)}{dZ} \tag{12.10}$$

where: J_A—the molecular diffusion flux of substance A in the Z direction, kmol/(ms);

dC_A/dZ—The concentration gradient of substance A, that is, the change rate of the concentration C_A of substance A in the Z direction, kmol/m^4;

D_{AB}—The molecular diffusion coefficient of the substance A in medium B, m^2/s.

The negative sign in the formula (12.10) indicates that the diffusion is carried out along the direction of the decrease of the substance A concentration.

According to the absorption process mechanism [2], the gas absorption process is simplified as a molecular diffusion process through the gas and liquid films. For a stable mass transfer process, the mass transfer rates of the gas and liquid films are the same. Therefore, the mass transfer rate in the effective films on either side of them can represent the absorption rate of that part. The absorption rate relation formula written separately based on the driving force and resistance of the gas film or liquid film is called the gas film or liquid film absorption rate equation.

Gas film absorption rate equation:

$$N_A = k_G(p - p_i) = (p - p_i)/k_G^{-1} \tag{12.11}$$

$$k_G^{-1} = Dp/RTZ_G p_{BM} \tag{12.12}$$

where: T—Absorption temperature, K;

p—Partial pressure of solute A in the main gas phase, kN/m^2;

p_i—Partial pressure of solute A at the phase interface, kN/m^2;

N_A—Absorption rate;

k_G—Gas film absorption coefficient;

p_{BM}—The logarithmic average of the partial pressure of the inert component B at the interface between the main flow and the phase in the gas phase, kN/m^2;

Z_G—Effective stagnation film thickness in gas phase, m;

D—Molecular diffusion coefficient.

The driving force of gas absorption is the gas film driving force $(p - p_i)$. When $(p - p_i)$ increases, the absorption rate increases, and the reciprocal of the gas film absorption coefficient k_G^{-1} means the transfer resistance of absorption substance passes through the gas film.

The exergy loss in the gas absorption process is also corresponding to the gas film driving force $(p - p_i)$. In order to create a large $(p - p_i)$, the usual method is to reduce the partial pressure of the solute in the lean solvent. It is generally realized by regeneration; therefore, it also consumes energy and exergy. In addition, increasing the operating pressure of the system can also increase the partial pressure of solute A, and increase $(p - p_i)$, which also consumes power energy. It can be seen that the absorption rate is also directly related to the process exergy loss.

The situation of chemical reactions is more complicated. The reaction rate equation is directly related to the n power of the reactant concentration. These equations are mostly empirical kinetic equations, which vary with experimental conditions and regression methods. There is no unified format, so it is difficult to uniformly express the relationship between driving force and resistance, but it is certain that the reaction rate equation can be expressed as general rate equation with just more complicated in form.

12.1.4 Dynamic Efficiency of Driving Force (Exergy Loss)

The exergy loss of the process is directly related to the thermodynamic parameters, and the driving force of the process rate equation is not only derived from the thermodynamic parameters such as T, p, C, but also reduced by the constraints of engineering arrangements, equipment structure, process parameters and other factors. There is an efficiency of the driving force, and the typical situation is as follows.

(1) The structure and operating parameters of the heat exchanger are different, resulting in different temperature difference correction factor F_t.

$$\Delta t_m = F_t.\Delta t_{\ln} \tag{12.13}$$

The temperature difference correction factor F_t is the dynamic efficiency of the temperature difference. The exergy loss is approximately proportional to the log-average temperature difference $\Delta t_{\ln}$, and the real driving force of the heat transfer process is Δt_m, which is F_t times smaller than $\Delta t_{\ln}$. Therefore, one of the important ways to reduce the exergy loss in the heat transfer process is to increase the heat transfer temperature difference correction factor F_t, so that all or most of the exergy loss of the temperature difference is used to promote the heat transfer process.

(2) The fluid flow rate is a function of pipe diameter and flow rate. For a certain flow rate, the piping system arrangement is very closely related to the required

delivery pressure drop, and there is a big difference. A considerable part of the pressure drop loss is in the control valve of the piping system.

Therefore, from the process exergy loss D_K, distinguish the driving exergy D_D that really promotes the process, and define the dynamic efficiency of the exergy loss [3].

$$\eta_{xk} = D_D / D_K \tag{12.14}$$

However, how to determine the D_D of a single process, and how to select the appropriate D_D and D_K, still needs a lot of work.

Generally speaking, the energy consumption of the plant is composed of two parts: one is the minimum energy (exergy) consumption of the process under reversible conditions, which is the thermodynamic energy consumption of the product in the form of physical energy and chemical energy. The rest are all lost to the production process as the driving force of the process (including external losses and irreversible losses). The exergy (energy) consumed by the process can be divided into the driving force exergy loss D_D that actually promotes the process and the exergy loss caused by improper operation and process equipment arrangement. This part of the exergy loss is not used to drive the process and can be reduced or even avoided.

12.2 Economical Insulation Thickness of Hot Fluid Pipeline

The process rate and the reduction of process exergy loss are a pair of complementary and opposite contradictions. Energy-saving is bound to reduce the driving force of the process, and the rate of the production process must remain unchanged, which requires an appropriate increase in equipment investment. There are too many variables and parameters to deal with this unity of opposites. Under current conditions, it is difficult to optimize the plant and even the entire production process. However, we can optimize the energy-saving measures of a certain unit process and receive significant results. The optimization of energy-saving measures for each unit process is helpful to the optimization of the entire process. Therefore, the optimization of energy-saving measures for unit processes is a problem that energy-saving workers should consider and solve. The optimization method is an independent subject, please refer to the related works [4–6].

12.2.1 Objective Function

Although there are various standards for evaluating a production plan or an energy-saving measure, in the final analysis, they are all the economic benefits. Usually used target values, such as maximum production profit, minimum heat exchange area, and minimum total energy consumption, are a standard for evaluating the objective function. However, the lowest energy consumption is not necessarily economical. The optimized energy consumption is the economic energy consumption, which means that this energy consumption is the optimal energy consumption. If continue to reduce energy, you will spend more capital but get less benefits. Therefore, most of the optimization goals of petrochemical processes are closely related to their economic benefits.

Characterizing the functional relationship of process optimization goal and it's the parameters is usually called the objective function. When starting to solve the problem, for a specific problem, there may be some constraints to ensure that the optimized value is obtained and these constraints are met during the optimization process. These constraints are the constraints of the optimization process.

With the objective function and constraint conditions, a complete optimization mathematical model is formed. The key to these mathematical models is to associate process variables with objective functions, including benefits and investments. Correctly establishing the objective function of the problem is an important part of the optimization process.

12.2.2 Classification of Optimization Methods [7]

There are various optimization methods, which can be divided into single-variable function optimization and multi-variable function optimization according to the number of variables. For multi-variable functions, to simplify the problem, simplifications and assumptions are often used to convert multi-variables into few variables, or even single variables; according to whether there are constraints, it can be divided into unconstrained optimization and constrained optimization.

However, the most commonly used classification is to divide optimization methods into indirect methods and direct methods.

When the mathematical model of the problem has a simple and clear mathematical expression, it is very convenient to use the differential method to solve the problem of unconstrained conditions. That is, find the first derivative (or partial derivative) of the objective function and set it to zero to obtain a set of equations. The stationary point of the function can be obtained by solving this equation set. By studying the second-order partial derivative or according to the actual meaning of the problem, it can be determined whether the function has an extreme value at the stagnation point. This kind of method is generally called analytic method or indirect method. It is used for the optimization of single variable or bi-variable function, and the optimization of economical pipe diameter and economical insulation thickness for high-temperature fluid transportation is the indirect optimization method of bivariate function.

The direct optimization method is to use an electronic computer as a tool to obtain an approximate solution to the problem within a given error range through a mathematical approximation method. This method is sometimes called numerical method and is mainly used in multivariable systems.

There are many other optimization methods, such as linear programming, nonlinear programming, and dynamic programming methods.

This section and the following sections take the optimization of equipment and unit processes commonly used in petrochemical processes as examples to introduce energy-saving measures optimization methods and ideas. The optimization of large-scale systems, such as the optimization of separation sequence, the optimization of heat exchange sequence, and the synthesis are not discussed here.

12.2.3 Economical Insulation Thickness of Hot Fluid Pipeline

Pipeline insulation is to reduce heat loss, and its objective function is:

$$S = S_E + S_I \tag{12.15}$$

where: S—Total annual cost, dollar/(am);

S_E—Annual operating cost (exergy loss cost for heat dissipation), dollar/(am);

S_I—Annual depreciation cost of thermal insulation investment, dollar/(am).

Therefore, the minimum sum of the heat dissipation cost of the thermal insulation pipeline and the annual depreciation cost of the thermal insulation investment is taken as the objective function. The optimization of this type of problem belongs to the optimization of the type of investment only and no output. Therefore, the minimum total annual cost is the optimization goal (if take the maximum annual benefit of the difference between the income of heat preservation and the investment depreciation cost as the optimization goal, the results will be the same).

After clarifying the optimization goal, it is necessary to determine the functional relationship between the thermal insulation investment cost and the heat dissipation cost as the insulation thickness changes. Obviously, the increase in insulation thickness will reduce the cost of heat dissipation exergy loss and increase the investment cost of insulation.

Annual operating expenses

$$S_E = C_h h\left(t_B - t_f\right)^2 \times 10^{-6} / \left[(1/2\pi\lambda)\left(\ln\frac{D}{d} + \frac{2\lambda}{D\alpha_T}\right)T_B\right] \tag{12.16}$$

Annual depreciation cost of insulation investment

$$S_I = (\pi/4)(D^2 - d^2)a.N \tag{12.17}$$

Substituting these two equations into Eq. (12.15), the relationship between the total annual cost and the outer diameter of the insulation can be obtained.

$$S = (\pi/4)(D^2 - d^2)a.N + C_h h\left(t_B - t_f\right)^2 \times 10^{-6} / \left[(1/2\pi\lambda)\left(\ln\frac{D}{d} + \frac{2\lambda}{D\alpha_T}\right)T_B\right] \tag{12.18}$$

Derivative dS/dD from the above formula on the heat dissipation outer diameter D, and set $\frac{dS}{dD} = 0$, after deriving and sorting out.

$$D\ln\left(\frac{D}{d}\right) = 0.002\left(t_B - t_f\right)\sqrt{\frac{\lambda C_h h}{aNT_B}}\sqrt{1 - \frac{2\lambda}{D\alpha_T}} - \frac{2\lambda}{\alpha_T} \tag{12.19}$$

This formula is the analytical formula of insulation thickness. However, this equation is an implicit function equation, and it is necessary to guess or iteratively calculate the outer diameter D of the insulation. It can also be done with the help of a programmable calculator or computer.

Generally, the factor value of $\sqrt{1 - \frac{2\lambda}{D\alpha_T}}$ is between 0.97 and 1.0. The following formula can also be used for approximate calculation.

$$D\ln\left(\frac{D}{d}\right) = 0.00198\left(t_B - t_f\right)\sqrt{\frac{\lambda C_h h}{aNT_B}} - \frac{2\lambda}{\alpha_T} \tag{12.20}$$

After calculating D, the insulation thickness δ can be calculated:

$$\delta = (D - d)/2 \tag{12.21}$$

The difference between exergy economical insulation thickness and thermal economical insulation thickness is that the calculation of heat dissipation energy cost is based on exergy rather than calculated based on heat, so as to distinguish the price of thermal energy of fluids at different temperatures, and use different insulation thicknesses for fluids of different temperatures. The thermal economy method is to take a uniform price regardless of the level of thermal energy parameters (temperature), which is obviously unreasonable.

The exergy price is determined by the cost equation of the energy conversion system [8, 9]. Thermal exergy is generally realized through energy conversion equipment, generally a fuel fired furnace, as shown in Fig. 12.1.

According to the cost balance:

$$C_h EX = C_F F + PN + C_m \tag{12.22}$$

For the heating furnace, the operating management cost C_m accounts for a small proportion and can be ignored, then:

$$C_h = C_F F/EX + PN/EX \tag{12.23}$$

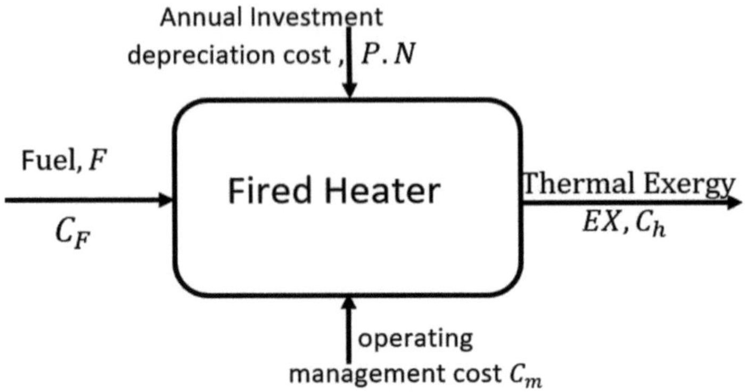

Fig. 12.1 Determination of the price of heat exergy

Generally, $\frac{EX}{F}$ is the exergy efficiency η_x of the heating furnace. When only heating one stream, the relationship between η_x and the energy efficiency η is:

$$\eta_x = \eta.\varepsilon$$

Set $C_{eq} = P.N/EX$, then:

$$C_h = C_F/(\eta.\varepsilon) + C_{eq} \tag{12.24}$$

The formula (12.24) is the exergy price of thermal energy. ε is the energy grade of the heat of the heating stream.

$$\begin{aligned} \varepsilon &= (1 - T_0/T_h) \\ T_h &= (T_1 + T_2)/2 \end{aligned} \tag{12.25}$$

where: D, d—Are the insulated outer diameter and the outer diameter of the pipeline, m;

a—Insulation materials and construction costs, dollar/m^3;

N—Insulation investment repayment rate, $N = \frac{(1+n)^m n}{(1+n)^m - 1}$ (m is the repayment period, n is the annual interest rate).

λ—Thermal conductivity of insulation material, 1.163 W/(mK);

h—Annual operating hours, h;

C_h—thermal exergy price, dollar/GJ;

C_F—Fuel price, dollar/GJ; if carbon tax is applicable for the project, it should include carbon tax;

F—Fuel demand, GJ/a.

EX—Thermal exergy supply, GJ/a.

ε—The average energy grade of the heated fluid of the conversion equipment;

η—Heating furnace thermal efficiency, %;

C_{eq}—Depreciation rate of heating furnace investment, dollar/(GJa);

P—Heating furnace investment, dollar;

C_m—Heating furnace operation and management fee, dollar/a.

Example 12.1 It is known that two D 250 outdoor pipelines, the conveying fluid temperature is 150 °C and 350 °C respectively, the ambient temperature is 15 °C, the average wind speed is 3 m/s, and the thermal conductivity of insulation materials is $\lambda = 0.046 + 0.0013t_p$, the number of operating hours is 8000 h, the price of insulation materials and construction is \$80/m^3, the annual interest rate n is 0.1, the repayment period m is 5 years, the thermal energy price and exergy price are \$0.9/GJ and \$2/GJ separately.

The economic insulation thickness to be calculated according to the thermal analysis method and the exergy analysis method respectively. The calculation results are summarized in Table 12.1 (based on a 100 m pipeline).

It can be seen from Table 12.1 that the investment/heat dissipation ratio of the exergy analysis method shows that the high-temperature fluid is 35% larger than the low-temperature fluid, and the insulation thickness at low temperature is less than the thermal method, while the insulation thickness at high temperature is greater than the thermal method. Obviously, it is more reasonable than the thermal analysis method, and the total cost is also lower [11].

12.3 Economical Pipe Diameter and Insulation Thickness for Fluid Transportation [10]

The pipeline investment and operating cost of fluid transportation occupy a certain position in the technical and economic field of petrochemical plants. In the past, in the design process of plants, pipeline diameters were often selected in a wide range of flow rates based on experience and technical specifications, and not paying attention to economic concepts, so the selected pipe diameter was not necessarily economical. In recent years, many literatures have proposed the concepts of economic pipe diameter and economic insulation thickness for fluid transportation, and proposed methods and steps for obtaining economic pipe diameter and economic insulation thickness of steam pipelines [18]; Table of economic pipe diameter of oil storage and transportation system has been also proposed under certain conditions [11]. This section only analyzes and discusses the selection of economic pipe diameters for incompressible fluid transportation.

Table 12.1 Example 12.1 calculation results

Description	150 °C		350 °C	
	Thermal analysis method	Exergy analysis method	Thermal analysis method	Exergy analysis method
Energy grade factor	0.32	0.32	0.538	0.538
Insulation thickness (mm)	81	69	129	139
surface temperature (°C)	19.1	19.9	22	21.4
Heat dissipation (GJ/a)	338.38	381.58	728.21	687.72
Heat dissipation exergy (GJ/a)	108.27	122.09	400.1	364.92
Heat loss cost (dollar/a)	215.5	243.0	779.8	726.4
Depreciation (dollar/a)	189.8	156.0	343.5	379.8
Total cost (dollar/a)	405.3	399.0	1123.3	1106.2
Investment/heat dissipation (dollar/GJ)	9.8	7.2	8.3	9.7
Investment/heat loss exergy (dollar/GJ)	30.7	22.4	15.1	18.2

12.3.1 Economical Pipe Diameter for Ambient Fluid Transportation

The economic pipe diameter of low temperature fluid transportation refers to the diameter of the pipeline with the smallest sum of annual pipeline investment depreciation costs and annual operating costs for fluid transportation; pipeline investment costs include pipeline materials and corresponding installation and construction costs; on the other hand, annual interest rates and operating costs should also be considered. For low temperature fluids that are not insulated, they are mainly due to the loss of kinetic energy for fluid transportation. These costs depend to a certain extent on materials, energy prices and financial policies in different regions and periods. The flow velocity corresponding to the economic pipe diameter is called the economic flow velocity.

For low-temperature fluid transportation that does not require heat preservation, the economical pipe diameter is only a function of pipeline investment costs and frictional kinetic energy costs.

12.3.1.1 Annual Depreciation Expense of Pipeline Investment

The annual depreciation cost of pipeline investment is a function of pipe diameter, material, depreciation rate, etc.; wall thickness can be selected according to process requirements and pipeline stress calculations. Assuming that the calculated pipeline is longer, the annual depreciation expense per meter of pipeline investment is:

$$S_p = 7.85\pi ABN(d + B) = 24.66ABN(d + B) \tag{12.26}$$

where: S_p—Annual depreciation expense of pipeline investment, yuan/(am);

A—Comprehensive indicators of pipeline investment and construction, dollar/t;

d—Inner diameter of pipeline, m;

B—Steel pipe wall thickness, m;

7.85—Density of steel, t/m^3;

N—Investment repayment coefficient,

$$N = \frac{(1 + i)^n i}{(1 + i)^n - 1}$$

i—Annual interest rate;

n—Service life, a.

12.3.1.2 Flow Energy Loss Cost of Fluid Transportation

The energy provided by the process pump to the fluid is mainly used to increase the position head of the fluid, the static pressure head required by the process, the pressure head loss to overcome the frictional resistance, and the dynamic pressure head at the corresponding flow rate. For incompressible fluids, the dynamic pressure head relative to the total energy is very small and can be ignored. The position head and the static pressure head are required by the process and cannot be changed, and the frictional resistance to overcome the fluid varies with different pipe diameters of the piping system. The annual operating cost of low-temperature incompressible fluids S_{op} is:

$$S_{op} = \frac{G\sum h_f}{367} C_{PF} h \tag{12.27}$$

$$\sum h_f = \lambda \frac{L + L_e}{d} \frac{u^2}{2g} \tag{12.28}$$

Friction coefficient under laminar flow conditions:

$$\lambda = \frac{64}{Re} \tag{12.29}$$

For turbulent flow conditions with Re greater than 2000, the following explicit function equation and Moody chart error are within 0.1%, which can meet the engineering error requirements:

$$\frac{1}{\sqrt{\lambda}} = -2.0\,\lg\left[\frac{K}{3.7d} + \left(\frac{6.81}{Re}\right)^{0.9}\right] \tag{12.30}$$

For petrochemical plants using pumped fluid transportation process, the Reynolds number Re is usually a turbulent situation, so formula (12.30) can be used to calculate the friction coefficient of the fluid transportation process.

$$\text{Before:}\quad u = \frac{G}{2.826d^2\rho}$$

$$\text{Therefore:}\quad Re = \frac{353.86G}{\mu d}$$

Assuming the absolute roughness of the tube wall $K = 0.1 \sim 0.3$ mm, then there is formula (12.31).

$$\lambda = \frac{1}{4.0\,\lg^2\left[\frac{5.4\times10^{-5}}{d} + 0.02856\left(\frac{\mu d}{G}\right)^{0.9}\right]} \tag{12.31}$$

Resistance loss of the pipeline

$$\sum h_f = \frac{(L + L_e)G^2}{626.76d^5\rho^2\,\lg^2\left[\frac{5.4\times10^{-5}}{d} + 0.02856\left(\frac{\mu d}{G}\right)^{0.9}\right]} \tag{12.32}$$

Then the flow energy loss cost per meter of tube length:

$$S_{OP} = \frac{G^3 C_{PF} h}{2.3\times10^5 d^5\rho^2\,\lg^2\left[\frac{5.4\times10^{-5}}{d} + 0.02856\left(\frac{\mu d}{G}\right)^{0.9}\right]} \tag{12.33}$$

In the above formulas:

$\sum h_f$—Pipeline resistance loss, m;

S_{OP}—Flow energy loss cost, dollar/(am);

μ—Fluid viscosity, 10^{-3} Pas;

λ—Coefficient of friction;

L, L_e—Respectively the length of straight pipe and the equivalent length of pipe fittings (valves, elbows, etc.), m;

u—Fluid flow velocity, m/s;

ρ—Fluid density, kg/m^3;

G—mass flow rate, t/h;

K—Absolute roughness, mm;

Re—Reynolds number;

h—Annual operating hours, h;

C_{PF}—Fluid flow energy price, dollar/(kWh).

Taking into account the conversion efficiency of the pump and the investment and maintenance costs of the pump to be allocated to the cost of the unit effective work. The price of fluid flow energy is:

$$C_{PF} = \frac{C_E}{\eta} + \frac{N.P_u}{W.h.\eta} \tag{12.34}$$

where:C_E—Electricity price, dollar/kWh;

η—Product of motor and pump efficiency;

N—Depreciation rate, (same as before);

P_u—Pump investment (including motor), dollar;

W—The actual power consumption of the pump (estimated), kW.

Since the latter term in formula (12.34) accounts for a small proportion (below 10%) of the total C_{PF}, even if the pump is selected in a wide range, it will not affect the accuracy of C_{PF}.

12.3.1.3 The Total Cost of Fluid Transportation and Economic Pipe Diameter

According to the above analysis, fluid transportation costs and pipeline investment costs. It can be treated as a function of diameter d. Therefore, the total cost per pipe length is:

$$S_T = 24.66ABN(d+B) + \frac{G^3 C_{PF} h}{2.3 \times 10^5 d^5 \rho^2 \lg^2 \left[\frac{5.4\times10^{-5}}{d} + 0.02856\left(\frac{\mu d}{G}\right)^{0.9} \right]} \tag{12.35}$$

It can be seen from Eq. (12.35) that after other known variables are determined, the total cost is only a function of the pipeline inner diameter d, and the pipe wall thickness B is not a fixed value, and varies with the size of the pipe diameter; to minimize the total cost, Eq. (12.35) takes the derivative of d and sets its derivative to zero, that is, $\frac{\partial S_T}{\partial d} = 0$, then the following mathematical model of economic pipe diameter is derived:

$$Cd^6 \lg^3[x] - 5\lg[x] - \frac{0.0471\left(\frac{\mu d}{G}\right)^{0.9}}{x} + 0.8686 = 0 \tag{12.36}$$

In the equation, C is a constant, and x is a function of d.

$$C = \frac{5.67 \times 10^6 \rho^2 ABN}{G^3 C_{PF} h}$$

$$x = \frac{5.4 \times 10^{-5}}{d} + 0.02856\left(\frac{\mu d}{G}\right)^{0.9}$$

Equation (12.36) is the single variable linear implicit function equation for finding the economic pipe diameter of low-temperature incompressible fluids. Because the actual pipeline series are relatively wide, the change of pipe diameter d is discontinuous. According to the Eq. (12.36), after calculating d, it still needs to be rounded to a series of similar pipe diameters. Therefore, two methods can be used to find the economic pipe diameter d: one is to select different pipe diameter schemes according to Eq. (12.35), and the pipe diameter with the smallest total cost is determined as the economic pipe diameter; the other is to follow Eq. (12.36). Iterative calculation, find d, and then round to the similar pipeline series diameter, which is the economic pipe diameter.

12.3.2 Economical Diameter and Insulation Thickness of Insulated Pipeline

Many of the oil transportation between equipment and equipment belong to the transportation of hot fluids that require insulation (such as atmospheric residual oil in oil refining equipment, catalytic cracking diesel, etc.); insulation thickness affects the determination of economic pipe diameter, and vice versa. The economical insulation thickness depends on the economical pipe diameter. In order to solve the economical pipe diameter of the insulated fluid transportation pipeline, the economical insulation thickness must also be considered. Therefore, the calculation of the economical pipe diameter of the insulated pipeline is more complicated than that of the low temperature fluid. The total cost includes the heat preservation cost of the pipeline and the cost of heat dissipation exergy in addition to the two costs of low temperature fluid transportation.

Hot fluid transportation and heat dissipation on the outer surface of the insulation will cause exergy loss, so there is thermal exergy loss caused by heat dissipation and electrical work exergy loss caused by friction. Due to the inequality in actual use, thermal exergy and work (exergy) are different in terms of actual use value or actual price [17]; if the price is based on unit exergy, the price of electrical work exergy is \$12/GJ, while the thermal exergy price is much less than this value. Therefore, here, we consider the cost of thermal exergy and work exergy separately.

1. Annual depreciation expense of pipeline investment per Eq. (12.26)

$$S_p = 24.66ABN(d + B)$$

2. Flow energy loss cost S_{OP} per Eq. (12.33)

$$S_{OP} = \frac{G^3 C_{PF} h}{2.3 \times 10^5 d^5 \rho^2 \lg^2 \left[\frac{5.4 \times 10^{-5}}{d} + 0.02856 \left(\frac{\mu d}{G} \right)^{0.9} \right]}$$

3. Annual depreciation expense of pipeline insulation investment S_I.

$$S_I = \frac{\pi}{4}\left[D^2 - (d + 2B)^2\right] A_1 N_1 \tag{12.37}$$

4. Heat loss exergy loss cost S_E.

$$S_E = \frac{C_H h (T_B - T_0)^2 \times 10^{-6}}{\frac{1}{2\pi\lambda}\left(\ln \frac{D}{d+2B} + \frac{2\lambda}{D\alpha_T}\right) T_B} \tag{12.38}$$

In the above formulas:

λ—Thermal conductivity of insulation material, 1.163 W/(mK);

D—Outer diameter of insulated pipeline, m;

d—Inner diameter of steel pipe, m;

B—Steel pipe wall thickness, m;

S_I, S_E—Are the annual depreciation cost of heat preservation and the exergy loss cost of heat dissipation loss, dollar/(am);

T_B, T_0—Respectively the fluid in the equipment and the ambient temperature, K;

N_1—Depreciation coefficient of insulation investment:

$$N_1 = \frac{(1+i)^{n_1} i}{(1+i)^{n_1} - 1} \tag{12.39}$$

n_1—Service life of insulation materials;

α_T—Radiative convection combined heat transfer coefficient of insulated pipeline to the environment, 1.163 W/(m^{2}K);

C_H—Heat loss exergy price, dollar/GJ.

The determination of C_H is based on the result of the exergy analysis of the plant. If the cost of the conversion equipment borne by the unit exergy value is ignored, then

$$C_H = \frac{C_F}{\eta_H \varepsilon} \tag{12.40}$$

C_F—Fuel price, dollar/GJ; if carbon tax is applicable for the project, it should include carbon tax;

η_H—Thermal efficiency of heat exchange equipment;

ε—The energy grade of the heated fluid of the conversion equipment;

$$\varepsilon = \frac{T_H - T_0}{T_H} \tag{12.41}$$

T_H—Average temperature of heated fluid, K;

5. Total cost of insulated pipeline and economic pipe diameter

The total annual cost of thermal fluid transportation is the sum of the above four costs:

$$\begin{aligned} S_{Total} &= 24.66ABN(d+B) + \frac{\pi}{4}\left[D^2 - (d+2B)^2\right]A_1 N_1 \\ &+ \frac{C_H h (T_B - T_0)^2 \times 10^{-6}}{\frac{1}{2\pi\lambda}\left(\ln\frac{D}{d+2B} + \frac{2\lambda}{D\alpha_T}\right)T_B} \\ &+ \frac{G^3 C_{PF} h}{2.3 \times 10^5 d^5 \rho^2 \lg^2\left[\frac{5.4\times 10^{-5}}{d} + 0.02856\left(\frac{\mu d}{G}\right)^{0.9}\right]} \end{aligned} \tag{12.42}$$

It can be seen from formula (12.42) that the pipeline total cost is a function of the outer diameter of the insulated pipe and the inner diameter of the steel pipe. This is an optimizing problem of a binary nonlinear equation. The economical pipe diameter and economical insulation thickness can be determined according to the method of the extreme value of the multivariate function. According to the actual meaning of the problem, there must be values of d and D that minimize the total cost. Formula (12.42) is used to obtain partial derivatives of d and D respectively, and set them to zero, and sort them into a binary nonlinear equation set:

$$C_1 - C_2(d+2B) + \frac{C_3}{(d+2B)\left(\ln\frac{D}{d+2B} + \frac{2\lambda}{D\alpha_T}\right)^2} - C_4\frac{5\lg x + 0.0471\left(\frac{\mu d}{G}\right)^{0.9} - 0.8686}{d^5 \lg^3 x} = 0 \quad (12.43)$$

$$C_2 D - C_3\frac{\frac{1}{D} - \frac{2\lambda}{D^2\alpha_T}}{\left(\ln\frac{D}{d+2B} + \frac{2\lambda}{D\alpha_T}\right)^2} = 0 \quad (12.44)$$

In the formula, C_1, C_2, C_3, C_4 are all the constants

$$C_1 = 24.66ABN$$

$$C_2 = \frac{\pi}{2}A_1N_1$$

$$C_3 = \frac{2C_H h\pi\lambda(T_B - T_0)^2 \times 10^{-6}}{T_B}$$

$$C_4 = \frac{G^3 C_{PF} h}{2.3 \times 10^5 \rho^2}$$

$$x = \frac{5.4 \times 10^{-5}}{d} + 0.02856\left(\frac{\mu d}{G}\right)^{0.9}$$

The simultaneous solution of Eqs. (12.39) and (12.40) is a relatively complicated task, which needs to be completed with the help of a computer. The calculation block diagram is shown in Fig. 12.2.

Example 12.2 Calculate the economical pipe diameter and economical insulation thickness for the transportation of 293 K (20 °C) water and a certain 513 K (240 °C) oil product. (For known conditions, see Tables 12.2 and 12.3) and compare with other different pipe diameter schemes.

Solution The calculation results of 293 K water are listed in Table 12.2. The calculation results of a certain oil product at 513 K are listed in Table 12.3.

Plot the calculation results as a curve, as shown in Fig. 12.3.

From the above example, it can be seen that the cost of kinetic energy loss (friction loss cost) has the most significant impact on the total cost. When the pipe diameter is smaller than the economic pipe diameter, the flow energy cost increases exponentially, while the investment cost decreases slightly, and the total cost increases.

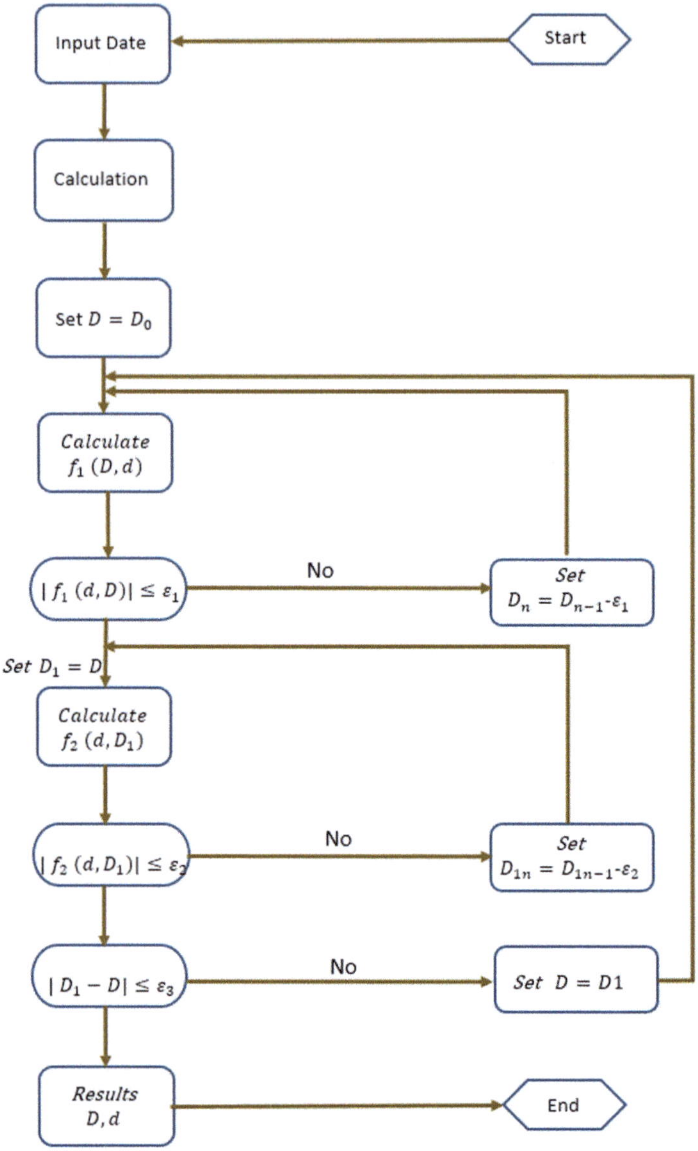

Fig. 12.2 The calculation block diagram of economical pipe diameter and economical insulation thickness

Table 12.2 20 °C water economical pipe diameter comparison

Known conditions	Description	Unit	Quantity	Description	Unit	Quantity
	Medium		293 K water	Annual interest rate	%	10
	Annual operating hours	h/a	8000	Service life	a	10
	Mass flow	t/h	100	Pump investment	dollar	833.3
	Fluid density	kg/m^3	998	Estimate shaft power	kW	60
	Fluid viscosity	PaS	0.001	Pump efficiency	%	80
	Comprehensive index of steel pipe	dollar/t	666.7	Electricity price	dollar/(kWh)	0.015
Calculation results	Pipe diameter × wall thickness	mm	168 × 5	114 × 4	219 × 7	
	Pipeline repayment	dollar/(am)	2.18	1.18	3.42	
	Operating cost	dollar/(am)	0.59	4.64	0.88	
	Total cost	dollar/(am)	2.77	5.82	3.57	
	Remarks		Economic pipe diameter	Smaller pipe diameter	Larger pipe diameter	

Therefore, when the economical pipe diameter is obtained, the pipe diameter should generally be selected in the nearest pipeline series, and when the difference between the pipeline series and the two pipe diameter series is roughly the same, the larger pipe diameter should be selected.

12.4 Optimization of Heat Exchange Equipment [5, 12]

Heat exchange equipment is an important equipment in the energy recovery link. For heat exchange equipment, the purpose is to recover energy, so there is a trade-off between the benefits of energy recovery and investment. For cooling equipment, the purpose is to cool the process streams to achieve the required temperature conditions, and there are also operating costs and investment costs, the objective function of its optimization is the minimum sum of investment and operating costs. The optimization problems of a single heat exchange equipment and cooling equipment are introduced below.

Table 12.3 Economic pipe diameter comparison of oil product transportation

Known conditions	Description	Unit	Quantity	Description	Unit	Quantity
	Mass flow	t/h	40	pump investment	dollar	833.3
	Annual operating hours	h/a	8000	Comprehensive index of insulation materials	$/m^3	80
	Fluid density	kg/m^3	850	Annual interest rate	%	4
	Fluid viscosity	Pas	0.0035	Insulation service life	a	5
	Fluid temperature	°C	240	Fuel price	Yuan/10^6 kJ	3.1
	Heated fluid temperature of heat exchange equipment	°C	400	Conversion equipment efficiency	%	90
	Ambient temperature	°C	15	Pump efficiency	%	80
	Average wind speed	m/s	3	Estimate shaft power	kW	50
	Comprehensive index of steel pipe	dollar/t	666.67	Electricity price	dollar/kWh	0.015
	Service life	a	10	Thermal conductivity of insulation material	W/(mK)	0.081
Calculation results	Description	Unit	Economic pipe diameter	Smaller pipe diameter	Larger pipe diameter	
	Pipe diameter × wall thickness	mm	108 × 4	89 × 3.5	159 × 5	
	Economical insulation thickness	mm	74	74	74	
	Pipeline repayment	dollar/(am)	1.05	0.87	1.56	
	Insulation cost	dollar/(am)	0.76	0.68	0.97	
	Heat loss exergy cost	dollar/(am)	1.64	1.45	2.15	
	Kinetic energy cost	dollar/(am)	0.62	1.69	0.09	
	Total cost	dollar/(am)	4.08	4.69	4.77	

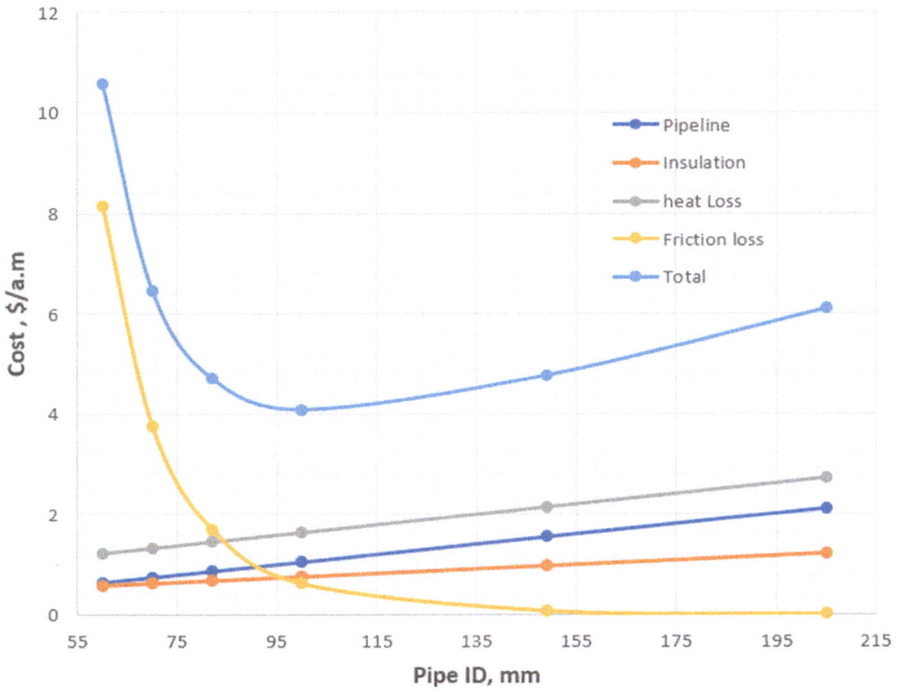

Fig. 12.3 Curve of relationship between annual cost and pipe diameter

12.4.1 Optimization of a Single Heat Exchanger

Using heat exchange equipment to exchange heat between the required cooling hot stream and the required heating cold stream to recover heat energy, which is a widely adopted energy-saving measure. However, a price must be paid for the recovery of heat energy, that is, the investment cost of heat exchange equipment must be paid. Generally, investment costs increase and heat recovery increases. There is an optimization problem between investment and energy-saving benefits. The objective function of a single heat exchange equipment optimization is the maximum difference between the annual energy-saving benefits of the recovered heat and the equivalent annual repayment cost of the equipment investment. Obviously, the economic effect of using heat exchange equipment is related to the price of energy and the cost of heat exchange equipment. We can use the following economic model to express the economic effect of using heat exchange equipment.

$$S = S_1 - S_2 = 3.6 C_h H Q - \beta A \tag{12.45}$$

In the formula:

S—the economic effect of using heat exchange equipment, dollar/a;

S_1—heat recovery benefit, dollar/a;

S_2—Annual repayment of equipment investment, dollar/a;

Q—The amount of heat recovery, MW;

C_h—the price of a unit of heat, dollar/GJ;

H—annual operating time, h/a;

A—The area of the heat exchanger, m^2;

β—Annual repayment cost per unit heat exchange area, dollar/(m^2a).

It goes without saying that the more heat recovered, the greater its economic value, and correspondingly, the equipment depreciation costs that need to be paid will inevitably increase. Obviously, there is an optimal heat exchange area to maximize the annual profit.

In order to solve the economic model Eq. (12.45), first determine the relationship between the heat recovery Q and the heat exchanger area A, establish the mathematical model of the non-phase change countercurrent heat exchanger shown in Fig. 12.4.

Assuming that the heat exchange area of the heat exchanger is A, after the heat exchange, the hot flow decreases from the initial temperature T_1 to the final heat exchange temperature T_2, and the cold flow rises from the initial temperature t_1 to the final heat exchange temperature t_2. Now, we take a micro-element dA on the heat exchanger (as shown in Fig. 12.4) and examine its heat exchange situation.

According to the heat transfer equation, you can write:

$$dQ = K(T - t)dA \tag{12.46}$$

According to the heat balance equation, it can be written (ignoring heat loss during heat transfer):

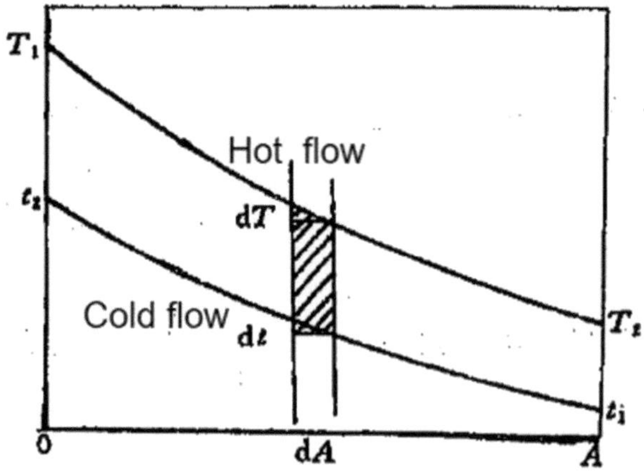

Fig. 12.4 Countercurrent heat transfer process

$$Hot\ flow\quad dQ = -W_h C_{Ph} dT \tag{12.47}$$

$$Cold\ Flow\quad dQ = -W_C C_{Pc} dt \tag{12.48}$$

In the above two formulas:

dQ—the heat exchange amount on the micro-element dA of the heat exchanger, kW;

K—total heat transfer coefficient, w/(m^2 °C);

T—the heat exchange temperature of the hot stream A, °C;

t—Heat exchange temperature of cold stream B, °C;

dA—The heat exchange area of the micro-element of the heat exchanger, m^2;

W_h—The hot stream flow rate, kg/s;

C_{Ph}—The specific heat capacity of the hot flow, W/(kg °C);

W_c—Flow rate of cold streams, kg/s;

C_{Pc}—Specific heat capacity of cold material, W/(kg °C);

dT—hot stream temperature difference on the micro-element dA of the heat exchanger, °C;

dt—cold stream temperature difference on the micro-element dA of the heat exchanger, °C.

Along the heat exchange area coordinate A, dT and dt are both negative, and the heat exchange Q must be positive, so the above equations are all negative.

According to Eqs. (12.47) and (12.48), the heat transfer amount Q is the same when heat loss is neglected, then:

$$W_h C_{Ph} dT = W_c C_{Pc} dt$$

Subtract $W_h C_{Ph} dt$ on both sides, then the above formula

$$W_h C_{Ph}(dT - dt) = (W_c C_{Pc} - W_h C_{Ph})dt$$

$$d(T - t) = \frac{W_c C_{Pc} - W_h C_{Ph}}{W_h C_{Ph}} dt$$

From the heat balance Eq. (12.48) of cold stream:

$$dt = \frac{-dQ}{W_c C_{Pc}}$$

From the heat transfer equation:

$$dQ = K(T - t)dA$$

$$d(T-t) = \left(\frac{W_c C_{Pc}}{W_h C_{Ph}} - 1\right)\frac{-K(T-t)dA}{W_c C_{Pc}}$$
$$\frac{d(T-t)}{(T-t)} = \frac{KdA}{W_c C_{Pc}}\left(1 - \frac{W_c C_{Pc}}{W_h C_{Ph}}\right) \tag{12.49}$$

We define $R = \frac{W_c C_{Pc}}{W_h C_{Ph}}$ as the heat capacity flow rate ratio, then the above formula is:

$$\frac{d(T-t)}{(T-t)} = \frac{K}{W_c C_{Pc}}(1-R)dA \tag{12.50}$$

After integrating the formula (12.50), we can get:

$$\frac{T_2 - t_1}{T_1 - t_2} = \exp\left[\frac{KA}{W_c C_{Pc}}(1-R)\right] \tag{12.51}$$

Considering $R = \frac{T_1 - T_2}{t_2 - t_1}$ and temperature efficiency

$$E = \frac{t_2 - t_1}{T_1 - t_1} \tag{12.52}$$

After mathematical processing, the formula (12.51) can be rewritten as:

$$E = \frac{1 - \exp\left[\frac{KA}{W_c C_{Pc}}(1-R)\right]}{R - \exp\left[\frac{KA}{W_c C_{Pc}}(1-R)\right]} \tag{12.53}$$

The above formula is the mathematical model of the temperature efficiency of the countercurrent heat transfer process.

From the heat balance equation and temperature efficiency model:

$$Q = W_c C_{Pc}(t_2 - t_1) = W_c C_{Pc}(T_1 - t_1)E \tag{12.54}$$

Substituting formula (12.53) into formula (12.54):

$$Q = W_c C_{Pc}(T_1 - t_1)\frac{1 - \exp\left[\frac{KA}{W_c C_{Pc}}(1-R)\right]}{R - \exp\left[\frac{KA}{W_c C_{Pc}}(1-R)\right]} \tag{12.55}$$

This establishes the relationship between the heat transfer area A and the heat transfer duty Q.

Substituting the above equation into the total cost equation, the Eq. (12.45) becomes:

$$S = 3.6C_h H W_c C_{Pc}(T_1 - t_1)\frac{1 - \exp\left[\frac{KA}{W_c C_{Pc}}(1 - R)\right]}{R - \exp\left[\frac{KA}{W_c C_{Pc}}(1 - R)\right]} - \beta A \tag{12.56}$$

This is the optimization problem of a single variable function. According to the necessary condition for the existence of the extreme value of the function, $dS/dA = 0$, take the derivative of Eq. (12.56) with respect to A and make it equal to zero, and then simplify:

$$(1 - E)(1 - ER) = \frac{\beta}{C_h H K(T_1 - t_1)} \tag{12.57}$$

In actual calculations, since C_h, β, H, $KT_1, t_1 W_h$, C_{Ph}, W_c, C_{Pc} are all known, the above formula is a quadratic equation in one variable with respect to temperature efficiency, and one of the irrational roots can be dropped, you can find E. Calculation steps for calculating the best heat exchange area of a single heat exchanger:

(1) Calculate the heat capacity flow rate ratio $R = \frac{W_c C_{Pc}}{W_h C_{Ph}}$
(2) Calculate the temperature efficiency E according to formula (12.57);
(3) Calculate the hot flow outlet temperature T_2

$$T_2 = T_1 - RE(T_1 - t_1)$$

(4) Calculate the outlet temperature of cold stream t_2

$$t_2 = t_1 + E(T_1 - t_1)$$

(5) Calculate the logarithmic mean temperature difference

$$\Delta T_{\ln} = \frac{(T_1 - t_2) - (T_2 - t_1)}{\ln \frac{T_1 - t_2}{T_2 - t_1}}$$

(6) Calculate the heat recovery amount Q by formula (12.55)
(7) Determine the best heat exchange area according to the following formula

$$A = \frac{Q}{K \Delta T_{\ln}}$$

(8) Calculate the annual repayment of equipment investment and heat recovery costs and the annual net energy-saving benefits.

$$S = S_1 - S_2, S_1 = 3.6C_h H Q, S_2 = \beta A$$

The optimization of the heat exchanger is suitable for the optimization of a single heat exchanger, and optimal matching between a single hot stream and a cold stream.

For heat exchange networks and sequences composed of many heat exchange equipment, synthesis and pinch technologies are used (such as Aspen energy analysis program).

12.4.2 Determination of the Best Outlet Temperature of the Cooling Water of the Cooler

The so-called optimal design of a single cooler is taking the lowest total annual cost as the criterion for optimization at given process cooling task, and. The total annual cost can generally be considered to be composed of two parts: the depreciation cost of the cooling equipment and the operating cost of the cooling water. In this way, the objective function of the optimal design of a single cooler can be expressed by (12.58):

$$C_t = C_A + C_W \tag{12.58}$$

where: C_t—is the total annual cost, dollar/a;

C_A—The depreciation cost of equipment, can be calculated using the formula (12.59):

$$C_A = aA \tag{12.59}$$

where: a—the depreciation cost per unit of heat exchange area, dollar/(am^2);

A—the area of the cooler, m^2;

C_W—cooling water operating cost, dollar/a;

$$C_W = \beta HW \tag{12.60}$$

β—cooling water price, dollar/t;

H—the total number of operating hours per year, h/a;

W—water consumption per hour, t/h.

According to the heat transfer equation:

$$Q = KA\Delta T \tag{12.61}$$

where: Q—is the heat exchange amount per unit time, kW;

K—total heat transfer coefficient, kW/(m^2 °C);

A—Heat exchange area, m^2;

ΔT—is a logarithmic average temperature difference (°C), when countercurrent heat transfer is used, it can be calculated according to formula (12.62);

$$\Delta T = \frac{(T_1 - t_2) - (T_2 - t_1)}{\ln \frac{T_1 - t_2}{T_2 - t_1}} \tag{12.62}$$

T_1—the inlet temperature of the hot stream, °C;

T_2—the temperature at the outlet of the hot stream, °C;

t_1—Cooling water inlet temperature, °C;

t_2—Cooling water outlet temperature, °C.
Heat balance equation:

$$Q = WC_P(t_2 - t_1)/3.6 \tag{12.63}$$

where: C_P—Specific heat of water, kJ/(kg °C).

According to the cost of the equipment and the price of cooling water to optimize the design of a single cooler, it can be described by the following mathematical model, that is, when the constraint conditions (12.64) are met, make the objective function $C_t = C_A + C_W$ with the minimum value.

$$Q = KA\Delta T,\ Q = WC_P(t_2 - t_1)/3.6 \tag{12.64}$$

This is an extreme value problem with constraints. We use the substitution method to turn the constrained extreme value problem into an unconstrained extreme value problem. After mathematical processing, you can get:

$$C_t = \frac{3.6H\beta Q}{C_P(t_2 - t_1)} + \frac{aQ\ln \frac{T_1 - t_2}{T_2 - t_1}}{K[(T_1 - t_2) - (T_2 - t_1)]} \tag{12.65}$$

The optimization design problem of a single cooler has become a problem of finding the best outlet temperature of the cooling water to minimize the total annual operating cost. Such a problem can be based on the necessary conditions for the existence of extreme values, namely: $dC_t/dt_2 = 0$.

After solving, the following algebraic equations can be obtained:

$$\frac{3.6KH\beta}{aC_P}\left[\frac{(T_1 - t_2) - (T_2 - t_1)}{t_2 - t_1}\right]^2 - \ln\frac{T_1 - t_2}{T_2 - t_1} + \left(1 - \frac{T_2 - t_1}{T_1 - t_2}\right) = 0 \tag{12.66}$$

Obviously, the cooling water outlet temperature that satisfies this algebraic equation is optimal t_2, which minimizes the cooler's total annual operating cost. However, this algebraic equation is difficult to solve analytically. Newton–Raphson method combined numerical solution method can be used to solve this equation.

12.5 Economical Thermal Efficiency of Heating Furnace

The smallest annual cost or the largest annual benefit is a common method for optimizing energy-saving measures. However, there is another optimization method that has been applied and valued: the energy-saving investment stagnation method, which is more convenient to apply. Take the determination of the economic thermal efficiency of the heating furnace as an example to illustrate the application of this method.

12.5.1 Background

At present, the efficiency of heating furnaces in petrochemical enterprises has reached above 85%. If the efficiency of the furnace continues to be improved, as the heat recovery system heat transfer temperature difference decreases, it will inevitably cost more equipment investment. That is, the profit rate of unit investment gradually decreases with the increase of thermal efficiency. How to determine the appropriate thermal efficiency of the heating furnace is a realistic topic for the improvement of energy saving of the heating furnace.

For heating furnaces, the main way to improve thermal efficiency is how to recover and utilize the low-temperature heat of exhaust flue gas. There are two main situations: one is to use its own flue gas or other process heat sources outside the furnace to preheat the furnace air to improve the furnace efficiency and save fuel; the other is using some low-temperature medium to exchange heat with the flue gas under the established process conditions (flue gas temperature), thereby reducing the exhaust flue gas temperature and improving the thermal efficiency (such as cold feed preheating, BFW preheating and steam superheating, etc.), and the fuel consumption and flue gas volume of the furnace remain unchanged.

We know that there are two limitations to improving the thermal efficiency of the heating furnace. One is the limitation of acid dew point corrosion of flue gas, in order to ensure that the waste heat recovery equipment is protected from the acid dew point corrosion, the exhaust flue gas temperature must be higher than the acid dew point temperature of the flue gas. With using sweeten gas and the application of glass tube air preheaters in recent years, this limitation is often broken. Another limitation is the limitation of economic exhaust flue gas temperature. The thermal efficiency of the heating furnace should not exceed the optimal thermal efficiency. The investment cost of energy recovery higher than the optimal thermal efficiency will inevitably be greater than the heat recovery benefit.

12.5.2 Method for Determining Economic Thermal Efficiency

There is an economic temperature approach [13] and economic thermal efficiency in the waste heat recovery of the heating furnace. The investment stagnation point method is used here to determine the economic flue gas temperature and economic thermal efficiency of the heating furnace [14, 15].

The so-called investment stagnation method is that as the exhaust flue gas temperature decreases, the benefits obtained by the unit investment of waste heat recovery gradually decrease, and when a certain point is reached, the depreciation cost of the investment is equal to the energy saving benefit, and continued investment "cannot make ends meet". We call this point the staging point of investment. This optimization method for determining the recovery system is the investment stagnation point method. The total investment before the stagnation point is less than the income. Therefore, this point is the limit of our investment. The waste heat recovery should be carried out within the scope of the investment stagnation point.

Take the waste heat recovery heat transfer area per square meter as the basis: For the case of recovering waste heat to reduce fuel, the energy saving benefit is the fuel price.

Energy-saving benefits

$$E_O = K\Delta T_m(C_F H)/Q_L \tag{12.67}$$

Unit investment cost

$$E_I = (P_{IH} + P_{ah})N + P_{EH}H \tag{12.68}$$

The stagnation point of investment is the point where income and expenditure are equal:

$$\frac{K\Delta T_m(C_F H)}{Q_L} = (P_{IH} + P_{ah})N + P_{EH}H \tag{12.69}$$

Sorted out

$$\Delta T_m = \frac{[(P_{IH} + P_{ah})N + P_{EH}H]Q_L}{KC_F H} \tag{12.70}$$

This formula is the economical heat transfer temperature approach at the end when the waste heat preheats the air.

When the waste heat recovery is not to preheat the air to save fuel, but to heat a certain low-temperature medium such as preheating water, etc., at this time, the revenue cannot be calculated by the fuel price, and the revenue should be calculated by the recovered heat exergy value.

$$E_O = K\Delta T_m C_h H\left(1 - \frac{T_0}{T_{cm}}\right)$$

And the economic heat transfer temperature difference changes as:

$$\Delta T_m = \frac{(P_{IH} + P_{ah})N + P_{EH}H}{KC_h H\left(1 - \frac{T_0}{T_{cm}}\right)}$$

Generally, the total heat transfer coefficient K does not change much during the heat recovery process, and a constant can be selected. T_{cm} is the arithmetic mean temperature of the heated low-temperature stream entering and leaving the furnace, K.

After determining the optimal end economic temperature difference, the economic exhaust flue gas temperature t_2 can be determined by the cold flow inlet temperature t_{c1}.

$$t_2 = t_{c1} + \Delta T_m \tag{12.71}$$

Economic exhaust flue gas heat loss q_2.

$$q_2 = (t_2 - 15.6)C_g(\alpha L_0 + W_s + 1)/Q_L \tag{12.72}$$

When the heat dissipation loss q_4 of the heating furnace has been measured, the general economic thermal efficiency is:

$$\eta = 1 - q_2 - q_4 \tag{12.73}$$

In the above formulas:

K—The heat transfer coefficient of waste heat recovery equipment, kW/m^{2}k;

ΔT_m—the average heat transfer temperature difference of the last square meter of investment, °C;

C_F—fuel price;

H—the number of operating hours per year;

Q_L—low heating value of fuel;

P_{IH}—The price of furnace tube or heat exchanger area, dollar/m^2;

P_{ah}—auxiliary equipment investment, dollar/m^2;

P_{EH}—Auxiliary system energy consumption fee (power and water of the boiler), dollar/m^2

N—investment repayment coefficient, see above for calculation;

T_{cm}—the average temperature of the heat recovered by the cold flow; K;

t_{c1}—cold flow inlet temperature, °C;

α—excess air coefficient;

L_0—Theoretical fuel air volume, taken as 14.2 kg/kg;

W_s—the amount of atomized steam, kg/kg;

C_g—specific heat capacity of flue gas, $C_g = 0.245 \times 4.18$ kJ/(kg °C);

C_h—exergy price of recovered heat, see above for calculation.

According to the investment stagnation method, the determination of economic thermal efficiency is based on the basis that the unit investment income is equal or greater than the investment depreciation expense. The average income should be much greater than the expenditure, and the income per square meter of the last investment equals the expenditure. The concept is clear, the process is simple and intuitive, and the equation is an explicit function. As long as the technical and economic price is determined, the economic thermal efficiency can be determined. The stagnation point of investment is the economic boundary for improving the thermal efficiency of the heating furnace, and the investment to improve the efficiency of the furnace should not exceed this point. The idea of the investment stagnation method can also be used to optimize other energy-saving measures.

References

1. Shanghai Institute of Chemical Engineering, *Basic Chemical Engineering* (Shanghai Science and Technology Press, 1978)
2. Tianjin University Chemical Engineering Principles Teaching and Research Office, *Chemical Engineering Principles* (Tianjin Science and Technology Press, 1983), pp. 78, 98
3. B. Hua, *Process Energy Analysis and Synthesis* (Hydrocarbon Processing Press, 1989), pp. 103–104
4. Y. Yang, *Practical Chemical Systems Engineering* (Chemical Industry Press, 1989)
5. S. Wei, *Petrochemical Production Process Optimization* (Refinery Design Editorial Department, 1988)
6. Y. Xu et al., *Petrochemical Practical Optimization Method* (Petroleum University Press, 1990)
7. H. Cao, K. Xie, Introduction to optimization methods. Refin. Des. **3** (1983)
8. X. Xiang, *Engineering Exergy Analysis Method* (Petroleum Industry Press, 1990), pp. 312–317
9. R.A. Gaggioli, *Second Law Analysis for Process and Energy Engineering* (American Chemical Society, 1983), pp. 34, 42
10. T.A. Chen, Economic pipe diameter and economic insulation thickness of fluid transport. Petrol. Refin. **10** (1986)
11. B. Lin et al., *Economic Flow Velocity and Pipe Diameter of Oil Storage and Transportation.* Oil Gas Storage Transp. **2** (1982)
12. S. Wei, Best heat exchanger area calculation method for a single heat exchanger. J. Beijing Chem. Inst. **1**, 42, 51 (1981)
13. W. Li, Economic temperature difference of fired heater waste heat recovery. Petrol. Refin. **1** (1984)

14. *Survey of the Technical Transformation of Refineries, Refinery Energy Conservation 10-Fired Heaters* (Science and Technology Information Institute of the Ministry of Petroleum, 1985), pp. 67–71
15. S. Zhao, Q. Chen, Heating furnace heat efficiency technical and economic analysis. Refin. Des. **1** (1982)
16. D. Yang, *Principles of Irreversible Process Thermodynamics and Engineering Applications* (Science Press,1989), p. 2
17. B. Hua, A. Chen, Practical calculation and exergy economic analysis of heat dissipation in refinery. Petrol. Refin. **2** (1984)
18. W.J. Weper et al., *Economics Sizing of Steam Piping and Insulation* (1980)

Chapter 13
Pinch Energy-Saving Technology and Its Application

Abstract Pinch energy-saving Technology is currently the most valued in practice, it is based on the first and second laws of thermodynamics, does not perform quantitative analysis and calculations, and uses the concept of optimization, but does not require advanced mathematics for optimization technology. This chapter first introduces the process dual subsystem system model, which is divided into the heat exchange subsystem and the process operation subsystem, the heat exchange subsystem includes three parts: cooling, heat exchange, and heating. the pinch energy-saving technology plays the most important role in the heat exchange subsystem; it includes 7 sections: the concept of pinch point and its determination; Pre-estimate the heat exchange network area and the optimal $\Delta T_{\min}$; Energy target determination; Pinch design of heat exchange network; Placement of the heat engine (pump) in the total energy system; The effect of cross heat transfer on heat exchange network area and energy; Energy saving principle of pinch technology; and in Sect. 13.7 provided the showcase of Pinch analysis examples using Aspen Energy Analyzer.

Keywords Pinch analysis · Vertical heat-transfer · Cross heat transfer · Engine location · Total-energy system · Energy analyzer · Composite curve

The pinch technology is a comprehensive heat recovery and utilization technology proposed by Professor B. Linnhoff of the University of Manchester in the United Kingdom. It has been widely used in the European and American petrochemical industries. Many scholars and energy-saving workers have also devoted themselves to this technology. The research and application of the work has made many research results and industrial application examples [1, 2].

13.1 The Concept of Pinch Point and Its Determination

The pinch concept is based on the second law of thermodynamics. The process energy integration technology developed based on the pinch has gone out of the field of thermodynamic analysis and formed a comprehensive synthesis technology for

T. A. Chen, *Energy Saving and Carbon Reduction*,
https://doi.org/10.1007/978-981-19-5295-1_13

energy recovery and utilization. The application of pinch technology is not limited to the analysis and synthesis of heat exchange networks, but also includes the optimization and synthesis of thermal energy power systems in process systems. But the most common application is still the synthesis and optimization of heat exchange networks.

13.1.1 Pinch Concept

The petrochemical process is mainly a thermal processing process. In the process of providing work and heat to satisfy the process operation, a large proportion of it is provided by the energy recovery system. For energy recovery systems that do not consider energy conversion and flow work, the dual subsystem process system model [3] is shown in Fig. 13.1.

According to the process system model in Fig. 13.1, the process system is decomposed into the heat exchange subsystem and the process operation subsystem. Raw material C enters the system and first exchanges heat with products A and B to recover energy. Then it is further heated to meet the process parameters and energy requirements, and then enter the process operating system (reaction, fractionation, etc.) to process the raw materials into products. In order to reduce the energy supplied, the product energy is recovered in the heat exchange subsystem to increase the temperature of the raw materials. After the heat exchange, the product is cooled to reach the target product temperature. Therefore, the heat exchange subsystem includes three parts: cooling, heat exchange, and heating. Between these three parts, there is obviously an interdependence between improving heat exchange and reducing heating

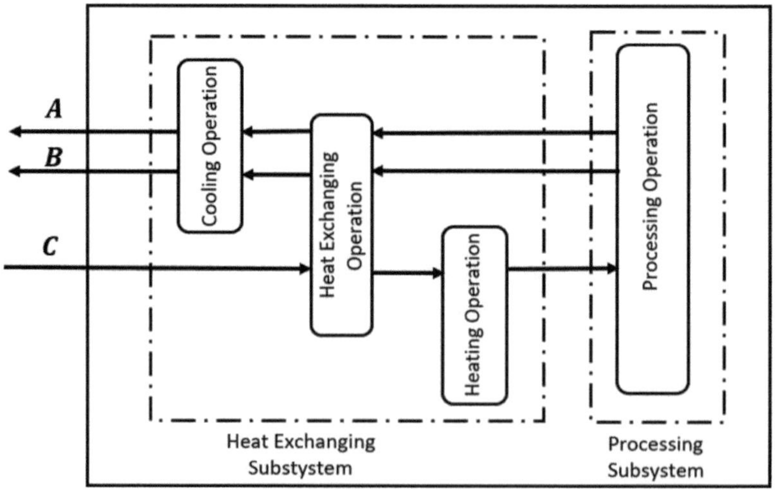

Fig. 13.1 Process dual subsystem model

and cooling. More comprehensively reflects the characteristics and laws of the energy recovery system.

13.1.1.1 Temperature Interval (Temperature Section)

In engineering design calculations, in order to ensure the heat transfer rate, it is usually required that the temperature difference between the cold and hot streams must be greater than a certain value (such as 20 K). This temperature difference is called the minimum allowable heat transfer temperature difference $\Delta T_{\min}$, and the determination of $\Delta T_{\min}$ should be Optimizing and determining based on technical economy and process parameters. This method of pre-determining the heat exchange area $A_{\min}$ and $\Delta T_{\min}$ based on the process parameters, streams data, and technical and economic conditions before the network is determined is called super-target optimization (see Sect. 13.2). The starting temperature of each hot flow and the target temperature subtract the minimum allowable temperature difference $\Delta T_{\min}$, then sorted from largest to smallest together with the initial target temperature of all cold streams, forming a temperature intervals of different temperature ranges [4], respectively using $T_1, T_2, \ldots T_{n+1}$ Indicates that the number of temperature intervals n can be calculated by the following formula:

$$n = 2Z - 1 - d \tag{13.1}$$

where: d—the number of repetitions of the stream with the same start temperature and end temperature (the hot stream start and end temperature should always be minus $\Delta T_{\min}$);

Z—Number of streams.

By dividing the original heat transfer problem into n temperature intervals (each temperature area includes all streams whose temperature fall within that interval), the original network synthesis problem can be decomposed into n sub-networks for synthesis, because all streams in the sub-networks are in the same temperature range, so the comprehensive problem is simplified.

Example 13.1 Given four hot and cold streams (see Table 13.1), if the minimum allowable temperature difference $\Delta T_{\min}$ is 10 °C, divide the temperature intervals.

Solution

From Table 13.1, we can see that $Z = 4, d = 2$, and from Eq. (13.1), the number of temperature intervals is 5. The endpoint temperature values of the temperature intervals (based on cold flow): 180 °C, 170 °C, 140 °C, 105 °C, 60 °C, 30 °C. These six temperatures can divide the original problem into five temperature zones, as shown in Fig. 13.2. It can be seen from the figure that the temperature zones SN_1 to SN_5 are respectively determined by the initial temperature T_I and the final temperature T_O

Table 13.1 Example 13.1 streams data sheet

Streams and types	Heat capacity flow rate (kW/°C)	Initial temperature (°C)	Final temperature (°C)	Heat load (kW)
(1) Cold	6	60	180	−720
(2) Hot	4	180	40	560
(3) Cold	5	30	105	−375
(4) Hot	8	150	40	880
Total				345

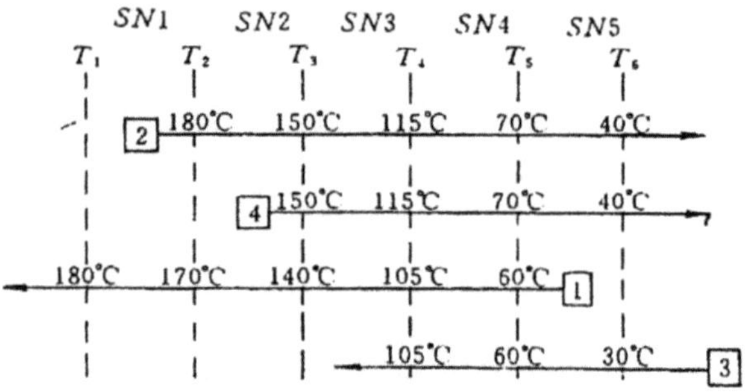

Fig. 13.2 Division of temperature intervals

of the stream. For example, SN_3 is determined by the initial temperature of stream 4 and the final temperature of stream 3.

13.1.1.2 Composite Curve

Add up the heat at the same temperature interval to obtain a composite temperature enthalpy line, which connects the temperature enthalpy lines of different temperature sections, and the curve plotted on the temperature enthalpy coordinate chart is called the composite curve. The conversion drawing method is (see Fig. 13.3).

(1) Plot the streams on the coordinates according to the temperature and enthalpy of each stream (temperature × flow × heat capacity) (the temperature and enthalpy coordinate system shows the relative value);
(2) Add the enthalpy of each stream between the temperature intervals to form the composite temperature—enthalpy line (because the heat capacity is assumed to be constant).
(3) According to the order of temperature from high to low or from low to high, the temperature enthalpy line between each temperature intervals can be connected to form a hot or cold composite curve.

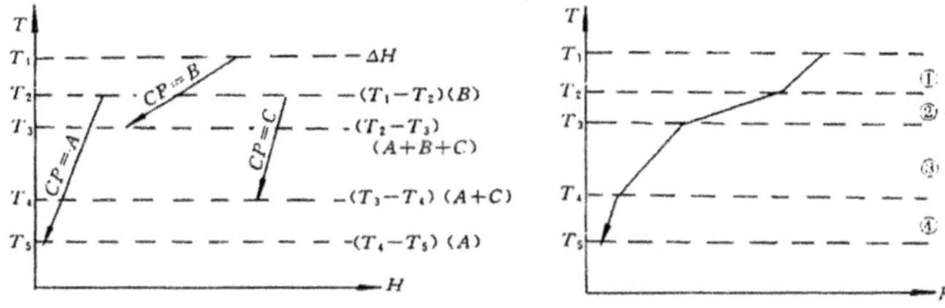

Note: CP is the heat capacity flow rate (heat capacity × flow), kW/°C

Fig. 13.3 Drawing method of composite curve.
Note CP is the heat capacity flow rate (heat capacity × flow), kW/°C

In drawing the composite curve, we find that the slope of the composite curve changes at the point where the flow enters and exits, that is, at the end temperature of the temperature section, forming an inflection point. Secondly, for the predetermined $\Delta T_{\min}$, when the hot flow composite curve is drawn, the position of the cold flow composite curve is determined accordingly. Or first determine the cold utility, and then draw the cold flow composite curve, check whether $\Delta T_{\min}$ is the same as the predetermined value, otherwise adjust the cold utility load until it matches the predetermined $\Delta T_{\min}$. The drawn composite curve is shown in Fig. 13.4.

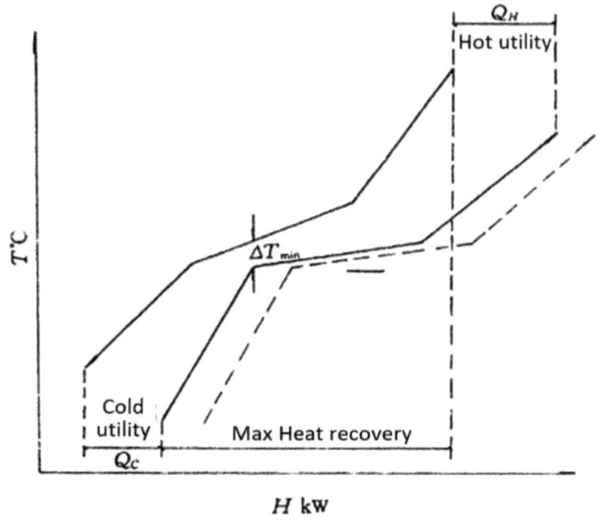

Fig. 13.4 Composite curve diagram of heat exchange network

13.1.2 How to Determine the Pinch Point

When the process parameters and streams data of a system are given, we can easily determine the pinch point position. There are generally three methods.

13.1.2.1 Graphical Method

In Fig. 13.4, the point with the smallest vertical distance between the cold and hot flow composite curves is the pinch point to be found, and the temperature difference at the pinch point is the network minimum temperature difference $\Delta T_{\min}$. From the point of view of heat recovery, there is a certain temperature in the heat exchange system. If heat energy passes through this temperature, energy will be wasted. This specific temperature is called the pinch point.

The composite curve on the $T - H$ diagram can be shifted horizontally along the H axis without changing the heat transfer load. Therefore, the cold flow composite curve can be shifted to the left along the $lines$ axis. At this time, the vertical distance between the two curves gradually decreases with the shift of the curve, that is, the heat transfer temperature difference ΔT gradually decreases. When the minimum vertical distance between the two curves is equal to the minimum allowable heat transfer temperature difference $\Delta T_{\min}$, the practicable limit position is reached. The geometric meaning of this limit position is the position where the vertical distance between the cold and hot flow composite curves is the smallest. It is not difficult to see from Fig. 13.4 that this narrowest position is the pinch point. At this time, the horizontal difference between the end points of the two curves represents the minimum cold and hot utility usage, and the maximum heat recovery (i.e., the maximum heat exchanged).

13.1.2.2 Problem Solving Table Method [5, 6]

The following example illustrates the method of finding the pinch point with the problem-solving table method (problem solving table).

Example 13.2 There are four streams in a heat exchange system, and the stream data are shown in Table 13.2.

The location of the pinch point of the heat exchange network and the minimum utility consumption can be calculated using the problem-solving table based on the selected $\Delta T_{\min}$. The example uses $\Delta T_{\min} = 20$ °C, and the calculation of the problem-solving table is shown in Table 13.3.

The streams data in the table are listed on the left, divided into 6 temperature zones, which are equivalent to 6 sub-networks SN1 to SN6. The temperature section is divided according to the starting and ending temperature of each stream. In order to ensure that all the heat exchanges between the cold and hot flows among the

Table 13.2 Example 13.2 streams data table

Streams and types	Heat capacity flow rate (kW/°C)	First temperature (°C)	Final temperature (°C)	Heat load (kW)
(1) Hot	3	160	60	300
(2) Hot	12	90	60	360
(3) Cold	3.8	20	125	−399
(4) Cold	4.5	25	100	−337.5

Table 13.3 Example 13.2 problem solving table

	Stream and temperature						1	2	3	4	5
Subnetwork	Cold stream			Hot stream			Deficit	Accumulated		Max. allowable heat flow (kW)	
	3	4			1	2	$(CP_c - CP_h) \times \Delta T_{sn}$	In	Out	In	Out
CP	3.8	4.5			3	12					
				160							
SN1			125	145			-45	0	45	**134**	179
SN2			100	120			20	45	25	179	159
SN3			70	90			159	25	-134	159	0
SN4			40	60			-201	-134	67	0	201
SN5			25				124.5	67	-57.5	201	76.5
SN6			20				19	-57.5	-76.5	76.5	**57.5**

various temperature sections are carried out at a temperature greater than or equal to $\Delta T_{\min}$, the upper limit of SN3 is 100 °C for cold flow 4 and 120 °C for hot flow 1. The complete heat exchange of each stream is an important feature of the problem-solving table, which means that the heat between each temperature section is either deficit or surplus. The first column in the problem-solving table is the deficit of heat of each sub-network, so the sign negative in the table is excess heat, and the symbol positive is indicating deficit. Another feature of the problem-solving table is that the hot flow transfers heat from the high temperature section to the low temperature, that is, the excess heat in the high temperature section can be used to supplement the deficit of heat in the low temperature section. The cumulative excess or deficiency value of each sub-network is shown in columns 2 and 3 in the table. The input of SN1 in column 2 is zero. This is a preliminary assumption before the design. The excess heat of SN1 is the output heat of SN1, that is, the input heat of SN2, and so on to calculate the entire sub-network. The negative value of SN2 and SN3 in the third column indicates that thermodynamics is impossible, so you must enter enough heat in SN1 to make SN3 equal to zero, that is, the minimum hot utility quantity of the network. This example requires 134 kW heat (That is, the input value of SN1 in column 4 of the table). The output value of SN6 is the minimum cold utility for this example.

The analysis results of Tables 13.3 can be shown in Fig. 13.5. The hot flow values on and below each sub-network correspond to the calculation results in the problem-solving table. The heat flow value between SN3 and SN4 is zero and the other heat

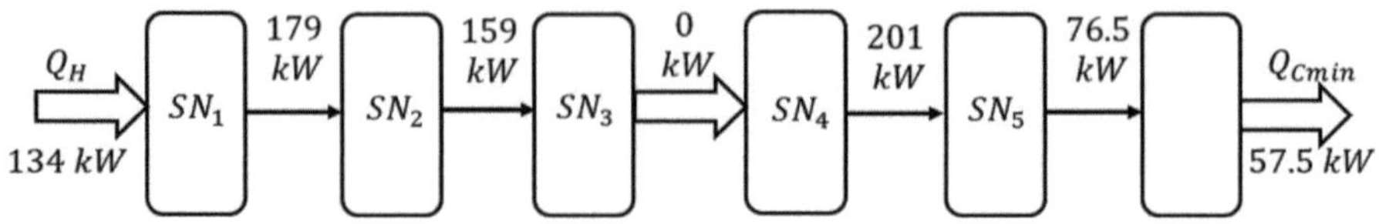

Fig. 13.5 Heat flow diagram of temperature intervals

flows are all positive values. The point where the heat flow is zero is the pinch point to find.

In Table 13.3, the 4th and 5th columns respectively represent the heat flow between temperature zones and the environment. This heat flow can be represented by a temperature zone heat flow diagram.

From Fig. 13.5, we can intuitively see the heat flow relationship between the temperature zones and the required minimum utility usage. The heat flow between SN3 and SN4 is zero, which means that no heat flows from SN3 to SN4. The point where the heat flow is zero is called the pinch point.

The pinch temperature is also not uniform in the literature. Some take the hot flow temperature, some take the cold flow temperature, and some take the average temperature of the cold and hot flow at the pinch point as the pinch temperature.

13.1.2.3 Heat Deficit Method [7]

The heat deficit method is a new method for solving pinch points proposed by Chinese scholars Lang Yidong [7] and Xu Yifang. That is, "except for the endpoints, the pinch point only exists at the inlet temperature of the cold and hot streams." The condition rule for the pinch point is that when each temperature section is combined from a single stream into a composite curve, the slope of the composite curve changes. The inflection point formed at the inlet temperature of the cold and hot streams in the composite curve is called the pre-pinch point or potential pinch point. Pinch points in the system network can only be generated from these breakpoints. The pinch point is determined by the maximum heat deficit method. Set the pre-pinch point as P, then the pinch point of the network P^*

$$P^* = \max\left\{Z_H^P\right\} \tag{13.2}$$

where: P—potential pinch point.

Z_H^P is the heat deficit at each potential pinch point, that is, from the high temperature end to each potential pinch point to do heat balance. The potential pinch point where the largest difference between the required heat of the cold flow and the heat release of the hot flow is the pinch point of the network. The heat deficit model can be graphically represented [7], which is not discussed here.

13.1.3 Grand Composite Curve

According to the data in the problem-solving table, the temperature section heat flow diagram can be drawn. From the temperature section heat flow diagram, the grand composite curve of the system network can be easily drawn. That is, with temperature as the ordinate and heat flow as the abscissa, in different temperature sections, mark the net heat flow and the coordinate values of the temperature section, and connect these coordinate points to form a Grand Composite curve.

From the problem-solving table, we can easily draw a grand composite curve, the temperature used in grand composite curve is the average temperature; that is, above the pinch, using hot stream temperature minus $1/2\Delta T_{\min}$, and below the pinch point, using cold stream temperature plus $1/2\Delta T_{\min}$; Fig. 13.6 is generated from Table 13.3, it depicts the heat flow diagram (left) and the grand composite curve of the temperature sections.

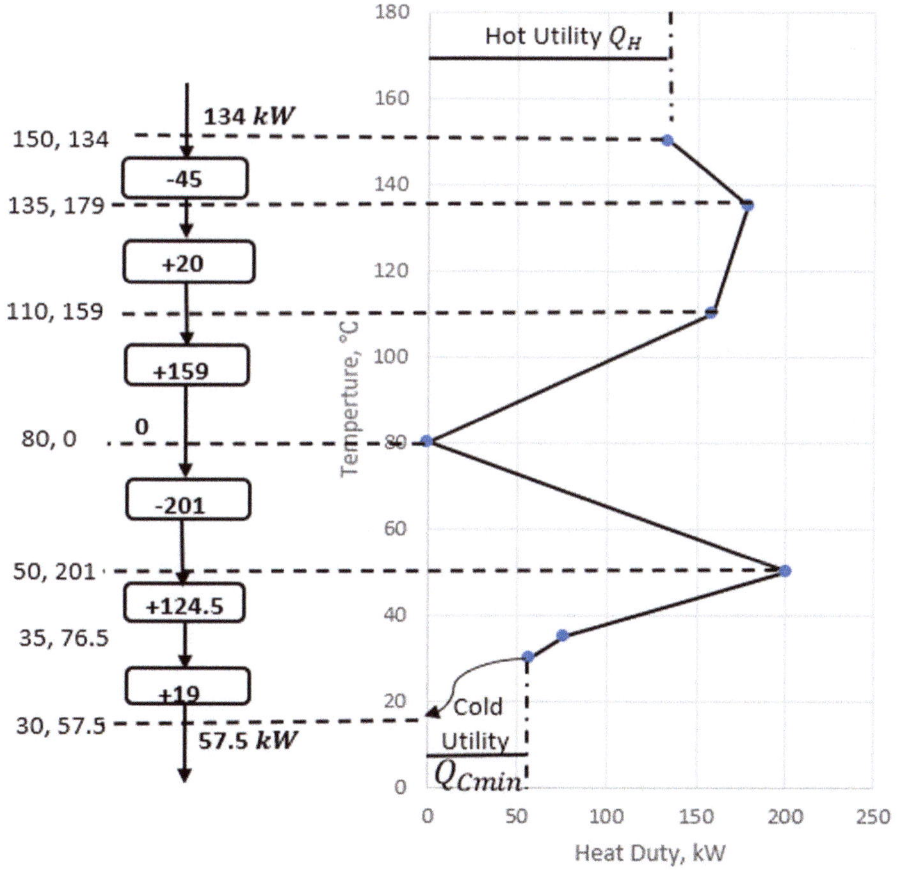

Fig. 13.6 Drawing of the grand composite curve

From Figure 13.6, you can find the key perimeter of the system: hot utility demand, cold utility demand, and pinch point. Based on the Grand Composite curve, a bridge between utility and process heat utilization can be built to use energy according to quality. For example, a hot utility uses steam as a heat source, and only uses medium-pressure steam in some high-temperature heating sections, while low-pressure steam can be used in low-temperature sections. The same is true for cold utility, through the grand composite curve, the cooled stream can extract heat at the highest possible temperature. This is a pinch point design method for multiple utilities [5, 14].

13.2 Pre-estimate the Heat Exchange Network Area and the Optimal ΔT_{min}

According to the stream data, we can draw the composite curve of the system (choose an initial value of ΔT_{min}). According to the vertical heat transfer method, determine the minimum area of the network, and then through technical and economic optimization, compare the total economic effects of different schemes, and determine the optimal ΔT_{min} (see Fig. 13.7).

13.2.1 Area Estimation Method

Under the premise of vertical heat transfer, the composite curve shown in Fig. 13.8 is divided into the vertical heat transfer temperature section (enthalpy section) according to the inflection point formed by the change in the slope of the curve

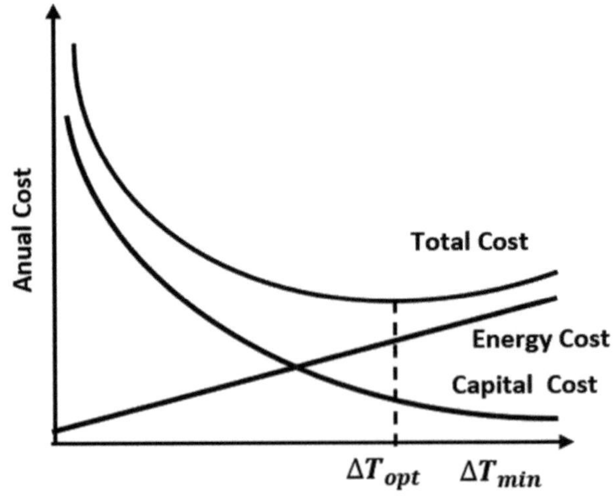

Fig. 13.7 Optimization of investment costs and energy costs

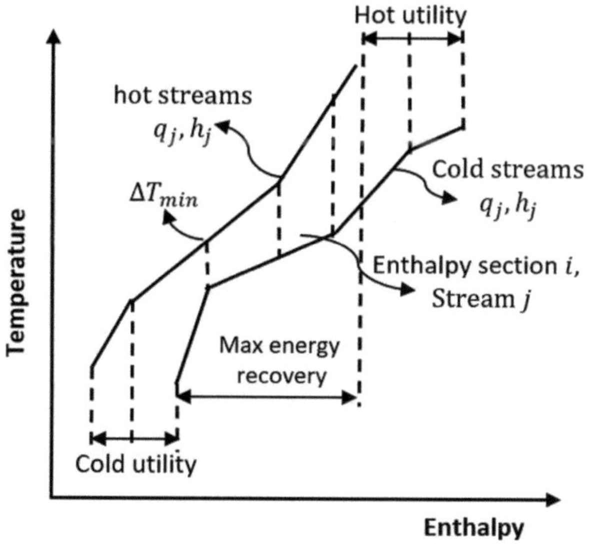

Fig. 13.8 Enthalpy section division of composite curve

caused by the inlet stream temperature. If the heat transfer coefficient of the i temperature section heat exchanger is the same and equal to a constant, there is no need to consider the number of flows. The heat transfer area of the i temperature section is:

$$A_i = \Delta H_i/(K.\Delta T_{lmi}) \tag{13.3}$$

where: ΔH_i—Width of the enthalpy of the ith temperature section;

ΔT_{lmi}—Logarithmic mean temperature difference at the ith temperature section;

K—Total heat transfer coefficient of heat exchanger.

Therefore, the minimum heat exchange area of the total network is:

$$A_{\min} = \frac{1}{K}\sum_{i=1}^{m}(\Delta H_i/\Delta T_{lmi})\partial \tag{13.4}$$

where: m—Is the number of temperature sections (sub-network) of the network.

The formula (13.4) is derived under the condition that the heat transfer coefficient is constant and the heat transfer coefficient K of each heat exchanger is the same.

In the case of different heat transfer coefficients, each temperature section has one or more parallel heat exchangers. The heat load of the heat exchanger and the temperature difference between the two end points can be read from Figure 13.19. If there is only a single hot flow or cold flow, we can estimate the film heat transfer coefficient and total heat transfer coefficient of the cold and hot flow of a single heat exchanger.

$$\frac{1}{K} = \frac{1}{h_i} + \frac{1}{h_o} \tag{13.5}$$

In the formula, h_i, h_o are the film heat transfer coefficient, including the fouling resistance coefficient. Then the area of the heat exchanger is:

$$A = Q/(K\Delta T_{lm}) \tag{13.6}$$

where: ΔT_{lm}—the logarithmic mean temperature difference between the two ends of the temperature section.

For the case of two hot flows and two cold flows, suppose that the hot flow 1 matches the cold flow 3, and the hot flow 2 matches the cold flow 4. Then the heat load and the logarithmic mean temperature difference of each heat exchanger are the same. Heat transfer coefficient of heat exchanger:

$$\frac{1}{K_{a1}} = \frac{1}{h_1} + \frac{1}{h_3}, \quad \frac{1}{K_{a2}} = \frac{1}{h_2} + \frac{1}{h_4} \tag{13.7}$$

Therefore, the total heat transfer area of this temperature section is:

$$\begin{aligned} A_1 &= A_{a1} + A_{a2} = \frac{Q}{K_{a1}\Delta T_{lm}} + \frac{Q}{K_{a2}\Delta T_{lm}} \\ &= \frac{Q}{\Delta T_{lm}}\left(\frac{1}{h_1} + \frac{1}{h_2} + \frac{1}{h_3} + \frac{1}{h_4}\right) \end{aligned} \tag{13.8}$$

Suppose streams 1, 4 are matched, streams 2, 3 are matched, and the above formula is also obtained; therefore, we can write the heat exchange area for multiple streams at any temperature section as

$$A_i = \frac{Q}{\Delta T_{lm}}\left[\sum_{i=1}^{h}\left(\frac{1}{h_i}\right) + \sum_{j=1}^{c}\left(\frac{1}{h_j}\right)\right] = \frac{Q}{\Delta T_{lm}}\sum_{j=1}^{n}\left(\frac{1}{h_j}\right) \tag{13.9}$$

where: h—number of hot streams,

c—number of cold streams.

When the load of each heat exchanger in the temperature section i is different, the area

$$A_i = \frac{1}{\Delta T_{lmi}}\sum_{j=1}^{n}\left(\frac{q_i}{h_j}\right) \tag{13.10}$$

The total area of the heat exchange network is estimated by the following formula:

$$A_{\min} = \sum \frac{1}{\Delta T_{lmi}} \sum_{j=1}^{n} \left(\frac{q_i}{h_j} \right) \tag{13.11}$$

where: n—Number of streams in the temperature section i;

m—Number of temperature sections of heat exchange network;

h_j—The film heat transfer coefficient of the stream (including fouling resistance).

For the corrosive materials in the heat exchange network, the investment cost of this type of heat exchanger must be greater than that of ordinary streams. Someone considered the amendment [9] to change the formula (13.11) into:

$$A_{\min} = \sum \frac{1}{\Delta T_{lmi}} \sum_{j=1}^{n} \left(\frac{q_i}{\varphi_j h_j} \right) \tag{13.12}$$

where: φ_j—Cost weighting factor for special streams.

13.2.2 *Determination of Total Annual Cost and* $\Delta T_{\min}$

The heat exchange network can produce different economic effects under different minimum allowable heat transfer temperature differences $\Delta T_{\min}$. Generally, when $\Delta T_{\min}$ decreases, the heat recovery increases and the area increases. As a result, the cold and hot utilities decrease and the investment increases, vice versa. If the minimum sum of energy consumption costs (fuel, cooling water, electricity) and the annual repayment of investment is the objective function, then $\Delta T_{\min}$ can be further optimized.

13.2.2.1 Annual Repayment of Investment of Heat Exchange Equipment [1, 2]

In the method of Linnhoff et al., the minimum heat transfer area $A_{\min}$ is obtained from the area target model, and then divided by the number of heat exchange units $U_{\min}$ to obtain the heat exchange area of each heat exchange unit.

According to the Euler composite principle, the number of heat exchange network system units (including heaters and coolers) is:

$$U_{\min} = N + L - S \tag{13.13}$$

where: U—Number of heat exchange units;

L—Number of independent heat load loops;

S—Is the number of independent subsystems in the network.

For the general heat exchange network $S = 1$, if the network reaches the minimum number of equipment (that is, all heat load loops are disconnected), the minimum number of heat exchange equipment is:

$$U_{\min} = N - 1 \tag{13.14}$$

Average heat transfer area of each heat exchange unit:

$$A'_{\min} = A_{\min}/U_{\min} \tag{13.15}$$

The investment of each heat exchange unit can be determined by the following power index form:

$$P_i = a\left(A'_{\min}\right)^b \tag{13.16}$$

Therefore, the total equipment investment of the heat exchange network is:

$$P = P_i \times U_{\min} \tag{13.17}$$

where: P_i, P—Equipment investment for unit i and network;

a, b—All are constants, determined by specific technical and economic conditions;

$A'_{\min}$—Average heat transfer area of heat exchange unit.

The disadvantage of the above method is that the area of each heat exchanger is considered to be the same and the unit area cost is the same. In fact, the area of each heat exchanger in the heat exchange network may be different, and different parts, pressure levels and materials will bring different equipment costs. Someone proposed the fitting equation of the heat exchange equipment cost according to 1.6, 2.5 and 4.0 MPa [12].

This method firstly determines the total heat transfer area ($A_{\min}$) of the ith temperature section from the heat of the ith temperature section, the temperature difference $\Delta T_{\min}$ and the heat transfer coefficient U in the composite curve shown in Fig. 13.8. According to the vertical heat transfer criterion and the number of streams, determine SH_i sets of heat exchange equipment, the average area A_o of each set is:

$$A_o = A_{\min i}/SH_i$$

The total investment cost of heat exchange network equipment is the sum of the costs under different pressure levels (assumed USD to RMB is 7, converted RMB to US dollar):

$$P_i = P_{16i} + P_{25i} + P_{40i}$$

$$P_{16i} = 2PRI \times \varphi \times SH_i \times X_{P16} \times 509.11 \times (A_o)^{0.7357} \times 1.428 \tag{13.18}$$

$$P_{1256i} = 2PRI \times \varphi \times SH_i \times X_{P25} \times 598.76 \times (A_o)^{0.729} \times 1.428 \tag{13.19}$$

$$P_{40i} = 2PRI \times \varphi \times SH_i \times X_{P40} \times 1171.9 \times (A_o)^{0.598} \times 1.428 \tag{13.20}$$

where: P_i—Investment in i temperature section heat exchange equipment, \$;

P_{16i}, P_{25i}, P_{40i}—Respectively, the cost of heat exchange equipment for the pressure level of 1.6 MPa, 2.5 MPa and 4.0 MPa in the ith temperature section, \$;

PRI—Is the price growth coefficient of heat exchanger (based on 1989);

φ—The ratio of the actual area to the minimum area of the ith temperature section, generally 1.0–1.3;

X_{P16}, X_{P25}, X_{P40}—They are the ratios of heat exchangers with three pressure levels in the i temperature section, satisfying $X_{P16} + X_{P25} + X_{P40} = 1.0$.

The total investment in heat exchange network equipment is:

$$P = \sum_{i=1}^{m} P_i \tag{13.21}$$

The equal annual repayment cost of the total equipment investment is the total investment multiplied by the equal annuity recovery coefficient N,

$$S_P = PN = P(1+i)^n / \left[(1+i)^n - 1\right] \tag{13.22}$$

where: S_P—Annual repayment of heat exchange network investment, \$/a;

n—Equipment service life, a;

i—Annual interest rate of investment loan.

13.2.2.2 Energy Consumption and Cooling Water Costs

Under a given $\Delta T_{\min}$, a minimum hot utility Q_H and cold utility Q_C can be obtained. It is generally believed that the load of cold utilities is mainly carried by the cooling water to the environment, and the cost is mainly the cost of cooling water. The hot utilities are divided into heat (fuel or external heating) and steam.

$$S_H = Q_H C_H \times H \times 3.6 \times 10^{-3} \tag{13.23}$$

where: S_H—The cost of hot utilities, \$/a;

Q_H—The amount of heat utilities, kW;

H—the number of operating hours per year, h;

C_H—The price of hot utilities, \$/GJ.

If steam heat source is used, the price of utilities can be converted from steam price (\$/t) to \$/GJ according to its enthalpy value (see Chap. 11). If it is an external heat source or provided by furnace fuel combustion, the heat price of utility is:

$$C_H = \frac{C_F}{\eta} + \frac{F_P N_F}{E} \tag{13.24}$$

In the formula:

C_F—fuel price, \$/GJ;

C_H—The price of heat energy, \$/GJ;

η—the efficiency of thermal energy conversion equipment;

F_P—Fixed asset investment of heat energy conversion equipment, \$;

N_F—investment annual repayment coefficient;

E—heat load of thermal energy conversion equipment, GJ/a.

Assuming that the cold utility Q_C is cooled by water cooling, the cooling water consumption is:

$$W = 3.6 Q_C / \Delta t \tag{13.25}$$

$$S_C = W H C_w \tag{13.26}$$

In the formula:

Q_C—cold utility load, kW;

S_C—cooling water cost, \$/a;

H—operating hours, h/a;

W—cooling water consumption, t/h;

C_w—cooling water price, \$/t;

Δt—the average temperature difference of cooling water, °C.

13.2.2.3 Estimation of Power Consumption Cost of Heat Exchange System

At different pinch point temperatures $\Delta T_{\min}$, the area of the heat exchanger increases, making the process more complicated, Both the number of equipment and the piping system have increased, correspondingly increased the power consumption of the system. Under the following two assumptions, a method for estimating system power consumption cost is proposed [12]

(1) In the heat exchange system, the total pressure drop of each heat exchanger is the same, and it is divided equally on both sides.
(2) The heat exchangers in the network are mainly in series. In the case of streams diversion, it is assumed that the flow of each diversion is equal.

The power N_{ij} consumed when the material flow i passes through the heat exchanger is:

$$N_{ij} = \frac{M_i \Delta P}{\eta \rho_i} \tag{13.27}$$

where: M_i—i stream mass flow, kg/s;

ΔP—the stream pressure drop of a single heat exchanger, MPa;

η—pump efficiency;

ρ_i—the relative density of the stream.

Calculate the number of heat exchange equipment SHS required for each stream from the heat exchange equipment number model, so the total power consumption of each stream heat exchanger is:

$$N_i = N_{ij} \times SHS = \frac{M_i \times \Delta P \times SHS}{\eta \rho_i} \tag{13.28}$$

The network total power required by each stream calculation is:

$$N = \sum N_i \tag{13.29}$$

The total cost of power consumption of the heat exchange network is:

$$S_N = NHC_E \tag{13.30}$$

where: S_N—heat exchange network power consumption cost, \$/a;

N—Power consumption of heat exchange network, kW;

C_E—Electricity price, \$/(kWh).

13.2.2.4 Total Annual Cost S_{all}

$$S_{all} = S_p + S_h + S_c + S_N \tag{13.31}$$

The optimization of $\Delta T_{\min}$ is to select different $\Delta T_{\min}$ to calculate the total annual cost. The $\Delta T_{\min}$ with the smallest total annual cost is the optimal ΔT_{opt}.

After pre-optimizing the heat exchange network and determining $\Delta T_{\min}$, the comprehensive design of the heat exchange network can be carried out.

Example 13.3 The heat exchange stream data and economic data of an atmospheric and vacuum unit of a refinery are shown in Table 13.4. The plant has a total of 21 hot streams and 5 cold streams. The results calculated by the super target program are shown in Table 13.5.

Main economic data and operating data: annual interest rate 10%, electricity price 0.03 \$/(kWh), equipment depreciation period of 5 years, fuel price 66 \$/t, heating furnace investment estimate 22 \$/kW, single heat exchanger total pressure drop is 0.076 MPa.

From the calculation results in Tables 13.5, it can be seen that when the heat transfer temperature difference $\Delta T_{\min}$ is 14 °C, the minimum total annual cost is 1.521 MM\$. However, if 20 °C is intuitively selected based on experience, the total annual cost is 1.559 MM\$. Optimizing the selection of $\Delta T_{\min}$ can save \$38,000 a year. Therefore, the optimal selection of $\Delta T_{\min}$ is not only necessary, but also has significant benefits. Comparing the calculation result of the program with the result of the actual synthesis network, the error of the number of heat exchangers is 2.8%, and the error of the total heat transfer area is 21%. It can be seen that the super-objective optimization of the heat exchange network is instructive and practical.

13.3 Energy Target Determination

If the heat recovery network is represented as a black box, the cold and hot flows through the black box absorb heat and release heat respectively, absorb heat Q_H from the heat source of the utility, and the heat discharged to the heat sink of the utility is Q_C, as shown in Fig. 13.9. When ignoring heat loss, for physical process operation (no chemical reaction), the energy difference ΔE between the cold and hot process streams is equal to the difference between the heat rates of the cold and hot utilities:

$$Q_C - Q_H = \Delta E \tag{13.32}$$

ΔE is a constant that only depends on the parameters of the process stream. Therefore, any increase in Q_H will inevitably lead to a corresponding increase in Q_C. Conversely, any savings in the heat source of utilities will inevitably be saved by the cold utilities, that is, a double saving of utilities consumption. As a result, there will be such a topic: in a given process, there must be a minimum of Q_H and Q_C. In

Table 13.4 Example 13.3 stream data table

Stream	Temperature (°C)		Flow rate (kg/s)	Oil characteristic factor	Relative density	Heat duty (kW)	Scale and fouling (m^2.°C/kW)	Remark
	Inlet	Outlet						
H1	98	44	7.62			3005.2	0.172	Phase change
H2	106	33	2.7			2847.2	0.258	Phase change
H3	211	136	33.33	12.1	0.793	6271	0.344	
H4	332	234	13.89	12.6	0.8242	4013.2	0.344	
H5	263	70	10.48	12.39	0.8144	5021.8	0.344	
H6	303	70	3.01	12.63	0.8223		0.344	
H7	323	64	1.57	12.74	0.8399		0.43	
H8	356	75	2.45	12.6	0.8482		0.43	
H9	106	80	8.89				0.344	Phase change
H10	55	31	2			3379.9	0.172	Phase change (steam)
H11	230	118	18.13	12.4	0.844	5004.8	0.344	
H12	307	188	12.05	12.5	0.8645	3918	0.516	
H13	166	95	4.68	12.39	0.8409	764.7	0.344	
H14	166	50	10.56	11.84	0.8149	2683.5	0.344	
H15	256	61	8.84	12.84	0.8555	4210.6	0.516	
H16	333	78	5.86	13.02	0.876	3914.2	0.688	
H17	346	84	2.13	13.23	0.8768	1497.6	0.688	
H18	380	130	0.77	13.25	0.8965	546.3	0.86	
H19	378	235	33.68	13.5	0.9219	14,592.9	1.291	
H20	235	128	32.88	13.5	0.9219	6862.9	1.291	
H21	128	89	15.29	13.5	0.9219	1324.6	1.291	
Cl	45	210	87.55	12.5	0.8551	34,673.8	0.344	
C2	206	365	81.96			44,259.7	0.516	
C3	166	180	7.71			831.5	0.258	
C4	97	100	38.39			482.2	0.344	
C5	103	150	2.97			592	0.344	
C6	150	150	2.97			6277.8	0.516	Phase change

Description of the stream data sheet: [1]Hot flow H19–H21, travel through the shell side, velocity 0.8 m/s, the rest travel through the tube side, velocity 1.4 m/s; [2]cold flow C1 travel through the shell side, velocity 0.8 m/s, other tube travel, the velocity is 1 m/s; [3]the heat loss of 5% has been considered for the hot flow

Table 13.5 Super target calculation results

Pinch temperature difference (°C)	Number of shell passes	Total annual cost (MM$)	Heating target (kW)	Cooling target (kW)	Heat transfer (kW)	Total heat transfer area (m^2)	Average heat transfer temperature difference (°C)	Average heat transfer coefficient (m^2 °C/kW)
10	140	1.541	15,505.77	7816.03	71,611.26	16,796.98	18	0.2629
11	132	1.531	15,834.91	8145.18	71,282.12	15,805.43	19.1	0.2643
12	124	1.523	16,164.59	8474.85	70,952.45	14,946.11	20.1	0.2640
13	119	1.522	16,494.79	8805.05	70,622.24	14,198.99	21.1	0.2646
14	112	1.521	16,825.51	9135.78	70,291.52	13,543.85	22.1	0.265
15	107	1.528	17,156.77	9467.03	69,960.26	13,227.88	22.7	0.265
16	101	1.530	17,488.56	9798.82	69,628.47	12,633.72	23.7	0.2656
17	97	1.534	17,784	10,094.26	69,333.03	12,155.41	24.6	0.2661
18	93	1.544	18,079.91	10,390.17	69,037.12	11,992.52	24.8	0.267
19	90	1.550	18,376.29	10,686.56	68,740.75	11,530.18	25.8	0.2674
20	88	1.559	18,673.14	10,983.41	68,443.69	11,128.59	26.7	0.2677
21	85	1.567	18,970.47	11,280.73	68,146.55	10,765.08	27.5	0.2681
22	82	1.576	19,268.26	11,578.52	67,848.77	10,437.21	28	0.2686

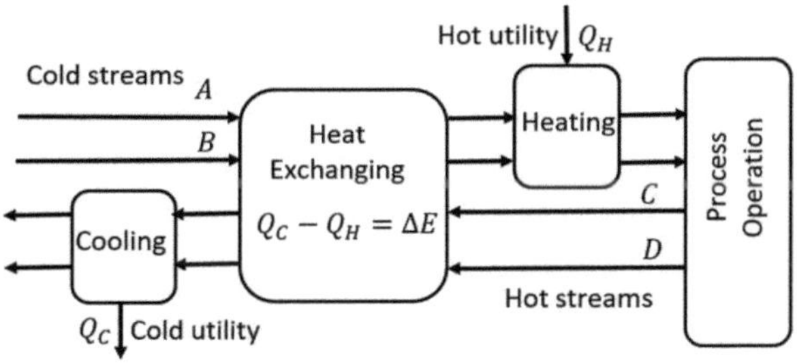

Fig. 13.9 Stream enthalpy difference and utilities

fact, such problems are not only in heat exchange, but also in other processes (such as reaction, separation, etc.), but at this time, the minimum requirement of utilities is not only heat, but also other forms of energy supply.

In formula (13.32), ΔE is directly determined by the energy difference between the cold and hot streams ($\Delta E = \sum \Delta H_i$).

Under ideal (reversible) conditions, the energy Q_H provided by the hot utilities should be equal to the heat required by the process ΔE. At this time, the heat transfer temperature difference of the cold and hot streams tends to zero, and $Q_C = 0$. This situation is obviously difficult to achieve in the actual production process. Then in order to complete the production task (the heat transfer rate required is constant), there must be a heat transfer temperature difference, and there must be energy loss. Therefore, energy consumption is inevitable, but there is an optimal energy consumption. For the heat transfer process, the optimized network minimum allowable heat transfer temperature difference $\Delta T_{\min}$ is determined based on the actual technical and economic conditions (see Sect. 13.2), and the minimum utility load is determined by $\Delta T_{\min}$. At this time, the energy consumption is inevitable, the energy consumption is the energy price that must be paid to implement the process task. The energy consumption (Q_H, Q_C) that is higher than the optimized $\Delta T_{\min}$ can be avoided, and it is the main object and potential for energy-saving improvement.

The idea of determining the minimum utility consumption is: under the optimized (or selected by experience) $\Delta T_{\min}$, complete the problem-solving table according to Linnhoff's problem table method based on the stream data and process parameters. In the last two columns of the problem-solving table (heat flow column), the input value in the first row is the minimum hot utility load, and the output value in the last row is the minimum cold utility load. This method can also be intuitively seen by drawing the composite curve of cold and hot from the problem-solving table. When the minimum allowable heat transfer temperature difference is the optimized $\Delta T_{\min}$, $Q_H \& Q_C$ (see Fig. 13.4) are the minimum hot utility and the minimum cold utilities.

Example 13.4 The test trail streams data of a refinery's atmospheric and vacuum unit are shown in Table 13.6. The plant has a total of 24 streams, of which 16 are hot streams and 8 are cold streams. Furnace efficiency is 85%, fuel heating value is 4.187 MJ/kg, when $\Delta T_{\min}$ is 18 °C and heat loss is 3%, determine the pinch temperature, fuel consumption and cooling load of the heat exchange system.

Table 13.6 A crude unit test trail stream information

Type of stream	Code	Stream name	Relative density	Characteristic factor (K)	Flow rate (kg/h)	Temperature (°C)		Remark
						Beginning	End	
Hot stream	HI	Vacuum bottom	0.9654	12.4	146,318	355	105	
	H2	Vacuum 3rd draw	0.8836	12.3	180,453	295	199	
	H3	Vacuum 3rd draw II	0.8836	12.3	51,703	199	70	Including vacuum 3rd middle reflux
	H4	Vacuum 2nd draw and middle reflux	0.8618	12.3	165,024	222	120	
	H5	Vacuum 2nd draw	0.8618	12.3	70,524	120	80	
	H6	Vacuum 1st draw and middle reflux	0.8393	12.2	30,573	155	42	
	H7	Vacuum overhead oil	0.8188	12.2	842	85	40	
	H8	Evacuate steam			3404	237	40	
	H9	Atm. 3rd draw	0.8208	12.4	30,157	299 (72)	40	72 °C is the data after preheating the air
	H10	Atm. 2nd middle reflux	0.8123	12.2	96,000	276	199	
	H11	Atm. 2nd draw	0.8113	12.2	27,436	245	60	
	H12	Atm. 1st middle reflux	0.8095	12.1	72,300	227	163	
	H13	Atm. 1st draw	0.8032	12	18,417	199	35	
	H14	Atm. top reflux	0.7826	12	76,000	156	69	
	H15	Atm. overhead oil	0.7636	12	27,288	142	32	Including Atm. top reflux
		Atm. overhead steam	1		2690	142	32	
	H16	Initial distillation overhead oil	0.7046	12.3	40,697	106	34	Including initial tower top reflux

(continued)

Table 13.6 (continued)

Type of stream	Code	Stream name	Relative density	Characteristic factor (K)	Flow rate (kg/h)	Temperature (°C)		Remark
						Beginning	End	
		Initial distillation overhead steam	1		675	106	34	
Cold stream	C1	Crude oil before desalination	0.8818	12.3	399,942	42	120	Including water content 15382 kg/h
	C2	Crude oil after desalination	0.8811	12.3	385,235	118	233	Including water content 675 kg/h
	C3	Initial distillation bottom	0.8892	12.3	371,423	225	362	
	C4	Atm. distillation bottom	0.9154	12.4	272,990	337	385	
	C5	Demineralized water			11,400	30	138	
	C6	0.35 MPa steam			5400	138	138	
	C7	1.0 MPa steam				138	181	
	C8	Preheating air						

Solution

The calculation result [2] is shown in Table 13.9. From Table 13.9, the hot utility load is 31,276.3 kW, and the cold utility (cooling load) is 15,470.52 kW.

$$\text{Fuel consumption}: \frac{31{,}276.3 \times 3.6}{0.85 \times 4.1868} = 3163.8\,\text{kg/h} = 3.164\,\text{t/h}$$

In the 8th sub-network, the output heat flow is 0, which is the pinch point. Take the average temperature of cold and hot flow as the pinch temperature, which is 234 °C (Table 13.7).

Table 13.7 Example 13.4 problem solving table

Subnetwork	Cold stream temperature (°C)	Hot Stream Temperature (°C)	Deficit (kW)	Accumulated		Maximum allowable heat flow (kW)	
				Inlet	Outlet	Inlet	Outlet
0	385						
1	362		5492.537	0.000	−5492.537	31,276.3	25,783.763
2	337	355.00	13,889.16	−5492.537	−19,381.69	25,783.763	11,894.61
3	281	299.00	10,892.37	−19,381.69	−30,274.06	11,894.61	1002.236
4	277	295.00	632.067	−30,274.06	−30,906.13	1002.236	370.17
5	258	276.00	138.641	−30,906.13	−31,044.77	370.17	231.529
6	233	251.00	−1575.183	−31,044.77	−29,469.59	231.529	1806.711
7	227	245.00	1382.068	−29,469.59	−30,851.66	1806.711	424.643
8	225	243.00	424.643	−30,851.66	−31,276.3	424.643	0
9	219	237.00	−250.519	−31,276.3	−31,025.78	0	250.52
10	209	227.00	−590.938	−31,025.78	−30,434.84	250.52	841.457
11	204	222.00	−567.342	−30,434.84	−29,897.5	841.457	1408.799
12	181	199.00	−5552.246	−29,897.5	−24,315.26	1408.799	6961.45
13	181	199. 00	2806.901	−24,315.26	−27,122.16	6961.45	4154.145
14	145	163.00	−2892.616	−27,122.16	−24,229.54	4154.145	7046.762
15	138	156.00	−197.473	−24,229.54	−24,032.07	7046.762	7244.234
16	138	156.00	2991.236	−24,032.07	−27,023.3	7244.234	4252.998
17	137	155.00	−71.248	−27,023.3	−26,952.05	4252.998	4324.246
18	124	142.00	−951.166	−26,952.05	−26,000.89	4324.246	5275.412
19	120	138.00	−316.306	−26,000.89	−25,684.58	5275.412	5591.719

(continued)

Table 13.7 (continued)

Subnetwork	Cold stream temperature (°C)	Hot Stream Temperature (°C)	Deficit (kW)	Accumulated		Maximum allowable heat flow (kW)	
				Inlet	Outlet	Inlet	Outlet
20	118	136.00	345.233	−25,684.58	−26,029.81	5591.719	5246.486
21	102	120.00	−1743.163	−26,029.81	−24,286.65	5246.486	6989.648
22	88	106.00	−1940.806	−24,286.65	−22,345.85	6989.648	8930.453
23	87	105.00	−65.277	−22,345.85	−22,280.12	8930.453	8996.18
24	77	95.00	325.176	−22,280.12	−22,605.3	8996.18	8671.004
25	76	94.00	−1895.156	−22,605.3	−20,710.14	8671.004	10,566.16
26	72	90.00	137.464	−20,710.14	−20,847.61	10,566.16	10,428.7
27	72	90.00	−3002.56	−20,847.61	17,845.05	10,428.7	13,431.25
28	67	85.00	155.763	17,845.05	−18,000.81	13,431.25	13,275.49
29	62	80.00	127.855	−18,000.81	−18,128.69	13,275.49	13,147.61
30	58	76.00	231.288	−18,128.69	−18,359.98	13,147.61	12,916.32
31	58	76.00	−1681.177	−18,359.98	−16,678.8	12,916.32	14,597.5
32	52	70.00	423.394	−16,678.8	−17,102.2	14,597.5	14,174.1
33	52	70.00	−424.37	−17,102.2	−16,677.83	14,174.1	14,598.47
34	51	69.00	80.617	−16,677.83	−16,758.45	14,598.47	14,517.86
35	42	60.00	1128.911	−16,758.45	−17,887.36	14,517.86	13,388.95
36	30	48.00	−1005.742	−17,887.36	−16,881.61	13,388.95	14,394.69
37		42.00	−563.315	−16,881.61	−16,318.3	14,394.69	14,958
38		40.00	−153.282	−16,318.3	−16,165.02	14,958	15,111.28

(continued)

Table 13.7 (continued)

Subnetwork	Cold stream temperature (°C)	Hot Stream Temperature (°C)	Deficit (kW)	Accumulated		Maximum allowable heat flow (kW)	
				Inlet	Outlet	Inlet	Outlet
39		35.00	−276.578	−16,165.02	−15,888.44	15,111.28	15,387.86
40		34.00	−44.702	−15,888.44	−15,843.74	15,387.86	15,432.56
41		32.00	−37.953	−15,843.74	−15,805.787	15,432.56	15,470.513

13.4 Heat Exchange Network Pinch Design

13.4.1 Pinch Design Concept

The pinch design method is developed from the design of the heat exchange network, but it is not only suitable for the integration of the heat exchange network itself, but also plays an important role in the integration of the general heat system.

The ultimate goal of heat recovery network design is to minimize the total annual cost of the sum of investment and operating costs. In order to achieve this goal, the pinch design method is carried out in two steps: First, find the network with the smallest number of heat exchangers, utility heaters, and coolers under the condition of minimum utility usage; then, adjust and optimize this network, appropriately reduce the number of heat exchange equipment (corresponding to the increase in the amount of utilities), to achieve the lowest total annual cost. As mentioned above, the problem-solving table algorithm can be used to calculate the utility requirements and pinch points before network design.

The pinch decomposes the problem into two independent subsystems: above the pinch point (hot end) there should only be heat exchangers and heaters, and no cooler is allowed; below the pinch point (cold end) should only have heat exchangers and coolers, and no heaters are allowed. There should not be a heat exchanger that transfers heat through the pinch point, that is, the cold and hot streams with the temperature across the pinch point directly exchanging heat are not allowed. However, what important is not the design of the upper (hot end) and lower (cold end) of the pinch point, but the selection of the matching of the cold and hot streams at the pinch point. The temperature of the cold and hot streams must be at least different by $\Delta T_{\min}$. Usually, several different matches can meet this constraint at the same time. The designer can make a choice according to process requirements (such as controllability, safety, etc.). However, due to the existence of $\Delta T_{\min}$ between all the hot and cold streams at the pinch point, it is the most constrained area in the design, and the number of reasonable matches is limited, and there is often only choice-a basic match. Otherwise, heat transfer through the pinch point will be resulted in. Therefore, the pinch point design method requires the design to start from the pinch point and develop to both ends respectively.

This is completely different from the intuitive and universal method of designing from the hot end to the cold end. If you start the design from the hot end, the design decisions you made earlier may make you have to make the choice of letting the heat flow across the pinch point later. However, the new method of designing from the pinch point allows the designer to determine the basic match in the most constrained area of the design, ensuring the minimum quantity of utilities. Once leaving the pinch point, the design is no longer so constrained, and the designer can freely choose the matching according to the process requirements.

In general, the pinch point design method of the heat recovery network has two important characteristics. First, recognize that the pinch point is the most constrained area of temperature, and the design starts from the pinch point and develops to both ends. Second, this method allows the designer to choose between alternatives.

13.4.2 Graphical Method of Heat Exchange Network

Because the comprehensive design of the heat exchange network follows certain rules. Choosing an appropriate heat exchange network expressive method is very important. Linnhoff et al. used the grid diagram to represent it, which is very convenient, this matching relationship between cold and hot streams is expressed in an intuitive and clear way, making it clear at a glance.

The grid diagram method is to arrange each stream in a horizontal line according to the hot and cold streams from top to bottom. The end points of the horizontal line are marked with the starting and final temperature; the stream number or name is added to the starting end of the hot and cold streams. And mark the heat capacity flow rate (CP) on one side of the horizontal line. The matching between the hot and cold fluids is connected by a vertical line, and the heat exchanger number is indicated by a circle, and the size of the heat exchanger load is indicated under it.

13.4.3 Pinch Point Design Method of Heat Exchange Network

Generally, the Pinch design should follow the following steps:

1. The minimum heat transfer temperature difference $\Delta T_{\min}$ of the heat exchange system is determined by optimization, or selected according to experience and specific conditions.
2. According to the stream data, divide the sub-network to complete the problem-solving table

The sub-network of the problem-solving table is arranged by subtracting $\frac{\Delta T_{\min}}{2}$ from the hot flow and adding $\Delta T_{\min}/2$ to the cold flow.

3. Mark the stream data, pinch points, and utility load on the grid diagram;
4. Design the heat-absorbing part (above the pinch point) and the heat-releasing part (below the pinch point). The main points are as follows.

(1) Start from the pinch point and expand to both ends for network design.
(2) According to the stream CP rules, match the cold and hot stream.
(3) The heat exchanger is designed according to the smaller heat load in the cold and hot streams, so that the heat exchange of the stream is matched from the initial temperature to the final temperature, and the number of heat exchange equipment is reduced.

(4) Only use external heating in the heat absorption part, and use an external cooling medium in the heat release part.
5. When necessary, according to the real minimum number of heat exchangers used in the entire system (that is, allowing heat to be transferred through the pinch point) as the goal, use the energy relaxation method to reintegrate the heat exchanger combination structure.
6. In actual design, the problems encountered are more complicated. Commonly used method is diversion. The number of process stream guidelines should be followed.
7. For complex systems, the CP difference rule should also be followed:

To implement above design approach, detail guidelines need to be followed, it will not be discussed here.

There are many programs and studies on pinch heat integration, however, Aspentech has integrated the Pinch analysis program with Aspen HYSYS and Aspen Plus, which makes it much easier use and transfers the process data directly from process simulator.

This section will introduce two examples of preheating train synthesis using Aspen Energy Analyzer (AEA).

13.4.4 Manually Input to Aspen Energy Analyzer

Example 13.5 Taking Crude unit test trail data from Example 13.4, propose a heat integrated Preheat train design;

In order to use the Aspen Energy analyzer for crude unit Heat integration, the relative density d_4^{20} and characteristic factor K of crude and its fraction in Table 13.14 will be used for calculating the heat duty per Chap. 2 provided calculation method of petroleum and its fraction, if you haven't set the HYSYS simulation yet.

Liquid petroleum fraction enthalpy

$$H_L = \{C_2 + C_3(T - 273) + 1.99 \times 10^{-3}(T - 273)^2\}C_1 \quad (2.44)$$

Gas phase enthalpy

$$H_V = C_4 + C_5(T - 273) + 1.657 \times 10^{-3}(T - 273)^2 \quad (2.46)$$

C1–C5—constant,

$$C_1 = 0.0533K + 0.3604$$

$$C_2 = 15.99 + 1.0396API - 1.1329 \times 10^{-2}API^2$$

$$C_3 = 1.5566 + 8.256 \times 10^3 API$$

$$C4 = 225.67 + 1.644API + 86.088\,\mathrm{K} - 6.925 \times 10^{-3}API^2 - 6.639\,\mathrm{K}^2$$

$$C_5 = 4.17 \times 10^{-3} API + 0.3161\,\mathrm{K} - 2.3295$$

API degree is calculated by the following formula.

$$API = \frac{141.5}{0.9942 d_4^{20} + 0.009181} - 131.5 \quad (2.41)$$

where: T—Petroleum fraction temperature, K;

d_4^{20}—Relative density of petroleum fraction at 20 °C.

H_L—Liquid phase petroleum fraction enthalpy, kJ/kg;

H_V—Enthalpy of gas phase petroleum fraction, kJ/kg;

K—Petroleum Fraction Characteristic Factor.

Please note that Eq. (2.44) is based on the reference temperature of −17.8 °C, when we apply the stream beginning and end temperature, the impact of equation reference temperature will be eliminated.

The calculated heat duties for each stream are summarized in Table 13.8. API degree and constants C_1–C_5 are calculated from Chap. 2 equations mentioned above.

Input the stream beginning and end temperatures, delta enthalpy, and flow rates information to the Aspen energy analyzer, you will have process stream information ready (Table 13.9).

The composite curve is generated automatically with process stream information, refer to Fig. 13.10.

Select the utilities from the program dropdown list, for this example, fired heater (1000 °C) and cooling water utilities have been selected; select the minimum Pinch temperature of 12 °C; the program default utility cost (fired heater and cooling water index) has been used for this example.

Annualization: set the rate of return as 8%, the plant equipment life is 10 years, results in annualization factor of 0.149 calculated by the program, the annual operating hours is 8765 hours.

The equipment capital cost is calculated by the program using the following formulas:

$$CapitalCostindex(heatexchanger) = a + b \times \left(\frac{heatexchangerarea}{shelles}\right)^C \times shells$$

$$CapitalCostindex(firedheater) = a + b \times (firedheaterduty)^C$$

where, a, b and c are constant, default data are used for this example.

Table 13.8 Heat duty calculation summary

Type of stream	Code	Stream name	Crude test trail data				
			Relative density	Characteristic factor (K)	Flowrate (kg/h)	Temperature (°C)	
						Beginning	End
Hot stream	H1	Vacuum bottom	0.9654	12.4	146,318	355	105
	H2	Vacuum 3rd draw	0.8836	12.3	180,453	295	199
	H3	Vacuum 3rd draw II	0.8836	12.3	51,703	199	70
	H4	Vacuum 2nd draw and mid reflux	0.8618	12.3	165,024	222	120
	H5	Vacuum 2nd draw	0.8618	12.3	70,524	120	80
	H6	Vacuum 1st draw and middle reflux	0.8393	12.2	30,573	155	42
	H7	Vacuum overhead oil	0.8188	12.2	842	85	40
	H9	Atm. 3rd draw	0.8208	12.4	30,157	299	40
	H10	Atm. 2nd middle reflux	0.8123	12.2	96,000	276	199
	H11	Atm. 2nd draw	0.8113	12.2	27,436	245	60
	H12	Atm. 1st middle reflux	0.8095	12.1	72,300	227	163
	H13	Atm. 1st draw	0.8032	12	18,417	199	35
	H14	Atm. top reflux	0.7826	12	76,000	156	69
	H15	Atm. overhead oil	0.7636	12	27,288	142	32
	H16	Initial distillation overhead	0.7046	12.3	40,697	106	34
Cold stream	C1	Crude oil before desalination	0.8818	12.3	399,942	42	120

(continued)

Table 13.8 (continued)

Type of stream	Code	Stream name	Crude test trail data				
			Relative density	Characteristic factor (K)	Flowrate (kg/h)	Temperature (°C)	
						Beginning	End
	C2	Crude oil after desalination	0.8811	12.3	385,235	118	233
	C3	Initial distillation bottom	0.8892	12.3	371,423	225	362
	C4	Atm. distillation bottom	0.9154	12.4	272,990	337	385

Stream name	Calculation results										
	API degree	Calculation constant					Gas enthalpy (kJ/kg)	Liquid enthalpy (kJ/kg)	ΔH (kJ/kg)	Heat duty (kW)	MCP (kW/C)
		C1	C2	C3	C4	C5					
Vacuum bottom	14.5	1.02	31.1	1.7					586.8	23,851.8	95.4
Vacuum 3rd draw	27.9	1.02	45.0	1.8					238.6	11,962.2	124.6
Vacuum 3rd draw II	27.9	1.02	45.0	1.8					313.6	4503.5	34.9
Vacuum 2nd draw and mid reflux	31.9	1.02	49.1	1.8					259.6	11,898.6	116.7
Vacuum 2nd draw	31.9	1.02	49.1	1.8					127.1	2490.4	62.3
Vacuum 1st draw and middle reflux	36.2	1.01	53.6	1.9					291.8	2478.4	21.9
Vacuum overhead oil	40.4	1.01	58.0	1.9	342.89	1.70	498.97	138.21	360.8	84.4	1.9

(continued)

Table 13.8 (continued)

Stream name	Calculation results										
	API degree	Calculation constant					Gas enthalpy (kJ/kg)	Liquid enthalpy (kJ/kg)	ΔH (kJ/kg)	Heat duty (kW)	MCP (kW/C)
		C1	C2	C3	C4	C5					
Atm. 3rd draw	40.0	1.02	57.5	1.9					694.1	5814.8	22.5
Atm. 2nd middle reflux	41.7	1.01	59.4	1.9					219.9	5863.8	76.2
Atm. 2nd draw	42.0	1.01	59.6	1.9					484.9	3695.3	20.0
Atm. 1st middle reflux	42.3	1.01	60.0	1.9					191.2	3839.0	60.0
Atm. 1st draw	43.7	1.00	61.4	1.9					429.3	2196.5	13.4
Atm. top reflux	48.2	1.00	66.1	2.0					251.3	5304.6	61.0
Atm. overhead oil	52.7	1.00	70.7	2.0	370.08	1.68	642.52	136.49	506.0	3835.7	34.9
Initial distillation overhead	67.9	1.02	86.6	2.1	359.83	1.84	573.65	163.41	410.2	4637.7	64.4
Crude oil before Desalination	28.2	1.02	45.3	1.8					−83.5	−9273.2	118.9
Crude oil after Desalination	28.4	1.02	45.5	1.8					−136.3	−14,585.3	126.8
Initial distillation bottom	26.9	1.02	44.0	1.8					−165.0	−17,021.6	124.2
Atm. distillation bottom	22.4	1.02	39.3	1.7					−40.6	−3076.5	64.1

Table 13.9 AEA process stream information

Description	Temperature (°C)		MCP (kJ)/(h.°C)	Enthalpy (kW)	Flow rate (kg/h)	Effective C_P (kJ)/(kg.°C)
	Inlet	Outlet				
Vacuum bottom	355	105	34,347	2385	146,318	0.235
Vacuum 3rd draw	295	199	448,575	11,962	180,453	2.486
Vacuum 3rd draw II	199	70	125,679	4504	51,703	2.431
Vacuum 2nd draw and middle reflux	222	120	419,951	11,899	165,024	2.545
Vacuum 2nd draw	120	80	224,136	2490	70,254	3.190
Vacuum 1st draw and middle reflux	155	42	78,958	2478	30,573	2.583
Atm. 3rd draw	299	40	80,823	5815	30,157	2.680
Vacuum overhead oil	85	40	6752	84	842	8.019
Atm. 2nd middle reflux	276	199	274,152	5864	9600	28.557
Atm. 2nd draw	245	60	71,909	3695	27,436	2.621
Atm. 1st middle reflux	277	163	121,232	3839	72,300	1.677
Atm. 1st draw	199	35	48,216	2197	18,417	2.618
Atm. top reflux	156	69	219,501	5305	76,000	2.888
Atm. overhead oil	142	32	125,532	3836	27,288	4.600
Crude oil before desalination	42	120	427,994	9273	399,942	1.070
Crude oil after desalination	118	233	456,583	14,585	385,235	1.185
Initial distillation bottom	225	362	447,283	17,022	371,423	1.204
Atm distillation bottom	337	385	230,738	3077	272,990	0.845
Initial distillation overhead oil	106	34	231,885	4638	40,967	5.660

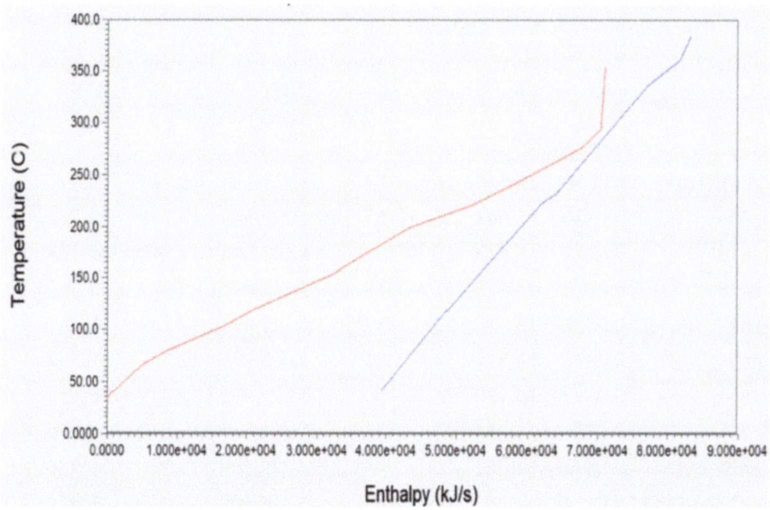

Fig. 13.10 Example 13.5 composite curves

Aspen energy analyzer can generate the recommended design, with optimization and adjusted using the retrofit mode, the following 23 designs have been generated (Table 13.10):

Among them, the A_Design16-0-0 has the lowest total cost index and lowest units and shells, therefore the A_design16-0-0 will be selected for recommendation. The grid diagram of A_design16-0-0 is shown in Fig. 13.11.

The grid diagram showing the pinch lines is shown in Fig. 13.12, this design has little duty heat transferred (278 kW) cross the pinch, but with minimum units and shells and lowest total cost index.

The design network cost indexes and performance are summarized in Table 13.11. This indicates this design is very close to the targeted design information.

13.4.5 Transfer Data from HYSYS File to Aspen Energy Analyzer

Example 13.6 A Diluent recovery unit of bitumen upgrader project configured with diluent recovery column, atm distillation tower and vacuum distillation tower, this unit processes 100 thousand barrels bitumen, the bitumen std mass density is 1016 kg/m^3; in order to reduce the bitumen feed viscosity, 40 thousand barrels of dilution solvent has been mixed with bitumen prior transferring to DRU unit for processing.

The simulated D2887 distillation data of bitumen and diluent feed is shown in Table 13.12;

Table 13.10 Comparison on generated designs

Design name	Total cost index (cost/s)	Area (m^2)	Units	Shells	Cap cost index (cost)	Heating (kW)	Cooling (kW)	Op. cost index (cost/s)
A_Design8	0.0830	16,440	24	53	4,319,107	12,749	39,782	0.0626
A_Design6	0.0826	15,664	26	55	4,228,869	12,749	39,782	0.0626
A_Design15	0.0822	15,613	22	52	4,142,093	12,749	39,782	0.0626
A_Design12	0.0822	15,613	22	52	4,142,093	12,749	39,782	0.0626
A_Design2	0.0822	15,613	22	52	4,142,093	12,749	39,782	0.0626
A_Design9	0.0818	15,374	24	51	4,068,027	12,749	39,782	0.0626
A_Design5	0.0817	15,120	25	50	4,031,932	12,749	39,782	0.0626
A_Design15-O	0.0811	14,234	22	44	3,731,910	12,942	39,976	0.0635
A_Design15-O-O	0.0811	14,222	22	44	3,728,509	12,942	39,976	0.0635
A_Design15-O-O-O	0.0811	14,222	22	44	3,728,509	12,942	39,976	0.0635
A_Design7	0.0811	14,441	25	54	3,944,226	12,703	39,737	0.0624
A_Design1	0.0808	14,524	24	50	3,894,581	12,703	39,737	0.0624
A_Design14	0.0808	14,524	24	50	3,894,581	12,703	39,737	0.0624
A_Design11	0.0808	14,524	24	50	3,894,581	12,703	39,737	0.0624
A_Design4	0.0806	14,291	25	49	3,850,308	12,703	39,737	0.0624
A_Design10	0.0805	14,138	24	48	3,794,594	12,742	39,775	0.0626
A_Design3	0.0804	14,016	23	52	3,801,067	12,703	39,737	0.0624
A_Design16	0.0804	14,016	23	52	3,801,067	12,703	39,737	0.0624
A_Design13	0.0804	14,016	23	52	3,801,067	12,703	39,737	0.0624
A_Design13-O	0.0799	13,834	23	48	3,690,019	12,704	39,737	0.0624
A_Design16-O	0.0797	13,655	22	48	3,665,698	12,704	39,737	0.0624
A_Design3-O	0.0795	13,350	22	49	3,623,901	12,704	39,737	0.0624
A_Design16-O-O	0.0791	14,084	22	46	3,706,688	12,508	39,542	0.0616

A simulation including the diluent recovery column, Atm. distillation tower and vacuum distillation tower with a simple configuration model has been established; in order to transfer the heat duty from simulation to Aspen energy analyzer easily, instead of configuring a detailed heat exchanger system, only a cooler for each hot stream to the targeted temperature and a heater for each cold stream to the targeted temperature. The HYSYS simulation sheet is shown in Figure 13.13.

The product yield and simulated Assay data of this DRU is shown in Table 13.13

The activated energy analysis in the HYSYS simulation can be used directly to generate the Aspen energy analyzer file by clicking the Analyze icon, and then clicking the Details icon to open the aspen energy analyzer file generated for your heat exchanger train synthesis.

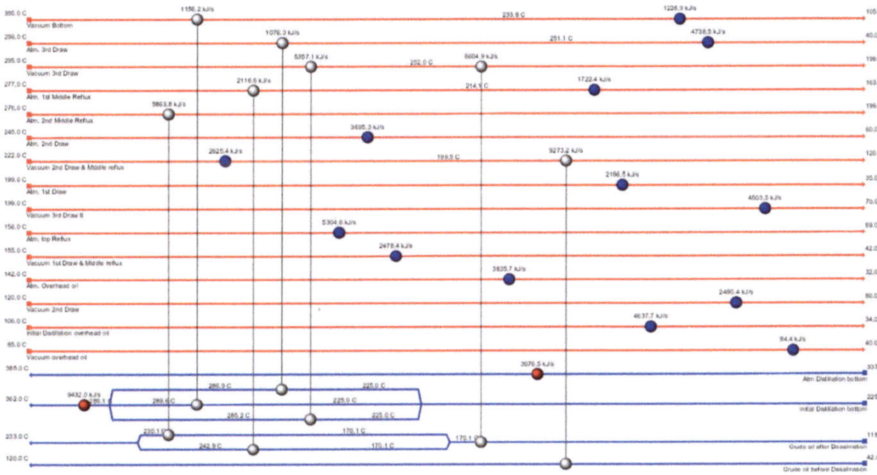

Fig. 13.11 Example 13.5 grid diagram

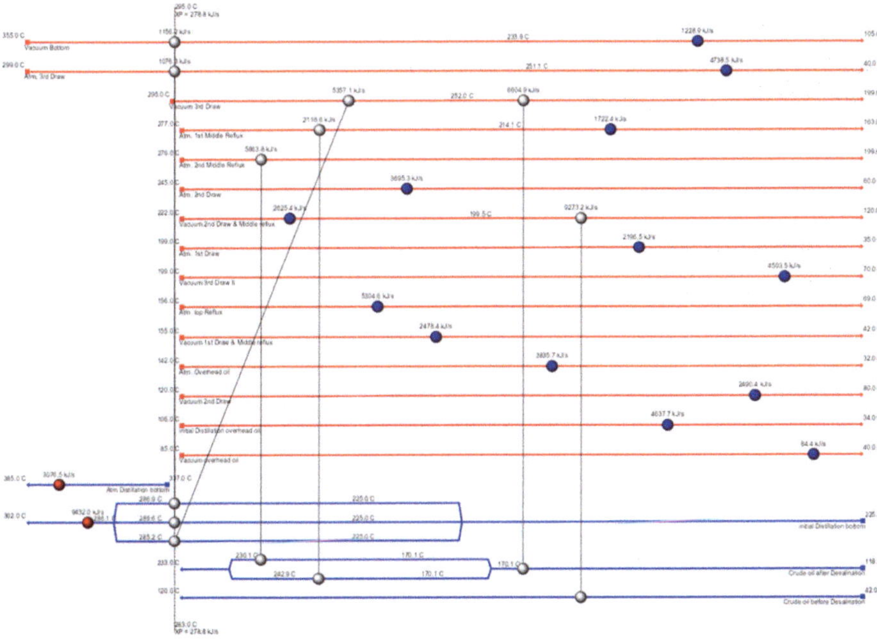

Fig. 13.12 Example 13.5 grid diagram showing pinch lines

Table 13.11 Network cost indexes and performance

Network cost index			Network performance		
	Cost index	% Target		HEN	% Target
Heating, Cost/s	0.0531	102.28	Heating, kW	12,508	102.28
cooling, Cost/s	0.0084	100.71	Cooling, kW	39,542	100.71
Operating, Cost/s	0.0616	102.06	Number of units	22	100.00
Capital, Cost	3,706,688	109.95	Number of shells	46	124.32
total cost, Cost/s	0.0791	103.71	Total area, m^2	14,084	117.33

Table 13.12 Simulated D2887 distillation data of bitumen and diluent

Description	Diluent	Bitumen
Liquid volume flow [barrel/day]	40,000	100,000
Liq. mass density @Std Cond [kg/m3]	715.99	1016.28
Molecular weight	100.10	452.31
(D2887 5%[Petrol]-Overall) [°C]	17.57	389.29
(D2887 10%[Petrol]-Overall) [°C]	30.93	427.38
(D2887 30%[Petrol]-Overall) [°C]	70.70	502.64
(D2887 50%[Petrol]-Overall) [°C]	104.01	569.84
(D2887 70%[Petrol]-Overall) [°C]	135.34	643.09
(D2887 90%[Petrol]-Overall) [°C]	161.32	707.29
(D2887 95%[Petrol]-Overall) [°C]	185.97	771.94

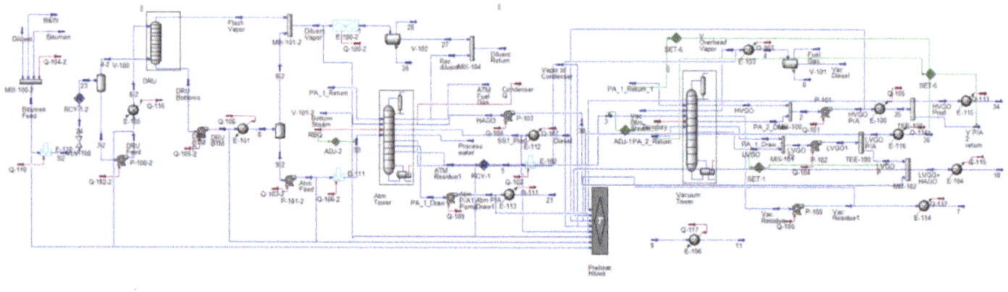

Fig. 13.13 DRU and distillation unit simulation sheet

For this example, the generated process stream information from the HYSYS simulation is shown in Table 13.14. In order to reduce cold utility usage, recover more heat, an additional cold stream of MP steam generation has been added as a cold stream.

The composite curve is generated automatically with process stream information, and is shown Fig. 13.14.

Select the utilities from the program dropdown list, for this example, fired heater (1000 °C) and cooling water utilities have been selected; select the minimum Pinch

Table 13.13 Simulated D2887 distillation data of products

Description	Diluent recovered	Diesel	HAGO	Vac diesel	LVGO	HVGO	Vac. residue
Liquid volume flow [barrel/day]	39,171	5332	4769	2734	13,530	15,763	58,720
Liq. mass density @Std Cond [kg/m3]	715.24	892.16	922.48	915.61	959.53	992.72	1067.09
Molecular weight	99.23	202.21	238.79	229.99	314.04	414.65	631.87
(D2887 5%[Petrol]-Overall) [C]	17.10	179.09	240.31	221.66	313.26	401.47	542.96
(D2887 10%[Petrol]-Overall) [C]	29.83	209.76	259.25	251.31	327.02	421.70	572.78
(D2887 30%[Petrol]-Overall) [C]	69.34	253.24	296.81	282.10	364.92	456.31	620.90
(D2887 50%[Petrol]-Overall) [C]	102.73	271.40	324.60	310.01	397.47	477.62	672.50
(D2887 70%[Petrol]-Overall) [C]	133.97	292.65	338.20	331.36	422.16	499.58	714.88
(D2887 90%[Petrol]-Overall) [C]	160.29	324.85	367.65	355.66	450.26	529.47	769.29
(D2887 95%[Petrol]-Overall) [C]	184.47	339.52	390.64	383.25	486.83	572.73	863.83

temperature of 10 °C; the program default utility cost (fired heater and cooling water index) has been used for this example.

Annualization: set the rate of return as 8%, the plant equipment life is 10 years, results in annualization factor of 0.149 calculated by the program, the annual operating hours is 8765 h.

The equipment capital cost is calculated by the program using the following formulas:

$$CapitalCostindex(heatexchanger) = a + b \times \left(\frac{heatexchangerarea}{shelles}\right)^C \times shells$$

$$CapitalCostindex(firedheater) = a + b \times (firedheaterduty)^C$$

Table 13.14 Process stream information

Description	Temperature (°C)		MCP (kJ/h.°C)	Enthalpy (kW)	HTC (kJ/m^2.°C)	Flow rate (kg/h)	Effective C_P (kJ/kg.°C)
	Inlet	Outlet					
V Overhead Vapor_To_4	169.6	35.0	–	9061	–	25,132	–
ATM Residue1_To_3	327.0	433.3	–	68,256	–	617,405	-
Atm Feed_To_33	215.1	338.0	–	71,126	–	743,712	–
DRU BTM_To_8	171.9	215.0	1,738,759	20,801	894	743,712	2.338
Vac. Residue1_To_7	386.8	130.0	–	72,773	–	412,261	–
HVGO Prod._To_34	239.7	80.0	–	10,096	–	103,019	–
HVGO P/A_To_25	323.3	239.7	1,756,581	40,805	2288	670,172	2.621
LVGO + HAGO_To_10	243.8	80.0	–	11,615	–	114,397	–
SS1_Prod_To_Diesel	195.6	80.0	–	2218	–	31,228	–
Diluent Vapor_To_1	175.3	50.0	–	21,642	–	122,067	–
Atm P/A Draw1_To_21	165.3	65.0	–	661	–	10,132	–
LVGO P/A_To_36	230.7	130.5	–	22,117	–	346,271	–
To Condenser@COL6	146.0	86.3	– -	38,710	–	241,512	–
Bitumen DRU Feed	76.7	180.0	2,077,057	59,600	720	865,800	2.399
MP steam Generation	174.0	183.0	5,000,000	12,500	720	17,000	

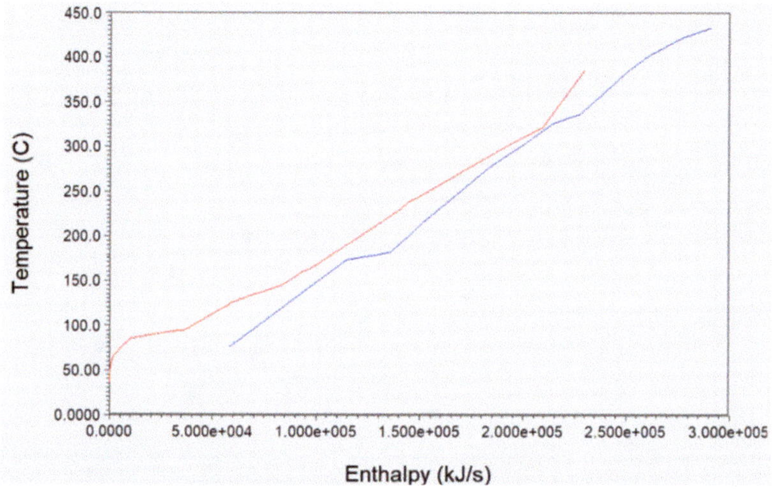

Fig. 13.14 Example 13.6 composite curves

where, a, b and c are constant, default data are used for this example.

Aspen energy analyzer can generate automatically the recommended designs, with optimization and adjusted using the retrofit mode, the following 12 feasible designs have been generated and summarized in Table 13.15.

Among the 12 designs, the designs, A_Design5-0-0-1 is recommended due to its lowest number of units and shells.

Table 13.15 Comparison on generated designs

Design name	Total cost index (Cost/s)	Area (m^2)	Units	Shells	Cap cost index (Cost)	Heating (kW)	Cooling (kW)	Op. cost index (Cost/s)
A_Design1	0.4655	115,077	34	272	34,175,517	68,256	65,671	0.3040
A_Design11	0.4056	60,275	39	162	21,510,973	68,256	65,671	0.3040
A_Design6	0.4056	60,275	39	162	21,510,973	68,256	65,671	0.3040
A_Design8	0.4056	60,275	39	162	21,510,973	68,256	65,671	0.3040
A_Design4	0.3973	53,019	36	143	19,741,000	68,256	65,671	0.3040
A_Design7	0.3972	52,742	38	145	19,735,035	68,256	65,671	0.3040
A_Design2	0.3898	46,766	31	125	18,167,438	68,256	65,671	0.3040
A_Design5	0.3883	50,268	31	88	17,838,158	68,256	65,671	0.3040
A_Design3	0.3856	47,217	34	85	17,281,993	68,256	65,671	0.3040
A_Design5-O	0.3848	47,337	28	79	17,109,641	68,256	65,671	0.3040
A_Design5-O-O-1	0.3847	47,335	28	78	17,089,462	68,256	65,671	0.3040
A_Design3-O	0.3818	43,251	34	81	16,460,506	68,256	65,671	0.3040

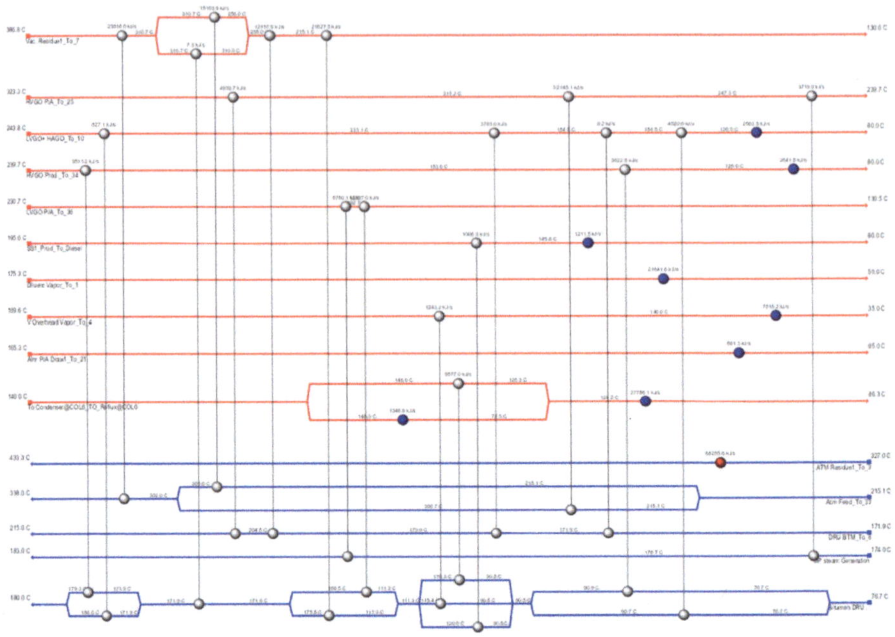

Fig. 13.15 Example 13.6 grid diagram

The grid diagram of A_Design5-0-0-1 is shown in Fig. 13.15.

The A_Design5-0-0-1 grid diagram showing the pinch lines is shown in Fig. 13.16, this design is in vertical heat transfer, no cross-pinch heat transfer, only one heat exchanger heat transfer near pinch, but in vertical heat transfer.

The design network cost indexes and performance are summarized in Table 13.16. This indicates that this design is very close to the targeted design information.

After completing preheat train synthesis, it is needed to convert the recommended design to HYSYS preheat train for cross-checking, and producing a project H&M balance, HYSYS simulation with optimized heat exchanger network-integrated will be the solid system design basis for heat exchanger, fired heaters, and cooling water.

The sub-flowsheet Preheat train of DRU HYSYS simulation is as Fig. 13.17.

Please note that the Aspen Energy Analyzer uses the simple MCP method, actual the petroleum and its fraction heat capacity is much different between high-temperature section and lower temperature section, therefore there is the error between HYSYS and aspen energy analyzer, convert the optimized heat exchanger train to HYSYS simulation will be great for cross-checking and accurately predicting the heat transfer train, it is also great to easily apply the EDR program for detail design.

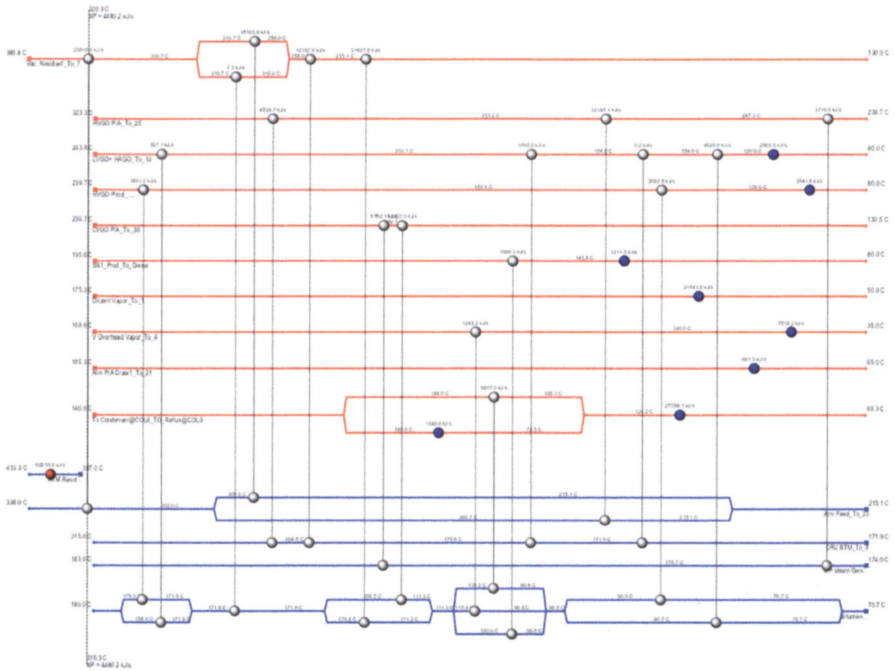

Fig. 13.16 Example 13.6 grid diagram showing pinch lines

Table 13.16 Example 13.6 network cost indexes and performance

Network cost index			Network performance		
	Cost index	% Target		HEN	% Target
Heating, Cost/s	0.2900	112.2	Heating, kW	68,256	112.2
Cooling, Cost/s	0.0140	112.7	Cooling, kW	65,671	112.7
Operating, Cost/s	0.3040	112.2	Number of units	28	164.7
Capital, Cost	17,089,462	102.0	Number of shells	78	144.4
Total cost, Cost/s	0.3847	109.9	Total area, m^2	47,335	75.1

13.5 Placement of the Heat Engine (Pump) in the Total Energy System [5, 14]

In order to effectively use energy, the total energy system that integrates the equipment that generates and uses power into the process flow is the hallmark of modern large-scale complex plants.

A process operation system requires heat input Q_P, input work W, a thermal engine will be needed to generate work W, the thermal engine requires heat input Q, Fig. 13.18a represents independent engine and process system, Fig. 13.18b reflect the

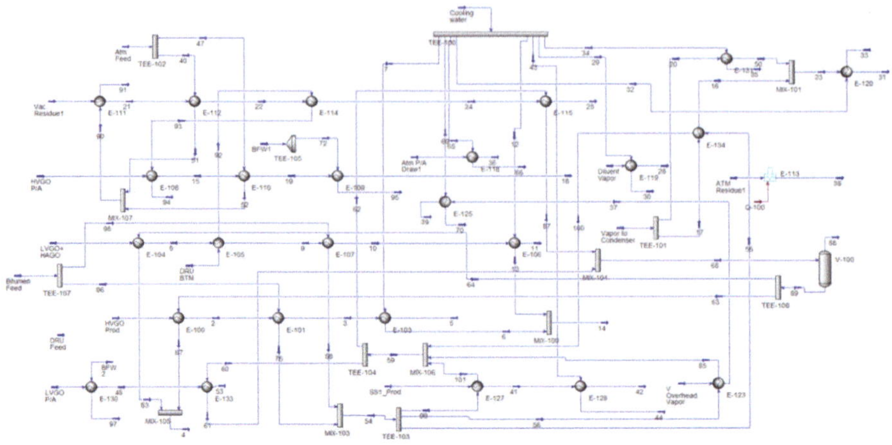

Fig. 13.17 HYSYS with optimized heat exchanger train incorporated

purpose of designing an integrated thermal system, the exhaust heat ($Q - W$) from thermal engine could be the heat input to process operation system, but it could not be fully recovered, the energy rejected is exhaust gas energy at discharge temperature (End temperature) of process operation, usually, a 60–70% of the exhaust heat can be recover.

From Fig. 13.18a, the independent thermal engine and process operation system require heat input of $Q + Q_P$, From Fig. 13.18b, the integrated engine and process system requires the total heat input of $Q + Q_P - Q_R$. The saved heat input is:

$$\Delta Q = (Q + Q_P) - (Q + Q_P - Q_R) = Q_R \tag{13.33}$$

where: ΔQ—Heat demand reduced by an integrated engine-process system;

Q_P—Heat input for an independent process system;

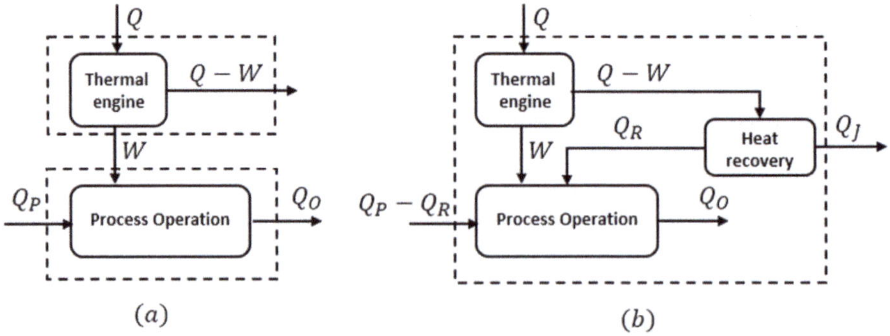

Fig. 13.18 Thermal power system

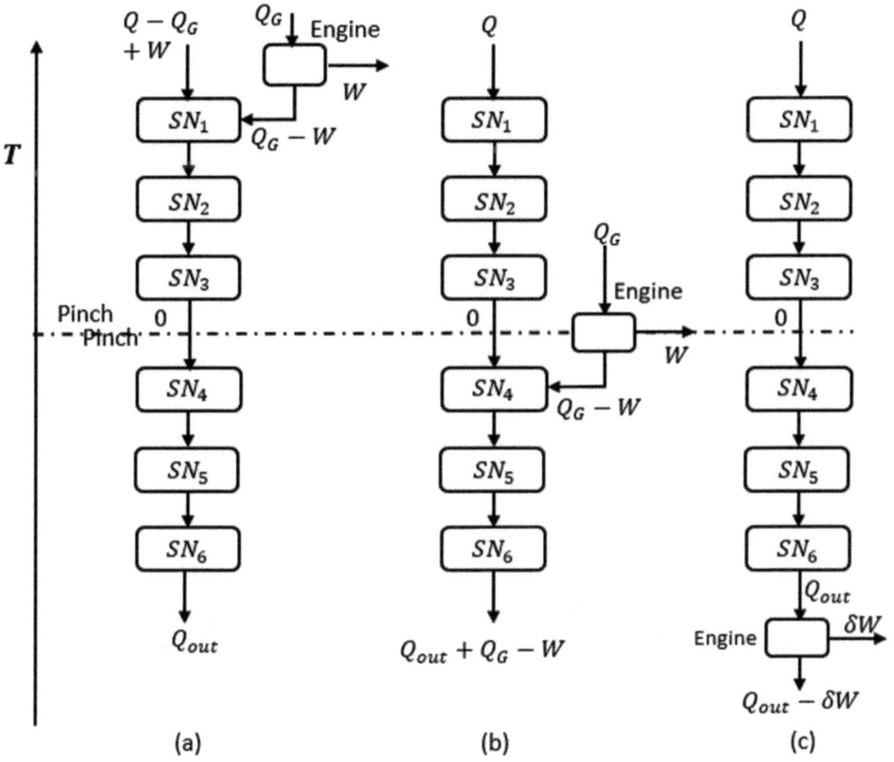

Fig. 13.19 The location of the heat engine

Q—Heat input for the thermal engine;

Q_R—Heat recovered from the exhaust gas of a thermal engine;

Q_J—Heat rejected from the exhaust gas of a thermal engine;

Q_O—Heat output from the process operation system;

W—Work performed by the thermal engine.

The most basic thermal power conversion equipment in the power system is the heat engine and the heat pump. When they are integrated with the heat recovery network, the most important issue is the relationship between these equipment and the process heat recovery pinch point, which depends on the location of the equipment relative to the pinch point. There are three situations: above the pinch, below the pinch, and across the pinch. According to the basic principles of pinch technology, the concepts of "appropriate placement" and "unsuitable placement" can be formed.

13.5.1 Heat Engine

As shown in Fig. 13.19a, the heat engine is placed above the pinch point. The original heat source from high-temperature utility to the heat recovery network has been changed to use hot utility and the engine exhaust heat supply to the heat recovery network in parallel; in case of the engine exhaust heat matches the hot utility demand, the heat engine to do work first, which reduces the energy loss. The integrated system has higher thermodynamic efficiency than the independent system. It can be seen from the energy relationship marked on the figure that due to integration, heat is converted into work (actually, waste heat is reduced), and hot utility reduced is $Q_G - W$, achieving the purpose of the system integration. Therefore, it is appropriate to place the heat engine above the pinch point.

On the contrary, if the heat discharged from the heat engine is supplied to the part below the pinch point, as shown in Fig. 13.19b, the heat engine crosses the pinch point position, and the heat ($Q_G - W$) has to "overflow" through all the temperature zones below the pinch point, until the waste heat is discharged from the lowest temperature zone. In this way, the heat engine obtains additional heat Q_G, after performing work W, discharges additional waste heat ($Q_G - W$). Compared with a separate heat engine, the integration has no benefit. Therefore, the placement of the heat engine across the pinch point is inappropriate.

Finally, as shown in Fig. 13.19c, the heat engine takes heat and exhausts heat from below the pinch point, that is, when the heat engine is placed below the pinch point, since there is no heat passing through the pinch point, the hot utility consumption Q remains unchanged. Part of the waste heat below the pinch point is converted into work, and the placement of the heat engine is also appropriate.

13.5.2 Heat Pump

The proper placement of a heat pump is the opposite of a heat engine. That is, only when the heat pump crosses the pinch point, energy can be saved.

In Fig. 13.20a, the heat pump takes heat from a certain temperature area above the pinch point and sends it to a higher temperature area for use. As a result, work is converted into heat. If the heat pump is placed below the pinch point, as shown in Fig. 13.20c, the net result is that work is transformed into heat and discharged to cold utilities. Therefore, the placement of the heat pump in Fig. 13.20a and b is not appropriate.

In Fig. 13.20c, the heat pump takes heat from below the pinch point and supplies it to the temperature zone above the pinch point. The net result is that the consumption of hot utilities is reduced by $Q + W$, and the consumption of cold utilities is reduced by Q_H. This placement of the heat pump across the pinch point is suitable.

In short, the heat engine is only suitable if it does not cross the pinch point, and the heat pump is only suitable if it is placed across the pinch point. The proper

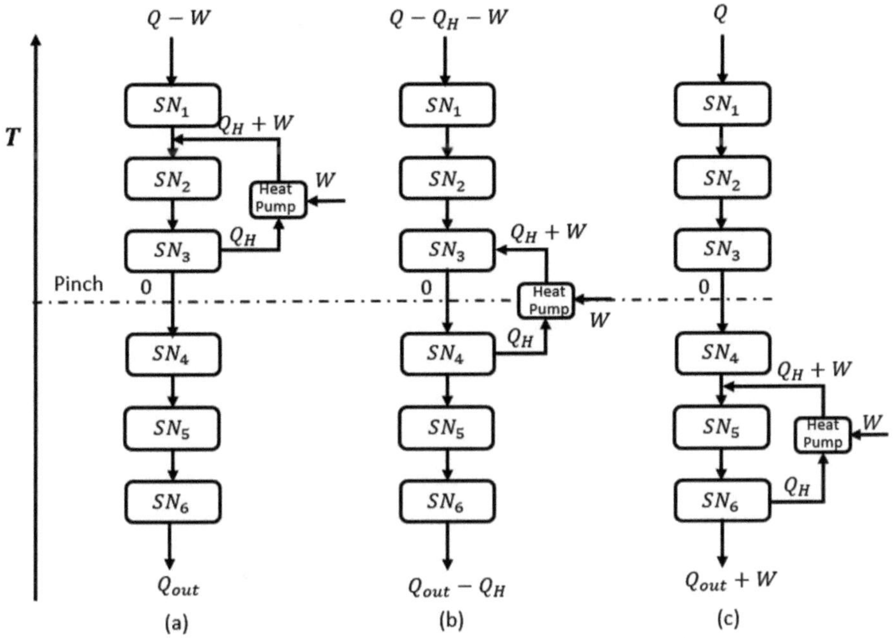

Fig. 13.20 The location of the heat pump

placement of the heat engine generates work from the heat at 100% efficiency (if takes all exhaust heat as effective heat), and the improper placement cannot generate work at a higher efficiency than when the heat engine is placed alone. The proper placement of the heat pump can achieve efficiency when the heat pump is placed alone, and improper placement is a waste of energy. The concept of the proper placement and inappropriate placement is simple, but it has a great effect. For example, heat pump distillation is an energy-saving distillation technology currently under development. However, according to the concept of proper placement of heat pumps, if the operating temperatures of the reboiler and condenser of the rectification tower are on the same side of the pinch point, this method will not save energy. Due to technical and economic reasons, heat pumps are mainly used in the rectification process where the temperature difference between the reboiler and the condenser is small. Therefore, this situation needs to be analyzed in particular, and the decision cannot be made based on intuition.

13.6 The Effect of Cross Heat Transfer on Heat Exchange Network Area and Energy

The pinch technology not only proposes the concept of a pinch point in the energy recovery system, but also requires that in the integrated design of the energy system, the guidelines of vertical heat transfer and avoiding cross heat transfer should always be followed, so as to use energy according to quality; This is the core of pinch technology. B. Linnhoff put forward the concept of vertical heat transfer in 1984 and used the driving force plots to discuss the heat transfer system for the first time [15]. According to the vertical heat transfer criterion, the "Super-target" method for determining the area of the heat exchange network is proposed according to the cold and hot streams conditions and the composite curve of the process. However, the impact of cross heat transfer on the heat transfer area and system energy target utility usage has not been quantitatively described. In the real process, there are many examples where cross heat transfer brings about an increase in energy and area targets. Discuss the impact of cross-heat transfer, transition from qualitative analysis to quantitative calculation, so that improvement of the process is not only necessary but also urgent.

13.6.1 Heat Transfer Model and Driving Force Diagram

Generally, the heat transfer mode of heat exchange equipment in the network can be divided into two types: vertical heat transfer and cross heat transfer, as shown in Fig. 13.21.

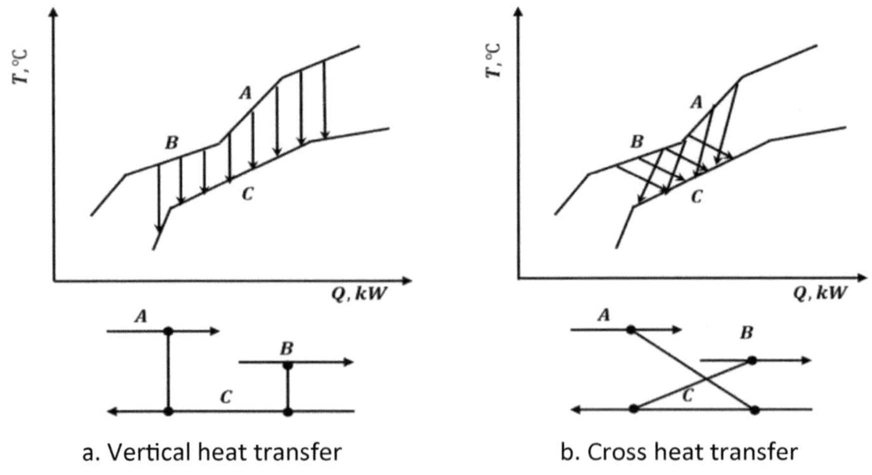

Fig. 13.21 Two heat transfer models

In order to characterize the relationship between the heat transfer temperature difference of the equipment and the stream temperature, it can be transformed into a driving force diagram [10] according to the composite curve; that is, taking the temperature of the cold stream as the abscissa and the heat transfer temperature difference as the ordinate, the relationship between the heat transfer temperature difference and the temperature of the stream is depicted. The graph curve is drawn based on the vertical heat transfer. The driving force plots is shown in Fig. 13.22.

For example, in the network design, the heat exchange equipment can achieve vertical heat transfer, which is consistent with the curve of the driving force plots. At this time, the system recovers the most energy and the heat exchange area is reasonable. Deviate from the vertical heat transfer, the temperature difference of the heat exchange equipment deviates from the driving force curve, and large fluctuations occur, as shown in Fig. 13.23.

From Fig. 13.15, it can be seen that cross-heat transfer increases the temperature difference of one heat exchange equipment, which will inevitably reduce the

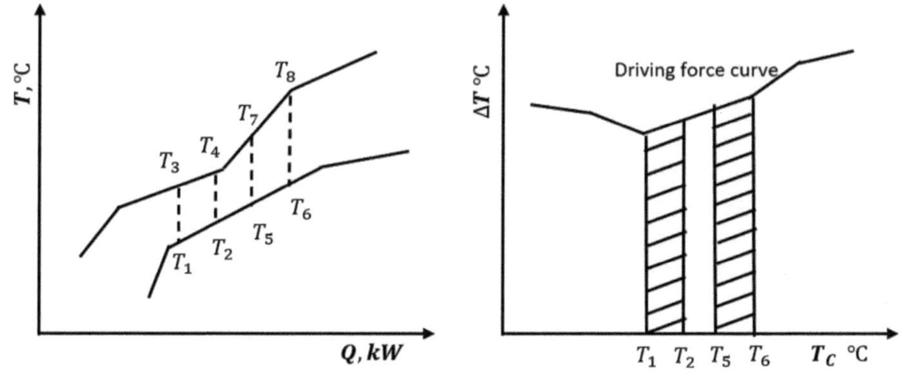

Fig. 13.22 Vertical heat transfer driving force Plots

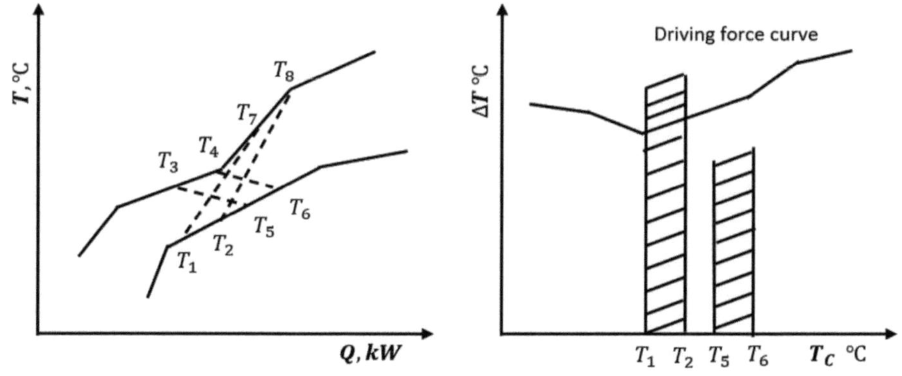

Fig. 13.23 Cross heat transfer driving force plots

temperature difference of another equipment. When the heat load is the same, the driving force curve shows that the increased temperature difference and the reduced temperature difference are equal. vice versa. Therefore, cross heat transfer affects the heat transfer area and energy consumption from two aspects.

(1) When two heat exchange equipment with similar heat transfer temperature difference (different stream temperature), the temperature difference of one equipment increases and the temperature difference of the other equipment decreases due to cross heat transfer. When the heat exchange equipment with increased temperature difference completes the same heat transfer load, the heat transfer area is correspondingly reduced. The area of equipment with reduced temperature difference will increase when the same heat transfer load is completed. The increased area is greater than the decreased area. When the temperature difference of the heat exchange equipment that reduces the temperature difference due to cross heat transfer decreases too much and tends to zero, the area of the equipment tends to infinity, which often forces the abandonment of the heat exchange load and increases energy consumption. Using the highest temperature heat source to heat a certain temperature section of the cold stream will inevitably break the balance of the original heat exchange network, reduce the final temperature of the cold stream after exchange, and affect the consumption of hot utilities.
(2) Since the heat transfer area is inversely proportional to the reciprocal of the average temperature difference of the equipment, cross heat transfer will reduce the total average temperature difference of the system and increase the area. When the area remains the same, the heat transfer load decreases, which affects the amount of utility demands.

In real industrial processes, cross heat transfer often affects both.

13.6.2 The Effect of Cross Heat Transfer on Area Targets

For the convenience of analyzing the problem, we assume that the heat load of each heat exchange equipment is the same, and the total heat transfer coefficient of cross heat transfer and vertical heat transfer are the same. From Fig. 13.22, take a certain vertical heat exchange equipment as the heat balance equation as:

$$Q_{vi} = KA_{vi}\Delta T_{mvi} \tag{13.34}$$

The area of the heat exchange network is

$$A_{vi} = \sum \frac{Q_{vi}}{K\Delta T_{mvi}} = K^{-1}Q_{vi}\sum \frac{1}{\Delta T_{mvi}} \tag{13.35}$$

In the same way, take a certain cross heat exchange equipment from Fig. 13.26:

$$Q_{ci} = K A_{ci} \Delta T_{mci} \tag{13.36}$$

$$A_{ci} = \sum \frac{Q_{ci}}{K \Delta T_{mci}} = K^{-1} Q_{ci} \sum \frac{1}{\Delta T_{mci}} \tag{13.37}$$

Since the heat load of the equipment before and after the cross-heat transfer is unchanged, that is, $Q_{ci} = Q_{vi}$, the ratio of the total heat transfer area in the case of cross heat transfer and vertical heat transfer is:

$$\alpha = \frac{A_{ci}}{A_{vi}} = \frac{\sum \frac{1}{\Delta T_{mci}}}{\sum \frac{1}{\Delta T_{mvi}}} \tag{13.38}$$

We define α as the cross-heat transfer factor, which reflects the influence on the total heat transfer area due to the uneven heat transfer temperature difference caused by cross heat transfer. To a certain extent, it reflects the utilization efficiency of network heat transfer temperature difference:

$$\alpha = \frac{\Delta T_{mv}}{\Delta T_{mc}} \tag{13.39}$$

In the formula: m—the number of network equipment.

For the same heat load of each heat exchange equipment:

$$\Delta T_{mv} = m \Big/ \frac{1}{\Delta T_{mvi}} \tag{13.40}$$

$$\Delta T_{mc} = m \Big/ \frac{1}{\Delta T_{mci}} \tag{13.41}$$

Generally, the cross-heat transfer factor $\alpha \geq 1$, obviously, for the vertical heat transfer $\alpha = 1$, the cross-heat transfer $\alpha > 1$. This can be proved from the following derivation.

As shown in Figs. 13.22 and 13.23, a set of (two) heat exchangers with vertical heat transfer and cross heat transfer, each heat transfer load is the same, and the heat exchange area can be calculated by the following formula:

Vertical heat transfer

$$A_v = \frac{Q}{K}\left(\frac{1}{\Delta T_{m1}} + \frac{1}{\Delta T_{m2}}\right) = \frac{Q}{K}\left(\frac{\Delta T_{m1} + \Delta T_{m2}}{\Delta T_{m1} \times \Delta T_{m2}}\right) \tag{13.42}$$

Cross heat transfer

$$A_c = \frac{Q}{K}\left(\frac{1}{\Delta T'_{m1}} + \frac{1}{\Delta T'_{m2}}\right) = \frac{Q}{K}\left(\frac{\Delta T'_{m1} + \Delta T'_{m2}}{\Delta T'_{m1} \times \Delta T'_{m2}}\right) \tag{13.43}$$

The heat transfer area ratio of cross heat transfer to total of vertical heat transfer:

$$\alpha = \frac{A_c}{A_v} = \frac{(\Delta T'_{m1} + \Delta T'_{m2})(\Delta T_{m1} \times \Delta T_{m2})}{(\Delta T'_{m1} \times \Delta T'_{m2})(\Delta T_{m1} + \Delta T_{m2})} \quad (13.44)$$

For two heat exchange equipment with the same heat load, cross heat transfer to increase the temperature difference from ΔT_{m1} to $\Delta T'_{m1}$. Assuming that the temperature difference increases by Δt, the heat transfer temperature difference of the other equipment must decrease by Δt. When used cold stream as a reference, there are:

$$\Delta T'_{m1} = \Delta T_{m1} + \Delta t$$
$$\Delta T'_{m2} = \Delta T_{m2} - \Delta t$$

So, the formula (13.44) becomes:

$$\alpha = \frac{\Delta T_{m1} \times \Delta T_{m2}}{\Delta T'_{m1} \times \Delta T'_{m2}} \quad (13.45)$$

For vertical heat transfer, obviously $\Delta T_{m1} = \Delta T'_{m1}$, $\Delta T_{m2} = \Delta T'_{m2}$, so $\alpha = 1$. For the case of cross heat transfer, the formula (13.45) can be written as:

$$\alpha = \frac{\Delta T_{m1} \times \Delta T_{m2}}{\Delta T'_{m1} \times \Delta T'_{m2}} = \frac{(\Delta T'_{m1} - \Delta t) \times \Delta T_{m2}}{\Delta T'_{m1} \times (\Delta T_{m2} - \Delta t)} = \frac{1 - \Delta t/\Delta T'_{m1}}{1 - \Delta t/\Delta T_{m2}} \quad (13.46)$$

For the cross-heat transfer of Fig. 13.23:

$$\Delta T'_{m1} = (T_7 + T_8 - T_1 - T_2)/2$$
$$\Delta T_{m2} = (T_7 + T_8 - T_5 - T_6)/2$$

Also, because:

$$(T_5 + T_6) > (T_1 + T_2)$$

Then:

$$\Delta T'_{m1} > \Delta T_{m2}, \frac{\Delta t}{\Delta T'_{m1}} < \frac{\Delta t}{\Delta T_{m2}}$$

$$\alpha = \frac{1 - \Delta t/\Delta T'_{m1}}{1 - \Delta t/\Delta T_{m2}} > 1 \quad (13.47)$$

That is, in the case of cross heat transfer, the cross-heat transfer factor $\alpha \geq 1$. For a heat exchange network with n groups of heat exchangers, there is obviously a conclusion $\alpha > 1$.

Equation (13.47) is derived under the assumption that the heat load of each heat exchanger is the same, and the number of heat exchanges before and after the crossover is unchanged. Generally (the total heat exchange load remains unchanged; the number m of heat exchangers remains unchanged):

$$\alpha = \sum_{i=1}^{m} \frac{Q_{ci}}{\Delta T_{mci}} / \sum_{i=1}^{m} \frac{Q_{vi}}{\Delta T_{mvi}} \tag{13.48}$$

where: ΔT_{mci}—the heat transfer temperature difference of the ith cross heat exchanger, K;

ΔT_{mvi}—the temperature difference of the ith vertical heat transfer heat exchanger, K;

Q_{ci}—heat load of the ith cross heat transfer heat exchanger, kW;

Q_{vi}—the heat load of the ith vertical heat transfer heat exchanger, kW.

13.6.3 *The Effect of Cross Heat Transfer on Energy (Heat Exchanger) and Exergy Loss*

Cross heat transfer affects the heat transfer load from two aspects: one is that the cross-heat transfer makes the temperature difference of one or more heat exchange equipment in the network too small, and is forced to abandon the heat transfer and reduce the heat transfer by pressing the heat to the cooling load; Under a given total heat transfer area, cross heat transfer will reduce the total heat transfer load. The former should be analyzed in conjunction with the specific process. The latter can be estimated by the crossover factor.

Assuming that the total heat transfer area is constant, the cross-heat transfer only affects the actual average temperature difference of the network. From the heat balance equation:

$$Q_v = AK\Delta T_{mv},\ Q_c = AK\Delta T_{mc}$$
$$\Delta Q = Q_v - Q_c = AK(\Delta T_{mv} - \Delta T_{mc})$$

From formula (13.40) and formula (13.41), we get:

$$\Delta Q = AK\left(\frac{m}{\sum(1/\Delta T_{mvi})} - \frac{m}{\sum(1/\Delta T_{mci})}\right) \tag{13.49}$$

Also, because:

$$\alpha = \sum \frac{1}{\Delta T_{mc}} \Big/ \sum \frac{1}{\Delta T_{mvi}}$$

Then:

$$\sum \frac{1}{\Delta T_{mci}} = \alpha \times \sum \frac{1}{\Delta T_{mvi}}$$

The formula (13.49) becomes:

$$\Delta Q = AK \frac{m}{\sum(1/\Delta T_{mvi})}\left(1 - \frac{1}{\alpha}\right) = Q_v\left(1 - \frac{1}{\alpha}\right) \tag{13.50}$$

Obviously, when the total heat transfer area is constant in the cross-heat transfer situation, the heat transfer amount is:

$$Q_c = Q_v - \Delta Q = Q_v - Q_v\left(1 - \frac{1}{\alpha}\right) = Q_v/\alpha \tag{13.51}$$

Generally speaking, when the total heat transfer area is constant, the cross-heat transfer load decreases. Since $\alpha \geq 1$, there will always be $\Delta Q \geq 0$, that is, $Q_c \leq Q_v$. A decrease in heat transfer by ΔQ means an increase in hot utilities ΔQ, cold utilities increase ΔQ, which is the energy cost of cross heat transfer.

In order to analyze the internal reasons of the impact of the cross-heat transfer process in utilities, we investigate the effect of cross heat transfer on the exergy loss of the heat exchange subsystem. The heat exchange subsystem exergy loss (see Fig. 13.29 in Sect. 13.7) is composed of three parts: exergy loss during heat exchange, exergy loss during heating, and exergy loss during cooling. For exergy loss in heat exchange process:

$$\Delta EX_2 = QT_0 \frac{T_H - T_C}{T_H T_C} = QT_0 \frac{\Delta T_{mv}}{T_H T_C} \tag{13.52}$$

where: $T_{H,} T_C$—the average temperature of the network hot and cold streams, K.

It can be seen that the exergy loss in the heat exchange process is proportional to the heat exchange amount Q and the vertical heat transfer temperature difference. When the heat exchange amount and process parameters remain unchanged, ΔEX remains unchanged [8]. In fact, some cross heat transfer cannot be compensated by the increase in area alone, such as the case where the temperature difference of a certain heat exchange equipment is too low due to cross heat transfer. In addition, in many cases, the area of the heat exchange network is unchanged, and the cross-heat transfer can be adjusted to receive better energy-saving effects. Therefore, we analyze the effect of the exergy loss of the entire heat exchange subsystem when the heat load changes with the same area.

Assuming that the heat transfer amount of cross heat transfer is reduced to ΔQ, the average temperature of the cold and hot streams of the heat exchange network will change to $T_C^{'}$ and $T_H^{'}$:

$$\begin{aligned}\Delta EX_2 &= Q_c T_0 \frac{T_H^{'} - T_C^{'}}{T_H^{'} T_C^{'}} - Q_v T_0 \frac{T_H - T_C}{T_H T_C} \\ &= Q_v T_0 \left(\frac{T_H^{'} - T_C^{'}}{\alpha T_H^{'} T_C^{'}} - \frac{T_H - T_C}{T_H T_C} \right)\end{aligned} \tag{13.53}$$

Generally, cross heat transfer increases the average temperature of the hot flow, $T_H^{'} > T_H$ (part of the low temperature section heat is pressed into the cooling load) and the temperature decreases after the cold flow exchange, the average temperature $T_C^{'} < T_C$, so there is $\left(T_H^{'} - T_C^{'}\right) > (T_H - T_C)$, but α is greater than 1, therefore, the change in process exergy loss only in terms of heat transfer depends on the cross heat transfer factor and the average temperature difference of the system streams. But it will inevitably increase the exergy loss during the cooling process and the heating process.

As far as the entire heat exchange subsystem is concerned, the reduction in heat exchange load will have to be compensated for by the increase of hot utilities, and the corresponding increase in cold utilities.

Heating exergy loss:

$$\Delta EX_3 = \Delta Q \left(1 - \frac{T_0}{T_{mc}} \right) / \eta_x \tag{13.54}$$

where: T_{mc}—the average temperature of the temperature section at which the final heat exchange temperature decreases, K;

η_x—heating furnace exergy efficiency.

Cooling exergy loss

$$\Delta \mathrm{EX}_1 = \Delta \mathrm{Q} \left(1 - \frac{\mathrm{T}_0}{\mathrm{T}_{\mathrm{cc}}} \right) \tag{13.55}$$

where: T_{cc}—the average temperature of the cooling medium, K.

The total process exergy loss change is:

$$\begin{aligned}\Delta EX = \Delta Q \left(1 - \frac{T_0}{T_{cc}} \right) + \frac{\Delta Q \left(1 - \frac{T_0}{T_{mc}} \right)}{\eta_x} \\ + Q_v T_0 \left(\frac{1}{\alpha} \frac{T_H^{'} - T_C^{'}}{T_H^{'} T_C^{'}} - \frac{T_H - T_C}{T_H T_C} \right)\end{aligned}$$

$$= Q_v(1 - 1/\alpha)\left[\left(1 - \frac{T_0}{T_{cc}}\right) + \left(1 - \frac{T_0}{T_{mc}}\right)/\eta_x\right]$$

$$+ Q_v T_0\left(\frac{1}{\alpha}\frac{T_H^{'} - T_C^{'}}{T_H^{'}T_C^{'}} - \frac{T_H - T_C}{T_H T_C}\right) \tag{13.56}$$

In fact, its energy level is close to 1 due to fuel heating. That is $\frac{1-\frac{T_0}{T_{mc}}}{\eta_x} \approx 1$, and the average temperature of the cooling medium is mostly close to the ambient temperature, $\frac{T_0}{T_{cc}} \approx 1$. At this time, the above formula can be simplified as:

$$\Delta EX = Q_v(1 - 1/\alpha) + Q_v T_0\left(\frac{1}{\alpha}\frac{T_H^{'} - T_C^{'}}{T_H^{'}T_C^{'}} - \frac{T_H - T_C}{T_H T_C}\right) \tag{13.57}$$

Therefore, the process exergy loss of the heat exchange subsystem can be determined by Q_v and the cross-heat transfer factor. It can be seen from above formula that cross heat transfer increases the irreversibility of the process, and the temperature difference distribution deviates from the driving force curve, which causes the average temperature difference between the cold and hot objects in the heat exchange part to change. The result is that the effective temperature difference decreases and the heat exchange rate decreases, and the hot and cold utilities increase.

13.6.4 Estimation of Cross Heat Transfer Factor α of Existing Heat Exchange Network

The key to using α to estimate the influence of cross heat transfer is to determine the average temperature difference of the vertical heat transfer network and the average temperature difference of the cross-heat transfer network.

(1) Vertical heat transfer temperature difference

In the actual process, it is impossible to determine the vertical heat transfer conditions before performing cross heat transfer. We can estimate the vertical heat transfer temperature difference based on the stream data, which is calculated from the average temperature of the hot flow and the cold flow:

$$\Delta T_{mv} = T_H - T_C \tag{13.58}$$

$$T_H = \sum_{i=1}^{n} \frac{T_{i1} - T_{i2}}{\ln(T_{i1}/T_{i2})}\frac{Q_i}{Q_A} \tag{13.59}$$

$$T_C = \sum_{i=1}^{p} \frac{t_{i1} - t_{i2}}{\ln(t_{i1}/t_{i2})}\frac{Q_i}{Q_A} \tag{13.60}$$

In the formula:

T_{i1}, T_{i2}—the population and outlet temperature of the heat flow heat exchange system (excluding heating and cooling) °C;

t_{i1}, t_{i2}—the outlet and inlet temperatures of the cold stream heat exchange system, °C;

n, p—the numbers of hot and cold streams, respectively;

Q_i, Q_A—respectively the heat load and total heat load of the ith heat exchange equipment, kW.

(2) Determination of cross heat transfer temperature difference

The formula (13.41) is based on the assumption that the heat load of each heat exchanger is the same, but the heat load of each heat exchange device in the actual network is not the same. At this time, the average heat transfer temperature difference of the cross-heat transfer can be estimated by the following formula:

$$\Delta T_{mc} = \frac{Q_A}{\sum(Q_i/\Delta T_{mci})} \tag{13.61}$$

Then the cross-heat transfer factor is calculated by formula (13.48), so as to estimate the influence of cross heat transfer on the area and energy target.

Example 13.7 For a heat exchange network composed of two cold streams and two hot streams, the stream data are shown in Table 13.17.

When $\Delta T_{\min}$ is 25 °C, the maximum energy recovery network (vertical heat transfer) is shown in Fig. 13.24. switch the heat exchangers (2) and (3) cold streams to make the network cross heat transfer, and estimate cross heat transfer effects. The adjusted cross heat transfer grid diagram and estimation results are shown in Fig. 13.25 and Table 13.18 respectively.

It can be seen from Tables 13.18 that the cross-heat transfer of the two heat exchangers increases the total network area by 45% at the same utility demand, and the results have been confirmed by using Pinch analyzer simulation results. Noted that the ΔT_{lm} of the heat changer 2 has been reduced to 10 °C from 30 °C, which

Table 13.17 Example 13.5 stream data table

Streams and types	Heat capacity flow rate (MW/°C)	Initial temperature (°C)	Final temperature (°C)	Heat load (MW)
(1) Hot	0.3	175	45	900
(2) Hot	0.15	156	30	260
(3) Cold	0.3	65	140	−440
(4) Cold	0.4	80	150	−575

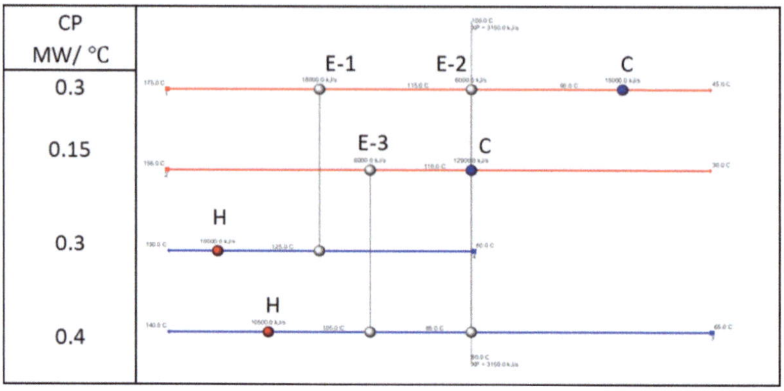

Fig. 13.24 Example 13.7 vertical heat transfer diagram

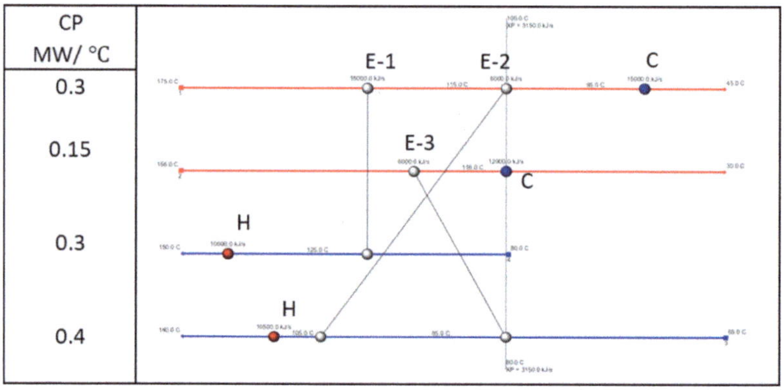

Fig. 13.25 Example 13.7 cross heat transfer diagram

Table 13.18 Cross heat transfer influence

Heat exchanger	Vertical heat transfer (°C)				Cross heat transfer			
	Inlet (°C)	Outlet (°C)	Heat duty (MW)	$\Delta T_{\min}$	Inlet (°C)	Outlet (°C)	Heat duty (MW)	$\Delta T_{\min}$
1	170	115	18	42.06	170	115	18	42.06
	80	125			80	125		
2	115	95	6	30	115	95	6	10
	65	85			85	105		
3	156	116	6	40.17	156	116	6	60.45
	85	105			65	85		
$\sum\left(\frac{Q_i}{\Delta T_m}\right)$				0.78				1.13
$\Delta T_{\min}$				38.59				26.61
α				1.00				1.45

causes an extreme impact on the heat transfer area. In reality of many existing heat exchanging networks, there may be a certain extent of cross heat transfer, it will be the potential to be saved.

If we keep the heat transfer area unchanged, the heat transfer duty will increase, and the consumption of hot and cold utilities will increase accordingly.

13.7 Energy Saving Principle of Pinch Technology

13.7.1 The Main Characteristics of Pinch Technology and Exergy Analysis

1. Features of pinch technology

Pinch technology has been practically applied in engineering design. Compared with other process thermodynamic analysis methods, it has the following characteristics:

(1) Practical. Pinch technology can be directly used in the design of new processes and process redesign. According to reports, the process synthesis team of the ICI Development Research Institute headed by Linnhoff completed dozens new engineering designs and old plant renovations using pinch technology. Each project saves energy, most of the project investment is close to or less than the original design. Pinch technology can also be combined with system optimization and other technologies to form a systematic process design method to solve quite complex process integration problems.
(2) Simplicity. The simplicity of the method is manifested in two aspects. First, only the data of material balance and energy balance are needed, and no other thermodynamic data (such as entropy) is needed. Second, the method focuses on the understanding of important physical phenomena such as pinch points, and form design criteria for various processes on this basis.
(3) Intuitive. Because the hot flow cascade model is used to express the characteristics of process energy degradation, the evaluation of the existing process and the synthesis of the new process are very intuitive and clear. In addition, the benefit is directly reflected in the amount of utilities, making it easy for engineering and technical personnel to understand and use.
(4) Flexibility. The design program compiled or Aspen Energy Analyzer according to this method can point out the key parts that affect energy consumption, and make specific designs for the key parts (such as near the pinch point) (such as basic matching), and the rest can be designed with flexibility.

2. The relationship between pinch technology and exergy analysis [17]

Exergy analysis and pinch technology are two energy-saving technologies developed based on the first and second laws of thermodynamics, but the methods used and development directions are different, forming their own characteristics.

In the energy recovery system, exergy analyzes the relationship between heat transfer equipment, process rate (heat transfer) and heat transfer driving force (temperature difference), finds out the location and cause of energy loss, points out the energy-saving potential, and turns the energy-saving potential into production, the reality of improvement must also be combined with technical, economic and engineering conditions to determine specific measures. The focus of exergy analysis is to put forward the concept of energy use according to quality, calculation analysis and evaluation methods.

The pinch technology (pinch design method) is through the establishment of the pinch concept and summarizes a series of empirical rules to ensure that energy is used according to quality during the integration process of the energy recovery system. In addition, the optimization problem between investment and benefit is introduced into the calculation of "super-target", and a series of computer application software has been developed, making the pinch technology a practical, convenient and economical energy comprehensive practical technology and has been widely used.

Exergy analysis can start from the process plant, equipment and a certain part of the equipment, and diagnose the current energy consumption from different perspectives such as energy conversion, process utilization, and recovery and recycling, find potential, and make improvements. It has a wide range of applications, including all energy-using processes in the production process.

The pinch technology is mainly used as a kind of integrated energy technology for energy recovery system. For the determination and matching of cold and hot utilities, the configuration of heat engine and heat pump position, a series of rules have been put forward and played a good role. However, there are few reports on the determination of operating conditions for process improvement and the improvement of energy conversion efficiency.

Through exergy analysis of process equipment (towers, reactors, etc.), it points out ways to improve process conditions, reduce exergy loss in the process, and improve the grade of "energy to be recovered" into the energy recovery system, thereby changing the pinch position of the energy recovery system. Synthesize a better energy recovery system. For example, the exergy analysis of the atmospheric and vacuum distillation unit of an oil refinery found that the return temperature of the middle reflux heat extraction can affect the process exergy loss. Increasing the middle reflux return temperature can reduce the exergy loss of the tower equipment process, increase the average temperature of the system heat source, and change the energy recovery, change the position of the system pinch points, thereby reducing the amount of hot utility demands.

13.7.2 Technical Characteristics of the Pinch

The energy recovery system usually contains a set of cold streams that need to be heated and a set of hot streams that need to be cooled. The cold and hot streams exchange heat to recover energy, but due to engineering and economic factors, the exhaust stream (hot stream) energy cannot be fully recovered, and there must be some cooling. The cold stream needs to be heated to reach the process operating conditions. When the energy recovery of the system is the maximum, the utility load for heating and cooling is the minimum, and the relationship between utilities (Q_H, Q_C) and heat recovery can be intuitively reflected on the streams composite T-H diagram (see Fig. 13.4).

It can be seen from Fig. 13.4 that the horizontal movement of the curve will not change the heat load of the cold and hot streams. When the cold flow curve shifts from the dotted line to the left, the heat transfer rate increases, the temperature difference decreases, and the amount of cold and hot utilities decreases. When the two curves approach a certain point, the minimum allowable heat transfer temperature difference $\Delta T_{\min}$ appears, the heat exchange load reaches the maximum, the cooling and heating load is the minimum, and the system is optimal. And the $\Delta T_{\min}$ appearing in the two curves becomes the checkpoint of the energy recovery system, which is a thermodynamic practical limit point, which we call the pinch point. Continue to shift to the left, the two curves intersect, the minimum heat transfer temperature difference tends to zero, and the process becomes a reversible process, recovering the most energy and the largest investment (towards infinity). This is the thermodynamic limit point, the limit of energy recovery, and reflects the objective law of the thermodynamic second law analysis.

Therefore, in actual design, we must determine the minimum allowable heat transfer temperature difference $\Delta T_{\min}$ according to technical and economic conditions. Figure 13.4 also objectively reflects the relationship between heat transfer, heating, and cooling. The pinch point of the heat exchange network has the following technical characteristics:

1. There are thermodynamic limitations

The heat transfer of a single heat exchanger has a heat transfer "pinch point", which is the limiting point that restricts the heat transfer of the equipment. Through the composite of cold and hot flows shown in Fig. 13.4, find the thermodynamic limit point of the system, that is, the pinch point. This point is the bottleneck restricting the further heat exchange of cold and hot streams, and minimizes the consumption of cold and hot utilities.

2. Position characteristics of the pinch point

The pinch point decomposes the heat exchange network problem into two independent subsystems, the hot end above the pinch point and the cold end under the pinch point. Above the pinch point, only hot utilities are needed, and below the pinch point, only cold utilities are needed. Matching the hot flow above the pinch point with the cold flow below the pinch point will inevitably lead to an increase in the amount of utilities. Therefore, the pinch technology proposes a design criterion, called the golden criterion, as shown in Fig. 13.26

.

(1) Don't transfer heat through the pinch point. If X units of heat are transferred through the pinch point, X units must be added to the hot utility.

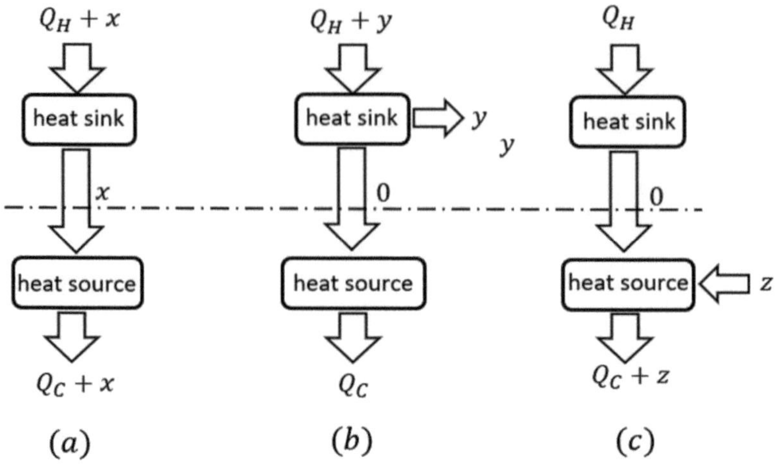

Fig. 13.26 The relationship between utility consumption and pinch points

(2) Hot utilities are not used below the pinch point. Otherwise, not only the hot utility will be consumed, but also the cold utility will increase.
(3) Cold utilities are not used above the pinch point. Otherwise, it will lead to an increase of the same number of hot utilities.

3. Heat transfer characteristics of the pinch point

The point where the slope of the composite curve changes in Fig. 13.4 is called a corner point (break point). Through observation, it can be found that any flow entering or leaving can cause the slope of the composite curve to change, thus forming corner point. It can be proved that the pinch points of the network can only appear at the corner points and the end points of the composite curve [8].

The pinch point is the point where the heat transfer driving force ΔT of the entire heat exchange network is the smallest. Therefore, extending from the vicinity of the pinch point to both ends, ΔT increases, that is, the temperature difference of the remaining parts is greater than the pinch temperature difference $\Delta T_{\min}$. Therefore, the hot and cold flow matching at the pinch point is the focus of the design. The heat exchanger at this point is called a pinch heat exchanger, and at least one of the two sides has a heat transfer temperature difference equal to $\Delta T_{\min}$. The temperature difference of other heat exchangers far from the pinch point is greater than the minimum temperature difference. Therefore, the feasibility criteria for matching pinch technical requirements are:

(1) On the side of the pinch point, the sum of the heat capacity flow rates flowing into the pinch stream should be less than or equal to the sum of the heat capacity flow rates flowing out of the pinch stream, that is:

$$\sum CP_{out} \geq \sum CP_{in} \tag{13.62}$$

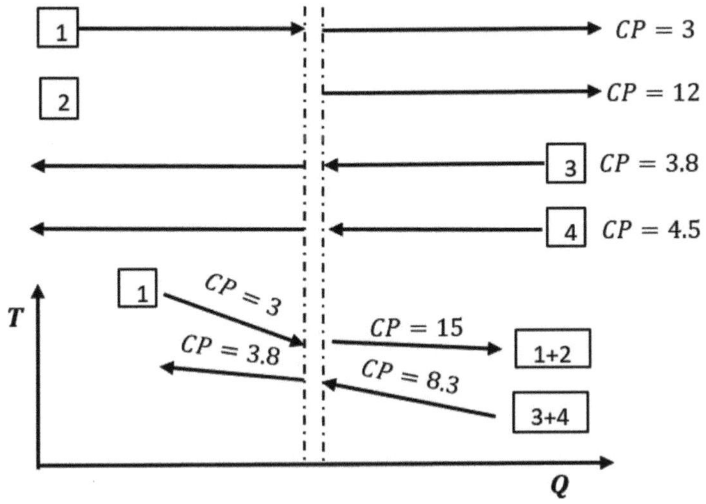

Fig. 13.27 Heat transfer characteristics of the pinch point

The definition of the pinch point of inflow and outflow is shown in Fig. 13.27. Figure 13.27 show the pinch point heat transfer characteristics of a certain heat transfer problem. For the stream with pinch point 2 we call the stream with the pinch point.

(2) The number of streams must meet the CP criterion of the above formula. It can be inferred that: the pinch point design technology should be followed, the number of streams flowing into the pinch point should be less than or equal to the number of streams flowing out of the pinch point, namely:

$$N_{in} \leq N_{out} \tag{13.63}$$

In order to meet this criterion in actual design, it is usually achieved by means of diversion.

13.7.3 Heat Transfer Exergy Loss

1. Thermodynamic analysis of the heat transfer process of a single heat exchanger

The actual heat exchange process is irreversible. The cold flow gets heat at a much lower temperature than the hot flow. Therefore, even if the heat loss is not considered, the released energy of the hot flow and the heat obtained of the cold flow are equal, but the exergy obtained by the cold flow is much lower than the exergy released by the hot flow, and the difference between the two is the irreversible exergy loss in the heat transfer process. From the integral formula of thermal physics exergy:

$$EX_h = \int_{T_1}^{T_2} C_p\left(1 - \frac{T_0}{T}\right)\mathrm{d}T \tag{13.64}$$

When the specific heat capacity can be regarded as a constant, we get:

$$EX_h = \Delta H\left(1 - \frac{T_0}{T_m}\right) \tag{13.65}$$

In the formula: T_m—the average temperature of the stream, $T_m = \frac{T_2 - T_1}{\ln(T_2/T_1)}$. Therefore, the exergy of hot stream is ($Q = G\Delta H$):

$$EX_h = Q\left(1 - \frac{T_0}{T_H}\right) \tag{13.66}$$

$$T_H = \frac{T_{H1} - T_{H2}}{ln(T_{H1}/T_{H2})}$$

The cold stream absorbs exergy as (ignoring heat loss):

$$EX_c = Q\left(1 - \frac{T_0}{T_C}\right) \tag{13.67}$$

$$T_C = \frac{T_{C2} - T_{C1}}{ln(T_{C2}/T_{C1})}$$

The exergy loss in the heat transfer process is:

$$EX_L = QT_0\frac{T_H - T_C}{T_H T_C} \tag{13.68}$$

Reflected in Fig. 13.28, the hot flow exergy is the area enclosed by 1256, the cold flow absorption exergy is the area enclosed by 3456, and the process exergy loss is the shaded part surrounded by the two curves in Fig. 13.28. When the ratio of the temperature difference between the two ends of the heat transfer device is $\Delta t_1/\Delta t_2 < 2$, the arithmetic mean temperature difference can be used. For pure countercurrent heat transfer:

$$\begin{aligned}\Delta T_m &= \frac{\Delta t_1 + \Delta t_2}{2} = [(T_{h1} - T_{c2}) + (T_{h2} - T_{c1})]/2 \\ &= [(T_{h1} + T_{h2}) - (T_{c2} + T_{c1})]/2\end{aligned}$$

When the specific heat capacity of the material is constant, there is: $T_h = \frac{T_{h1} + T_{h2}}{2}$, $T_c = \frac{T_{c2} + T_{c1}}{2}$

$$\Delta T_m = T_h - T_c \tag{13.69}$$

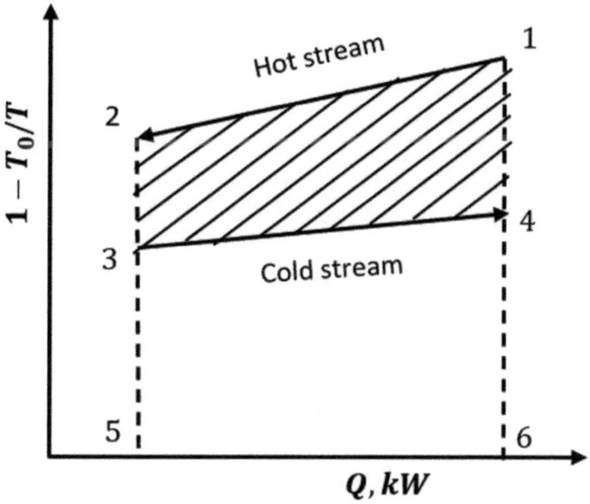

Fig. 13.28 Exergy loss of heat exchanger

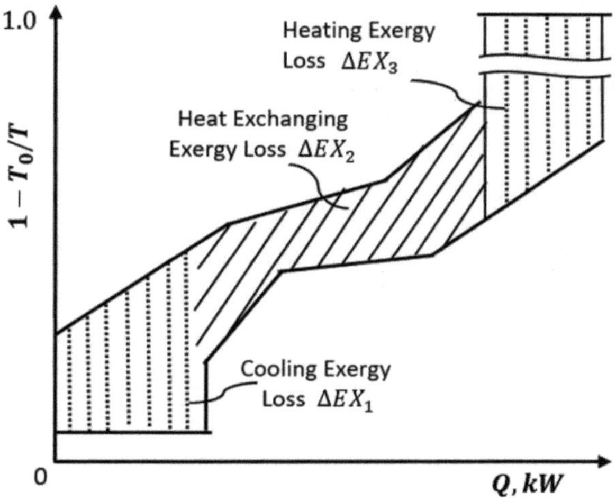

Fig. 13.29 Exergy loss distribution diagram of heat exchange subsystem

Therefore, the heat transfer process loss is:

$$EX_L = QT_0 \frac{\Delta T_m}{T_H T_C} \tag{13.70}$$

It can be concluded from the above formula that the heat transfer exergy loss is proportional to the heat transfer Q and the heat transfer driving force ΔT_m, and inversely proportional to the average temperature of the streams, that is, under the condition of certain heat transfer, the greater ΔT_m, the greater exergy the loss; the higher the T_h, T_c, the smaller the exergy loss. It reflects the objective attribute that the low temperature heat transfer exergy loss is greater than the high temperature heat transfer exergy loss under the same heat transfer temperature difference.

Reducing the heat transfer temperature difference within a reasonable range of technology and economy can reduce heat transfer exergy loss, thereby increasing the energy recovery rate and reducing energy consumption. This can be explained from the pinch point diagram shown in Fig. 13.4. The increase in the temperature difference of a single heat exchange device will affect the average temperature difference of the heat exchange network. When the temperature difference of the heat exchange network increases, that is, the cold flow curve in Fig. 13.4 shifts to the right (such as the position of the dotted line). As a result, the heat exchange amount is reduced, and the heating load increases, which leads to an increase in energy consumption.

2. Exergy loss of heat exchange network

Exergy loss is an important parameter to measure the change of energy quality in the heat transfer process. Increased exergy loss will inevitably lead to an increase in energy consumption. For the heat exchange subsystem shown in Fig. 13.1, it includes three parts: heat exchange, cooling and heating. The exergy loss of the heat exchange system is shown in Fig. 13.29.

It can be seen from Fig. 13.29 that the exergy loss in the heat transfer process of the heat exchange subsystem is composed of three parts: exergy loss in heat transfer equipment, exergy loss in the heating process, and exergy loss in the cooling process. The exergy loss distribution diagram can be developed from Composite curve diagram, using the energy grade ($\varepsilon = 1 - T_0/T$) as ordinate instead of T in composite curve diagram. The so-called energy grade is the proportion of exergy in heat, a parameter that characterizes energy quality, and the maximum value of the ordinate is 1. exergy loss EX_L in the process is:

$$EX_L = \Delta EX_1 + \Delta EX_2 + \Delta EX_3 \tag{13.71}$$

The exergy loss in the heat transfer process ΔEX_2 is a function of the heat transfer temperature difference and the amount of heat transfer:

$$\Delta EX_2 = \sum_{i=1}^{n} Q_i T_0 \frac{\Delta T_{mi}}{T_{hi} T_{ci}} \tag{13.72}$$

Reducing the heat transfer temperature difference can not only reduce the heat transfer exergy loss, but also reduce the system temperature difference, thereby increasing the heat transfer load and reducing the heating and cooling load.

The heating load of the system (hot utilities) is mostly converted into heat by the combustion of fuel in the heating furnace, and then transferred to the heated fluid. The exergy loss is the difference between the fuel chemical exergy and the exergy obtained by the heated fluid, which can be calculated as follows:

$$\Delta EX_3 = Q_H\left(1 - \frac{T_0}{T_m}\right) \Big/ \eta_x - Q_H\left(1 - \frac{T_0}{T_m}\right) = Q_H\left(1 - \frac{T_0}{T_m}\right)\left(1 - \frac{1}{\eta_x}\right) \tag{13.73}$$

where: T_m—is the logarithmic mean temperature of the heated fluid;

η_x—exergy efficiency of the heating furnace, for only one heated fluid, $\eta_x = \eta\varepsilon$;

η—the thermal efficiency of the heating furnace.

For the cooling load, the low-temperature hot stream that cannot match for heat exchange is cooled and discarded with the cooling medium. Generally speaking, the temperature carried by the cooling medium is also close to the ambient temperature, and the energy taken away by the cooling medium is finally rejected into the environment. The exergy loss of the heat transferred from the hot flow to the cooling medium is:

$$\Delta EX_1 = Q_c\left(1 - \frac{T_0}{T_{mc}}\right) - Q_c\left(1 - \frac{T_0}{T_c}\right) = Q_c T_0 \frac{T_{mc} - T_c}{T_{mc} T_c} \tag{13.74}$$

where: T_{mc}—the average temperature of the cooling hot fluid, K;

T_c—the average temperature of the cooling medium, K.

Generally, in the heat exchange subsystem, the heating load of the system is the driving force to complete the process operation and energy recovery. Without the heating load, the process operating conditions cannot be reached, and an energy recovery and recycling cannot be formed. The heating load is generally greater than the cooling load. This is due to the heat dissipation of process equipment, energy recovery equipment and pipelines, and most of the rest enters the cooling load.

References

1. X. Yifang et al., Multi-pinch points of multi-effect evaporation system and analysis of furfural solvent recovery system. Pet. Refinery **4** (1988)
2. S. Xiaowen, S. Meisheng, Application of pinch point design method in atmospheric and vacuum distillation unit. Refinery Des. **5**, 55–65 (1990)
3. H. Yaowen, in *New Chemical Technology—Thermodynamic Analysis of Chemical Process*, ed. by Education Committee of Chinese Chemical Society (1982)
4. Y. Jihong, *Ma Dexian* (Hydrocarbon Processing Press, Introduction to Process Systems Engineering, 1989)
5. B. Linnhoff et. al., User guide on process integration for the efficient use of energy. I. chem. E. Rugby, Engl.
6. S. Meisheng, Pinch design method. Refinery Des. **3**, 29–40 (1989)
7. Y.D. Lang et. al., Simultaneous optimization and heat integration with process simulators. Comput. Chem. Eng. **12**(4), 311–327 (1988)
8. L. Jiong, Research on pinch point theory and heuristic rules of crude oil distillation heat transfer network. Master's Thesis of Petroleum University, p. 12, 1992

9. S.G. Hall et al., Capital cost targets for heat exchanger networks comprising mixed materials of construction, pressure rating and exchanger types. Comput. Chem. Eng. **14**(3), 319–335 (1990)
10. B. Linnhoff, S. Ahmad, Cost optimum heat exchanger networks-part 1, minimum energy and capital using simple models for capital. Comp. Chem. Eng. **14**(7), 729–750 (1990)
11. J.M. Douglas, in *Conceptual Design of Chemical Processes* (McGraw-Hill Book company, U.S., 1988), pp. 216–289
12. G. Wenhao et al., in *Super-Target Application of Pinch Point Technology—Advanced Optimization of Pinch Point Temperature Difference in Heat Exchange System*. The 6th National Conference on Thermodynamic Analysis and Energy Saving, Guangzhou (1992), p. 12
13. Y. Taiping, Designing heat exchanger composite to save energy by bottleneck location method. Pet. Quart. **23**(3) (1987) (Taiwan)
14. J. Chusheng et al., in *Thermodynamic Basis and Application of Industrial Energy Saving* (Chemical Industry Press, 1990)
15. B. Linnhoff, in *Pinch Technology Has Come Age* (C. E. P, July 1984)
16. T.N. Jioe, B. Linnhoff, in Using pinch technology for process retrofit. Chem. Eng. (1986)
17. C. Anmin, L. Kun, in *Thermodynamic Analysis of Energy Recovery System*. Paper of the Seventh National Conference on Thermodynamic Analysis and Energy Saving (Dalian, 1994), p. 8

Chapter 14
Key Energy-Saving Technologies

Abstract This chapter is focusing on the key energy-saving technologies, which have played very important roles in energy saving and carbon emission reduction; It includes the following topics, (1) Pump and compressors speed control technology and its energy-saving principle; (2) Gas turbine energy-saving principle and its selection; (3) Combined gas turbine cycle and its application in LNG plant with simulation support; (4) Flue Gas Turbine Energy Saving of FCCU, including energy-saving principle, loading impacts, evaluation approach, and approaches to improve Flue gas power recovery efficiency; (5) FCCU regenerating gas CO combustion outside of regenerator, CO ignition rate equation, CO incinerator and energy recovery; (6) Flue gas energy recovery system optimization, including CO duct pre-combustion, pressure combustion, flue gas energy and power recovery system optimization; (7) Low-Temperature Heat Recovery and Utilization, including direct use as process heating, upgrading use of compression heat pump, absorption heat pump, refrigeration, and power generation, direct use and upgrading use system integration.

Keywords Frequency control · Combined-cycle gas-turbine · CO combustion · Flue-gas turbine · Low-temperature heat · Heat pumps

The energy-saving technologies (including energy-saving techniques) used in the oil and gas and petrochemical process have a wide variety of types and extremely rich content, which can be found throughout the entire process of energy conversion, energy utilization, and energy recovery. Many of them are closely integrated with the production process, and they have both energy-saving benefits (including Carbon tax reduction) and the benefits of increased product quantity and improved quality brought about by improved processes. Based on the contents of the foregoing chapters and combining the characteristics of petrochemical processes, this chapter lists some of the key energy-saving technologies that have exerted benefits in actual production or have broad application prospects in the petrochemical, chemical, and oil and gas field. The purpose of energy-saving technology improvement in the energy conversion process is to improve the efficiency of conversion equipment including energy efficiency and exergy efficiency. The second is to provide the process with energy that is compatible with the required energy grade (such as the cascade utilization of steam and the technology of pump/compressor speed regulation). The energy-saving

T. A. Chen, *Energy Saving and Carbon Reduction*,
https://doi.org/10.1007/978-981-19-5295-1_14

technology of the technological process aims to improve the technological process, reduce the amount of energy used in the process, and improve the grade of energy to be recovered.

14.1 Pump Speed Control Technology

The power consumption of pumps and compressors in petrochemical plants accounts for about 80% of the total power consumption of the whole plant. Because electricity is a high-quality energy source, the electric energy obtained in the power plant has already paid a large price—energy loss. Saving electric energy not only saves energy and mitigates CO_2 emissions but also has significant benefits.

Petrochemical equipment is often operating at low load due to production plans, raw material and product transportation, market conditions, and other reasons. In addition, the safety margin and the expansion and development of the plant are considered during the design of the plant, which makes many pumps operate at a low load. Moreover, some pumps are affected by factors such as plant processing plan, weather, season, etc., and the load rate fluctuates greatly. The "common practice" is to use regulating valve throttling control under low load, causing a large amount of energy loss. To overcome this situation, the change in the number of revolutions per minute (RPM) of the pump should be used to adjust the flow rate to avoid the throttling loss of the pump.

14.1.1 Variation Law of Pump Flow Rate with Speed

The stable operating point of the centrifugal pump is determined by the piping system characteristic equation and the pump characteristic equation:

$$PipelineH = H_0 + kQ^2 \tag{14.1}$$

$$PumpH = f(Q, n) \tag{14.2}$$

In the equation:

H—the head of the pump;

H_0—the static pressure head of the piping system, which is the operating pressure required by the process;

Q—the flow of the piping system;

n—the number of revolutions of the pump;

k—is the resistance reduction coefficient of the piping system.

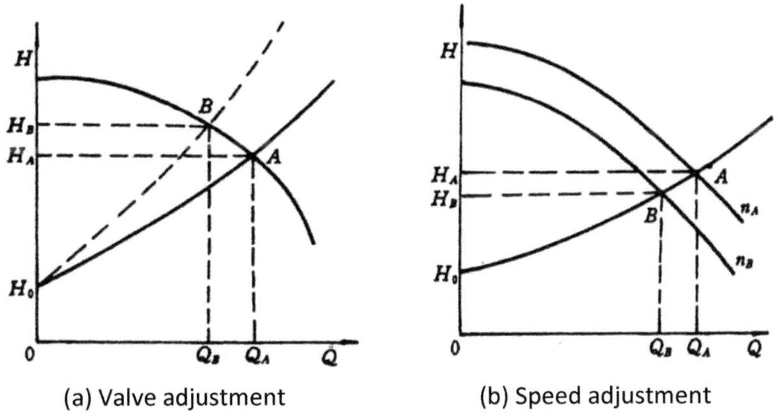

Fig. 14.1 Pump and piping system characteristics

The head of the pump and the pressure of the piping system (the sum of the system resistance drop and the operating pressure required by the process) are balanced at the pump outlet (before the regulating valve). There are two ways to achieve balance. One is to adjust the piping system (using a regulating valve). When the flow of the pump decreases, the head rises and the regulating valve closes so that the pump and the piping system achieve pressure balance. Another adjustment method is to open the pipeline adjustment valve. The pump speed is used to adjust the flow change, and its head is adapted to the required pressure of the piping system. This is the pump speed control technology. The pump and piping system characteristics of the two regulating systems are shown in Fig. 14.1.

For the valve adjustment of Fig. 14.1a, when the flow rate decreases from Q_A to Q_B, the operating point moves from A to B, and the head increases from H_A to H_B. With speed adjustment, the system pipeline characteristics remain unchanged, and the pump speed changes from n_A to n_B, forming a new pump characteristic curve. At this time, the flow rate decreases, and the pump head decreases. The characteristic curves of the two ways of adjustment are shown in Fig. 14.2.

It can be seen from Fig. 14.2 that the control method is different, and the pump outlet head is different. Under low load, the regulating valve control increases the pump head, while the pump speed control reduces the pump head. The two adjustment methods have different shaft power:

$$\Delta N = \frac{\rho Q \Delta H}{102}\ \text{kW} \tag{14.3}$$

where

N—shaft power, kW;

ρ—the relative density of the fluid;

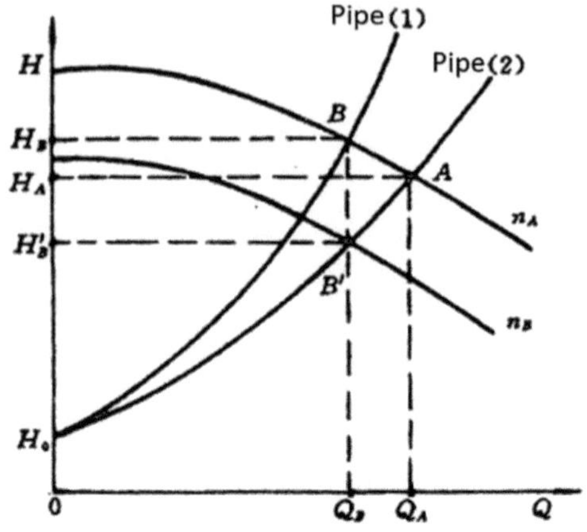

Fig. 14.2 Comparison of characteristic curves of the two adjustment methods

H—pump head, m;

Q—the flow rate of the pump, m^3/s.

On the other hand, the relationship between the change of the centrifuge pump speed and the pump flow, head, and power consumption is based on similar laws:

$$\frac{Q'}{Q} = \frac{n'}{n} = r \tag{14.4}$$

$$\frac{H'}{H} = \left(\frac{n'}{n}\right)^2 = r^2 \tag{14.5}$$

$$\frac{N'}{N} = \left(\frac{n'}{n}\right)^3 = r^3 \tag{14.6}$$

Therefore, if the flow is adjusted by the number of revolutions, and the load is $Q'/Q = L$, then the Eq. (14.6) is: $\frac{N'}{N} = L^3$, that is, adjust using the number of revolutions, the shaft power change is proportional to the cube of the load ratio. The shaft power change per unit flow is:

$$\frac{N'/Q'}{N/Q} = L^2 \tag{14.7}$$

The Eq. (14.7) is the characteristic of shaft power change when the speed is adjusted: the shaft power per unit flow is proportional to the square of the load ratio *L*. Theoretically, during low-load operation, there is not only no fixed energy consumption of electricity, but also, as the load ratio decreases, the unit consumption of electricity decreases. In the actual process, the flow adjustment method is more complicated. In addition to the speed adjustment, other adjustment methods are often supplemented. Especially when the flow rate decreases and the pump head decreases, which cannot meet the needs of process operation and transportation, control valves should be combined.

In addition, the flow rate changes using speed regulation technology, and the efficiency of the pump is generally unchanged.

14.1.2 Speed Control Method and Classification [1]

There are two main types of speed control methods for pump or compressor systems. One type is the direct speed regulation by the electric motor. The second is that the speed of the motor remains unchanged, and the purpose of speed regulation is achieved by changing the speed of the transmission facility between the motor and the pump.

Pump speed regulation device	Motor speed regulation	Pole-changing
		Frequency Changing
		SCR cascade speed regulation
	Mechanical speed regulation	Fluid coupling speed regulation
		Oil film clutch speed regulation

1. Motor Speed Regulation

According to electrical engineering, the speed equation of an AC motor is:

$$n = 60(f/P)(1 - S) \tag{14.8}$$

where: f is the power supply frequency of the stator, P is the number of pole pairs, and S is the slip ratio. Changing the number of rotations of the AC motor can be achieved by changing f, P, S.

(1) Changing the number of pole pairs P of the motor, this variable-speed method is simple to control and can only run at two or three fixed speeds, so it is difficult to meet the requirements of the production process.
(2) Change the slip rate S, connect a silicon control inverter in series in the rotor winding as the back EMF, this kind of speed regulation is thyristor cascade speed regulation. Used for winding induction motor speed range is small, the inverter draws reactive power from the grid, so in the case of high system efficiency, the total power factor of this speed control system is not high, its effect is lower than Frequency.
(3) Change the power supply frequency f, this kind of speed regulation to variable frequency, with low loss and high efficiency. Many applications have been obtained in petrochemical plants. The current problem is that the imported or assembled frequency converter does not have a series of products for large and

medium pumps (200–1000 kW) in petrochemical plants. The quality of domestic equipment needs to be improved.

2. Mechanical speed regulation

(1) Fluid coupling is slip-type transmission equipment that uses liquid (20# mechanical oil) kinetic energy and potential energy (mainly kinetic energy) to transmit power. It steplessly changes the pump's speed and output parameters under the motor's constant speed. The amount of liquid discharged by the fluid coupling increases or decreases according to the change of the discharge pressure of the pump, so that the sliding speed is changed and the speed is changed automatically, so the input power of the motor can be reduced. In this way, it can be started in a no-load state. Therefore, the motor drive current can be reduced, no special starting device is needed, the speed change range is wide, and the responsiveness is better. Due to sliding, the conduction rate decreases, but it also has the advantages of a small maintenance workload and high reliability.

(2) Oil film clutch [2]

The speed regulating oil film clutch works by adjusting the working oil pressure in the working oil chamber between the cylinder and the piston to change the degree of compaction of the driving and driven friction plates. When fully compressed, the master and slave friction plates run synchronously. When the pressing force is low to a certain critical value, a relative sliding slip occurs between the driving and driven friction plates, making the output shaft speed lower than the input shaft speed. Changing the pressing force can change the relative slip, to achieve the purpose of step-less speed regulation.

The piston displacement is controlled by adjusting the oil pressure, thereby pushing the driving and driven friction plates to move relative to each other. Change the oil film gap between the driving and driven friction plates. Because the transmitted torque and speed difference are related to the size of the oil film gap. Therefore, when adjusting the speed, you only need to change the oil pressure to push and control the displacement of the piston. Adjusting the oil film gap between the driving and the driven plate can realize stepless speed regulation, the transmission ratio is λ:

$$\lambda = W_2/W_1 \tag{14.9}$$

W_2 is the speed of the driven shaft, and W_1 is the speed of the driving shaft. The actual working condition λ is generally in the range of 1–1/3. In addition to the friction loss of the friction plate, the efficiency of the oil film clutch also includes the transmission loss of the oil supply pump and gear and the loss of bearings and seals.

The comparison of various speed control schemes is shown in Table 14.1.

Table 14.1 Comparison of various speed control schemes

Description		Applications	Reliability	Maintenance difficulty	Advantage	Disadvantage	Efficiency	Remarks
1	Hydraulic coupler	High-speed, large-capacity, squirrel-type motor	High	General	Less investment	1. Slip loss 2. No backup operation plan 3. Require more space	$\frac{\eta_T}{\eta_B}$	η_T, η_B are the output of the coupler and the input speed respectively. Positive pressure ventilation should be adopted for the automatic control actuator in explosion-proof places
2	TL clutch	High-speed, large-capacity, squirrel-type motor	Control system hydraulic proportional valve failure	General	1. Less investment 2. Can switch to synchronous operation when the control system fails	1. Slip loss 2. No backup operation plan when rotating parts fail		
3	SCR cascade speed regulation	Synchronous speed 1500 rpm and below the winding motor	higher	Requires a higher level of technology	1. It can automatically switch to full speed operation in case of failure; 2. The slip power generated during speed regulation can be recovered; 3. The price is only 30–40% of the frequency speed regulation	1. Low power factor 2. The harmonics generated in the process of speed regulation have a certain degree of pollution to the grid 3. Only suitable for winding motors	90%	Self-aligning machine trigger, strong anti-interference performance, simple circuit

(continued)

Table 14.1 (continued)

Description		Applications	Reliability	Maintenance difficulty	Advantage	Disadvantage	Efficiency	Remarks
4	Frequency	Production plant, power system, squirrel motor	higher	Requires a higher level of technology	1. Strong protection function; 2. Much faster response 3. No-slip loss 4. High power factor	1. Large investment 2. Complicated line 3. Pollution to the power grid	98%	According to the current price, it is more suitable for a load of 15–160 kW

14.2 Gas Turbine and Its Selection

14.2.1 Energy Saving Principle

The operating temperature used in petrochemical processes is mostly between 200 and 400 °C, and these heats are all converted from fuel through the furnaces; the boiler exergy efficiency also highly depends on the steam generation conditions. In recent years, people have devoted themselves to improving the efficiency of heating furnaces and boilers, so that the thermal efficiency of many large and medium-size furnaces and boilers has reached greater than 85%. Continuing to improve the thermal efficiency of the furnace and boiler can save energy, but it is restricted by the following factors: First, the acid dew point of the flue gas is limited, and the flue gas temperature shouldn't be lower than the acid dew point temperature; the acid dew point varies with the type of fuel and the sulfur content; secondly, the flue gas heat recovery needs to be traded off between the heat recovery benefits and investment, which limits the low temperature of the flue gas heat recovery, the recovered energy grade is low, with the heat transfer driving force reduced, the general investment will be higher so that the economical flue gas temperature cannot reach the acid dew point limit point, so there is no need to spend great effort on flue gas stack and duct anti-corrosion. At the same time, the application of combined cycle gas turbines integrated with the heating furnace in the industry to improve furnace exergy efficiency has begun to attract people's attention.

The gas turbine-boiler or heating furnace combined system is characterized by first using high-temperature and high-pressure gas expansion to do work, the exhaust gas heat and excess air can be taken as the heat source of the furnace or HRSG (heat recovery steam generator), it achieves a result of co-generation of work and heat, which improves the energy utilization efficiency of the fuel consumed. The result of the thermodynamic analysis of the system shows that the exergy efficiency can be increased by about 10% [16]. However, a simple heating furnace/boiler burns fuel under normal pressure and does not have the function of expansion. There are also some disadvantages to the combined cycle gas turbine system. One is that the excess air volume is extremely large, and the energy lost at the same exhaust temperature is more than that of the heating furnace/boiler. The heat dissipation area of the system is increased, and the heat dissipation loss is also increased. The cogeneration of heat and work overcomes this disadvantage by using up the excess oxygen in a duct fired boiler or furnace.

In the gas turbine cycle, the compressor consumes power, and the expansion turbine generates power. The amount of work done is:

$$W_s = \int_{p_1}^{p_2} V dp \tag{14.10}$$

If the gas working fluid of the gas turbine cycle is regarded as an ideal gas, then (14.10) can be written as:

$$W_s = nRT \ln \frac{p_1}{p_2} \tag{14.11}$$

For the pressure of the working fluid (air), because it is compressed at a low temperature (close to the ambient temperature), it consumes less work. When the working fluid is burned with fuel in the combustion chamber, the temperature increases, and the expansion does more work. It is assumed that the working fluid of expansion and compression pressure has an ignorable difference. Ignoring the system pressure drop, the ratio of expansion work and compression consumption work is:

$$\frac{W_{s2}}{W_{s1}} = \frac{nRT_2 \ln \frac{p_1}{p_2}}{nRT_1 \ln \frac{p_1}{p_2}} = \frac{T_2}{T_1} \tag{14.12}$$

In the equation, $T_2(K)$ is the combustion gas average temperature during the expansion process, and $T_1(K)$ is the air average temperature during the compression process. W_{s2} is expansion work, W_{s1} is compression power consumption. Generally, the expansion process T_2 is 500–1000 °C (773–1273 K), while the compression process temperature is 20–150 °C (293–423 K). The ratio of expansion work and compression consumption work is over 3; due to turbine efficiency and other losses, the actual work ratio is slightly lower, typically, about 50% of the work developed by the turbine sections is used to power the axial flow compressor.

A schematic diagram for a simple cycle, a single-shaft gas turbine is shown in Fig. 14.3a. Ambient air enters the axial flow compressor, since air inlet conditions vary from day to day and from location to location, it is convenient to use ISO conditions that are 15 °C, 1.013 bar, and 60% relative humidity.

Air is compressed to some higher pressure; compression raises the air temperature. The compressed air enters the combustion system, where fuel is supplied and combustion occurs. The combustion process occurs at essentially constant pressure; the combustion system is designed to provide mixing, burning, dilution, and cooling. Thus, by the time the combustion mixture leaves the combustion system and enters the turbine, it is at a mixed average temperature.

In the turbine section of the gas turbine, the energy of the hot gases is converted into work. This conversion takes place in two steps [3]. In the nozzle section of the

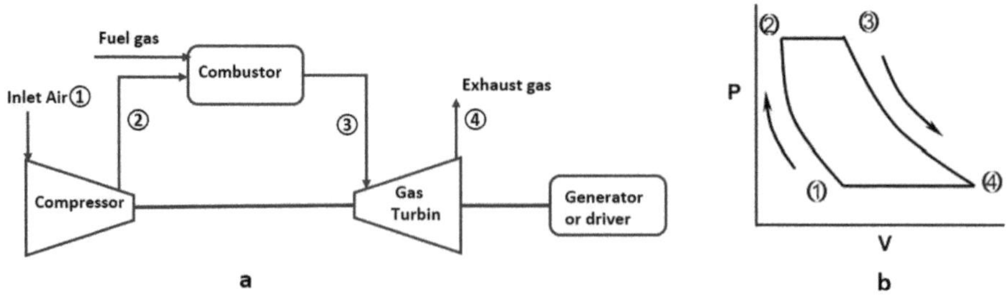

Fig. 14.3 Schematic diagram of single-shaft gas turbine and PV diagram

turbine, the hot gases are expanded and a portion of the thermal energy is converted into kinetic energy. In the subsequent bucket section of the turbine, a portion of the kinetic energy is transferred to the rotating buckets and converted to work. Some of the work developed by the turbine is used to drive the compressor, and the remainder is available for useful work at the output flange of the gas turbine.

The thermodynamic cycle upon which all gas turbines operate is called the Brayton cycle, Fig. 14.3b shows the classical pressure-volume (PV) diagram, The numbers on this diagram correspond to the numbers also used in Fig. 14.3a. Path 1–2 represents the compression occurring in the compressor, path 2–3 represents the constant-pressure addition of heat gain in the combustion systems, and path 3–4 represents the expansion occurring in the turbine.

The path from 4 back to 1 on the Brayton cycle diagrams indicates a constant-pressure cooling process. In the gas turbine, this cooling is done by the compression air, which provides fresh, cool air at point 1 continuously in exchange for the hot gases exhausted into the atmosphere at point 4. The actual cycle is an "open" rather than a "closed" cycle, as indicated.

Every Brayton cycle can be characterized by two significant parameters: pressure ratio and firing temperature. The pressure ratio of the cycle is the pressure at point 2 (compressor discharge pressure) divided by the pressure at point 1 (compressor inlet pressure). In an ideal cycle, this pressure ratio is also equal to the pressure at point 3 divided by the pressure at point 4. However, in an actual cycle, there is some slight pressure loss in the combustion system and, hence, the pressure at point 3 is slightly less than at point 2.

The other significant parameter, firing temperature, is thought to be the highest temperature reached in the cycle. GE defines firing temperature as the mass-flow mean total temperature sent as firing temperature by point 3 in Fig. 14.3b.

Typically, more than 50% of the work developed by the turbine sections is used to power the axial flow compressor.

The gas turbine will be derated up to 8% due to the following main factors.

Air intake conditions

Derating is caused by gas turbine Air intake conditions. The heat rate of the gas turbine provided by the vendor is per ISO condition (15 °C and 1.013 atm, 60% of humidity), a heat rate deviation is caused by different conditions from ISO conditions since its performance is changed by anything that affects the density and/or mass flow of the air intake to the compressor.

Humidity

Since humid air is less dense than dry air it directly affects the performance of gas turbines, for the same volume entering the gas turbine, it reduces mass flow rate and reduces the power output.

Inlet and exhaust losses

Derating is caused by gas turbine Inlet and exhaust losses. installing air filtration, silencing, evaporative coolers, or chillers into the inlet or heat recovery devices in the exhaust causes pressure losses in the system, eventually, cause lowering the power output.

Derating caused by gas turbine (including inlet compressor) fouling (1–4%).

Derating caused by gas turbine (including inlet compressor) aging (1–4%).

14.2.2 Gas Turbine Selection

Generally, two types of gas turbines are widely used in power generation, LNG plants, and the energy industry, Aeroderivative gas turbines, and heavy-duty gas turbines. Its efficiency or heat rate varies from type to type of gas turbine, and its compression ratio, firing temperature (T_3), and exhaust temperature (T_4); the T_3 temperature is controlled and highly dependent on materials used and manufacturing level.

Generally speaking, gas turbine selection is targeted on the max profit (NPV), including operating cost (Fuel), target product revenue, and capital cost. The target product could be either power generation or compressor driving power, which can be estimated per power price. The basic considerations of gas turbine selection are listed here:

1. Decide the capacity of the gas turbine set based on actual operation needs. The total capacity could be power generation capacity or mechanical driving power needed depending on the actual application with a reasonable design margin.

For an LNG mixed refrigerant gas compressor, the compressor power demand may be supplied by the gas turbine, helper steam turbine, or motor helper, depending overall on the steam and utility system arrangement.

2. Select gas turbines based on the efficiency (heat rate), gas turbine sparing policy, turndown capacity, guaranteed continuous operation hours, of course, the equipment price, etc.
3. The heat rate from manufacture is at ISO condition, which is 15 °C and 1.013 atm, 60% of humidity.
4. Actual operation will be different from ISO conditions and will be impacted by multiple factors, so a derating factor of up to 8% should be applied, it is noted that the derating factor will be different at the start of run and end of run, but for design, the expected max derating factor should be applied.
5. The gas turbine loading factor will impact the operation cost and capital cost greatly, selecting an optimal operation loading will contribute to lowering the total cost greatly. A typical gas turbine heat rate impacted by loading is shown in Fig. 14.4.

Figure 14.4 is estimated based on aero-derivative gas turbine performance data, it is for understanding how the gas turbine loading impact the efficiency or heat rate, when you evaluate the gas turbine projects, it is strongly recommended to contact the vendor to get the loading impacts for specified gas turbine. In case of lack of vendor information, you can use the following equations for a rough estimate:

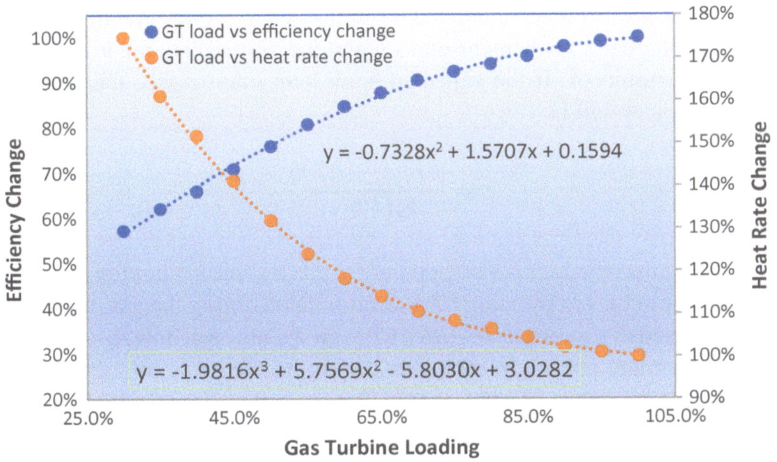

Fig. 14.4 GT loading impact on GT efficiency

Gas turbine efficiency changes by loading

$$E_{ef} = -0.7328L^2 + 1.5707L + 0.1594 \tag{14.13}$$

Gas turbine heat rate changes by loading

$$H_R = -1.9816L^3 + 5.7569L^2 - 5.803L + 3.0282 \tag{14.14}$$

where

E_{ef}—Gas turbine efficiency change at loading L, %;

L—Gas turbine loading, %;

H_R—Gas turbine heat rate change at loading L, %;

In addition to derating air intake conditions, inlet and exhaust losses, GT fouling, and aging, the predicted heat rate (efficiency) should be the derated heat rate (efficiency) multiple the heat rate (efficiency) changes.

6. Capital cost, the gas turbine cost is per vendor quotation, plus the cost for instruments and control system which falls in client scope.
7. Operating cost is mainly a fuel cost, which can be figured using actual power demand and derated heat rate at predicted loading. The fuel cost is calculated using total heat demand multiple by heat price $/GJ;
8. The main product of the gas turbine set could be either the power generated or the power driving a compressor, which eliminates/reduces the power demand of the compressor, therefor the power sales will increase the revenue.
9. A simplified dynamic payback period (ref. to Chap. 11) can be used for gas turbine selection in a simple economic evaluation.

When the GT project investment can be simplified as a one-time investment I (or discounting each year's investment into I), and the annual income after production is equal to the annuity A, If the expected bank loan interest is i, then the payback period (t) of project will be

$$t = \frac{-lg(1 - Ii/A)}{\lg(1 + i)} \tag{11.19}$$

10. NPV (net present value) is also popular for gas turbine evaluation, simple evaluation can be achieved by using NPV accumulated during the gas turbine lifetime at the targeted discounted rate (*IRR*), Convert the net cash flow $R_t = (CI - CO)_t$ of an investment project in each year during its life cycle into the current time (usually the project start time, that is, 0 years), the sum of which is called the net present value (NPV) of the project. The calculation equation is:

$$NPV = \sum_{t=0}^{n} (CI - CO)_t \left[\frac{1}{(1 + i)^t}\right] \tag{11.20}$$

where

n—Project construction and production service years;

i—Set discount rate or benchmark investment profit rate;

t—years;

$(CI - CO)_t$—Net annual cash flow, year.

11. The GT model with maximum accumulated NPV will be the best suitable gas turbine for your project. It can be recommended for detailed evaluation.

Example 14.1 Some oil gas facility requires installing a power plant with a net power output of 100 MW, considering power system loss of 2%, the gas turbine net output requires 102 MW, this facility is required to install a simple cycle gas turbine power generation.

To initial screening the suitable gas turbines, GE heavy-duty gas turbines and Aeroderivative gas turbines were considered, the gas turbine outputs and heat rates are taken from GE 2021–2022 GE Gas Power catalog [15].

The prices of the gas turbines are the best guess for this showcase example, when conducting a gas turbine selection project, it is strongly recommended to have a vendor input on each gas turbine price, net output, ISO heat rate, etc.

The gas turbine fuel used is the sweet natural gas with LHV of 47,200 kJ/kg, the gas turbine total derating factor is 8%; the targeted internal rate of return or discount rate for this project is 8%, the electricity price is \$0.12/kW-h, and natural gas price is \$5/GJ in the local area of this facility.

Table 14.2 Example 14.1 basic input data and GT selection results

Gas turbine information	GT model	GE heavy duty		GE aeroderivative			
		6F.03	6F.01	LM 6000 PF	LM 2500 + G4 DLE	LM 2500 + DLE	LM 2500 DLE
	SC net output, MW	88	57	44.7	33	30.5	22.2
	ISO heat rate, kJ/KWh	9788	9469	8702	9802	9871	10,466
	GT price (best guess), $ MM	29.2	26.4	21.3	17.7	16.6	16.1
Power generation, MW		102					
Required IRR (discount rate) %		8.0%					
Fuel gas LHV, kJ/kg		47,200	47,200	47,200	47,200	47,200	47,200
Expected GT performance	N units calculated	1.2	1.8	2.3	3.1	3.3	4.6
	N + 1 number of units	3	3	4	5	5	6
	N gen capacity (MW)	176.0	114.0	134.1	132.0	122.0	111.0
	N actual gen margin (%)	73	12	31	29	20	9
	Derated heat rate, kJ/kWh	10,639	10,292	9459	10,654	10,729	11,376
	Derated efficiency (%)	33.8	35.0	38.1	33.8	33.6	31.6
	Gas turbine loading (%)	58.0	89.5	76.1	77.3	83.6	91.9
	GT efficiency change (%)	82.4	97.8	93.0	93.6	96.0	98.4
	Predicted efficiency (%)	27.9	34.2	35.4	31.6	32.2	31.1

(continued)

Table 14.2 (continued)

Gas turbine information	GT model	GE heavy duty		GE aeroderivative			
		6F.03	6F.01	LM 6000 PF	LM 2500 + G4 DLE	LM 2500 + DLE	LM 2500 DLE
	SC net output, MW	88	57	44.7	33	30.5	22.2
	ISO heat rate, kJ/KWh	9788	9469	8702	9802	9871	10,466
	GT price (best guess), $ MM	29.2	26.4	21.3	17.7	16.6	16.1
	Predicted heat Rate, kJ/kW-h	12,918.4	10,522.7	10,169.0	11,388.2	11,172.0	11,561.5
	Fuel gas rate, kg/h	27,917	22,740	21,975	24,610	24,143	24,985
	Heat demand, GJ/h	1318	1073	1037	1162	1140	1179
Economic Evaluation	Fuel cost at $5/GJ, $/h	6588	5367	5186	5808	5698	5896
	Annual fuel cost, $MM	$57.71	$47.01	$45.43	$50.88	$49.91	$51.65
	Power generation revenue $MM, @ $0.12/kW-h	$107.22	$107.22	$107.22	$107.22	$107.22	$107.22
	N unit costs, $MM	$58.40	$52.80	$63.90	$70.80	$66.40	$80.50
	N + 1 unit cost, $MM	$87.60	$79.20	$85.20	$88.50	$83.00	$96.60
	Total capital $MM @ 50% client scope cost	$131.40	$118.80	$127.80	$132.75	$124.50	$144.90
	Dynamic payback period, year	3.10	**2.23**	**2.35**	2.71	2.48	3.04

(continued)

Table 14.2 (continued)

Gas turbine information	GT model	GE heavy duty		GE aeroderivative			
		6F.03	6F.01	LM 6000 PF	LM 2500 + G4 DLE	LM 2500 + DLE	LM 2500 DLE
	SC net output, MW	88	57	44.7	33	30.5	22.2
	ISO heat rate, kJ/KWh	9788	9469	8702	9802	9871	10,466
	GT price (best guess), $ MM	29.2	26.4	21.3	17.7	16.6	16.1
	NPV @8% discount rate, $MM	$455.76	**$595.30**	**$605.04**	$535.49	$555.19	$514.16

Solution Following the gas turbine selection steps above, the basic data input and calculation results are summarized in Table 14.2.

From Table 14.2, Both GT 6F.01 and LM 6000 PF have the shortest payback period; with further NPV evaluation, the LM 6000 PF has maximum accumulated NPV, aero-derivative gas turbines LM 6000 PF is the best suitable gas turbines for this application, therefore, it is recommended; that is: Set N + 1 operation model, normal 3 LM 6000 PF gas turbines running with about 30% margin, one gas turbine is stand by.

14.3 Combined Gas Turbine Cycle and Its Application

14.3.1 Combined Gas Turbine Cycle

A typical simple-cycle gas turbine (ref. to Fig. 14.3) has one engine—A gas turbine, and it might have heat recovery for process use, the power produced efficiency is relatively low, 30–40% of the fuel input into shaft output.

A typical combine cycle [3] is generally defined as two or more engines working in tandem, 40–50% of efficiency could be achieved, it includes that installing the heat-recovery steam generators (HRSGs) in the exhaust gas duct, using backpressure steam turbines to drive the compressors or pumps, the steam turbine generator, providing process heating, or a combination thereof.

A typical combined gas turbine cycle is shown in Fig. 14.5.

From Fig. 14.5, the combined cycle gas turbine has a high energy utilization efficiency of the fuel input to the gas turbine, the efficiency could reach 50–60%.

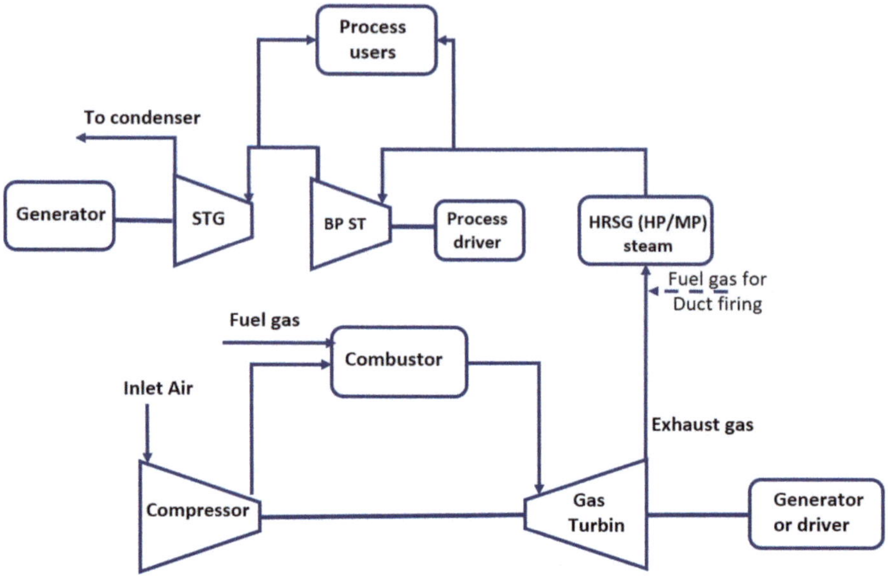

Fig. 14.5 A typical combined cycle gas turbine

The exhaust flue gas from a gas turbine has a higher temperature and high excess air, the exhaust gas heat can be recovered to generate an MP or HP steam, the excess air in the exhaust gas is good for using in duct firing for steam balance or superheating the HP steam as needed, duct firing in HRSGs provides quality superheated steam and can ramp-up to full capacity in case of need.

The steam generated from HRSGs can be used for driving the compressor/pumps as needed using a backpressure steam turbine or being directed to a steam turbine generator for power generation, it also provides steam to process heating and BFW deaerator.

14.3.2 Application of LNG Plant

Take a typical two-train LNG plant as an example, and illustrate the combination of gas turbines, steam turbines, boilers, and compressors, this plant has a nominal capacity of 6 MTPA for each train. It consists of:

- Mercury Removal Units, remove mercury from feed gas to 5 ng/Nm^3.
- Acid Gas Removal Units using amine solvent, the treated gas meets LNG feed gas specification of CO_2, H_2S, and COS.
- Dehydration Units remove water from feed gas to <1 ppmv
- LNG/Refrigeration Units using a licensed process with two refrigerants of propane and Mixed Refrigerant.

- The Refrigerant Compressors are driven by Frame 7 Gas Turbines, HRSG followed to recover exhaust gas heat.
- Common LNG storage and Jetty Loading.
- Common utility and general facilities, Common power generation to provide all the power needed.

The power and steam system can be highly integrated, the compressor power and steam system consist of:

- Gas Turbine driven system, provides main power to MR compressors;
- Followed with heat recovery steam generators (HRSGs), package boilers will provide any deficit steam if needed;
- Steam helper turbines (backpressure steam turbines) provide the balanced power demand for MR compressors;
- Back pressure steam turbines use the high-pressure steam generated from GT-HRSGs to provide the required power to LNG BOG (boil off-gas) compressors.
- LP steam from Back pressure steam turbines is used for power generation via STG (steam turbine generator), this STGs are LP steam condensing turbine generators via exhaust steam condenser.
- Steam/Condensate/BFW distribution and collection system.

14.3.2.1 Process Sketch and Process Description

The gas turbines, steam turbines, HRSGs, boilers, and compressors Integration for a typical LNG train are shown in Fig. 14.6.

Compressor driven system

There are 2 Refrigerant compression strings for each train, each of them requires about 95–100 MW of Power; Refrigerant compressors were co-driven by Frame 7 gas turbines and HP steam back pressure turbines.

Frame 7 Gas turbines provide about 75–85% of power demand; Steam helper turbines are driven by HP steam turbines and provide the remaining power needed for MR Compression trains. The boil-off gas (BOG) compressor is driven by the HP steam turbine;

Back pressure LP steam from HP steam turbines is supplied to condensing steam turbine generators (STGs) to generate power, and supply to the plant and utility system;

Steam system

There are 2 HRSGs and 1 Package boiler to generate HP steam (69 bar, 525 C) per train; The package boiler also supplies steam for startup and maintaining the plant steam balance; HP (high pressure) steam generated from HRSGs to recover exhaust gas heat from gas turbines; the exhaust gas from gas turbines will have a temperature of around 535 °C.

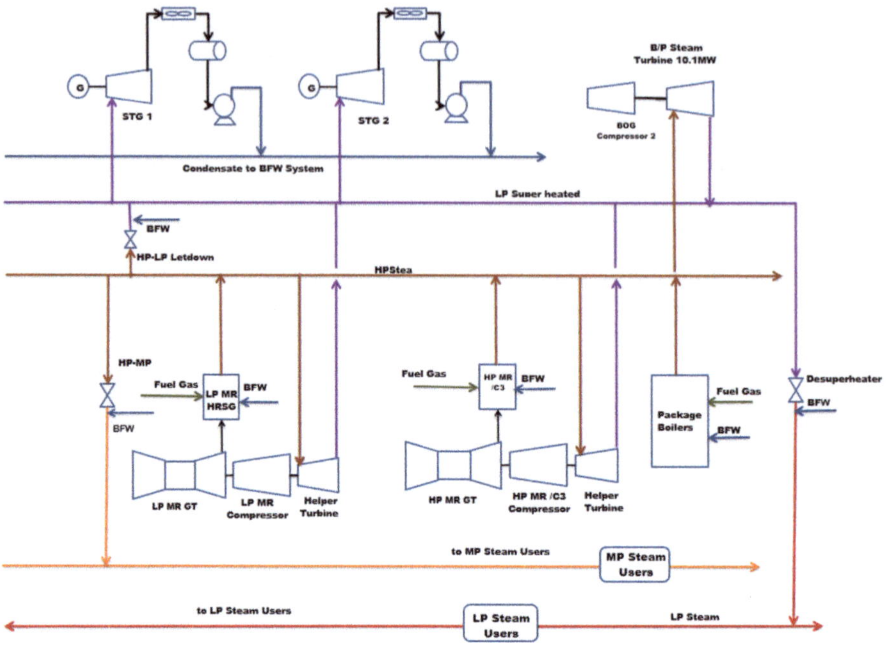

Fig. 14.6 A typical GT-HRSGs+STGs integration in an LNG plant

HP steam generated from HRSGs and package boilers will be used in driving Helper turbines and BOG steam turbines; according to steam balance, a small part of HP steam will be let down to MP steam for process heating; Partially LP steam (3–5 bar) from backpressure steam turbines will be used in process heating and deaeration steam.

Duct firing

Exhaust flue gas from gas turbines contains about 15% oxygen content, the exhaust gas temperature from gas turbines is about 520–540 °C, this exhaust gas temperature is not high enough to generate a 525 °C of superheated HP steam, so the minimum duct firing in HRSGs is needed to provide quality superheated steam and capability to ramp-up to full capacity in case if need; with duct firing, we can fully use the excess air in the exhaust gas for additional firing without installing supplemental air blower.

Increase duct firing duty to generate more HP steam, which changes pinch position and is more efficient. Duct firing duty can be used for steam generation at near 100% of efficiency, and can be ramped up as needed rapidly; with duct firing, HRSG capacity can be increased to minimize the number of package boilers for the entire LNG plant, all 2 HRSGs can supply all the steam needed at normal operation.

Power generation

All the power demand except the compressor power will be supplied by steam turbine generation (STGs), all the unbalanced LP steam will be directed to steam turbine power generation to generate the power needed.

14.3.2.2 Process Simulation

A simulation model has been built for this showcase of GT-HRSG+STG integration of the LNG plant, the simulation sheet is shown in Fig. 14.7.

GE 7E-03 heavy-duty turbines are selected to drive the refrigerant compressors, providing about 75–85% of the power required; HP steam drives backpressure steam turbine starters/helpers to provide the remaining power demand of refrigeration compressors.

GE 7E-03 gas turbine has an ISO heat rate is 10664 kJ/kWh per GE 2021–2022 gas power catalog. The simulation uses the derated heat rate, the derated heat rate can be adjusted by the exhaust gas temperature and efficiency of air compressors and gas turbines.

Full heat recovery from GT exhaust flue gas; HP (high pressure) steam generated from HRSGs to recover exhaust gas heat; duct firing in HRSGs has been designed to generate superheated HP steam and utilize excess air in the gas turbine exhaust gas.

Gas turbines, helper steam turbines, and steam turbine generators (STGs) work in tandem from the same source of heat.

HP steam (525 °C, 69 bar) will mainly be used for backpressure steam turbines including MR compressor steam turbine helpers, BOG compressor steam turbine, the backpressure of LP steam is 5 bar; about 6–8% HP steam will be a letdown to meet the steam balance of all levels of the steam networks.

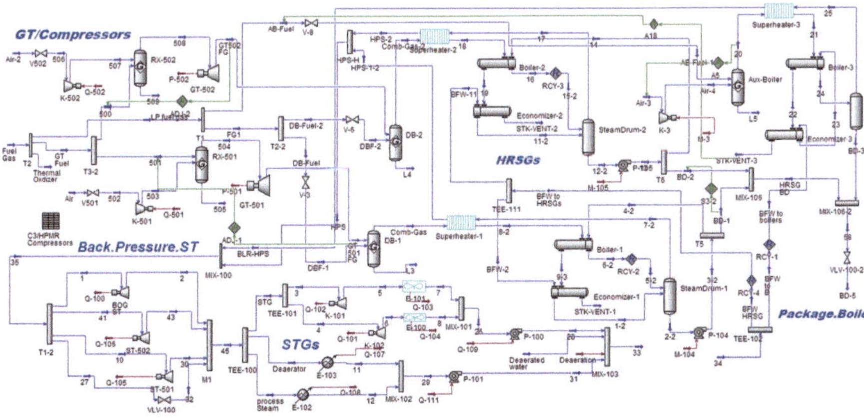

Fig. 14.7 GTs, HRSGs, STGs integration simulation sheet

Process steam used in process heating and the reboilers accounts for 25% of total LP steam amount, the LP steam used for BFW deaeration is about 5%, the reminding LP steam will be directed to the LP steam turbine generator for power generation.

All the flash steam from boilers blowdown water and is directed to the deaerator.

14.3.2.3 Simulation Results and Discussion

Per the simulation results, the GT-ST-HRSG integration system efficiencies are summarized in Table 14.3.

Table 14.3 GT-ST-HRSG integration efficiencies summary

Description		Units	Gas turbine 1	Gas turbine 2
Gas turbine	Selected turbine		GE 7E.03	GE 7E.03
	ISO heat rate	kJ/kWh	10,664	10,664
	ISO efficiency	%	33.76	33.76
	Gas turbine power	MW	201.3	198
	Gas turbine compressor	MW	123.2	121.2
	Fuel LHV	kJ/kg	45,570	45,570
	GT fuel	kg/h	19,886	19,559
	GT fuel heat	kJ/h	906,205,020	891,303,630
	Derated heat rate	kJ/kWh	11,800	11,420
	Derated efficiency	%	30.51	31.52
Supply fuel	GT fuel heat	kJ/h	906,205,020	891,303,630
	HRSG fuel heat	kJ/h	38,287,914	38,287,914
	Package boiler fuel heat	kJ/h	145,960,710	
	Sub total	kJ/h	2,020,045,188	
Power generation	GT net power	MW	78.1	76.8
	Steam turbine helper	MW	19.97	16.99
	BOG steam turbine	MW	10.5	
	STG1	MW	14.28	
	STG2	MW	14.28	
	Subtotal	MW	230.92	
Process heat	Process heat	MW	54.25	
	Deaerator heat	MW	10.85	
	Subtotal	MW	65.1	
Evaluation results	Power efficiency	%	41.15	
	Energy efficiency	%	52.75	

From Table 14.3, the simple gas turbine ISO efficiency is about 33.8%, with considering a derating, the actual efficiency is about 31%. With this type of integration, the total gas turbine and steam turbine power generation efficiency reach 41.2%, and overall energy efficiency is 52.8%.

Please note, for this integration of using GE 7E-03 heavy-duty gas turbine, the ISO efficiency is only 33.8%, some latest heavy-duty gas turbine models and aero-derivative gas turbines have much higher efficiency, if using these highly efficient heavy-duty turbines or aero-derivative gas turbine replace this GE 7E-03 heavy-duty gas turbine, Additional efficiency increase of up to 10% can be achieved.

The performance of HRSGs and package boiler are summarized in Table 14.4.

From Table 14.4, the HRSGs have an efficiency of 82.52% at 220 °C flue gas temperature, the duct firing is about 2%, it mainly provides the capability of superheating the steam to 525 °C.

The package boiler has an efficiency of 91%, this package boiler provides the capability of steam system balance, in case of steam deficit, it generates HP steam to make up the steam deficit.

Maximizing up the HRSG capacity by adjusting the duct firing rate will be an attractive option, it will save the package boiler investment, it is recommended to have further investigation to optimize the steam generation system further.

Instead of using gas turbine exhaust gas to generate steam, a gas turbine also can be integrated into a process heat furnace for power and heat supply system integration, its energy-saving principle is the same, it will not be discussed here.

14.4 Flue Gas Turbine Energy Saving of Fluid Catalytic Cracking Unit

Since the catalytic cracking flue gas energy recovery unit was put into use in Canada in 1962 [4], flue gas turbines have been widely used in the catalytic cracking unit of oil refineries. At present, the total number of gas turbines in the world is over 160 units [5]. China had a rapid development period of FCCU flue gas turbines, the first flue gas turbine was put into use in the Fushun Second Petroleum Plant in November 1978, due to the low price of domestically-produced flue gas expander units, various plants have followed and installed. At the end of 1995, the total number of flue gas turbines reached 66, which are almost in all large and medium-sized catalytic cracking plants. The speed of its development is much higher than that of other countries. Flue gas turbines can play an important role in the energy saving of the plant and the entire oil refinery. However, the investment in flue gas turbine units is relatively high. The relationship between investment and energy-saving benefits must be considered, and the technical and economic effects must be taken as an important basis for project decision-making.

Table 14.4 HRSGs and package boiler performance

Description		Units	Gas turbine 1	Gas turbine 2	Notes
HRSG performance	HRSG fuel	kg/h	840	840	
	HRSG fuel heat	kJ/h	38,287,914	38,287,914	
	Exhaust gas temperature	°C	535	535	
	Exhaust gas rate	kg/h	1,086,882	1,069,556	
	Exhaust gas heat capacity	kJ/kg.°C	1.17	1.17	
	Exhaust gas heat	kJ/h	477,000,000	469,200,000	
	HRSG total heat	kJ/h	515,287,914	507,487,914	
	Superheater	kJ/h	99,630,000	97,470,000	
	Boiler coil	kJ/h	223,600,000	219,300,000	
	Economizer	kJ/h	102,000,000	102,000,000	
	Flue gas temperature	°C	220	220	
	HRSG steam rate	kg/h	142,869	139,770	525°C, 69 bar
	HRSG steam heat	kJ/h	425,230,000	418,770,000	
	HRSG efficiency	%	82.52	82.52	
Package boiler performance	Package boiler fuel	kg/h	3,203		
	Package boiler fuel heat	kJ/h	145,960,710		
	Superheater	kJ/h	31,060,000		
	Boiler coil	kJ/h	85,700,000		
	Economizer	kJ/h	16,150,000		
	Flue gas temperature	°C	203		
	Available heat	kJ/h	132,910,000		
	Steam rate	kg/h	44,550		
	HRSG efficiency	%	91.06		
	HP to LP let down	kg/h	20,250		

14.4.1 Energy-Saving Principle and Expansion Work Estimate

14.4.1.1 Energy Saving Principle

The energy-saving principle of a catalytic cracking flue gas turbine is similar to that of a gas turbine, including an air compressor (main air compressor), a combustion chamber (regenerator), and an expansion turbine (gas turbine). The difference is that the coke burning in the catalytic cracking unit is the aimed process, to recover the catalyst's activity, what is burned in the combustion chamber (regenerator) is not the usual fuel, but the coke attached to the catalyst; and also, the air compressor used is providing the regeneration air. In addition, because the regenerated flue gas is the "waste" of the oil refining process, it often carries catalyst dust, which has a certain impact on the energy recovery and investment of the flue gas turbine.

The power recovery of the flue gas turbine is not only related to the pressure ratio of the flue gas turbine but also related to the temperature of the flue gas. The main air compression is carried out at low temperatures. Under the same pressure ratio, the compressor consumes less power than the flue gas turbine expansion. In the expansion work of the flue gas, not only the pressure energy of the flue gas is recovered, but also part of the enthalpy of the flue gas is converted into work. In the actual process, the efficiency of the flue gas turbine, and pressure drop of equipment and system are considered. The design work recovery rate (the ratio of the expansion work of the gas turbine to the power consumption of the main air compressor) is often between 0.7 and 1.4.

14.4.1.2 Expansion Work Estimate

The expansion work of the flue gas turbine can be calculated by the following equation:

$$N = P_1 V_1 \frac{k}{k-1}\left[1 - \left(\frac{P_2}{P_1}\right)^{\frac{k-1}{k}}\right]\eta \tag{14.15}$$

Accord g to $\left(\frac{P_2}{P_1}\right)^{\frac{k-1}{k}} = \frac{T_2}{T_1}$, $PV = nRT$, a simple equation for calculating expansion work can be derived, and the calculation results are the same.

$$N = nRT_m ln\left(\frac{P_1}{P_2}\right) \tag{14.16}$$

where

T_m—is the logarithmic mean temperature of the flue gas turbine inlet and outlet, K;

$$T_m = (T_2 - T_1)/ln\left(\frac{T_2}{T_1}\right) \tag{14.17}$$

where

N—power output of the flue gas turbine, kW;

η—flue gas turbine efficiency;

n—amount of flue gas entering the flue gas turbine, kmol/s;

R—gas constant, 8.314 J/(mol K);

k—adiabatic expansion index;

V_1—flue gas flow at the inlet of flue gas turbine, m^3/s;

P_1—the flue gas inlet pressure, kPa;

P_2—The outlet pressure of the flue gas turbine, kPa;

T_1, T_2—the temperature of the inlet and outlet of the flue gas turbine, K;

T'_m—the logarithmic average temperature of flue gas turbine inlet and outlet temperature under low load, K.

14.4.1.3 Load Ratio Impact on Expansion Work

High-efficiency flue gas turbine units installed may not be in high-efficiency operation in actual operation. There are some high efficiencies in single-set units, the actual work (generation) is even less than 50% of the design value in the operation of the plant. The main reason is that the flue gas turbine is operating at a low load. The reasons for the low-load operation of flue gas turbines generally include:

(1) Considering the development of the enterprise and the process unit, the unit's actual production and processing volume are lower than the designed processing volume, that is, the flue gas volume change caused by the unit running at a low load.
(2) The actual coke production rate is different from the designed coke production rate, which is generally manifested as the nature of the processed raw materials deviates from the designed raw material properties.
(3) The operating parameters, such as flue gas carbon monoxide content, excess oxygen concentration, etc., all affect the flue gas volume changes, resulting in flue gas turbine load rate fluctuations.
(4) The choice of flue gas turbine is too large, resulting in low-load operation of flue gas turbine.
(5) When the bypass double-acting slide valve is used to control the pressure of the reactor and regenerator, the flue gas volume often cannot fully pass through the flue gas turbine system. In some FCCUs, the amount of flue gas running loss reached 20–30%, and the inlet pressure of the butterfly valve was reduced by 0.05–0.06 MPa [6].

For the flue gas turbine under the design load, the cross-sectional area of flow between the flue gas turbine blades is constant. When the flue gas turbine and the AC motor (generator) are on the same shaft system, the speed of the flue gas turbine is restricted by the frequency of the AC grid during operation, and it runs at a constant speed. When the regeneration flue gas flow decreases, the flue gas turbine is objectively required the inlet butterfly valve to be more closed with increased throttling, and the inlet pressure is reduced to increase the volume flow, to ensure that the operating pressure of the regenerator is not fluctuated [6]. At this time, the power recovery rate of the flue gas turbine is significantly reduced, and the output power of the flue gas turbine drops sharply along with the flue gas turbine characteristic curve. For example, when the flue gas volume of a plant is 1800 m^3/min, the recovery power of the flue gas turbine is 5.91 MW, and when it is reduced to 1100 m^3/min, the recovery power is only 1.99 MW.

According to the characteristics of the flue gas turbine, the cross-sectional area of the flue gas flow through the flue gas turbine is constant. When the load rate decreases, to keep the volume flow rate of the flue gas inlet constant, the flue gas inlet pressure must be reduced. Regarding the flue gas as an ideal gas, there are: when the load rate decreases and the volume flow rate remains unchanged, the flue gas inlet pressure drops to

$$P_1' = P_1 L \tag{14.18}$$

And the expansion work under low load is:

$$N\prime = nRT_m'\eta\prime \ln\left(\frac{P_1'}{P_2}\right) \tag{14.19}$$

Suppose the expansion work under full load is:

$$N = nRT_m\eta \ln\left(\frac{P_1}{P_2}\right)$$

The ratio of expansion work under low load to full load expansion work is:

$$\varphi = N\prime/N = \frac{n\prime RT_m'\eta\prime \ln\left(\frac{P_1'}{P_2}\right)}{nRT_m\eta \ln\left(\frac{P_1}{P_2}\right)} \tag{14.20}$$

Before and after the load is reduced, the flue gas pressure at the outlet of the flue gas turbine is generally unchanged, and the load rate: $L = n\prime/n$, then Eq. (14.20) is:

$$\varphi = L \times \frac{RT_m'\eta\prime \ln\left(\frac{P_1'}{P_2}\right)}{nRT_m\eta \ln\left(\frac{P_1}{P_2}\right)} \tag{14.21}$$

Therefore, when the load rate deviates from the design load, the amount of work done is:

$$N' = N\varphi \tag{14.22}$$

In equation s (14.18)–(14.22).

N'—power output of the flue gas turbine after the load rate change, kW;

η'—flue gas turbine efficiency after the load rate change;

n'—amount of flue gas entering the flue gas turbine after the load rate changes, kmol/s;

L—load rate of the flue gas turbine;

φ—the change in the ratio of the work done by the flue gas before and after the load rate changes;

P_1'—the flue gas inlet pressure after the flue gas turbine load rate changes, kPa;

T_m'—the logarithmic average temperature of flue gas turbine inlet and outlet temperature under low load, K.

14.4.2 Energy-Saving Benefit Estimation

The use of flue gas turbines to recover kinetic energy is an important energy-saving technology. Correctly estimating the energy-saving amount and value is important for project decision-making. The energy-saving of a flue gas turbine is mainly reflected in the power saving benefit of the reduction of power consumption of the air compressor (or more generated electricity). At the same time, it should be noted that after the flue gas turbine is used, the temperature of the regenerated flue gas decreases, and the heat enthalpy decreases, so that the amount of heat recovery is reduced. In addition, there is water, electricity, and steam consumed by the flue gas turbine system.

In the ideal adiabatic expansion process, the expansion work is equal to the enthalpy drop. The enthalpy reduction of the flue gas after passing through the flue gas turbine can be determined by the expansion work. The work efficiency loss of the flue gas turbine is mainly heat dissipation and flue gas leakage. Turbine expansion work (considering the heat dissipation of equipment and pipelines) estimates the flue gas enthalpy drop and then estimates the steam reduction. For the commissioned flue gas turbine, the steam reduction can be directly estimated from the flue gas temperature drop, but the actual turbine intake of flue gas volume changes must be considered. The flue gas turbine itself consumes water, electricity, and steam; consumed steam is mainly used for flue gas turbine wheel, gas seal injection steam, and cooling steam injected to inlet pipeline to ensure that the inlet temperature of the flue gas turbine is within specification, but does not include steam injected to prevent secondary combustion and safety operation. In addition, some production facilities were modified to set up a flue gas turbine to recover kinetic energy, which increased the regeneration operating pressure, and the power consumption of the air compressor increased correspondingly, which should be deducted from the energy-saving calculation. Of course, after increasing the regeneration pressure, since the temperature of the flue gas turbine is much higher than the temperature of the air compressor, the extra work recovered by the flue gas turbine expansion is much greater than the extra work consumed by the air compressor. The increased energy consumption is listed in Eq. (14.23) item 4.

The calculation of energy-saving benefits after adding a flue gas turbine system can be calculated by Eq. (14.23)

$$M = \left[E_N C_E - \frac{E_N C_S}{3.6(H_S - H_0)} - \left(E_S C_S' + E_E C_E + E_W C_W\right) - \Delta N C_E\right] \times H \times 10^{-4} \quad (14.23)$$

In the equation:

M—annually energy-saving benefit, $MM/a,

E_N—Flue gas turbine expansion work, kW;

E_S—Steam consumption of flue gas turbine system, t/h;

E_E—Electricity consumption of flue gas turbine system, kW;

E_W—water consumption of flue gas turbine system, t/h;

H_S, H_0—are steam enthalpy and feed water enthalpy, kJ/kg;

C_S'—1.0 MPa steam price, $/t;

C_E, C_S, C_W—are electricity, $/kWh, steam and cooling water prices, $/t;

ΔN—power increases with main air pressure increase, kW;

H—annual operating hours, h.

The energy savings of the flue gas turbine system are:

$$\Delta E = (E_N - E_E - \Delta N)\varphi_E + \frac{E_N \varphi_s}{3.6(H_S - H_0)} - E_S \varphi_S' - E_W \varphi_W \quad (14.24)$$

where

φ_E—Electric energy conversion coefficient, MJ/kWh;

φ_s—3.5 MPa steam energy conversion coefficient, MJ/t;

φ_S'—1.0 MPa steam energy conversion coefficient, MJ/t;

φ_W—the cooling water energy conversion coefficient, MJ/t;

ΔE—integrated energy saving of flue gas turbine system, M]/h.

Reduced value of energy consumption per unit of processing capacity:

$$\Delta E_C = \Delta E / G \tag{14.25}$$

where

G—the raw material processing capacity of the device, t/h;

ΔE_C—is the reduction value of energy consumption per unit of processing capacity, MJ/t.

Depending on the project location, Car tax reduction may be applied according to national regulations on the carbon tax.

14.4.3 Technical and Economic Evaluation

In the current market economy environment, an energy-saving project mainly depends on whether it is economically feasible and whether it can create economic benefits for the enterprise under the premise of satisfying technical and engineering feasibility. Investment in the flue gas energy recovery system has been on the rise in recent years. In addition to the continuous development of the automatic control system, the price of raw materials and price factors are also very important aspects. Therefore, when using flue gas turbines to recover flue gas kinetic energy projects, it is more important to conduct technical and economic analysis and evaluation of flue gas turbine systems.

The investment in the flue gas turbine system usually includes the flue gas turbine, the third-stage cyclone separator, high-temperature control valves, high-temperature pipelines, and supporting electrical equipment and instrument control system. In addition to the one-time investment, we must also consider the maintenance cost, the replacement cost of spare parts including blades, unforeseen costs, and proprietary technology costs. After the flue gas turbine recovers the power, the waste heat boiler investment is slightly reduced, and the impact is not significant. When the one-time investment of the flue gas turbine is determined, the service life of the flue gas turbine becomes an important factor; It is generally 10–15 years.

According to the service life, the total investment of the flue gas turbine can be converted into the annual investment for technical and economic evaluation. There are two types of technical and economic evaluations: dynamic and static. The static evaluation does not consider the time value of funds, and only calculates the annual converted investment amount and the annual energy-saving benefits, which deviates from the objective economic law. Most of the capital for infrastructure projects was bank loans, and there was time value for funds, so dynamic evaluation methods should be used.

According to the cash flow method, the net cash flow value (cash inflow minus cash outflow) in different years is multiplied by the present value coefficient, so that the year when the accumulated discounted cash flow is zero is the dynamic investment payback period (ref. to Eq. 11.18 in Chap. 11).

For stable operation conditions, the power generation and consumption of the flue gas turbine system remain stable, that is, the annual energy-saving benefits remain unchanged, and the flue gas turbines are mostly one-time investments (for the renovation project, it is an increase in investment and new benefits), which can be derived the following simplified empirical equation for finding the payback period of investment:

$$t = \frac{-\ln(1 - I.i/A)}{\ln(1 + i)} \tag{14.26}$$

From the above equation, as long as the total investment I, the annual energy-saving benefit A and the capital discount rate i(annual interest rate) are known, the dynamic investment payback period can be easily calculated, and in many cases, this simplified method can be used. This equation can still be used when it is not a one-time investment but can be discounted as a one-time investment, and the annual benefit can be an average annual benefit.

Therefore, when evaluating flue gas turbine projects, a dynamic payback period should be adopted. Flue gas turbine is a major energy-saving project, and a benchmark investment payback period should be determined by the higher authority department as the basis for project decision-making. The flue gas turbine system should not be blindly installed on the ground, and a feasibility demonstration should be carried out.

14.4.4 *Ways to Improve the Power Recovery Rate of Flue Gas Turbine*

The expansion work of the flue gas turbine is a function of flue gas temperature, pressure, and flue gas flow rate. Changing these three parameters has a great influence on the output power of the flue gas turbine. The following only proposes methods and approaches to improve the power recovery rate of flue gas turbines:

14.4.4.1 Increase the Flue Gas Temperature

Increasing the flue gas temperature at the inlet of the flue gas turbine can increase the expansion work, but due to the restriction of the flue gas turbine material, it should not exceed 730 °C. However, within a limited range, increasing the flue gas temperature can increase the power recovery rate. Generally, for every 10 °C in temperature increase, the output power increases by about 1%. At present, the inlet temperature of the domestic flue gas turbine design is 580–680 °C, most of which are operated between 600 and 650 °C, and the inlet flue gas temperature of the flue gas turbine in some refineries is as low as 580 °C. Therefore, from the process and system improvement, the flue gas turbine inlet temperature can be increased to 680–730 °C, which will undoubtedly increase the flue gas turbine power recovery rate. How to increase the flue gas temperature should start from the following aspects.

(1) Strengthen the insulation of pipelines and equipment. The operating temperature of the regenerator dilute phase is mostly above 650 °C. However, when entering the flue gas turbine through the second and third stage cyclone separators and flue gas pipes, the flue gas temperature drops by around 30 °C, and some as much as 50 °C even more. Such a large temperature drop in flue gas temperature not only reduces the power recovery rate but also increases the heat energy loss and reduces the amount of steam generation correspondingly. The improvement method is to use materials with good thermal insulation effect and low thermal conductivity in the construction of the flue gas duct lining design, such as non-tortoise shell mesh lining. When site and engineering conditions permit, appropriate external insulation may be considered. In this way, not only the power recovery rate can be increased, but also the steam output of the waste heat boiler can be increased accordingly. Particular attention should be paid to repairing the falling parts of the lining.
(2) Pay attention to reasonably reducing the amount of water spray and steam usage in the flue gas pipeline.
(3) Increasing the regeneration operating temperature is beneficial to increase the flue gas inlet temperature, but it is determined by the process operation.

14.4.4.2 Adjust the CO Combustion Process to Increase the Inlet Flue Gas Temperature

For the two-stage regeneration process and the single-stage incomplete regeneration process, the regeneration flue gas contains 3–10% of CO, and where the CO is burned has a greater impact on the inlet temperature of the flue gas turbine and the flue gas volume of the flue gas turbine. Burning CO before the flue gas turbine, the flue gas flow rate increases, and the flue gas temperature increases. Generally, the pressure drop of the CO incinerator is small, and the impact on the power is ignorable. However, the concern is that after the CO is completely burned, the flue gas temperature reaches 900–1200 °C, and the material of the flue gas turbine is unbearable, so the flue gas heat removal facility should be installed to reduce the flue gas temperature to 700–730 °C.

14.4.4.3 Increase the Flue Gas Flow

(1) When the CO-containing flue gas is combusted under positive pressure in the two-stage regeneration process or the single-stage incomplete regeneration process, the CO combustion flue gas and the second regenerator flue gas of the two-stage regeneration process can be sent to the flue gas turbine to increase the flue gas flow, thereby increasing the expansion work of the flue gas turbine.

(2) Increase the flue gas load rate of the flue gas turbine. Make the amount of regeneration flue gas well matched with the flue gas turbine, and do not leave a big design margin for flue gas turbines selection. Otherwise, part of the flue gas energy will be lost to the throttling of the inlet valve, and the actual work will be less than the small flue gas turbine, which reduces the energy-saving effect of the actual operation of the flue gas turbine, and the economy becomes worse. The basic principle of selecting a flue gas turbine is: fully consider the actual factors of the process and production, and select the long-term stable operation flue gas volume as the basis for selecting and designing the flue gas turbine; under the determined actual flue gas load, full consideration of leakage and the flue gas emitted by regulating pressure. In the process operation, it is advisable to adopt two-step control to reduce the amount of flue gas leakage and increase the actual operating load rate of the flue gas turbine.

14.4.4.4 Increase Regeneration Operating Pressure

The increase of the regeneration pressure is restricted by the operating conditions of the Reactor- Regenerator system and even the fractionation system. From the energy point of view, increasing the regeneration pressure is conducive to the increase of the flue gas turbine's work capacity. The pressure of the fractionation tower is also increased correspondingly, which can reduce the compression ratio of the off-gas compressor, thereby reducing the power consumption of the gas compressor, which is conducive to energy saving. Secondly, with the same amount of coke burning rate, the flue gas linear velocity of the equipment is reduced, and the processing capacity of the plant can be increased. In addition, after increasing the regeneration pressure, the oxygen partial pressure increases, which is also conducive to the coke burning of the regenerator. However, increasing the operating pressure also brings some unfavorable factors, mainly due to the increase in the partial pressure of the oil and gas in the reaction system and the increase in the coke yield. The distillate catalytic cracking unit can be compensated by injecting dry gas or increasing the amount of lifting steam.

14.5 FCCU Regenerating Gas CO Combustion Outside of Regenerator

14.5.1 Background

The catalytic cracking regeneration process has two forms: complete regeneration and incomplete regeneration. For the incomplete regeneration process, the flue gas contains 3–10% carbon monoxide, and its recovery and utilization are important issues for energy conservation and environmental protection. For the complete regeneration process, due to the limitation of the heat balance and the operating gas velocity of the regeneration equipment, it is often necessary to modify the regeneration system and set up internal and external heat extraction facilities; in addition, the heavy metal content in the feedstock of the heavy oil catalytic cracking is high, and the production operation causes the catalyst to deactivate and needs to be unloaded. With a certain amount of promoted catalyst, the loss of platinum metal combustion promoter will increase accordingly. Many refineries set up carbon monoxide boilers to recover the CO chemical energy and sensible heat of flue gas. However, due to various reasons, the afterburning of carbon monoxide boilers is relatively large, and the amount of introduced CO flue gas in actual operation is not much. Therefore, the development of carbon monoxide combustion and utilization technology outside the regenerator provides a new method for the utilization of excess coke energy, which has important and practical significance.

(1) Incomplete regeneration is used to introduce carbon monoxide to the outside of the regenerator for combustion, which increases the coke burning capacity by 20–30% compared with the complete regeneration method, and alleviates the contradiction between the insufficient capacity of the main air system and the excessively high linear velocity of the regenerator.
(2) It satisfies the heat balance of the reactor- regenerator, does not need to install the heat removal facilities inside (outside) the regenerator, simplifies the equipment structure, and facilitates operation.
(3) In the case of heavy metal pollution in residue catalytic cracking and high unit consumption of catalysts, the use of regeneration flue gas carbon monoxide external combustion technology can avoid the consumption of platinum combustion promoters and save costs.
(4) Compared with the existing carbon monoxide boiler option, the regeneration flue gas carbon monoxide external combustion technology changes the carbon monoxide combustion conditions (the CO boiler is burning and heat transfer at the same time), and it can use very little supplementary fuel to ensure stable carbon monoxide combustion and protect the environment.

14.5.2 Laboratory Research Results

According to the research results of the carbon monoxide combustion of the regenerated flue gas in the laboratory, it is found that:

(1) Regenerative flue gas carbon monoxide combustion has two stages: ignition and deflagration; under laboratory conditions, the ignition temperature is about 639 °C, and the deflagration temperature is 700–710 °C. When it reaches above the deflagration temperature, the combustion process can be quickly completed within a residence time of 0.01 s to burn out CO.
(2) Regression experimental data found that the reaction rate during the ignition phase had a 1.0-level to the oxygen concentration; the activation energy during

the ignition phase was 267.57 kJ/mol, and the apparent frequency factor was 2.72×10^{19} ml/(mol s).

The combustion rate equation in the ignition phase is:

$$\frac{d[CO]}{dt} = 2.72 \times 10^{19} \exp(-32183/T)[\mathrm{CO}][\mathrm{O_2}][\mathrm{H_2O}]^{0.5} \tag{14.27}$$

In the equation:

T—combustion flue gas temperature, K;

t—burning time, s;

〔CO〕—CO concentration in the reaction mixture, mol/ml;

〔O_2〕—O_2 concentration in the reaction mixture, mol/ml;

〔H_2O〕—H_2O concentration in the reaction mixture, mol/ml.

14.5.3 Flue Gas CO Duct Pre-Combustion

CO pre-combustion spans two combustion stages: ignition and deflagration. It is difficult to regress strictly to the ignition velocity equation. However, the FCCU has tens of meters of flue gas ducts, the residence time of the flue gas is between 2 and 3 s, which is long enough to complete the pre-combustion process, the author had carried out a regeneration flue gas duct pre-combustion test trial, and the test trial data shows that under the appropriate oxygen concentration and initial pre-combustion temperature; the pre-ignition process can be completed within 0.7–1.5 s. This pre-combustion time is smaller than the residence time of the flue gas in the flue gas duct, so it is completely feasible to complete the pre-combustion process using the existing flue gas duct.

The regeneration flue gas CO pre-combustion test trail schematic flow is shown in Fig. 14.8, The test trial was carried out on the flue gas duct with an inner diameter of 500 mm and a total duct length of about 50.8 m.

The exhaust flue gas of the catalytic cracking regenerator is injected with the industrial air on the flue gas duct about 9.0 m from the outlet of the secondary cyclone separator of the regenerator, and the flow rate was measured with a rotameter. At this time, the oxygen concentration in the flue gas containing carbon monoxide has increased, and the test trial was carried out by changing the air injection volume and the mixing temperature after the air injection. Under the different mixing temperatures and oxygen concentrations of flue gases, the temperature and flue gas composition changes in different measuring locations of the flue gas are observed.

Different ignition conditions, the required pre-ignition time is different, the test trial data are shown in Table 14.5.

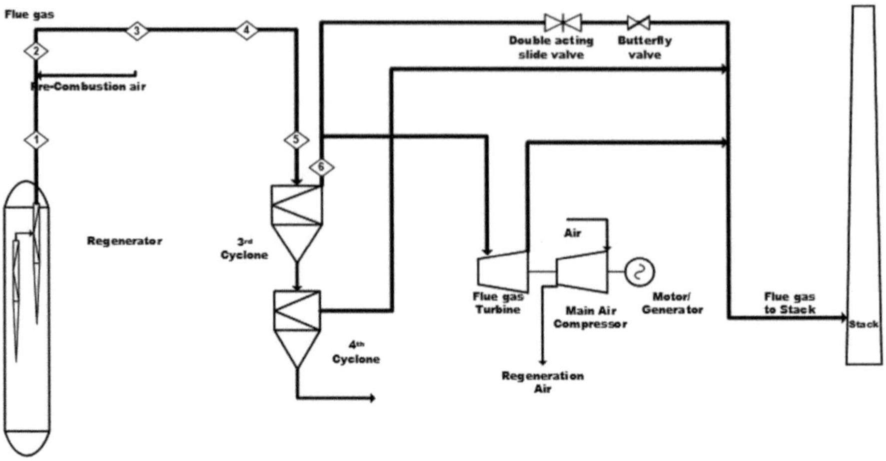

Fig. 14.8 Regeneration flue gas CO pre-combustion test trail schematic flow. Measuring location 1—second cyclone outlet; 2—immediately downstream of air injection; 3—middle of flue gas duct I; 4—middle of flue gas duct II; 5—3rd cyclone inlet; 6—3rd cyclone outlet

Table 14.5 Residence time and pre-ignition temperature datasheet

Description	I		II		III	
	Temperature (°C)	Residence time (s)	Temperature (°C)	Residence time (s)	Temperature (°C)	Residence time (s)
After air injection	630	0	634	0	675	0
Mid of duct I	638	1.39	672	1.32	754	1.28
Mid of duct II	664	2.11	796	2.02	761	1.93
3rd cyclone inlet °C	695	3.2				
Initial concentration CO (%)		4.4		3.8		3.6
Initial concentration 0_2 (%)		1.4		2.6		1.4

As can be seen from Table 14.5, under different ignition conditions, the time required to complete the pre-combustion process is also different, and in data III, at the ignition temperature of 675 °C, it takes only 1–1.3 s to complete the pre-combustion process. In the II with higher O_2 concentration, pre-combustion from 672 to 796 °C requires only a residence time of 0.7 s. After deducting the heat dissipation temperature drop of the flue gas duct, the 2nd regenerative flue gas mixture is also between 670 and 685 °C, so it is safe to complete the pre-combustion process within 1.5 s.

In general, the pre-combustion time may be estimated by Eq. (14.28).

$$t = C\int_0^x \frac{dx}{(1-x)^2 \exp\frac{-32183}{T^0+81.3[CO]_0 x}} \tag{14.28}$$

where

C—constant; $C = \frac{8.64\times10^{-19}\left(\frac{T^0}{P}\right)^{1.5}}{[O_2]_0[H_2O]_0^{0.5}}$.

x—Carbon monoxide pre-combustion conversion (determined by the required pre-combustion end temperature);

t—pre-combustion time, s;

T^0—the initial combustion temperature of the mixed flue gas, K;

P—flue gas pressure, kg/cm^2 (a);

$[O_2]_0$—initial concentration of oxygen in mixed flue gases, volume fraction (wet);

$[H_2O]_0$—initial concentration of water vapor in mixed flue gases, volume fraction (wet);

14.5.4 Technical Process Flow of CO Combustion Outside Regenerator [8]

An oil refinery's heavy oil catalytic cracking unit adopts the external combustion technology of regenerated flue gas, and the process is shown in Fig. 14.9.

The FCCU 1st regeneration flue gas contains 5–7% carbon monoxide. It is discharged from the 1st regenerator to the third stage cyclone, separated from the catalyst dust, and then enters the flue gas expansion turbine. After the pressure energy of the flue gas is recovered, it enters the carbon monoxide incinerator through a three-way butterfly valve and a water seal. The containing CO flue gas mixes with the combustion flue gas of the afterburning fuel gas of the CO incinerator to increase the temperature of the first stage regeneration flue gas to the ignition condition. The secondary air distribution supplies the required air volume for fuel combustion. The carbon monoxide is burned in the CO incinerator, and the temperature at the end of the CO incinerator reaches 900–1000 °C, the carbon monoxide content achieves 50–280 ppm depending on the fired fuel rate, then mixed with the 2nd regeneration flue gas (about 650 °C), the temperature is reduced to about 880–930 °C, then enters the waste heat boiler to generate steam, then the exhaust flue gas is discharged to stack at about 180 °C. When the carbon monoxide incinerator and the waste heater boiler are overhauled, the first and second stage regeneration flue gas is directly discharged into the flue gas stack through the three-way butterfly valves.

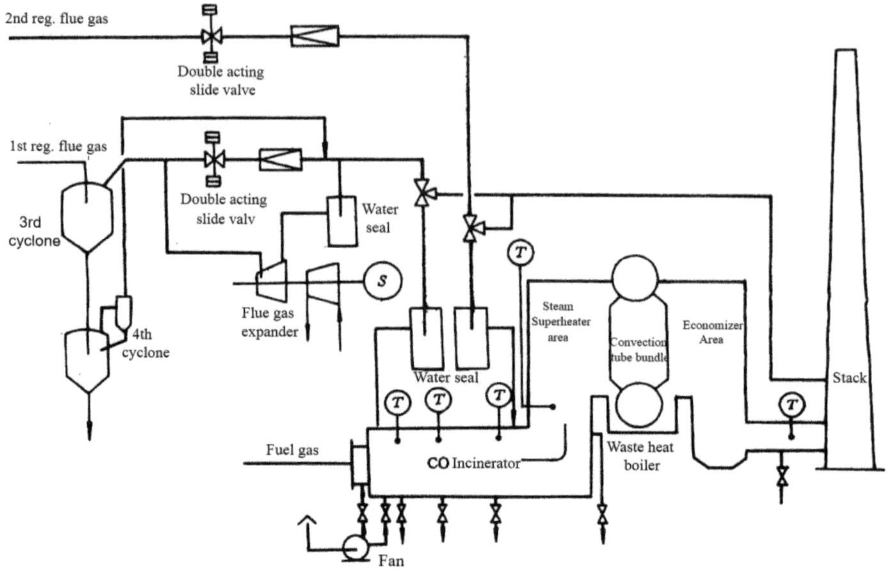

Fig. 14.9 Typical FCCU CO combustion flow scheme

14.5.5 Test Trial Results

The combustion status and waste heat recovery under different supplementary combustion fuel rates are shown in Table 14.6.

The supplementary fuel is refinery gas with a LHV of 31.4 MJ/Nm^3 (7500 kcal/m^3). According to the process shown in Table 14.6 and Fig. 14.9, when the temperature of the regenerated flue gas is reduced to 430–440 °C after the flue gas Turbine does work, only 310 Nm^3/h of supplementary fuel is required. After the supplementary combustion, the average flue gas temperature is about 600 °C (the temperature of the flame zone in the center of the furnace is higher than the ignition temperature), and complete combustion is achieved in the incinerator. The incinerator outlet temperature is above 900 °C, and the CO content in outlet gas is less than 300 ppm, which meets the environmental protection requirements at the test trial period. The combustion flue gas enters the waste heat boiler to generate medium-pressure steam and recover the chemical energy of carbon monoxide and the sensible heat energy of the flue gas. In case of requiring CO content less than 50 ppm, it is expected to increase the fuel gas to about 500 Nm^3/h.

According to the test trail results of the plant, when the production load rate of the plant is 80–90% and the 1st regeneration flue gas is fully fed into the CO incinerator, the total plant superheated steam volume of the waste heat boiler is 45–48 t/h. Except for the steam generated by the slurry, the flue gas waste heat boiler steam generated by the waste heat boiler is 37 t/h, and the net recovered flue gas sensible heat and carbon monoxide chemical energy are about 27 MW. Assumed 90% of the efficiency

Table 14.6 Minimum fuel demand and performance summary

Description	Fuel consumption (Nm^3/h)	230	310	453	715
Regenerative system operating conditions	Dense phase temperature of 1st regeneration flue gas (°C)	640.5	641.5	656.5	640.5
	Dilute phase temperature of 1st regeneration flue gas (°C)	615	616	617	620
	A regenerative flue gas outlet flue gas temperature (°C)	614	614	613	616
	Flue gas temperature at the outlet of 3rd cyclone	569	571	565	561
	Dense phase temperature of 2nd regeneration flue gas (°C)	707.5	707.5	708	709
	Dilute phase temperature of 2nd regeneration flue gas (°C)	805	808	787	798
	2nd regeneration flue gas outlet flue gas temperature (°C)	704	706	712	704
	1st regeneration flue gas volume (Nm^3/h)	45,980	47,080	45,430	46,420
	2nd regeneration flue gas volume (Nm^3/h)	21,726	24,276	30,600	21,216
CO incinerator	The inlet temperature of 1st regeneration flue gas (°C)	438	439	433	429
	The inlet temperature of 2nd regeneration flue gas (°C)	632	633	653	636
	Total air distribution (Nm^3/h)	11,852	14,176	16,213	15,629
	One time air distribution (Nm^3/h)	5040	5040	5256	5868
	The temperature in front of the incinerator (°C)	615	615	688	
	The temperature in the middle of the incinerator (°C)	702.3	752.5	843.6	715.5
	The temperature at the back of the incinerator (°C)	867	907	965	942
	Mixed flue gas (°C)	883	894	876	925
	Flue gas CO at the end of the incinerator (ppm)	280	264	57	112
Waste heat boiler operating data	Drum water supply (t/h)	31.5	37.5	36.5	37.7
	Steam output from the waste boiler (t/h)	30.7	33.7	36.3	35.2
	Steam generated from hot oil slurry (t/h)	11.6	11.3	9.9	10.7
	Total superheated steam (t/h)	42.25	45	46	45.8
	Steam pressure (MPa)	3.82	3.82	3.82	3.82

(continued)

Table 14.6 (continued)

Description	Fuel consumption (Nm^3/h)	230	310	453	715
	Deaeration water temperature <°C)	92	91	94	92
	Superheated steam temperature (°C)	441	441	441	441
	Exhaust flue gas temperature (°C)	175	175	177	173
	Heat recovery from waste boilers (kW)	28,374	30,585	31,363	31,333
	Waste boiler efficiency (%)	81.65	81.8	81.0	82.23

of the waste heat boiler, the annual fuel saving is 20,776.3 tons and the annual energy saving benefit is $MM1.4.

For the process where the regenerative flue gas first passes through the flue gas turbine to expand and do work, the flue gas temperature at the inlet of the CO incinerator is as low as 500 °C, and it needs to be burned fuel to increase the temperature to reach the ignition temperature. According to the on-site test trial results:

(1) There is a flame initiation phenomenon in the afterburning carbon monoxide incinerator. That is, there is a radial temperature gradient in the front of the incinerator, and the flame area in the center of the incinerator (about 800 °C) can ignite the surrounding flue gas and spread the combustion. Therefore, it is not necessary to make the average temperature of the flue gas reach the ignition temperature, and the fuel consumption can be reduced. A certain refinery has a calibrated minimum fuel consumption of 310 Nm^3/h (standard), and a supplementary combustion/heating ratio of 7–8%.
(2) The carbon monoxide incinerator has a large air distribution volume, which reduces the temperature of the flue gas, but increases the oxygen concentration; therefore, there is a suitable air distribution volume for carbon monoxide. Generally, based on the results of site test trial, the excess air coefficient should be maintained at 1.5–1.8.
(3) According to the site test trial results, the residence time of the carbon monoxide incinerator is between 0.5 and 1.3 s, which can ensure the complete combustion of CO. The length to diameter ratio of the carbon monoxide incinerator should be determined according to the specific conditions of the site.
(4) Due to the restriction of the CO equilibrium conversion rate, the furnace outlet temperature should not exceed 1200 °C.

This technology has been widely used in catalytic cracking units of oil refineries. It is also suitable for energy recovery of CO-containing flue gas and process waste gas.

14.6 CO Combustion and Flue Gas Energy Recovery System Optimization

With the continuous increase of oil refining processing depth, heavy oil catalytic cracking units have been continuously developed and expanded in oil refineries, which have played an important role in improving the economic benefits of enterprises. A prominent feature of heavy oil catalytic cracking is high coke yield, serious heavy metal pollution, and a large excess of heat, which has a fundamental impact on the energy use and consumption of catalytic cracking units. For low-quality raw materials, especially heavy oil catalysis, the two-stage regeneration process is a better choice. It not only avoids the expensive cost of using platinum accelerants but also relieves the limitation of the regeneration capacity of the main compressor, especially for slowing down the high-temperature hydrothermal deactivation of catalysts; therefore, it has been widely used. It changed the past practice of adding combustion accelerants in the regenerated catalyst bed to completely burn CO, and led CO to the outside of the regenerator for combustion and utilization, which improved the heat balance of the reactor-regenerator system. In the two-stage regeneration process, the first stage is usually incomplete regeneration, burning 65–90% of carbon and most of the hydrogen in the coke, and the flue gas contains about 5–9% CO. The second-stage regeneration is high-temperature oxygen-enriched completing regeneration, burning off the remaining 10–35% of carbon and all the H_2. The exhaust flue gas temperature is above 720 °C with an oxygen content of about 3–5% (v).

As for the combustion of CO in flue gas, many kinds of literature [8, 9] have carried out research and development. In the research and development, the author discovered the combustion law of the regenerated flue gas CO, and proposed the ignition temperature and deflagration temperature of regenerated flue gas CO combustion; the empirical kinetic equation of the ignition process has been also regressed, and the investigation of flue gas pipeline pre-combustion and CO high-temperature combustion is also carried out. It has realized industrial application in the heavy oil catalytic cracking units in China. However, the current industrial application is the atmospheric CO combustion at atm pressure downstream the flue gas turbine, and supplementary combustion is needed to increase the temperature. In the case of excess heat and steam generation in the refinery, it is uneconomical to use a large amount of fuel for supplementary combustion of CO to increase the amount of steam generated. The volume and space of the CO incinerator are also too large, which needs to be improved and perfected. In addition, by combining flue gas energy recovery and sensible heat recovery, there are different combustion methods and flue gas energy recovery processes. Studying these processes, finding reasonable CO combustion methods, and developing and finding non-supplementary CO combustion technologies are major profound and energy-saving subjects.

14.6.1 Two-Stage Regeneration Flue Gas Mixed Pre-Combustion

For the two-stage regeneration process, since the flue gas contains 5–8% of CO, it is an important subject to burn and utilize the chemical energy of CO. The choice of CO combustion and energy utilization approach is not only the CO combustion itself but also the combination with flue gas kinetic energy recovery and flue gas sensible heat recovery. Generally speaking, there are three CO combustion methods for different processes, pre-combustion in the flue gas duct, CO combustion before the flue gas turbine, and CO combustion after the flue gas turbine; it is also divided into two types of CO combustion according to the combustion pressure: atmospheric pressure combustion and positive pressure combustion. In the operation of industrial plants, we have already mastered the CO combustion ignition law and the operation experience of atmospheric pressure combustion of CO. it will not repeat here, but focus on pre-combustion and positive pressure combustion.

Before entering the flue gas turbine, the flue gas CO can be pre-burned using the flue gas duct, and make full use of the characteristics of excess O_2 and high temperature of 2nd flue gas to increase the flue gas temperature. The purpose of CO duct pre-combustion is to appropriately increase the inlet temperature of the flue gas turbine and increase the power recovery rate; reduce the amount of supplementary combustion and the structure size of the CO incinerator; at optimal conditions, CO can be completely burned without supplementary combustion, which is conducive to energy saving and environmental protection.

14.6.1.1 CO Pre-combustion and 2nd Regeneration Flue Gas Mixing Ratio

In general, the 1st regeneration flue gas leaving the second cyclone temperature is about 680 °C, all above the ignition temperature of CO, as long as the flue gas contains oxygen, combustion can be carried out, the flue gas duct pre-combustion test trail had been carried out in a demonstration FCCU. Oxygen is provided by injecting industrial air, the injection of air increases the oxygen concentration, but the initial temperature of the flue gas drops, there is the best point, the test trial indicated that after the air injection, 630–640 °C of flue gas temperature, the excess oxygen concentration of 1.1–2.6%, in 1–1.5 s residence time can make the flue gas temperature rise to 750 °C, the pre-combustion process can be completed; for two stages regenerative flue gas mixing, the flue gas mixing temperature is higher than required ignition temperature, and the O_2 in mixed flue gas will be high enough to complete the pre-combustion, there is no temperature drop concern caused even with air injection, the pre-ignition speed is faster, and the control of pre-combustion and its flue gas temperature should be mainly achieved by controlling excess O_2.

Excess O_2 can be controlled in two ways: the control of the 2nd regeneration operation is mainly determined by the process operation and production, and it is not appropriate to make substantial adjustments for this purpose. The second is at stabilizing the 2nd regeneration operation, controlling the mixing amount of the 2nd regeneration flue gas, for which it is still necessary to make a control scheme and equip with a good control system.

The degree of pre-combustion is how much CO conversion, how high the flue gas heats up, it mainly depends on the refractory material used in flue gas duct lining and the flue gas energy recovery process, for the process of pre-combustion flue gas directly flowing into the flue gas turbine, the pre-combustion temperature increase to 710–730 °C is more appropriate, for the operating temperature of 1st regeneration dense phase of 670–680 °C, considering the flue gas temperature drop, the flue gas turbine inlet is about 630–650 °C, the temperature increase is about 80–100 °C, the power recovery rate of the flue gas turbine is increased by about 8–10% (excluding the recovery power of the 2nd regeneration flue gas); in general, the 1st regeneration flue gas contains excess O_2 of less than 0.2%, the O_2 concentration is too low to cause CO combustion, CO combustion oxygen is mainly through the mixing of 2nd regeneration flue gas so that the mixed flue gas oxygen concentration increases. The CO pre-combustion rate and the ratio of 2nd regeneration flue gas rate to 1st regeneration flue gas rate are determined by heat balance and oxygen balance using the following equation set:

$$CO = \frac{C_P(\Delta T_1 + (V_2/V_1)\Delta T_2)}{126.36} \tag{14.29}$$

$$\frac{V_2}{V_1} = \frac{1/2CO}{O} \tag{14.30}$$

where

C_P—Average specific heat of flue gas, kJ/(m^3 °C);

CO—Pre-ignited CO percentage in 1st regeneration flue gases, %(v);

O—Excess oxygen concentration in 2nd regeneration flue gas, %(v);

V_1—1st regeneration flue gas volume, Nm^3/h;

V_2—2nd regeneration flue gas mixed with 1st regeneration flue gas, Nm^3/h;

T_1—1st regeneration temperature, °C;

T_2—2nd regenerator temperature, °C;

T—flue gas turbine inlet temperature, °C;

ΔT_1—1st regeneration flue gas total temperature rise, °C;

ΔT_2—2nd regeneration flue gas total temperature rise, °C;

Equations (14.29) and (14.30) can be solved to determine the percentage of pre-combustion CO completing the pre-combustion process, and the amount of 2nd regeneration flue gas required for supplying enough oxygen of the pre-combustion process; it is suitable for the higher pressure on the 2nd regeneration flue gas; for the 2nd regeneration flue gas pressure is equal to or lower than the pressure of 1st regeneration, the pressure of 1st regeneration flue gas should be lowered and mix under the 2nd regeneration flue gas pressure.

Example 14.2 For a heavy oil catalytic cracking unit, the 1st regeneration flue gas is 132,200 Nm^3/h, the CO concentration is 6.95%, and the regenerated flue gas temperature is 680 °C; the 2nd regeneration flue gas volume is 31,280 Nm^3/h, the excess O_2 concentration is 4.39%, the regenerated flue gas temperature is 720 °C, and the inlet temperature of the flue gas expander can be 730 °C. How much to mix the 2nd regeneration flue gases and how much CO is to be burned off in the 1st regeneration flue gases? The flue gas takes a specific heat of 1.465 kJ/(m^3 °C).

Solution

$$\Delta T_1 = T - T_1 + 40 = 730 - 680 + 40 = 90$$

$$\Delta T_2 = T - T_2 + 10 = 730 - 720 + 10 = 20$$

From Eqs. (14.29) and (14.30)

$$CO = \frac{1.465\left(90 + \frac{V_2}{V_1}20\right)}{126.36}$$

$$\frac{V_2}{V_1} = \frac{0.5CO}{4.39} = 0.1139CO$$

$$CO = \frac{1.465(90 + 20 \times 0.1139CO)}{126.36} = 1.072\%$$

$$\frac{V_2}{V_1} = 0.1139CO = 0.1139 \times 1.072 = 0.1221$$

$$V_2 = 132200 \times 0.1221 = 16142\,\text{Nm}^3/\text{h}$$

Therefore, to meet the requirements of the inlet temperature of the flue gas turbine of 730 °C, it is required to mix the 2nd regeneration flue gas of 16,142 Nm^3/h.

For the case of setting a flue gas heat remover before the flue gas turbine or directly entering the positive pressure CO incinerator after completing the pre-combustion, the first and second regeneration flue gases can be all mixed and pre-ignited, and the excess O_2 of the 2nd regeneration flue gas can be fully used, its pre-combustion temperature T can be calculated by the following equations (deducting the 40 °C temperature drop of the pipeline).

$$\Delta T = \frac{63.18 V_3 O}{C_P(V_1 + V_3)}$$

$$T_m = \frac{T_1 V_1 + T_2 V_3}{V_1 + V_3}$$

$$T = T_m + \Delta T - 40$$

where

V_3—1st regeneration flue gas rate, Nm^3/h;

T_m—Flue gas mixing temperature, °C;

T—Pre-ignition temperature, °C.

Other symbols are the same as above.

The pre-combustion time required can be estimated by Eq. (14.28).

14.6.1.2 The Effect of Pressure on the CO Combustion Reaction and CO Incinerator Size

After the flue gas passes through the flue gas turbine, CO is burned in the atmospheric pressure furnace, and the prominent contradiction is that the equipment is large, required a larger space, and the inlet temperature of the flue gas is as low as 500 °C, which must be replenished with fuel to increase temperature. Instead, they use of positive pressure combustion is conducive to reducing the volume and footprint of CO incinerator, saving investment, optimizing the flue gas energy recovery process, and eliminating the fuel replenishment. Effect of positive pressure on CO combustion reaction and CO incinerator size:

1. **Effects on chemical equilibrium**

Carbon monoxide combustion reaction $CO+0.5O_2=CO_2$ is a volume reduction reaction, according to the principle of chemical equilibrium movement, increase the pressure, equilibrium moves in the direction of volume reduction, which is conducive to improving the equilibrium conversion rate of CO combustion.

2. **Effect of pressure on reaction rate**

(1) The influence of pressure on the reaction rate is mainly to change the concentration of reactants per unit volume and change the reaction rate. When the pressure is high, the concentration of reactants increases, the reaction rate is accelerated, and the flame propagation rate is also affected, but the impact is not large.
(2) When the pressure is increased, the concentration of inert components (CO_2, N_2, etc.) per unit volume in the flue gas also increases, which is not conducive to improving the reaction rate. More importantly, after the concentration of these inert components is increased, the free radicals collide with the inert components in the CO combustion chain reaction, which slows down the rate of chain reaction.
(3) Positive pressure combustion can reduce the ignition energy of the fuel [10], thereby reducing the ignition temperature, which can be seen from the test trial pre-combustion data, the atmospheric pressure experimental data ignition temperature is 639 °C, while in the 0.25 MPa pipeline pre-experiment, the ignition temperature is only 634 °C. However, the temperature of our regenerated

flue gas is about 680 °C, which is higher than the ignition temperature, and there is no significant impact on reducing the ignition energy.

Therefore, positive pressure combustion is advantageous for chemical equilibrium, so that the reaction moves in the direction of CO into CO_2, but the chemical reaction rate is advantageous to each factor, and the concentration of reactants, products, and non-reactants increases at the same time, so that the reaction concentration changes, and the reaction rate is increased by the reaction rate of 2.0 level. At the same time, the concentration of inert components increased, which slowed down the chain reaction rate. In short, for the regenerative operating pressure of the existing catalytic cracking unit, positive pressure combustion is more advantageous than atmospheric pressure.

3. **Effect of pressure on CO incinerator equipment**

The flow volume of the flue gas is inversely proportional to the pressure, the flue gas volumetric flow at positive pressure combustion is greatly reduced, it is roughly the $\frac{P}{P_0}$ times of the original volume, the general regenerative pressure is 0.2–0.4 MPa, such as the use of positive pressure combustion, the incinerator volume will be reduced to 1/2–1/4 of the atmospheric pressure combustion volume at the same combustion time, which reduces the equipment size greatly, the shortage of the occupying larger space of CO incinerator at atmospheric pressure has been overcome.

Positive pressure combustion is completely feasible from the process technology, but it is necessary to further study the design and production of the positive pressure CO incinerator, formulate operating procedures different from the atmospheric pressure furnace, and avoid leakage affecting safe operation.

14.6.1.3 Determination of the CO Burning Time

There are two keys to the design of the CO incinerator, one is to determine a reasonable combustion residence time, which can ensure the full combustion of CO and make the equipment compact and reasonable; the second is that the combustion is stable, neither de-ignition nor tempering, so that the propagation speed of the flame and the gas flow velocity is balanced; generally, under turbulent conditions, the flame propagation speed is related to the pulsation speed of the gas flow and current it is difficult to accurately determine it. According to the experimental data and operating experience, the operating velocity of gas flow in the CO incinerator should not be more than 10 m/s. The determination of the combustion time is very important in the design of the CO incinerator.

(1) **CO incinerator with supplementary combustion**

When the inlet temperature of flue gas to the CO incinerator is lower than the ignition temperature, it is necessary to fire fuel and raise the flue gas temperature, according to the experimental and on-site test trial results, the determination of the CO incinerator residence time should be divided into three parts: the re-ignition temperature raising area, the high-temperature combustion area, and the stability area. The total residence time is the sum of the three parts.

1. Re-ignition temperature raising area

This area is mainly a supplementary combustion temperature raising, including air injection, air and flue gas mixing, and pre-combustion, so that the flue gas temperature rises to 750 °C or higher; due to the existence of flame ignition flue gas CO phenomenon in this area, the flue gas quickly heated up to deflagration point, accounting for the area of the residence time of 0.3–0.4 s.

2. Combustion zone

In the re-ignition temperature zone, the flue gas has been raised to 750 °C, at this temperature the CO combustion comes into the high-temperature deflagration zone, and the CO combustion rate in this area is extremely fast, and the CO can be fully burned in about 0.1 s. The residence time in this area is considered as 0.1–0.2 s.

3. Stabilization zone

The area is mainly considered from the perspective of safe operation, avoiding operational fluctuations, and transferring the combustion area to the downstream equipment. Generally, the residence time is 0.3–0.6 s. This area may not be considered when considering high-temperature refractory material lining for the flue gas duct between the CO incinerator and the waste heat boiler.

The total residence time is the sum of the three parts. The total residence time is between 0.9 and 1.15 s, and 0.4 to 0.6 s for high-temperature refractory material lining duct downstream of CO incinerator; and the characteristics of these three areas should be considered accordingly to the furnace structure.

(2) **A high-temperature combustion incinerator that does not replenish the fuel**

For the case of using the flue gas pre-combustion to reach the deflagration temperature, it is only necessary to inject the oxidizer (air) into the CO incinerator, there is also a mixing time between air and flue gas. However, at this time, the mixing and combustion are carried out at the same time, and the length of the combustion time depends on the degree of homogeneity of the mixture of two, the total combustion time, including mixing and combustion, stabilizing the three regions, mixing and burning time to take 0.4–0.6 s, the total residence time will be within 1.1 s.

(3) **Ignition type of CO incinerator**

The ignition time is longer, the initial temperature reaches 640 °C after the air injection, it takes about 1.5 s to complete the ignition process, and the residence time required for the ignition process can be estimated by Eq. (14.28) (pre-ignition final temperature is 710 °C). However, at this time, the incinerator does not have a premix time, so the total furnace residence time is between 2.0 and 2.2 s.

At this time, it is obvious that the volume and footprint are large, the project is uneconomical, and it is generally more combined with flue gas duct pre-combustion to reduce the size of the equipment. Pre-combustion of CO in the flue gas duct completes the ignition process. Make the CO incinerator just a high-temperature combustion process, the equipment will be compact.

14.6.2 *Flue Gas Energy Recovery Processes and Characteristics*

For the FCCU two-stage regeneration process, the rational arrangement of CO combustion approaches and the combination of the most suitable flue gas energy recovery process is a major area of energy saving for the process unit. In general, according to the confluence and diversion of the two stages of flue gas, flue gas kinetic energy recovery and CO combustion mode, the location and number of segments of the waste heat boiler layout, 6 different processes can be composed, and the energy-saving and economic effects are also different.

14.6.2.1 CO Atmospheric Combustion Without 2nd Regeneration Flue Gas Turbine (Process 1)

Process Flow diagram of CO atmospheric combustion without 2nd flue gas turbine is shown in Fig. 14.10.

The characteristics of this process are that the flue gas is directed to 1st regeneration flue gas turbine first, then flows into the CO incinerator for combustion, the 2nd regeneration flue gas flows into the rear of the incinerator or is mixed with the 1st regeneration flue gas flows into the CO incinerator, and then into the waste heat

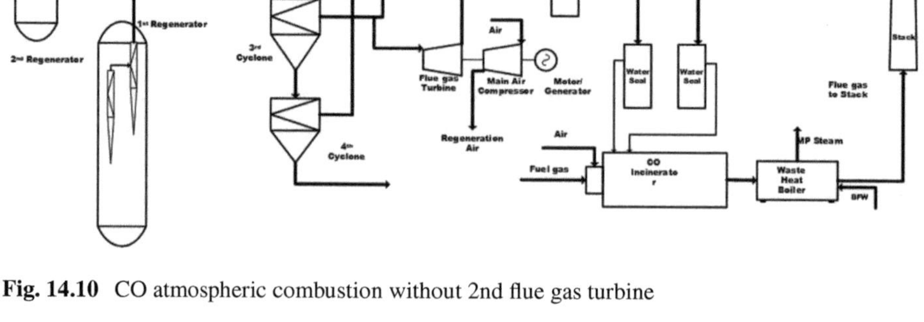

Fig. 14.10 CO atmospheric combustion without 2nd flue gas turbine

boiler to recover the heat energy to generate steam. The disadvantage is that the kinetic energy of the 2nd regeneration flue is not recovered. The CO incineration at atm pressure, which can completely burn CO but requiring replenish fuel, it is low in the supplementary combustion ratio compared with the traditional CO boiler.

The 2nd regeneration flue gas enters the rear or front of the CO incinerator, and there is no essential difference in the process. However, the mixture with the 1st regeneration flue gas from the front makes full use of the excess O_2 and high-temperature characteristics of the 2nd regeneration flue gas, which is conducive to reducing the amount of fuel replenishment, and the unfavorable factor is the reduction of the CO concentration in the mixed flue gas.

14.6.2.2 2nd Flue Gas Cooling and Mixing with 1st Flue Gas into the Flue Gas Turbine (Process 2)

Process Flow diagram of 2nd flue gas cooling and mixing with 1st flue gas into the flue gas turbine is shown in Fig. 14.11.

The 2nd regeneration flue gas pipeline is equipped with a steam superheater to reduce the temperature of the 2nd regeneration flue gas, to prevent secondary combustion when the high-temperature oxygen-enriched flue gas of the second regenerator mixed with the 1st regeneration CO flue gas, to ensure that the mixed flue gas temperature does not exceed 600 °C, the 2nd regeneration flue gas needs to be reduced to 380 °C to achieve the mixing flue gas temperature of 598 °C. Considering the flue gas enters the flue gas turbine at 588 °C after the temperature drop, the exhaust flue gas is incinerated in the CO incinerator or burned by the CO boiler, but the former has a low amount of fuel replenishment, and the latter requires a fuel replenishment ratio of about 30%.

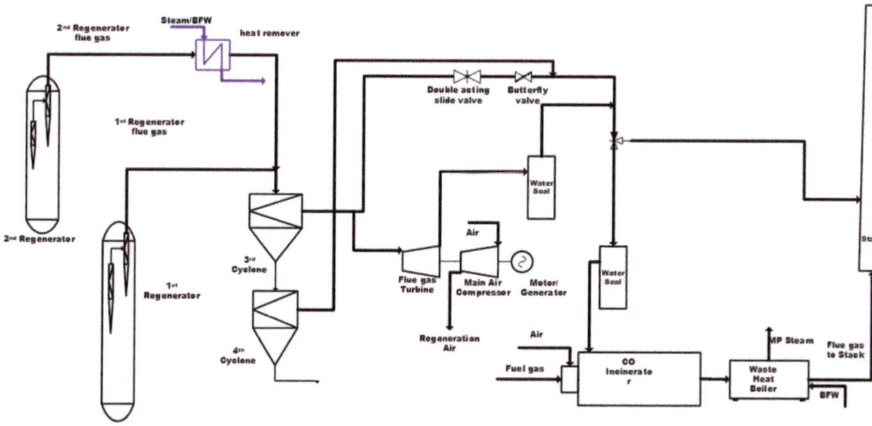

Fig. 14.11 Second flue gas cooling and mixing with 1st flue gas into the flue gas turbine

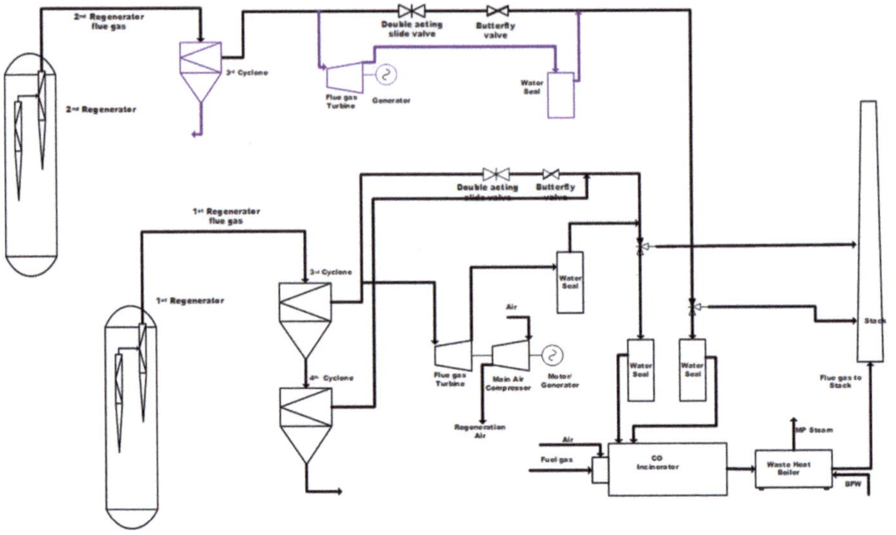

Fig. 14.12 Recover both flue gas kinetic energy, CO atm pressure combustion

The characteristics of this process are: reducing the flue gas temperature to prevent tail combustion, recovering 2nd regeneration flue gas kinetic energy, but there is a shortage that the temperature of the flue gas entering turbine is low, which requires replenish fuel in the CO incinerator, and the expansion work is reduced.

14.6.2.3 Recover Both Flue Gas Kinetic Energy, CO atm Pressure Combustion (Process 3)

Process Flow diagram of Recover both flue gas kinetic energy, CO atm pressure combustion is shown in Fig. 14.12.

The process is roughly the same as process 1, except that a small flue gas turbine is added to the 2nd regeneration flue gas to recover the flue gas kinetic energy, and the recovered power is increased, but the steam production is correspondingly reduced. Investment in flue gas turbine, third cyclone, and related systems has been increased, and the control and operability should be considered. It is recommended that a technical-economic evaluation should be conducted before applying this process.

14.6.2.4 CO Pre-combustion, CO atm Pressure Combustion (Process 4)

Process Flow diagram of CO pre-combustion, CO atm pressure combustion is shown in Fig. 14.13.

This process takes the characteristics of 2nd regeneration flue gas at high temperature with excess O_2 to pre-ignite 1st regeneration flue gas CO. The temperature of

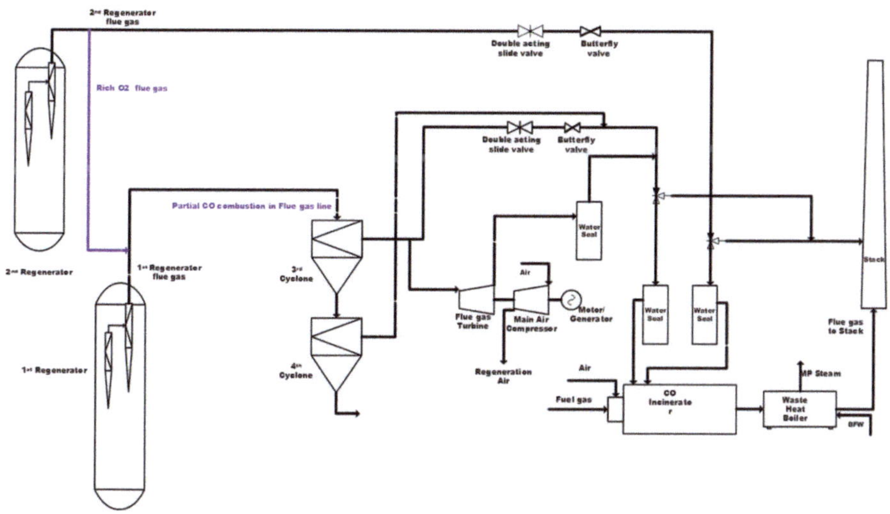

Fig. 14.13 CO pre-combustion, CO atm pressure combustion

flue gas entering the flue gas turbine can be controlled by adjusting the amount of 2nd regeneration flue gas (oxygen content), there is no flue gas heat remover to be provided.

The process is characterized by duct CO pre-combustion, a flue gas turbine recovers both 1st regeneration and 2nd regeneration flue gas kinetic energy, the CO in the flue gas is burned at atmospheric pressure, and the following waste heat boiler recovers the regenerated flue gas CO chemical energy and the flue gas sensible heat.

When the pressure of the 2nd regeneration flue gas is equal to or lower than that of the 1st regeneration flue gas, the 1st regeneration flue gas needs to be slightly throttled to produce a small amount of kinetic energy loss, but the amount of kinetic energy recovery is still increased than that of process 1.

14.6.2.5 Flue Gases Are Mixed and Pre-ignited, Cooler, CO atm Pressure Combustion (Process 5)

Process Flow diagram of Flue gases are mixed and pre-ignited with a cooler followed; CO atm pressure combustion is shown in Fig. 14.14.

The process uses high-temperature oxygen-enriched 2nd regeneration flue gas to mix with 1st regeneration flue gas, the mixed flue gas temperature and excess oxygen increase, CO concentration decreases, and the mixed flue gas temperature are higher than the CO ignition temperature, approaching or even reaching the deflagration temperature.

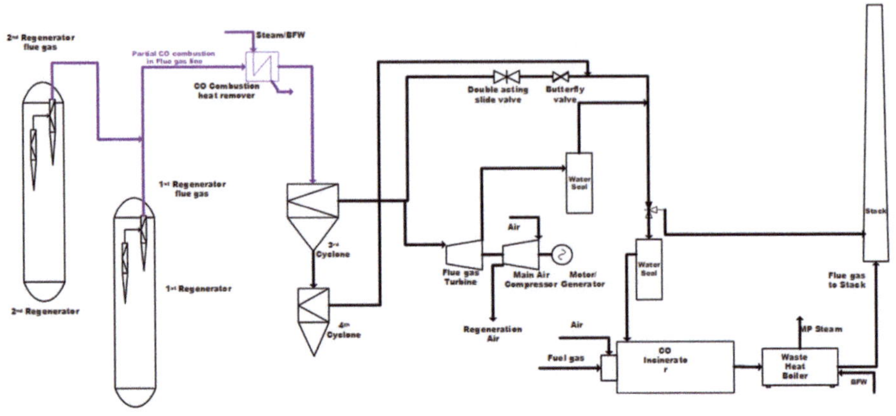

Fig. 14.14 Flue gases are mixed and pre-ignited, cooler, CO atm pressure combustion

The characteristics of this process are the use of flue gas duct CO pre-combustion, making full use of the advantages of high temperature and high excess oxygen of 2nd regeneration flue gas. The degree of pre-combustion mainly depends on the oxygen concentration in the mixed flue gas. Generally, there is a residence time of 2–3 s in the flue gas duct, the reaction rate is very fast at high-temperature, therefore, the combustion result will make the oxygen in the flue gas used up, for the case of 1st regeneration coke burning of 85–90%, the 2nd regeneration flue gas contains about 2–5% O_2, accounting for 20–40% of the total flue gas rate, the oxygen content in the mixed flue gas is about 1%, it can consume 1.5–2% of CO, the total temperature rise between 130 and 180 °C, After deducting the temperature drop of the flue gas duct caused by duct heat dissipation, the flue gas temperature will be in 800–880 °C; at such a high temperature, the flue gas turbine material cannot withstand, must be set up a cooling facility, but this heat is not much, generally can be used for superheated steam.

Although the two flue gases are mixed, 1st regeneration flue gas has also a throttling loss, but the 2nd regeneration flue gas kinetic energy is recovered, and the work recovery rate is also improved according to the inlet temperature of the 730 °C entering the flue gas turbine, and the total kinetic energy recovery is increased.

14.6.2.6 Two Flue Gases Mixing and Pre-combustion, CO Positive Pressure Combustion (Process 6)

Based on process 5, relocate the CO incinerator upstream of the flue gas turbine to burn CO at positive pressure, and since the pre-combustion can make the flue gas reach higher than 800 °C, it is not necessary to replenish fuel and increase the temperature, and the flue gas will be directly mixed with combustion air to burn CO out, there will be no fuel gas replenishment, but designed for startup and flexibility. The outlet of the CO incinerator is about 1100 °C, design and control the first stage

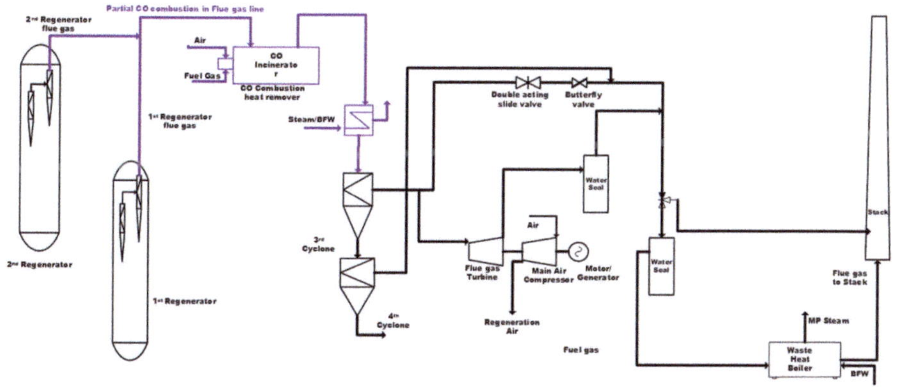

Fig. 14.15 Two flue gases mixing and pre-combustion, CO positive pressure combustion

waste heat boiler flue gas outlet temperature to be 730 °C, then flow through the third cyclone to further separate the catalyst dust, then enters to flue gas turbine for power recovery, and the 550–600 °C of exhaust flue gas from flue gas turbine is discharged into the second stage waste heat boiler to recover heat energy.

Process Flow diagram of Two flue gases mixing and pre-combustion, CO positive pressure combustion is shown in Fig. 14.15.

The process features positive pressure CO combustion so that the size of the CO incinerator equipment is reduced to 1/2–1/4 of the atmospheric pressure furnace. First, pre-ignite the flue gas to above the deflagration point, there is no need to replenish fuel to raise the temperature in the CO incinerator so that the CO external combustion technology from a small amount of replenishing fuel to no fuel gas firing at all. The flue gas from 1st waster heat boiler meets the inlet temperature requirements of the flue gas turbine, and the flue gas flow rate at the entrance of the flue gas turbine increases and the work increases.

In the engineering application, it is necessary to do a good job of pressure balance and develop a positive pressure CO incinerator.

14.6.3 Comparison of Energy-Saving Effects of Various Energy Recovery Processes

To compare the energy-saving effect and economy of various flue gas energy recovery processes, take a 1.4 Mt/a scale heavy oil catalytic cracking unit as the basis for comparison, and the regeneration method is a two-stage regeneration process, 1st stage regenerator is incomplete regeneration, burning 85–90% of carbon and most of H_2, and the second stage is complete regeneration, burning off the remaining carbon and H_2, and the data of the energy recovery system are shown in Table 14.7. The calculation results of the various processes are also listed in Table 14.7.

Table 14.7 Evaluation summary of various energy recovery processes

Process		Units	1	2	3	4	5	6	Notes
System data	1st flue gas	Nm^3/h	132200						1st FQ/2nd FG
	2nd flue gas	Nm^3/h	31280						
	1st Reg./2nd Reg. temperature	°C	630/ 720						
	Reg. pressure	MPa	0.35 / 0.32						
	Coke yield	%	9						
	Air consumed Index	kg/kg	9.6	9.6	9.6	9.6	9.6	10.99	
	Flue gas comp., CO	%	6.95 / 0						
	C02	%	9.93/13.91						
	02	%	0.17/4.39						
CO combustion	Pre-combustion CO	%				1.13	2.08	2.08	After deducting 40 °C pipeline temperature drop
	Injected 2nd FG	Nm^3/h				17001	31280	31280	
	Pre-Comb Temp.	°C				730	801	801	
	CO incinerator pressure	MPa	Atm	Atm	Atm	Atm	Atm	0.3	
	Inlet temp.	°C	533.6	460.7	511.9	604.3	594.9	801	
	Inlet 02	%	0.98	0.98	0.98	0.52	0.14	0.14	
	Fuel rate	kg/h	1810	1810	1810	500	500		
	Air rate	Nm^3/h	56396	56396	56396	320763	30520	21895	

(continued)

Table 14.7 (continued)

Process		Units	1	2	3	4	5	6	Notes
	Air pressure	MPa	Atm	Atm	Atm	Atm	Atm	Pressure Air	
	Outlet temp.	°C	998.7	944.4	982.5	922.8	867.14	1031.1	
	Total flue gas	Nm^3/h	219876	219876	219876	194809.3	192627	184002	
Expander energy recovery	Flue gas rate	Nm^3/min	2090	2600	2090/510	2362	2566.7	2913.4	
	Inlet pressure	MPa	0.31	0.28	0.28/0.28	0.28	0.28	0.28	
	Outlet pressure	MPa	0.106	0.106	0.106/0.106	0.106	0.106	0.106	
	Inlet temp.	°C	640	588	640 / 680	730	730	730	
	Enthalpy drop	kJ/kg	261.75	221.77	238.9/238.6	260.2	257.5	257.5	
	expansion work	kw	8862	9450.2	8084/ 2085.3	10084.8	10939.4	12405	
	Outlet temp.	°C	499	460.7	511.5/515	597	594.9	594.9	
Waste heat boiler	Before expander	Inlet temp. °C		690			801	1031.1	
		Outlet temp. °C		380			730	730	
		Flue gas, Nm^3/h		31280			162107	184002	
		Heat recovery, kW		3749.7			4685	21424.2	
		Usage		Superheat steam			Superheat steam	Generate superheat steam	
	After expander	Usage	Generate superheat steam					Generate steam	
		Inlet temp. °C	993.7	944.4	982.5	922.8	867.14	594.9	
		Outlet temp. °C	200	200	200	200	200	200	

(continued)

Table 14.7 (continued)

Process		Units	1	2	3	4	5	6	Notes
		Flue gas, Nm^3/h	219876	219876	219876	194809.3	192627	184002	
		Recovered heat, kW	67909.9	63293	66532.4	55450.2	49696.2	23098.3	
	Steam temp.	°C	450	450	450	450	450	450	
	Steam pressure	MPa	3.8	3.8	3.8	3.8	3.8	3.8	
	Steam enthalpy	kl/tg	3334.78	3334.78	3334.78	3334.78	3334.78	3334.78	
	BFW temp.	°C	103	103	103	103	103	103	
	Steam generation	t/h	84.2	83.1	82.5	63.75	67.42	61.4	
	Heat loss	kW	3573.9	3528.56	3501.6	2865	2615.6	2606.5	
Energy utilization summary	Energy supply	Air compressor, kW	9920	9920	9920	9920	9920	11310	
		Airfan, kW	110	110	110	63	60	0	
		Coke, kW	173687.5	173687.5	173687.5	173637.5	173687.5	173687.5	
		Fuel, kW	21050.3	21050.3	21050.3	5815	5815	0	
		Subtotal	204767.8	204767.8	204767.8	189485.5	189482.5	184997.5	
	Energy utilization and recovery	Coke desorption and catalyst takeout, kW	838154.3	88154.3	88154.3	88154.3	88154.3	88154.3	
		Heat removed, kW	18840	18840	18840	18840	18840	18840	
		Power recovery , kW	8862	9450.2	10169.3	10084.8	10939.4	12405	
		Steam gen, kW	67909.9	67042.7	66532.4	55450.2	54381.2	49522.5	

(continued)

Table 14.7 (continued)

Process		Units	1	2	3	4	5	6	Notes
		Subtotal, kW	183766.2	183487.2	183696	172529.3	172314.9	168921.8	
	Energy rejection	Heat loss, kW	5676.5	6065.5	5856.7	3475.7	3838.1	3343	
		Flue gas loss, kW	15215.1	15215.1	15215.1	13480.5	13329.5	12732.7	
		Subtotal, kW	20891.6	21280.6	21071.8	16956.2	17167.6	16075.7	
	Energy utilization efficiency, %		89.7	89.6	89.7	91.1	90.9	91.3	
	Thermal energy recovery , %		89.8	89.4	89.1	90.5	89.9	90.11	
	Power recovery, %		88.4	94.2	101.4	101	109.6	109.7	
	Primary energy utilization efficiency, %		93.62	94.08	94.93	95.73	96.56	96.9	
Compare results with process 1	Reduced FCCU energy consumption	MJ/t		21.74	61.91	93.04	129.36	143.83	Steam \$8.6/t, fuel \$128/t
	Energy saving benefits	\$/a		139,700	361530	735140	956571	1594000	Power \$0.05/kWh

For the 1st regenerator burning coke ratio of 65–70%, the proportion of 2nd regenerator flue gas to total flue gas increases, and the energy recovery results of the above processes will be more favorable than the 1st regenerator 85–90% coke burned situation.

Table 14.7 lists the energy-saving effect and technical and economic data of various processes, process 1 has been industrialized since the 1990s, many of the FCCUs are designed to use this process, and processes 2–5 are processes formed by making different improvements based on process 1.

Process 2 adopts the method of reducing the 2nd regenerator flue gas temperature to recover the 2nd regenerator flue gas kinetic energy, although the amount of flue gas increases, at the same time, the inlet temperature of the flue gas turbine decreases, coupled with the pressure drop when the flue gas is mixed, the flue gas turbine expansion work does not increase much. The process is also slightly more complicated. the CO incinerator is required to replenish the fuel firing capacity, the energy-saving and economic benefits will be inferior to process 1.

Process 3 adds a 2nd regenerator flue gas kinetic energy recovery system, which is the same as process 1 and saves energy 61.91 MJ/t, the annual energy-saving benefit of $0.36 MM, the estimated capital of $1.1 MM, the net investment payback period is about 3 years, with the recovery of 2nd regenerator flue gas kinetic energy, the operability and controllability also been complicated.

Process 4 adopts controlled flue gas duct pre-ignition, which increases the inlet temperature of the flue gas turbine, increases the inlet flow of the flue gas turbine, and increases the work recovery rate by 12.6% compared with process 1. The investment did not increase, the energy consumption of the unit was reduced by 93 MJ/t, and the annual energy-saving benefit was $0.74 MM. The CO incinerator inlet temperature is high, and the amount of replenishment can also be reduced.

Process 5 uses fully pre-ignited flue gas with a flue gas temperature of 800–850 °C. the flue gas has been cooled by the flue gas heat remover, the 2nd regenerator flue gas full flows to the flue gas turbine, although 1st regeneration flue gas has small throttling losses, when the temperature of the regenerated flue gas intake is increased to 730 °C, the power recovery rate is increased by 21.2%, the energy consumption of the unit is reduced by 129.36 MJ/t, and the annual energy saving benefit is $0.96 MM. After the adoption of Process 5, the 2nd regenerator flue gas duct is reduced by about 2/3 or more, a flue gas heat remover is added, and the investment remains almost unchanged.

Process 6 adopts flue pre-combustion, CO positive pressure combustion process. The inlet temperature of the CO incinerator is the pre-ignition final temperature, which has reached above the deflagration temperature, so there is no need to replenish fuel firing the combustion temperature. The CO incinerator design changes qualitatively from a small amount of replenishment to no replenishment at all. However, the CO combustion air of process 6 requires the main air, the power consumption increases, the flue gas turbine does more work, and the work recovery rate reaches 109.7%. As a result, the energy consumption of the FCCU was reduced by 143.83 MJ/t, and the annual energy saving benefit was $1.6 MM. Waste heat boilers are set up in sections, a waste heat boiler is required upstream of the flue gas turbine to reduce the flue gas temperature, it is expected that the investment will increase, deducting the cost of shortening the 2nd regenerator flue gas duct, and the investment will remain almost unchanged.

In the above 6 processes, processes 4, 5, and 6 have their characteristics, the investment changes little, the energy-saving effects are remarkable, and engineering applications can be carried out according to specific conditions and user requirements. Among them, process 6 with the best result is recommended.

14.7 Low-Temperature Heat Recovery and Utilization

After the optimization and adjustment of the heat exchange process of the refinery process unit, there is still a lot of low-temperature waste heat (up to 150 °C) that is cooled and rejected. From the perspective of the system and the entire plant, this waste heat may be a useful resource, which can be provided to many users who need low-temperature heat to replace steam or other high-grade energy sources. Currently, carbon reduction and energy conservation continue to deepen. To reduce the carbon emission and energy consumption of equipment and the entire plant, the recovery of low-temperature heat is essential. However, due to the low-temperature characteristics of low-temperature heat, there are objectively a difficulty of recovery technology and low economic benefits.

The recovery and utilization of low-temperature heat require the establishment of a supporting recovery system throughout the entire plant, and the heat is taken out of the plant through an energy-carrying working medium (usually water). To ensure the safe and reliable operation of the plant, a constant heat extraction method will be considered, and the heat load fluctuation can be adjusted using setting a cooler in the recovery system.

The utilization of low-temperature heat in the entire plant is to collect the low-temperature heat that cannot be used in the plant but can be used outside of the plant. Most of the recovered heat is the high-temperature part of air cooling and water-cooling duty, and many of them have the condensing phase change heat. This requires that the safety and the reliability of operation should be fully considered when the plant is recovering the low-temperature heat. There are usually the following methods: First, to solve the problem of unsynchronized startup and shutdown of the different plants; the existing cooling system can be retained, once the upstream unit fails, it can be switched over in time to ensure normal operation. The second is to use reliable heat exchange equipment to prevent leakage of process streams, such as U-tube heat exchangers. To ensure a constant temperature of hot water entering the plant, a cooling and heating system should be installed in the hot water station.

14.7.1 Direct Use as a Heat Source for Heating

Low-temperature heat is prioritized for continuous and stable heat load users, which not only changes the excessive energy transmission loss caused by users when using high and medium-temperature heat sources but also replaces high and medium-temperature heat sources. This is the most attractive solution for low-temperature heat utilization.

Heating the plant Low-temperature streams

Using low-temperature heat to replace the high and medium temperature heat sources used in production can not only directly reduce production energy consumption but also because most of the production heat is continuous and the load is stable, the energy-saving range is large and the benefit is high. Therefore, when arranging the low-temperature heat program, it should be given priority. Such heat users include (1) heating of feed stream and tower reboiler for gas fractionation, MTBE, etc.; (2) heating of washing water in catalyst plant; (3) heating of chemical water and freshwater for power and boiler system; (4) heating of oil tank of oil storage and transportation system, etc.

Provide Heating for offices and living quarter

For the plant located in some cold weather district, there is a lot of steam heating energy consumption of office buildings, production control rooms, laboratory, maintenance, firefighting, education and training centers, living quarters, and scientific research and design institutes. This part of the energy consumption affects the overall energy consumption of the entire plant. These types of steam energy consumption usually can be replaced with low-temperature heat sources.

After low-temperature heat is prioritized for continuous and stable heat load users, it should be considered for upgrading utilization of low-temperature heat. There are four ways of such use, heat pump, heat transformer (second type of absorption heat pump), refrigeration, and power generation.

14.7.2 Heat Pump

The use of heat pump technology to increase the temperature of the process streams and reuse the streams in the production process is a technical means to effectively utilize low-temperature heat energy. There are two types of heat pumps: compression type and absorption type. Many petrochemical plants have successfully used compression heat pumps and achieved good energy-saving effects. Heat pumps are generally used on occasions where the temperature of the required heat is not much different from the temperature of the low-temperature heat stream. The larger temperature difference will be uneconomical, so be cautious when making plans.

14.7.2.1 Compression Heat Pump

Compression heat pumps are commonly used in gas fractionation units, it replaces the original heating medium such as steam with a small amount of high-quality electric energy; please note that with the optimization of the low-temperature heat utilization program, the use of low-temperature hot water instead of steam as the heat source of the bottom reboiler is more economical than the compression heat pump system.

A simple schematic of the compression heat pump is shown in Fig. 14.16.

Heat pump principle and performance index

A heat pump is a device that transfers heat energy from a lower temperature system to a higher temperature system. The ideal heat pump heating cycle is a reversed Carnot cycle, such as "pumping" heat from the T_1 temperature to T_2 temperature, The reversed Carnot cycle, and temperature-entropy diagrams are shown in Fig. 14.17.

For the heat pump used in the fractionation process, T_1 represents the evaporation temperature of the condenser at the top of the tower, and T_2 is the temperature of the reboiler at the bottom of the tower, and the heat pump (generally compressor)

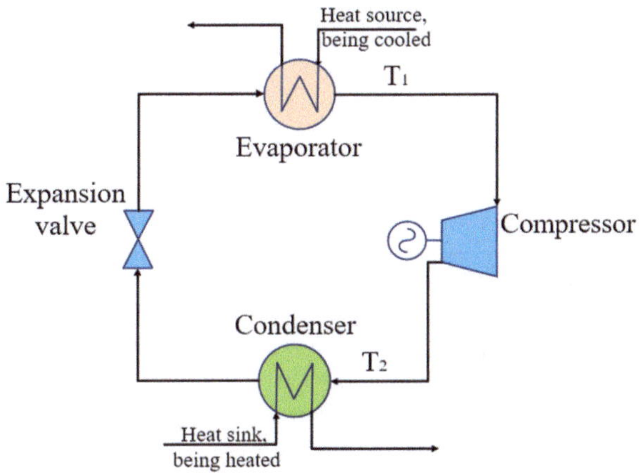

Fig. 14.16 Simple schematic of compression heat pump

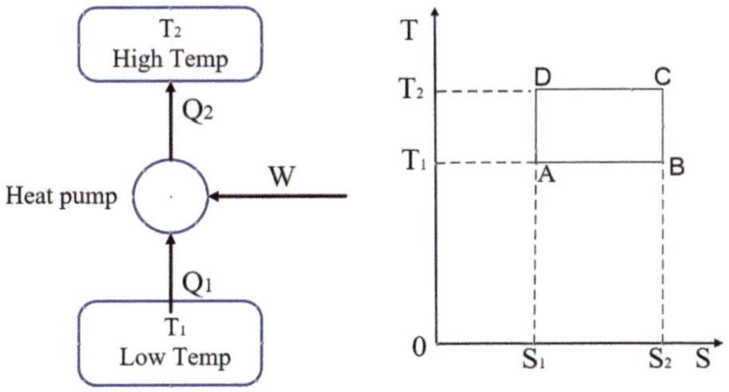

Fig. 14.17 Ideal heat pump cycle

consumes electric power W (kW), absorbs heat from the low-temperature heat source as Q_1, and releases heat Q_2 to the high-temperature heat source. Obviously, $Q_2 = Q_1 + W$. Coefficient of performance of heat pump cycle:

$$COP = \frac{Q_2}{W} = \frac{Q_2}{Q_2 - Q_1} = \frac{T_2}{T_2 - T_1} \tag{14.31}$$

Equation (14.31) is an ideal heating coefficient of performance based on a constant temperature heat source and in a reversible condition. It is only related to the temperature of the two heat sources and has nothing to do with the performance of the working fluid. For example, for the case of $T_1 = 313$ K, $T_2 = 333$ K, the coefficient of performance is $COP = 16.65$.

Most of the process heat sources are not under strict constant temperature conditions, and the actual process is irreversible. The heat pump heating performance coefficient is far lower than the ideal reverse Carnot cycle. But it suffices to show that the heat pump uses a small amount of high-quality electric energy to booster the low-grade energy which is going to be rejected to a desired energy grade and put it into for reuse. For plants that require low-grade heat, the use of heat pump technology can stop the use of high-grade fuel and steam energy, and obtain significant energy-saving benefits. The actual heat pump process heating performance coefficient is:

$$(COP)_{real} = \frac{Q_2}{W}\eta_C(1 - \eta_D) = \frac{T_{m2}}{T_{m2} - T_{m1}}\eta_C(1 - \eta_D) \tag{14.32}$$

where

T_{m2}, T_{m1}—are the average temperature of the two heat sources respectively;

η_C—compression mechanical efficiency, %;

η_D—heat pump system heat dissipation loss, %.

Compression heat pumps are generally divided into two types, closed and open, as shown in Fig. 14.18. Open heat pumps are divided into two types: the tower top product is the working fluid and the bottom product is the working fluid. A closed heat pump adopts external working fluid circulation.

The open heat pump avoids the loss of temperature difference in the heat exchange process, and the efficiency is greatly improved. The COP of the open heat pump of a gas separation plant in an oil refinery reached 7.52 [12].

The actual heating coefficient of the closed heat pump [11] (as shown in Fig. 14.18a):

$$COP = \frac{T_2 + \Delta T_R}{T_2 - T_1 + \Delta T_R + \Delta T_C}\eta_C(1 - \eta_D) \tag{14.33}$$

where

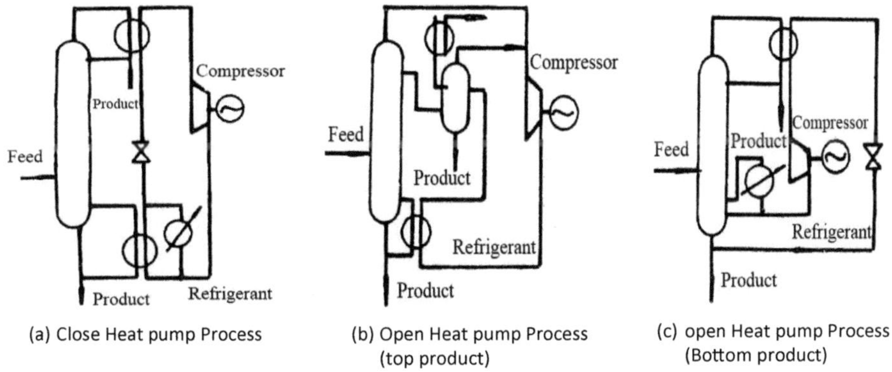

Fig. 14.18 Three basic processes of heat pump distillation

ΔT_R—column bottom reboiler and heat pump circulating working fluid heat transfer temperature difference, K;

ΔT_C—the temperature difference between the heat transfer between the top condenser and the circulating working fluid, K.

In Eq. (14.33), T_2 and T_1 can be the average temperature of the bottom and top of the tower respectively.

Applications

There have been many successful applications of heat pumps in petrochemical enterprises: the typical application is in the propylene-propane fractionation tower of the gas fractionation unit of a refinery [11, 12].

Since both propane and propylene are good working fluids in the heat pump cycle, the propylene-propane separation heat pump generally adopts an open cycle (Fig. 14.19). Open heat pumps are divided into two types: propylene heat pumps with the top product circulating of the tower and propane heat pumps with the bottom product circulating of the tower. Figure 14.19b shows the propylene heat pump cycle. The propylene gas at the top of the tower is directly fed into the heat pump compressor, and the pressure and temperature are increased as the heat source of the reboiler at the bottom of the tower. After the propylene is condensed, part of it is used as reflux liquid at the top of the tower, and part of it is exported as a product. Figure 14.19c shows the propane heat pump cycle. The propane liquid at the bottom of the tower is partly exported as a product and partly used as a heat pump cycle working fluid. Propane pressure is reduced via the control valve, then flows into the condenser/evaporator at the top of the tower, flashed and evaporated, absorbs the low-temperature waste heat of the propylene product at the top of the tower, and then enters the compressor to increase the pressure and increase the temperature, then directly enter the bottom of the tower as the bottom heat source.

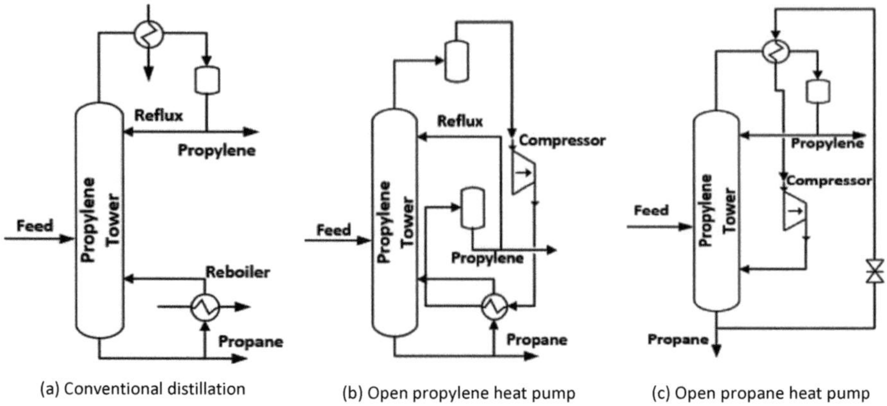

Fig. 14.19 Propylene rectification process heat pump

14.7.2.2 Absorption Heat Pump (AHP)

There are two types of absorption heat pumps [13], Type I absorption heat pump and Type 2 absorption pump. Type 1 absorption heat pump takes waste heat of 30–70 °C water as a low-temperature heat source and uses high-temperature heat (>150 °C) as driving heat source, and it produces 70–130 °C of heat for process or utility system use. Its thermal performance coefficient will be always great than 1. The purpose of installing AHP is to upgrade the waste heat to be reused in process and utility heating.

The simple Type I absorption heat pump sketch is shown in Fig. 14.20.

Type I absorption heat pump requires a small amount of driving heat source with a higher temperature, to produce mid-temperature heat, which may not be the desired temperature for use due to so much surplus heat at this temperature range. It may be useful for individual users depending on the low-grade heat balance.

For type I absorption heat pump, Generator, and condenser are at lower pressure, absorber and evaporator are at higher pressure.

Type II absorption heat pump (Heat transformer) takes 60–100 °C of the hot water heat as the driving heat source, produces 100–150 °C hot water or steam, and portion of supplied will be rejected; which is an effective way of low-grade heat upgrading, upgraded heat can be used in process heating or utility system. It is different from a type I absorption heat pump, Generator and condenser are at higher pressure, absorber and evaporator are at lower pressure. The simple Type II absorption heat pump sketch is shown in Fig. 14.21.

14.7.2.3 Heat Pump Heat Balance and Performance Index

Based on the heat input and heat output of the absorption heat pump system, the heat balance is shown in Eq. 14.34

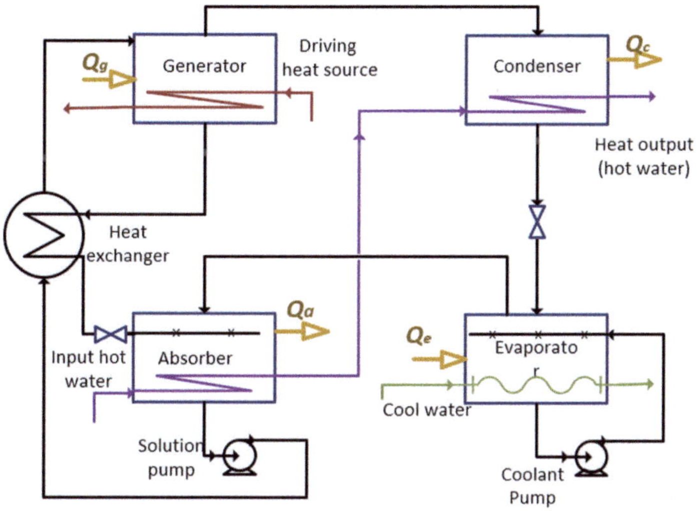

Fig. 14.20 Type I absorption heat pump sketch

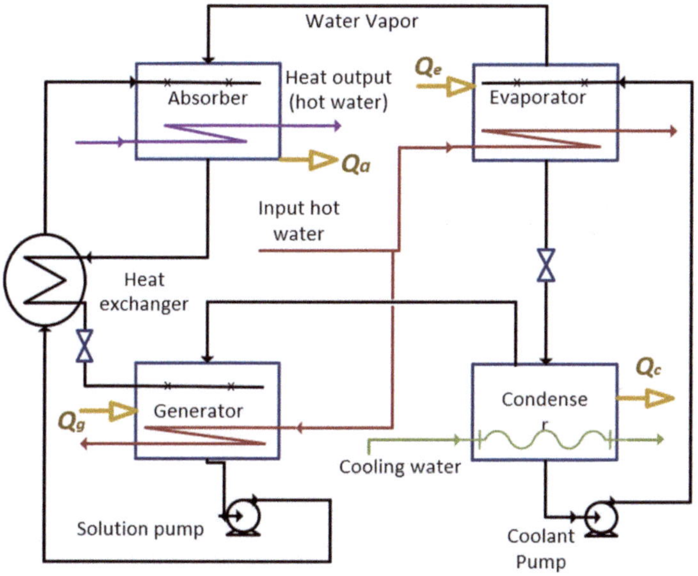

Fig. 14.21 Type II absorption heat pump sketch

$$Q_e + Q_g + W_p = Q_a + Q_c \tag{14.34}$$

where

Q_e—Evaporator input heat, for Type I heat pump, heat is supplied from the environment;

Q_g—Generator input heat;

W_p—Power consumed by solution pump and coolant pump;

Q_a—Absorber output heat;

Q_c—Condenser output heat.

The absorption heat pump efficiency can be characterized by the performance indexξ

$$\xi = \frac{Q_a + Q_c}{Q_e + Q_g + W_p} \tag{14.35}$$

For type I absorption heat pump, Q_e is from environment, it is not energy input, so

$$\xi = \frac{Q_a + Q_c}{Q_g + W_p} \tag{14.36}$$

The power consumed by the solution pump and coolant pump is relatively smaller, and it can be ignorable,

$$\xi = \frac{Q_a + Q_c}{Q_g} \tag{14.37}$$

Generally speaking, the performance index of Type I absorption heat pump is always great than 1, type II absorption heat pump performance index performance calculation should follow the Eq. 14.35 or Eq. 14.36, and the performance index will be always less than 1.

14.7.2.4 Heat Transformer

The absorption heat transformer is a low-temperature heat recovery technology, it is also called the type II absorption heat pump, where the low-temperature heat (such as 90 °C hot water) is converted into two parts by the heat transformer, one part is converted into higher temperature heat, which will be used as a heating heat source in process or utility system; the other part is degraded into waste low-temperature heat, which will be rejected through cooling.

A typical single-stage absorption heat transformer process and work principle are shown in Fig. 14.22.

The working fluid is composed of refrigerant and absorbent. In the generator (G), part of the refrigerant absorbs low-grade heat with a temperature of T1 (such as 70–100 °C) Q_G, evaporates to form saturated steam with a higher pressure. The

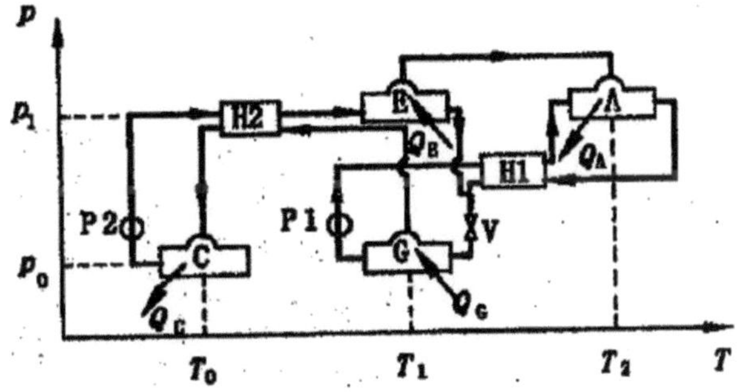

Fig. 14.22 Single-stage absorption heat transformer process. C—condenser; G—generator; E—evaporator; A—absorber; H1, H2—heat exchanger; P1, P2—solution pump; V—throttle valve

steam is pre-cooled in the heat exchanger (H2) and then flows into the condenser (C) and condenses here, releasing lower-grade heat Q_C to the environment at the ambient temperature T_0. The condensed liquid is boosted by the solution pump (P2) and absorbed heat by the heat exchanger (H2) to form a supercooled liquid with a pressure of P1 and a temperature of T1, which is sent to the evaporator (E), where it absorbs the low-grade heat Q_E and evaporates at the temperature of T1. The evaporated vapor flows into the absorber (A), while the non-evaporated liquid is depressurized through the throttle valve (V) and flows back into the generator (G). The dilute solution (the main component is absorbent) flowing out of the generator (G) is boosted by the solution pump (P1) and preheated by the heat exchanger (H1) before entering the absorber (A), where it absorbs the evaporation refrigerant vapor flowing from the evaporator. Since the absorbent has a thermal effect when it absorbs refrigerant vapor, the absorber (A) releases a higher grade of heat with a temperature of T2 (such as 120–150 °C), and the concentrated solution in the absorber (A) is heat exchanged in the heat exchanger (H1) releases heat and merges with the liquid from the evaporator through the throttle valve (v) to reduce the pressure and then flow into the generator (G).

Current, using LiBr solution, a two-stage absorption heat pump can achieve 150 °C of heat output (hot water or steam), two-stage absorption heat pump actual running results indicated that: based on the cooling water of 40 °C, take low-grade heat temperature of 80 °C as the heat source, the outlet temperature of first stage absorption heat pump is 115 °C, and output temperature of 2nd stage absorption heat pump is about 150 °C.

The heat transformer technology is still under development. The focus of the research is to select suitable working fluids, the target of this research should be upgrading the low-grade heat to around 200 °C. At present, it is generally believed that a mixture containing TFE (2,2,2-trifluoroethanol) is more suitable as a working fluid.

14.7.3 Refrigeration

There are generally two technologies for low-temperature heat refrigeration. One is lithium bromide refrigeration technology, which is used for shallow refrigeration, generally producing cold water at 5–10 °C, and the second is ammonia absorption refrigeration technology, which is used for deeper refrigeration in industrial processes. The evaporation temperature is generally between −10 and −35 °C. The working heat source includes low-pressure steam (exhausted steam) of the system and low-grade heat.

14.7.3.1 Lithium Bromide Refrigeration Technology

Working principle and refrigeration cycle [14]

At a certain temperature and concentration, the saturation pressure of lithium bromide (LiBr) aqueous solution is much lower than the saturated vapor pressure of water at the same temperature. Therefore, utilizing the pressure difference, the lithium bromide solution can be used to absorb water vapor and reduce the water vapor pressure. The lower the pressure, the lower the evaporation temperature, so that low-temperature chilling water can be obtained. The lithium bromide refrigeration uses steam or low-temperature heat as the heat source, lithium bromide solution as the absorbent, and water as the refrigerant to obtain cold energy.

A simple absorption refrigeration sketch is shown in Fig. 14.23.

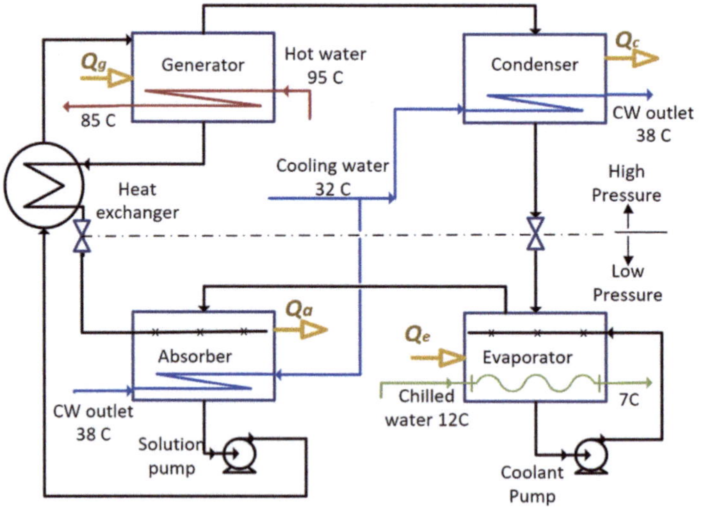

Fig. 14.23 A single effect absorption refrigeration sketch

The weak solution in the absorber is pumped through the heat exchanger to the generator. Then, it becomes the intermediate solution with water vapor before flowing into the generator separating the water and lithium bromide solution.

The generator utilizes the heat energy from waste heat such as flue gas, exhaust steam, or hot water. In the Generator, the water will be vaporized and leave the lithium bromide solution, it will be formed as a liquid and drop to the bottom of the generator. The liquid flows down to preheat the weak solution in the heat exchanger and becomes the strong solution. Then, it flows back to spray over the absorber to absorb the vapor for the next process again.

Meanwhile, the vapor which left from the lithium bromide solution in the generator flows into the condenser. The solution will be cooldown by the cooling water, then it will be condensed to the refrigerant water. Then it flows down to the evaporator, the water will be evaporated due to the vacuum condition which made the water boiling temperature becomes low.

In the Evaporator, 12 °C chilled water which returns from the process units or utility system flows into the evaporator. The chilling water will be chilled by spraying refrigerant water over the chilled water pipe in the evaporator. Therefore, the chilled water temperature will be decreased to 7 °C.

Finally, after the refrigerant water absorbs the heat of chilled water, it becomes the vapor again at an as low boiling temperature under the vacuum condition in the evaporator. After that, the vapor will be absorbed by a strong solution and become a weak solution for the new cycle again.

The principle of double-effect lithium bromide refrigeration is the same as that of single-effect refrigerators, except that one generator is added. It consists of a high-pressure generator, a condenser, a low-pressure generator, an evaporator, and an absorber.

Application

1) Background

In the process operation of the catalytic cracking plant of a petrochemical complex, it was found that the catalytic dry gas contained a large amount of liquefied gas components, and the components above C_3 accounted for 13.4%. There are many ways to solve the problem of "no dry gas", such as increasing the operating pressure, increasing the number of trays, and increasing the absorption agent recirculation rate. However, the catalytic cracking unit is affected by objective factors such as the gas compressor, and the above methods are restricted.

2) Solution

A low-temperature heat refrigeration approach was recommended to achieve the purpose of increasing the absorption rate by lowering the absorption temperature. a 1.75 MW lithium bromide refrigeration unit was installed and test trialed, the concentration of C_3 and heavier components in the dry gas of 5–7% had been achieved and recovered 8500 t/a of liquefied gas (including 4500 t/a of propylene).

14.7.3.2 Ammonia Absorption Refrigeration

Ammonia absorption refrigeration is a refrigeration method that consumes heat energy to obtain cold energy. The principle of ammonia absorption refrigeration is the same as that of lithium bromide absorption refrigeration (ref. to Fig. 14.23). The working fluid is changed from lithium bromide aqueous solution to ammonia aqueous solution.

Ammonia absorption refrigeration is a device that uses ammonia as a refrigerant and water as an absorbent to form a solution circulation system. Because ammonia is used as the refrigerant, ammonia absorption refrigeration is suitable for refrigeration conditions with an evaporation temperature of +5 to −40 °C, which can be used in the refrigeration system of the refining and petrochemical plant, which greatly reduces the power consumption.

The heating source of ammonia absorption refrigeration is usually steam. It is best to use various waste heat (such as high-temperature water, and high-temperature gas) rejected during the production process, and sometimes direct combustion gas fuel can be used to heat the generator.

14.7.4 Power Generation

When a large amount of excess low-temperature heat is difficult to find a suitable utilization scheme at the same level and upgraded level, power generation should be considered for further utilization of low-grade heat. If the low-temperature heat is only used for power generation, it will result in a large investment, lower power generation efficiency, and a long payback period. Therefore, it is recommended to combine the production heating, heat pumps, refrigeration, and power generation into an integrated system; arranging low-grade heat users based on the needs of energy quantity and quality.

The low-grade heat power generation Rankine cycle technology had been developed in China using water or low-boiling organic substance as the working medium. Among them, the "power generation and heat supply" unit in refineries using water as the working medium was put into operation in the 1980s, since then, it has been operating normally and has received great energy-saving effects.

The low-temperature heat power generation and heating system using water as the working medium flow sketch is shown in Fig. 14.24.

14.7.5 Low-Grade Heat Integration System

The flow chart of Power generation and low-grade heat integration system in summer is shown in Fig. 14.25.

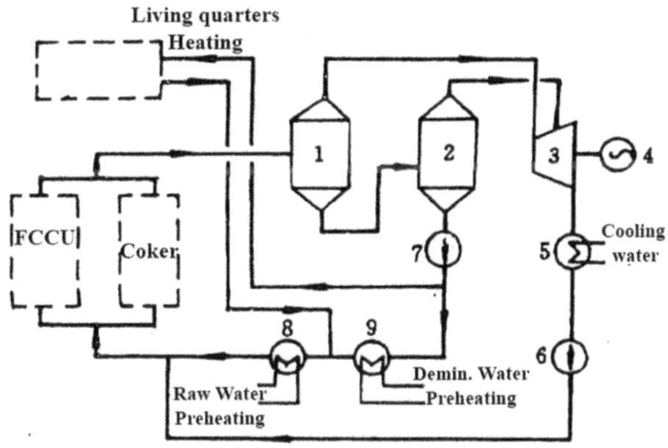

Fig. 14.24 Flow chart of low-temperature heat power generation and heating system. 1—1st expansion vessel; 2—2nd expansion vessel; 3—steam turbine; 4—generator; 5—condenser; 6—cooling water pump; 8—freshwater preheater; 9—desalinated water preheater

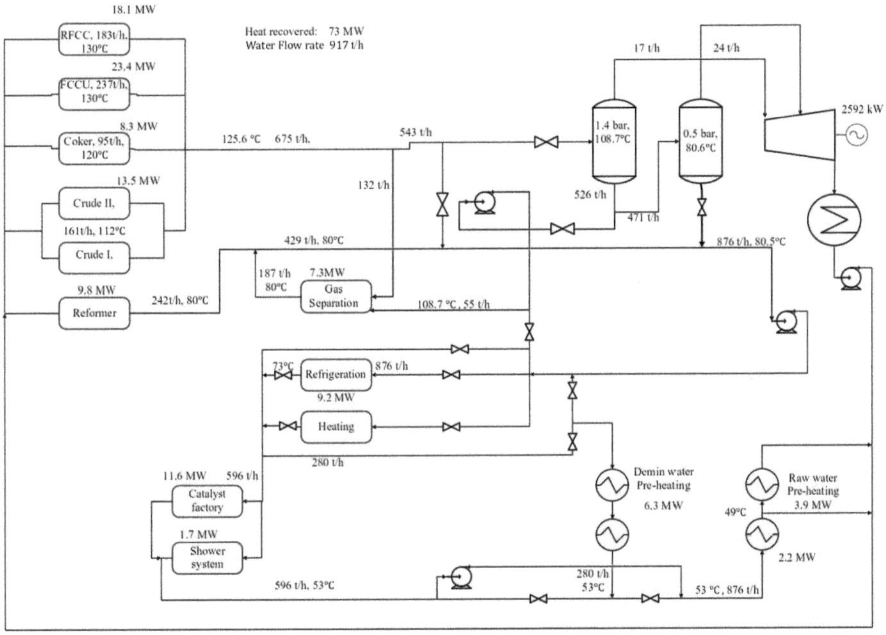

Fig. 14.25 Flow chart of power generation and low-grade heat integration system in summer

Figure 14.25 indicates a typical low-grade heat power and heating integration in an oil refinery, for oil and gas production facilities, chemical or petrochemical plants, a similar low-grade heat integration system could be established, it is expected to have significant carbon mitigation and energy-saving benefits.

From Fig. 14.25, this system uses demineralized water as a working fluid to collect low-grade heat from 6 process units of a refinery, the total low-grade heat recovered is 73.1 MW.

The water temperature from RFCC and FCCU is 130 °C, 120 °C from the delayed Coker unit, 112 °C from two crude units, the mixed water has a temperature of 125.6 °C, a small portion of mixed water (20%) was directed to gas separation unit to replace steam (7.3 MW) used in reboiler, remaining of mixed how water was directed a two-stage of flash drums, two flash steams were directed to a steam turbine generator to generate power of 2592 kW.

Hot water collected from the reformer unit is 80 °C, it merges with the hot water from the second flash drum, and is pumped to a low-grade heat refrigeration system (9.2 MW), the refrigeration system uses lithium bromide absorption refrigeration units, and provides chilled water to production units and office's air conditioning.

The hot water from the refrigeration system has a temperature of 73 °C, it flows to a low-grade heat utilization system to provide heat of 25.7 MW to the catalyst factory, demineralized water preheating, raw water preheating, and shower system.

The following suggestions are provided for establishing low-grade heat integrated utilization system:

(1) Identify and collect low-temperature heat sources, focusing on the recovery of heat from cooling streams above 80 °C in the production equipment; the production equipment should consider the practicability of heat extraction projects and ensure safety and reliability.
(2) Identify suitable users. The conditions of low-temperature heat users are very different, and they must be adapted to local conditions. Some companies have a lot of users, and the amount of low-temperature heat demand is large, and the recovered heat cannot meet the needs of users during the peak heat demand period; some companies are the opposite, there are more low-temperature heat sources and fewer users, so it is necessary to develop ways of upgrading and utilization.
(3) Develop a feasible plan for low-grade heat integrated utilization system. There are generally three schemes: pure heating type; heating-refrigeration combined type and heating- refrigeration -power generation type.
(4) Develop new technologies for the upgrading and utilization of low-temperature heat, such as the development of absorption heat transforming technology to increase the temperature level and facilitate utilization.

References

1. P. Jie, Discussion on variable speed energy saving methods for pumps. Refinery Des. **4**, 28 (1987)
2. C. Delian, in *Using Oil Film Clutch Speed Regulation to Achieve Energy Saving in Circulating Water Systems, Petroleum Refining*. Proceedings of the 2nd Energy Conservation Experience Exchange Conference (1989), p. 396

3. F.J. Brooks, in *GE Gas Turbine Performance Characteristics (GER-3567H)* (GE Power Systems Schenectady, NY)
4. W. Xiaozhong, Petrol. Refining Des. **5**, 22–25 (1985)
5. D Daoji 1996 Refining Des. 26 3 17
6. X. Hui, in *Collected Papers of the Annual Meeting of Catalytic Cracking* (1993), pp. 341–345
7. L. En, Operating a CO boiler and saving energy. Refinery Des. **2**, 70 (1981)
8. C. Anmin, Z. Fuyi et al., Catalytic cracking regeneration flue gas CO external combustion technology and its application. Petrol. Refinery Eng. **2** (1995)
9. F. Wei, S. Lin, G. Yang, Kinetics of CO vapor phase slow oxidation. J. Univ. Petrol. **4**(5), 10, 115 (1988)
10. Z. Qiongjiao et al., in *Combustion Engineering* (Coal Industry Press, 1988)
11. G. Qu, Open heat pump successfully used in refinery gas propane. Propylene Sep. Petrol. Refinery Eng. **5** (1988)
12. L. Deping, Analysis of the impact of heat pump power on energy saving and economic benefits. Petrol. Refining **3**, 37 (1993)
13. Z. Yuqing, W. Jinfa, G. Huibin, in *Two-Effect Lithium Bromide Absorption Refrigerator, and Its Application* (China Machine Press, 1990), p. 4
14. Lithium Bromide Absorption Chiller for waste heat recovery. https://www.gmsthailand.com/product/lithium-bromide-absorption-chiller/ShuanglingProduct brochure
15. GE Gas Power products & services catalog. https://www.ge.com/gas-power/resources/catalog
16. C Anmin L Kun 1988 Thermodynamic analysis of gas turbine-furnace combined system Pet. Refin. Eng. 4 55 62

Appendix
Frequently Used Data

See Tables A.1, A2, A3 and A4.

Table A.1 Critical constants and eccentric factors for some substances (cited from Chemical thermodynamics, Zhang Lianke, Chemical Industry Press, 1980. 12)

Substance name	T_C(K)	P_C(atm)	$V_C(\text{cm}^3/\text{mol})$	Z_C	ω
Argon	150.8	48.1	74.9	0.291	0.00
Helium	5.2	2.24	57.3	0.301	0.00
Krypton	209.4	54.3	91.2	0.287	0.00
Hydrogen	33.2	12.8	65	0.305	0.00
Nitrogen	126.2	33.5	89.5	0.29	0.04
Oxygen	154.6	49.8	73.4	0.288	0.021
Chlorine	417	76	124	0.275	0.074
Ammonia	405.6	111.3	72.5	0.242	0.25
Carbon monoxide	132.9	34.5	93.1	0.295	0.04
Carbon dioxide	304.2	72.8	94	0.274	0.225
Hydrogen chloride	324.6	82.6	81	0.25	0.27
Hydrogen sulfide	373.2	88.2	98.5	0.284	0.10
Sulfur	1314	116			0.273
Sulfur dioxide	430.8	77.8	122	0.268	0.51
Sulfur trioxide	491	81	130	0.26	0.51
Water	647.1	217.6	56	0.23	0.348
Methane	190.6	45.4	99	0.288	0.007
Ethane	305.4	48.2	148	0.285	0.091
Propane	369.8	41.9	203	0.281	0.145
N-Butane	425.2	37.5	255	0.274	0.193

(continued)

T. A. Chen, *Energy Saving and Carbon Reduction*,
https://doi.org/10.1007/978-981-19-5295-1

Table A.1 (continued)

Substance name	T_C(K)	P_C(atm)	V_C(cm^3/mol)	Z_C	ω
Isobutane	408.1	36	263	0.283	0.176
Ethylene	282.4	49.7	129	0.276	0.086
Propylene	365	45.6	181	0.275	0.148
Acetylene	308.3	60.6	113	0.271	0.184
Benzene	562.1	48.3	259	0.271	0.21
Toluene	591.7	40.6	316	0.264	0.257
Methanol	512.6	79.9	118	0.224	0.556
Acetic acid	594.5	57.1	171	0.2	0.45
Acetone	508.2	46.4	209	0.232	0.318

Table A.2 Chemical exergy and temperature correction coefficients for major inorganic compounds (cited from Exergy analysis of energy systems, Zhu Mingshan, Tsinghua University Press, 1988. 4)

Substance	ex_c, (kJ/mol)	ξ, J/(mol.k)	Substance	ex_c, (kJ/mol)	ξ, J/(mol.k)
$AlCl_3$	229.83	892.07	HCl(g)	45.77	173.72
$Al_2(SO_4)3$	308.36	539.44	Na_2S	962.86	−798.31
Ar	11.67	39.16	$NaHCO_3$	44.68	−39.33
BaO	261.04	−596.01	MgO	50.79	−218.78
$BaSO_4$	32.55	−415.3	$MgCl_2$	73.39	343.21
$BaCO_3$	63.01	−356.77	$MgCO_3$	22.59	62.30
C	410.53	57.07	$MgSO_4$	58.24	−67.78
CaO	110.33	−227.74	MnO	100.29	−115.94
$Ca(OH)_2$	53.01	−201.46	Mn_2O_3	47.24	−113.51
$CaCl_2$	11.25	349.74	Mn_3O_4	108.37	−213. 13
$CaOSiO_2$	21.34	−228.15	N_2	0.67	2.34
$CaOAl_2O_3$	88.03	−251.08	Ne	27.07	90.83
CO	275.35	−25.61	NO	88.91	−4.60
CO_2	20.12	67.40	NH_3(g)	336.69	−154.22
Fe	368.15	−147.28	Na_2O	346.98	−585.76
FeO	118.66	−71.76	NaCl	0	0.38
Fe_3O_4	96.90	−70.29	Na_2SO_4	62.88	−413.5
Fe $(OH)_3$	30.29	43. 85	Na_2CO_3	89.96	−364.22
Fe_2SiO_4	220.41	−125.35	Na_3AlF_6	581.95	138.95
$FeAl_2O_4$	103.18	−66.36	SO_2 (g)	306.52	−114.64

(continued)

Table A.2 (continued)

Substance	ex_c, (kJ/mol)	ξ, J/(mol.k)	Substance	ex_c, (kJ/mol)	ξ, J/(mol.k)
H_2	235.22	−169.74	SO_3 (g)	239.70	−14.02
H_2O(g)	8.62	−118.78	H_2S(g)	804.46	−329.66
He	30.12	101.08	ZnO	21.09	−745.63
HF	152.42	46.61	$ZnSO_4$	73.68	−587.52
O_2	3.93	13.22	$ZnCO_3$	22.34	−503.46

Table A.3 Chemical exergy and temperature correction coefficients for major organic compounds (Cited from Exergy Analysis of Energy Systems, Zhu Mingshan, Tsinghua University Press, 1988. 4)

Substance	ex_c, (kJ/mol)	ξ, J/(mol.k)	Substance	ex_c, (kJ/mol)	ξ, J/(mol.k)
CH_4(g)	830.15	−201.96	$C_{11}H_{24}$ (1)	7361.33	−238.03
C_2H_6(g)	1493.77	−221.63	$C_{12}H_{26}$(l)	8013. 03	−247.07
C_3H_8 (g)	2148.99	−238.36	$CH_3C_6H_5$ (1)	3928.36	61.63
C_4H_{10}(g)	2801.06	−255.1	CH_3OH (l)	716.72	−33.26
C_4H_{10}(l)	2803.20	−540.11	C_2H_5OH (l)	1354.57	−43.68
C_5H_{12}(g)	3455.61	−270.29	C_3H_7OH (l)	2003.76	−52.26
C_5H_{12}(l)	3454.52	−152.38	C_4H_9OH (l)	2659.10	−61.55
C_6H_{14}(g)	4109.48	−286.14	$C_5H_{11}OH$ (l)	3304.69	−67.07
C_6H_{14}(l)	4015.38	−193.84	HCHO(g)	537.81	−86.02
C_7H_{16}(g)	4763.44	−355.81	CH_3CHO(g)	1160.18	−107.95
C_7H_{16}(l)	4756.45	−209.58	$(CH_3)_2CO$ (l)	1783.85	−20.59
C_2H_4(g)	1359.63	−172.38	HCOOH (l)	288.24	112.84
C_3H_6(g)	1999.95	−196.27	CH_3COOH (l)	903.58	105.52
$CH_2CHCH_2CH_3$(g)	2654.29	−211.33	C_6H_5OH (s)	3120.43	224.05
C_2H_2(G)	1265.49	−114.43	C_6H_5COOH (s)	3338.08	372.5
CH_3CCH(g)	1896.48	−138.2	$HCOOH_3$ (g)	998.26	−35.81
C_5H_{10} (l)	3265.11	−86.44	$CH_3COOC_2H_5$ (l)	2254.25	53.09
C_6H_{12} (l)	3901.16	−62.93	$(CH_3)_2O$ (g)	1415.78	−150.04
C_6H_6 (l)	3293.18	85.65	$(C_2H_5)_2O$ (l)	2697.26	−88.91
C_8H_{16} (l)	5243.89	−73.51	CH_3Cl (g)	723.96	206.86
C_4H_6 (g)	2522.53	−130.08	CH_2Cl_2 (l)	622.29	480.03
C_8H_{10} (l)	4580.1	50.92	CCl_4 (l)	441.79	1367.83
C_8H_{18} (l)	5407.78	−211.42	$C_6H_{12}O_6$ (s)	2966.92	590.7
C_9H_{20} (l)	6058.81	−220.87	$C_{12}H_{22}O_{11}$ (s)	5968.52	1136.21
$C_{10}H_{22}$ (l)	6710.05	−743.08	$(NH_2)_2CO$	686.47	182.67

Table A.4 Coefficients for ideal gas enthalpy, entropy and heat capacity equations (cited from API technical data book-petroleum refining, 4th Ed. 1983)

Compound name	*A*	*B*	$C \times 10^4$	$D \times 10^7$	$E \times 10^{11}$	$F \times 10^{15}$	*G*
Non-hydrocarbons							
Oxygen	−0.981760	0.227486	−0.373050	0.483017	−1.852433	2.474881	1.124314
Hydrogen	12.326740	3.199617	3.927862	−2.934520	10.900690	−13.878670	−3.938247
Water	−2.463420	0.457392	−0.525117	0.645939	−2.027592	2.363096	0.660170
Hydrogen sulfide	−0.617820	0.238575	−0.244571	0.410673	−1.301258	1.448520	0.954068
Nitrogen	−0.934010	0.255204	−0.177935	0.158913	−0.322032	0.158927	1.042363
Ammonia	−0.946950	0.480156	−0.862580	1.749520	−6.542850	8.558870	0.715489
Carbon monoxide	−0.975570	0.256524	−0.229112	0.222803	−0.563256	0.455878	1.092470
Carbon dioxide	4.778050	0.114433	1.011325	−0.264936	0.347063	−0.131400	1.343357
Sulfur dioxide	1.394330	0.110263	0.330290	0.089125	−0.773135	1.292865	1.194796
Paraffins							
Methane 58.40160	58.401600	0.571700	−2.943122	4.231568	−15.267400	19.452610	0.343962
Ethane	163.059600	0.264878	−0.250140	2.923341	−12.860530	18.220570	1.082172
Propane	165.723800	0.172601	0.940410	2.155433	−10.709860	15.927940	1.206577
N-Butane	164.444000	0.098571	2.691795	0.518202	−4.201390	6.560421	1.351649
2-Methylpropane	162.081100	0.046682	3.348013	0.144230	−3.164196	5.428928	1.561697
N-Pentane	173.460900	−0.002795	4.400733	−0.862875	0.817644	−0.197154	1.736161
2-Methylbutane	169.016300	−0.031504	4.698836	−0.982825	1.029852	−0.294847	1.871908
2,2-Dimethylpropane	145.603200	0.004372	4.064654	−0.276464	−2.174529	4.685030	1.567876
N-Hexane	133.193900	0.229107	−0.815691	4.527826	−25.231790	47.480200	0.577037
2-Methylpentane	152.967100	0.041484	3.116936	0.622252	−6.385257	12.592750	1.445766

(continued)

Table A.4 (continued)

Compound name	A	B	$C \times 10^4$	$D \times 10^7$	$E \times 10^{11}$	$F \times 10^{15}$	G
3-Methylpentane	148.171800	0.095013	1.605403	2.399189	−16.146920	32.727430	1.218716
2, 2-Dimethylbutane	166.769200	−0.119500	6.005360	−1.780500	2.920860	−1.344890	2.173019
N-Heptane	134.125900	0.180209	0.347292	3.218786	−18.366030	33.769380	0.746003
2-Methylhexane	140.722700	0.096697	1.770980	2.121378	−14.382430	28.531670	1.139842
3-Methylhexane	149.500800	0.039990	3.001828	0.786641	−7.296178	14.037090	1.421940
Olefins							
Ethene	173.778500	0.144963	1.710121	0.761974	−4.503085	6.664928	1.748330
Propene	193.229300	0.030810	3.512242	−0.494661	−0.226171	1.125539	1.965351
I-Butene	187.766400	−0.018519	4.263451	−0.940582	1.072240	−0.349830	1.999354
CIS-2-butene	210.450300	−0.042795	4.034318	−0.684280	0.134493	0.878860	2.146788
Trans-2-butene	187.242400	0.037032	3.551222	−0.560440	0.158471	0.444673	1.673007
Diolefins and acetylenes							
Propadiene	229.263800	0.033745	3.715168	−1.062607	1.864623	−1.435039	1.886725
1, 2-Butadiene	197.381000	0.039560	3.479813	−0.756071	0.840374	−0.265154	1.703710
1,3-Butadiene	208.818500	−0.100603	5.651872	−2.123463	4.830541	−4.738449	2.339874
Ethyne (acetylene)	274.099600	0.022636	5.459217	−2.976556	8.738401	−10.045390	2.226331
Propyne	172.513400	0.080387	3.240066	−0.879745	1.599146	−1.400978	1.662198
1-Butyne	185.609000	0.053336	3.461058	−0.803301	1.052757	−0.533994	1.610754
Aromatics							
Benzene	225.051800	−0.122662	4.310824	−1.138140	1.494985	−0.564766	2.178204
Methylbenzene	206.736400	−0.101151	4.225723	−1.061438	1.337653	−0.484075	2.054833

(continued)

Table A.4 (continued)

Compound name	A	B	$C \times 10^4$	$D \times 10^7$	$E \times 10^{11}$	$F \times 10^{15}$	G
Ethylbenzene	193.426300	−0.093633	4.390639	−1.126299	1.458215	−0.543200	1.974441
1,2-Dimethylbenzene	186.293600	−0.014950	3.342431	−0.484082	−0.460172	1.705569	1.551212
1,3-Dimethylbenzene	194.820200	−0.068902	3.995003	−0.924756	1.059785	−0.345267	1.847583
1,4-Dimethylbenzene	184.781200	−0.030090	3.300522	−0.394355	−0.821163	2.173243	1.646826
N-Propylbenzene	187.364700	−0.099907	4.668002	−1.269276	1.767887	−0.694428	1.974834

MIX
Papier aus verantwortungsvollen Quellen
Paper from responsible sources
FSC® C105338

If you have any concerns about our products,
you can contact us on
ProductSafety@springernature.com

In case Publisher is established outside the EU,
the EU authorized representative is:
Springer Nature Customer Service Center GmbH
Europaplatz 3, 69115 Heidelberg, Germany

Printed by Libri Plureos GmbH
in Hamburg, Germany